D1775347

HOLOGRAMS & HOLOGRAPHY

Design, Techniques, & Commercial Applications

LIMITED WARRANTY AND DISCLAIMER OF LIABILITY

THE CD WHICH ACCOMPANIES THE BOOK MAY BE USED ON A SINGLE PC ONLY. THE LICENSE DOES NOT PERMIT THE USE ON A NETWORK (OF ANY KIND). YOU FURTHER AGREE THAT THIS LICENSE GRANTS PERMISSION TO USE THE PRODUCTS CONTAINED HEREIN, BUT DOES NOT GIVE YOU RIGHT OF OWNERSHIP TO ANY OF THE CONTENT OR PRODUCT CONTAINED ON THIS CD. USE OF THIRD PARTY SOFTWARE CONTAINED ON THIS CD IS LIMITED TO AND SUBJECT TO LICENSING TERMS FOR THE RESPECTIVE PRODUCTS.

CHARLES RIVER MEDIA, INC. ("CRM") AND/OR ANYONE WHO HAS BEEN INVOLVED IN THE WRITING, CREATION OR PRODUCTION OF THE ACCOMPANYING CODE ("THE SOFTWARE") OR THE THIRD PARTY PRODUCTS CONTAINED ON THE CD OR TEXTUAL MATERIAL IN THE BOOK, CANNOT AND DO NOT WARRANT THE PERFORMANCE OR RESULTS THAT MAY BE OBTAINED BY USING THE SOFTWARE OR CONTENTS OF THE BOOK. THE AUTHOR AND PUBLISHER HAVE USED THEIR BEST EFFORTS TO ENSURE THE ACCURACY AND FUNCTIONALITY OF THE TEXTUAL MATERIAL AND PROGRAMS CONTAINED HEREIN; WE HOWEVER, MAKE NO WARRANTY OF ANY KIND, EXPRESS OR IMPLIED, REGARDING THE PERFORMANCE OF THESE PROGRAMS OR CONTENTS. THE SOFTWARE IS SOLD "AS IS " WITHOUT WARRANTY (EXCEPT FOR DEFECTIVE MATERIALS USED IN MANUFACTURING THE DISK OR DUE TO FAULTY WORKMANSHIP);

THE AUTHOR, THE PUBLISHER, DEVELOPERS OF THIRD PARTY SOFTWARE, AND ANYONE INVOLVED IN THE PRODUCTION AND MANUFACTURING OF THIS WORK SHALL NOT BE LIABLE FOR DAMAGES OF ANY KIND ARISING OUT OF THE USE OF(OR THE INABILITY TO USE) THE PROGRAMS, SOURCE CODE, OR TEXTUAL MATERIAL CONTAINED IN THIS PUBLICATION. THIS INCLUDES , BUT IS NOT LIMITED TO, LOSS OF REVENUE OR PROFIT, OR OTHER INCIDENTAL OR CONSEQUENTIAL DAMAGES ARISING OUT OF THE USE OF THE PRODUCT.

THE SOLE REMEDY IN THE EVENT OF A CLAIM OF ANY KIND IS EXPRESSLY LIMITED TO REPLACEMENT OF THE BOOK AND/OR CD-ROM, AND ONLY AT THE DISCRETION OF CRM.

THE USE OF "IMPLIED WARRANTY" AND CERTAIN "EXCLUSIONS" VARY FROM STATE TO STATE, AND MAY NOT APPLY TO THE PURCHASER OF THIS PRODUCT.

HOLOGRAMS & HOLOGRAPHY

Design, Techniques, & Commercial Applications

John R. Vacca

CHARLES RIVER MEDIA, INC.
Hingham, Massachusetts

Copyright 2001 by CHARLES RIVER MEDIA, INC.
All rights reserved.

No part of this publication may be reproduced in any way, stored in a retrieval system of any type, or transmitted by any means or media, electronic or mechanical, including, but not limited to, photocopy, recording, or scanning, without prior permission in writing from the publisher.

Production: Publishers' Design and Production Services
Cover Design: The Printed Image
Cover Image: AD 2000 Inc.
Printer: Phoenix Color Printing

CHARLES RIVER MEDIA, INC.
20 Downer Avenue, Suite 3
Hingham, MA 02043
781-740-0400
781-740-8816 (FAX)
www.charlesriver.com

This book is printed on acid-free paper.

Holograms & Holography: Design, Techniques, & Commercial Applications
John Vacca
ISBN: 1-886801-96-7

Library of Congress Cataloging-in-Publication Data

Vacca, John R.
    Holograms and holography : design, techniques, & commercial applications / John Vacca.
      p. cm.
    1. Holography.   I. Title.
TA1540 .V33 2001
621.36'75—dc21
                                  2001000922

All brand names and product names mentioned in this book are trademarks or service marks of their respective companies. Any omission or misuse (of any kind) of service marks or trademarks should not be regarded as intent to infringe on the property of others. The publisher recognizes and respects all marks used by companies, manufacturers, and developers as a means to distinguish their products.

Printed in the United States of America
01 7 6 5 4 3 2 First Edition

CHARLES RIVER MEDIA, INC. titles are available for site license or bulk purchase by institutions, user groups, corporations, etc. For additional information, please contact the Special Sales Department at 781-740-0400.

Requests for replacement of a defective CD must be accompanied by the original disc, your mailing address, telephone number, date of purchase and purchase price. Please state the nature of the problem, and send the information to CHARLES RIVER MEDIA, INC. 20 Downer Avenue, Suite 3, Hingham, MA 02043. CRM's sole obligation to the purchaser is to replace the disc, based on defective materials or faulty workmanship, but not on the operation or functionality of the product.

About the Cover Hologram

Cover Hologram produced by A D 2000, Inc. of New Haven, CT. Both the Earth/Grid embossed type hologram on the cover and the Brain/Skull photopolymer type hologram in the CD-ROM sleeve are stock images available through the HoloBank® line (see accompanying CD-ROM for full information).

The hologram on this cover is known as an embossed type rainbow hologram. After concept, a model maker was hired to create the model of the world, and an artist for the grid and stars. The model was "flattened" to facilitate better viewing in ambient lighting environments. Blue Ridge Holographics was hired to shoot the H1 and H2 photoresist holograms at an overall image size of 6" x 6" (later cropped to the 2¾" size you see here). This image is an excellent example of the combined use of an actual 3D object with two-dimensional flat artwork, together causing a dramatic appearance of depth.

The photoresist master was then silvered, re-tooled (for replication efficiency), and production shims were created. The images were then micro-embossed into 2 mm metallized polyester, coated with pressure sensitive adhesive and a release liner, and kiss cut to final size. Finally, the hologram labels were applied by hand to the finished book cover and laminated for protection.

HoloBank® offers the world's largest selection of stock image holograms that can be customized for use in any promotion, advertisement, or security authentication. The Earth with Grid is one of almost a thousand varied stock hologram images; many of which are offered at a variety of sizes and on a multitude of products. Please contact www.holobank.com for further information.

To Tullio G. Bortoletto, for his inspiration, support, interest, and friendship over the past few years.

—John R. Vacca

CONTENTS

Foreword		xi
Acknowledgments		xiii
Introduction		xv

Part I Overview of Hologram Technology: Practical Uses — 1

1	Practical Holography	3
2	Holographic Principles	23
3	Using Light to Shape Empty Space	37
4	Holographic Interferometry	59
5	Laser Electronics	83
6	Holography Marketing Implications	91

Part II Commercial Applications — 107

7	A Universe of Commercial Holograms	109
8	Design: Anatomy of a Hologram	119
9	Mastering Holograms	197
10	Electroforming	205
11	Holographic Embossing	215
12	Metallizing	225
13	Converting	233

	14	Products	241
	15	Patterns	251

Part III		Integral and Portrait Holography	269
	16	Motion Picture Holographic Image Production	271
	17	The Market for Holographic Art: How Well Are Holograms Represented?	285
	18	Selling Holographic Art	305

Part IV		Computer-Generated Holography	319
	19	Computer-Generated Holograms on the Web	321
	20	Internet Holography	339
	21	Holographic Storage Systems: Converting Data into Light	359

Part V		Electro- and Electron Holography	397
	22	Electro-Holography	399
	23	Electron Holography	461

Part VI		Custom Holography, Security, Results, and Future Directions	479
	24	How to Make Your Own Holograms	481
	25	Holographic Security	531
	26	Summary, Conclusions, and Recommendations	561

Part VII		Appendixes	597
	A	International Hologram Manufacturers Association (IHMA) and List of Holographic-Related Sites	599

B	Holographic Research Condensed Report	605
C	Hologram Reference Wave Noise	617
D	Quantum Holography	623
E	Holographic Images	625
F	Frequently Asked Questions	627
G	ISAR Imaging: A Generalized Geometry	633
H	Hologramic Theory	641
I	Holographic Games	645
J	Guide to the CD-ROM	647
	Glossary of Terms and Acronyms	655
	Index	663

FOREWORD

Holograms and Holography
Design, Techniques,
and Commercial Applications

By Michael Erbschloe
Vice President of Research
Computer Economics
Carlsbad, CA

One of the key building blocks of the digital future is holographic technology. Holographs will take many shapes and forms, and will be used in a wide variety of digital applications. This means that developers of digital media and entertainment will be using holographs to create, deliver, protect, enhance, enrich, and even mystify their work. It also means that there will be a growing market for holographic development skills. This book will help the digital developer and the digital artist gain an understanding of this fascinating technology and some of the many ways they can use holographs.

Digital media and entertainment developers who are not yet using holographs are missing out on a fantastic opportunity to move their work to a new height and add a new edge to their creations. Digital artists who have not begun to work with holographs are now pushing themselves beyond their old limits and into one of the most promising digital realms where technology can help them grow. Those who are not using holographs need to start.

Holographs are the next best things to life itself, because of their realistic effects. They go far beyond flat art, far beyond flash art, and will even exceed today's animation and special effects methods and techniques. Holographs will be used to enrich Web sites, liven up digital presentations, and bring superb effects to video productions. In 10 years, holographic technology will make the virtual reality applications created in the late 1990s look like the work of first-grade computer users.

As a developer or artist, adding holographs to your future will help advance your career and set you apart from others. The digital media field is very attractive and is drawing in many young developers and artists. Don't be left behind the pack. You need to jump ahead, and understanding and using holographic technology is one of the best ways to propel your work into the digital future. The rewards will be numerous and will include money, fame, and a level of self-satisfaction and self-realization that others will not even be able to imagine.

ACKNOWLEDGMENTS

There are many people whose efforts contributed to the successful completion of this book. I owe each a debt of gratitude, and want to take this opportunity to offer my sincere thanks.

To my publisher, David Pallai, whose initial interest and support made this book possible, and for his guidance and encouragement over and above the business of being a publisher. To production coordinator, Courtney Jossart, and editorial assistant Kelly Robinson, who provided staunch support and encouragement when it was most needed. To my copyeditor, Beth A. Roberts, whose fine editorial work is invaluable. To Michael Erbschloe, who wrote the Foreword. Finally, to all of the other people at Charles River Media whose many talents and skills are essential to a finished book.

To my wife, Bee Vacca, for her love, help, and understanding of my long work hours.

Finally, to the organizations and individuals who granted me permission to use the research material and information necessary for the completion of this book.

I thank you all so very much.

INTRODUCTION

Attention magnet. Impossible to put down. Visual magic. These are a few of the phrases used to describe holograms, a technology that combines physics and optics to create a three-dimensional image on a two-dimensional surface.

Holography is used to create visual impact on everything from trading cards to pharmaceutical promotions and packaging. The reason for this is simple: it sells. This has been proved with virtually every use of holographic imagery.

Created by the recording of light waves from a laser source reflected off an object, a holographic recording encapsulates an infinite number of views of an object. Photographs merely record one view. Placed under ordinary light, the hologram *replays* the image so an observer may see the recreated light waves and perceive the image of the object as if it were still there.

So fascinating to hold, touch, and move about, holograms are growing in recognition and practicality. This is good news for business products professionals who can apply them to a myriad of products for a number of uses.

The hologram market itself has been estimated at between $500 and $590 million; tallies that are more accurate have been forestalled by its newness.

History

Unlike a description of the complicated mechanics used to create holograms, the history of the product itself is quite simple. Holograms were invented in 1947, but it wasn't until 1968 that they were seen outside the laboratory. Before that time, a laser was required to make (as well as see) a hologram. A Polaroid engineer who invented and patented the white-light, rainbow transmission hologram laid the groundwork for today's mass-production holograms.

Holography dates from 1947, when British/Hungarian scientist Dennis Gabor developed the theory of holography while working to improve the resolution of an electron microscope. Gabor, who characterized his work as "an experiment in serendipity," coined the term *hologram* from the Greek words *holos*, meaning whole, and *gramma*, meaning message. Further development in the field was stymied during the next decade because light sources available at the time were not truly *coherent* (monochromatic or one-color, from a single point, and of a single wavelength).

Another major advance in display holography occurred in 1968 when Dr. Stephen A. Benton invented white-light transmission holography while researching holographic television at Polaroid Research Laboratories. This type of hologram could be viewed in ordinary white light, creating a *rainbow* image from the seven colors that make up white light. The depth and brilliance of the image and its rainbow spectrum soon attracted artists who adapted this technique to their work and brought holography further into public awareness.

Benton's invention is particularly significant because it made possible mass production of holograms using an embossing technique. These holograms are *printed* by stamping the interference pattern onto plastic. The resulting hologram can be duplicated millions of times for a few cents apiece. Consequently, embossed holograms are now being used by the publishing, advertising, and banking industries.

Another method for the mass production of holograms (the photo polymer) was developed by the Polaroid Corporation. Unlike *embossed* holograms (which are, in fact, transmission holograms with a mirror backing), the photo polymer hologram is a reflection hologram that can be viewed in normal room ambient light. The Mirage hologram has been used successfully in advertising, direct mail, product packaging, and point-of-sale displays.

Uses

Today's holograms are used for three main reasons: *security applications* for financial documents (they can't be realistically photocopied and are difficult to simulate), credit cards, and tamper-evident stickers; *commercial/consumer applications* for publication covers, greeting cards, direct-mail literature, collectibles, trading cards, packaging, and displays; and *promotional pieces* such as T-shirts, hang-tags on merchandise, stickers, POGs, and even key chains.

Generally, holograms can be applied to a product via hot-stamp foil, laminate film, or in a pressure-sensitive format, and may then be run through an impact or ink-jet printer for imaging, if desired. The jury is still out on use in laser printer engines, so caution should be used if a hologram-enhanced product is slated for a laser printer.

Holograms are becoming increasingly familiar to the layperson, which is likely to help distributors in sales—especially as they match product with purpose. There are three key reasons why holograms are used in today's market:

- **Impact**: Holograms are eye-catching and will be looked at significantly longer than other graphic mediums. If a customer's name or slogan is on the hologram, the message is enforced.
- **Pass-around value**: People are impressed by good holograms, and are likely to bring them to the attention of colleagues and associates.
- **Retention**: People tend to keep holograms. It's a well-known fact that people have holograms on their desks that were created more than 10 years ago. Some people carry business cards sporting holograms 12 years after they were received.

What other promotional product has that kind of return? A pen? Probably not!

The customer now generates holographic applications. They are the ones who are pulling holograms into the general market; specifically, the printed business product market. The business products professionals will be pushing holograms to satisfy those needs.

Hologram distributors faced with end-user interest can sell accounts on their value-added aspects. These aspects can help business products professionals ease their clients over the holographic price hurdle, which can be substantial.

Holograms are not for price shoppers; some budgets may find holograms financially inaccessible.

Custom jobs typically cost $9,000 before reproduction. Stock art, available from some manufacturers, can bring the prices for the entire job down to $6,900–$7,000. With that in mind, someone set on holograms who cannot afford a holographic project should consider holographic patterns rather than images. Patterns are designed with the dramatic color change of holography without the up-front art charges, because they are available from inventory.

There are ways to work around budgets and still get the look you want. The end user's perception of holography as being expensive and complex is a misconception that should not necessarily be alleviated, as it adds value and cachet to a distributor's service. Distributors should know, however, that holograms are as easy to use and apply as regular hot-stamp foil, laminate, and pressure-sensitive materials.

Applications

Holography's unique ability to record and reconstruct both light and sound waves makes it a valuable tool for industry, science, business, and education. The following are some applications:

- Double-exposed holograms (holographic interferometry) provide researchers with crucial heat-transfer data for the safe design of containers used to transport or store nuclear materials.
- A telephone credit card used in Europe has embossed surface holograms that carry a monetary value. When the card is inserted into the telephone, a card reader discerns the amount due, and deducts (erases) the appropriate amount to cover the cost of the call.
- Supermarket scanners read the bar codes on merchandise for the store's computer by using a holographic lens system to direct laser light onto the product labels during checkout.
- Holography is used to depict the shock wave made by air foils to locate the areas of highest stress. These holograms are used to improve the design of aircraft wings and turbine blades.
- A holographic lens is used in an aircraft *heads-up display* to allow a fighter pilot to see critical cockpit instruments while looking straight ahead through the windscreen. Several automobile manufacturers are researching similar systems.
- Magical, unique, and lots of fun, candy holograms are the ultimate snack technology. Chocolates and lollipops have been transformed into holographic works of art by molding the candy's surface into

tiny, prism-like ridges. When light strikes the ridges, it is broken into a rainbow of brilliant iridescent colors that display 3D images.

- Researchers at the University of Alabama in Huntsville have developed subsystems of a computerized holographic display. While the work focuses on providing control panels for remote driving, training simulators, and command and control presentations, researchers believe that TV sets with 3D images might be available for as little as $1,000 within the next 10 years.

- Holography is ideal for archival recording of valuables or fragile museum artifacts. For example, the form of a 2,300-year-old Iron Age man unearthed at Lindow Moss, a peat bog in Cheshire, England, was recorded by a pulsed laser hologram for study by researchers. The Forensic Science Department of Scotland Yard made a reconstruction model of the Lindow Man.

- Scientists at Polaroid Corporation have developed a holographic reflector that promises to make color LCDs whiter and brighter. The secret lies in a transmission hologram that sits behind an LCD and reflects ambient light to produce a white background.

- Comic books began using dimensional images and prismatic patterns on the covers and reported astronomical sales. The collectibles market has embraced dimensional imagery because of the value it adds to the item. Anything with a dimensional image is seen as a premium piece.

- People will stop whatever they are doing to look at a hologram. Sales tripled for Ghostbusters cereal when Kellogg's put an image on the package. The best-selling *National Geographic* never sold as well as it did because of its holographic cover. The appeal of holography is widespread.

- An international confectionery company decided to use holography as in-pack premiums for their bubble gum. The success of this promotion was so great that their market share doubled and they expanded into two countries—all because of the appeal of holography.

- Cadbury, Sa in South Africa, used a dimensional wrapper to launch a promotional campaign for its Crunchie chocolate bars that resulted in an immediate 30-percent increase in sales. They reported this as "a fantastic achievement in the current economic climate."

- But is it just for kids? NO! When Molson Breweries prepared to launch a new product, Molson Dry, they considered the packaging with extreme caution. Imagery is very important in this market. Beer is an image product, so any time you change an image, it is of critical importance to see how it plays among your customers. After careful consideration, Molson decided to go with a dimensional label that was an undeniable success!
- Miller Breweries also used a dimensional label. The combination of a clever promotion and the label led to the most successful promotion for the Halloween season (second-biggest promotional period next to St. Patrick's Day) ever! They reported sales of 180 percent of the previous year's Halloween volume.

As you can see, there is plenty of proof that dimensional imagery can help boost sales of nearly any product. The best part is that it has not been overdone—the surface has merely been skimmed.

Purpose

The purpose of this book is to show experienced (intermediate to advanced) holography professionals how to design and create holographic applications for experimental, commercial, military, and private use. It also shows through extensive hands-on examples how you can gain the fundamental knowledge and skills you need to create, install, and configure holograms. This book also provides the essential knowledge required to deploy and use holographic applications that integrate data, voice, and video. Fundamental holographic concepts are demonstrated through a series of examples in which the selection and use of appropriate holography technologies are emphasized.

In addition, this book provides practical guidance on how to design and implement holograms. You will also learn how to optimize and manage a complex hologram.

In this book, you will learn the key operational concepts behind holograms. You will also learn the key operational concepts behind the major holography services. You will gain extensive hands-on experience designing and creating resilient holograms. You will also develop the skills needed to plan and design large-scale hologram projects.

In addition, you will gain the knowledge of concepts and techniques that allow you to expand your existing holographic systems. This book provides the advanced knowledge that you'll need to design and configure holographic solutions for the Internet.

Finally, through extensive hands-on examples (field and trial experiments), you will gain the knowledge and skills required to master the implementation of advanced holograms. In other words, you will gain the knowledge and skills necessary for you to take full advantage of how to deploy advanced holographic applications.

Scope

Throughout the book, extensive hands-on examples will provide you with practical experience in creating, installing, and configuring holographic applications. In addition to advanced holographic application technology considerations in commercial organizations and governments, the book addresses, but is not limited to, the following line items as part of creating and installing holograms:

- The CD contains the latest and best holographic images and the software for their creation and manipulation.
- Hundreds of commercial applications include anti-counterfeiting, stationery, entertainment, trading cards, virtual reality, 3D Web pages, premiums, and more.
- Chapters on the holographic creation include discussions of motion picture holographic production, electro and electron holography, holographic inteferometry, and laser electronics.

This book leaves little doubt that a new architecture in the area of advanced hologram creation and installation is about to be constructed. No question, it will benefit organizations and governments, as well as their holography professionals.

Target Audience

With regard to holograms, the book is primarily targeted at academia, scientists, laypersons, IT managers, system administrators, government computer security officials, Internet administrators, and Web developers—basically, all types of people and organizations around the world who manufacture holograms.

xxii INTRODUCTION

Organization of This Book

This book is organized into seven parts, including the appendixes (which include a glossary of hologram terms and acronyms).

PART I: OVERVIEW OF HOLOGRAM TECHNOLOGY: PRACTICAL USES

Part One discusses practical holography, basic hologram principles, shaping empty space with light, holographic interferometry, laser electronics, and the holography marketplace.

Chapter 1, " Practical Holography," provides an overview of hologram design types and techniques, and their commercial applications.

Chapter 2, "Holographic Principles," covers the basic hologram principles, including but not limited to, vibrations, light, and interference.

Chapter 3, "Using Light to Shape Empty Space," covers how to shape empty space with light. This includes, but is not limited to, experiments, spectral harmonics, holographic optical elements (HOES), techniques, dreamscapes, and reflections and projections.

Chapter 4, "Holographic Interferometry," answers many questions about holographic interferometry and the procedures that are used in holographic photography.

Chapter 5, " Laser Electronics," continues the theme of Chapter 4 by showing you how to set up a transmission and reflection hologram using laser electronics. It presents a different approach to setting up transmission and reflection holograms than what is discussed in Chapter 4. The chapter also continues to examine the practical uses of hologram technology.

Chapter 6, " Holography Marketing Implications," concludes the discussion on the practical uses of hologram technology. It examines the holography marketplace through the eyes of the father of Australian Holographics (AH), Dr. David Ratcliffe. The chapter also examines the practical uses of holographic art in the electronics marketplace (commerce).

PART II: COMMERCIAL APPLICATIONS

The second part of this book identifies intranet security trends on the Web: client and server, procedures and tools, and system and intranet administration currently in place within most organizations.

Chapter 7, "A Universe of Commercial Holograms," covers holographic production, conversion, and commercial applications, as well as a number of other related topics.

Chapter 8, "Design: Anatomy of a Hologram," addresses the design and anatomy of a hologram by looking at computer graphics in general.

Chapter 9, "Mastering Holograms," explains how to master holograms by showing you how to create and combine the necessary photographic and holographic elements to produce a *master* hologram on special photosensitive emulsions. It also shows you how a multidimensional image is recorded as a unique *interference pattern* that is created using laser light and precision optical techniques.

Chapter 10, "Electroforming," shows you how the *master* hologram can be copied repeatedly on a variety of formats, depending on your final application. To produce embossed holograms appropriate for high-volume runs, manufacturing facilities that specialize in electroforming equipment are able to generate metallized shims that preserve the holographic *interference pattern* exactly.

Chapter 11, "Holographic Embossing," shows you how to affix the metal shims to a high-speed embossing machine that stamps the holographic pattern into rolls of very thin plastic or foil. This chapter also shows you how various backings and laminations can be applied to the rolls. When properly illuminated, the embossed patterns focus light waves in specific ways to produce a multidimensional image—a hologram!

Chapter 12, "Metallizing," shows you how metallized films are routinely used for holographic applications. Some of these might be for security and anti-counterfeiting applications, while others are for graphical use.

Chapter 13, "Converting," examines the converting stage of the holographic process that manufacturers go through to create commercial holograms.

Chapter 14, "Products," deals with the *products* stage of the commercial holograms process. We look at the creation of many unique holographic projects for products and customers around the globe.

Chapter 15, "Patterns," deals with the *final* stage of the commercial holograms process: *patterns*. We look at how the defocusing of the correlation plane decreases the shift invariance of the correlators, thus increasing the number of patterns that can be stored in each correlation operation.

PART III: INTEGRAL AND PORTRAIT HOLOGRAPHY

Part Three covers integral holography as a two-step process. Images are first shot on 16mm or 35mm motion picture film and then transferred with a laser to hologram film. Integral holograms can be displayed flat, curved, or as a cylinder viewable from 360 degrees. Displaying the hologram on a curved or cylindrical surface makes the hologram viewable to a larger audience than normal. Integral holograms can be created from living people in motion, stop-motion animation, video, and/or computer-generated imagery.

Chapter 16, "Motion Picture Holographic Image Production," shows you how to produce holographic motion pictures; in other words, how to transfer motion picture film or video directly to the holographic image.

Chapter 17, "The Market for Holographic Art: How Well Are Holograms Represented?" discusses the holographic art market, including holographic gifts and novelties, worldwide wholesale distribution service, limited manufacturing, mail order, and custom production of holograms.

Chapter 18, "Selling Holographic Art," covers the problems encountered when trying to sell holographic art: holographic art and the conventional world, business considerations, improving relationships with the contemporary art scene, and ideas for increased success and recognition.

PART IV: COMPUTER-GENERATED HOLOGRAPHY

Part Four discusses the ability to turn your work into actual three-dimensional holography displays. The process presented is quick, easy, and surprisingly affordable. Some of the computer-generated holography customers mentioned here are the NYU Medical Center, Mitsubishi, The American Museum of Natural History, Michigan State University, and Allied.

Chapter 19, "Computer-Generated Holograms on the Web," covers the engine memory, interface design, and implementation issues required to create computer-generated holograms.

Chapter 20, "Holography on the Internet," reviews several holography sites and presents a case study of the development of one of those sites.

Chapter 21, "Holographic Storage Systems: Converting Data into Light," discusses the use of holograms to store data in memories that are both fast and vast. It also covers the replacement of hard disks with data holograms that have 2,000 times more capacity.

PART V: ELECTRO- AND ELECTRON HOLOGRAPHY

Part Five discusses a new visual medium—electro-holography—capable of producing realistic 3D holographic images in real time. It also examines the use of electron holography to record and reconstruct off-axis object wave (exit surface waves) electron holograms in conjunction for use with the electron microscope.

Chapter 22, "Electro-Holography," discusses a new visual medium (electro-holography) capable of producing realistic 3D holographic images in real time. It also covers subsequent research in *holovideo*, which led to computation at interactive rates; full-color images; synthetic images and real-world input; and, most recently, a scale-up to a 36-MB display system capable of producing images as large as approximately 100mm in width, height, and depth.

Chapter 23, "Electron Holography," examines off-axis electron holograms through a virtual electron microscope; artifacts in electron holography; holographic neural networks; reconstruction of electron holograms using simplex algorithms; high-resolution electron microscopy; and interference experiments with energy filtered electrons.

PART VI: CUSTOM HOLOGRAPHY, SECURITY, RESULTS, AND FUTURE DIRECTIONS

Part Six discusses practical holography, or how to make your own holograms. It also examines silver halide hologram emulsions, a medium particularly suited for custom pieces. Transmission and reflection holography are also covered. Transmission holograms are rear lit and can be produced as either laser viewable or rainbow-colored white light viewable images. Reflection holograms are lit from the same side as they are viewed, and are viewable in white light as a single-color image.

Chapter 24, "How to Make Your Own Holograms," shows you how to make your own hologram right in the confines of your home.

Chapter 25, "Holographic Security," examines the world of holography security.

Chapter 26, "Summary, Conclusions, and Recommendations," looks at holographic results and future directions. It looks at holographic neural networks, aeronautical engineering, the dynamic structure of holographic space, autostereoscopic holographic displays and computer graphics, digital holography systems, and interactive virtual reality holodecks.

PART VII: APPENDIXES

Ten appendixes provide additional resources that are available for holography. Appendix A is a list of holographic-related sites, and a discussion of what the International Hologram Manufacturers Association (IHMA) really is. Appendix B is a condensed report on holographic research. Appendix C discusses hologram reference wave noise. Appendix D describes quantum holography. Appendix E is a discussion of holographic images. Appendix F is a list of frequently asked questions. Appendix G deals with a generalized geometry for ISAR imaging. Appendix H discusses the hologramic mind. Appendix I covers holographic games. Appendix J provides a guide to the included CD-ROM. The book ends with a glossary of holography-related terms.

Conventions

This book uses several conventions to help you find your way around, and to help you find important facts, tips, and cautions.

You see eye-catching icons in the left margin from time to time. They alert you to critical information and warn you about problems.

This icon highlights a special point of interest about a holographic topic.

This icon highlights a warning that URLs and/or contact information contained in the book can change without notice.

This icon gives you good advice and tips — things you should do to maintain a hologram.

This icon alerts you to little-known, but potentially very valuable, information about intranet security.

> ■ **SIDEBARS**
>
> We use sidebars like this one to highlight related information, give an example, discuss an item in greater detail, or help you make sense of the swirl of terms, acronyms, and "initialisms" so abundant to this subject. The sidebars are meant to *supplement* each chapter's topic. If you're in a hurry on a cover-to-cover read, you can skip the sidebars. If you're quickly flipping through the book looking for juicy information, read *only* the sidebars.

John R. Vacca
jvacca@hti.net

PART I

OVERVIEW OF HOLOGRAM TECHNOLOGY: PRACTICAL USES

This part of the book discusses practical holography, basic hologram principles, shaping empty space with light, holographic interferometry, laser electronics, and the holography marketplace. It provides an overview of hologram design types and techniques, and their commercial applications.

Also discussed are the basic hologram principles, including but not limited to vibrations, light, and interference. Shaping empty space with light is also covered. Part One answers many questions about holographic interferometry and the procedures used in holographic photography. It presents a different approach to setting up transmission and reflection holograms. Finally, it concludes the discussion on the practical uses of hologram technology.

CHAPTER 1

PRACTICAL HOLOGRAPHY

In This Chapter

- Hologram types and commercial applications
- Holographic formats
- Hologram production
- Holographic displays
- How a hologram works

Artists have been trying to achieve 3D design on a 2D surface since the first cave dweller drew on walls. It's finally available. *Holography* is a recording of light made with lasers in laboratories. It is real three-dimensionality, not a gimmick or an illusion. Therefore, holography can achieve visual effects that cannot be produced by any other medium.

Webster's Dictionary defines holography as "the process of making or using a three-dimensional picture that is made on a photographic film or plate without the use of a camera, that consists of a pattern of interference produced by a split coherent beam of radiation." In other words, a hologram is a three-dimensional picture produced by the interference of two laser light beams on a photosensitive plate. The interference is caused by the same beam being split into two beams, one of which is reflected off an object and onto the plate. The other beam is directed straight at the plate as a reference beam for the first beam to interfere with, thus creating the three-dimensional interference pattern.

Holography is also the process of recording total information onto high-resolution film, resulting in a three-dimensional *picture* when reconstructed. The recording, called a *hologram*, is actually of an interference pattern that is created by an object beam (light from the object) and a reference beam (light from the laser). The interference pattern recorded on the emulsion is a record of phase and amplitude information. For constructive interference, the beams are in phase. For destructive interference, the beams are completely out of phase (relative phase 1/2 wavelength). The degree of blackening of an interference fringe recorded on the plate depends on the amplitudes of light from various parts of the object as well as on phase. The amplitude of light depends on how intense the object beam is when it meets the reference beam at each point on the plate.

The hologram has properties of a lens, except that the light wave is intercepted and stored (on the film) before being allowed to continue its propagation to form an image. It is only when the hologram is suitably illuminated that the information contained in the hologram can be decoded and the scene reconstructed or made visible. The emulsion reacts strongly where the light is intense, and less strongly elsewhere. Thus, the constructive interference fringes will become surfaces where there has been strong activation of the silver halide grains in the emulsion. After being processed chemically, the exposed plate will be a photographic negative that contains a permanent record of the interference pattern.

Holograms have an interesting characteristic, redundancy, in which each piece of a hologram is a hologram. It's like looking through a window; if the window is broken, you can still look through each piece, but with a smaller view of what's outside. Thus, if a hologram (film) is cut into smaller pieces, each piece will still show the same holographic image, but with a limited view (smaller view than the original hologram).

Still a mystery to most of the population, holography is poised on the verge of exploding into a part of daily life. One of the most remarkable inventions in graphics history, true three-dimensionality on a two-dimensional surface makes all but the bravest media and art directors, advertising agencies, and marketing vice presidents squeamish.

Increases in sales and recognition have been reaped by the few who have dared to be different. This beginning chapter brings you their stories in the form of an overview of hologram design types and techniques, and their commercial applications. Also, take a look at who's using holography in the sidebar, "Who's Using It?"

New and Mixed Dimensions

With holography, expensive items such as jewelry, artifacts, and money can be displayed without security or risk. Dangerous items such as nuclear fuels and toxic items can be shown in three dimensions. Substances that are extremely delicate or sensitive to environmental changes can be portrayed.

For example, the November 1985 issue of *National Geographic* featured a hologram of the Taung Child, a 2-million-year-old skull of a five-year-old child, on its cover. This skull could not have been moved. The only way that an individual could have viewed it was by going to South Africa. This issue of *National Geographic* had the most advertising pages for that publication since 1960. The magazine's first use of holography in March 1984 resulted in 500,000 new subscriptions—a tremendous increase.

To expose the internal components of equipment in the same hologram, multiple images of objects can be manipulated. This peeling away of the center skin is viewed simply by changing perspectives slightly.

Since the optical qualities of lenses are preserved in holograms, microscopes and magnifying glasses included in holograms actually function. The 1987 *Cole-Palmer Catalogue* of electronic equipment had an embossed hologram on its cover. The image was of a circuit board, including capacitors and resistors. In the lower midsection of the board was a magnifying glass. Looking through it revealed the company's logo, which appeared enlarged or magnified.

Some other examples include direct mail pieces sporting holograms from Fingerhut. The company went into a rollout of 3 million holograms after a test of 40,000 pieces of a small eagle.

After a test run of 440,000, Montgomery Ward went into a rollout of 2 million pieces for its auto club membership. The hologram was used as an interactive technique that served as a validation device, adding value to the temporary membership card, like the hologram on MasterCard and Visa cards.

In another example, Merck, Sharp, & Dohme hot stamped a hologram of a stomach and intestine in a letter they sent to doctors announcing a new drug (Primaxin). When announcing the new drug Xanax (an improvement over Valium), the Upjohn Company used a hologram of a brain and the chemical formulation affixed to a three-page foldout.

■ **WHO'S USING IT?**

Uddeholm Steel Corporation recently placed a five-page insert including a hologram of a toy soldier in *Iron Age* magazine. Of the respondents to a Chilton Company study for *Iron Age*, 90.6 percent noticed the ad. The closest competitive ad scored 77.9 percent. More important, 72.1 percent started to read the ad, and 48.5 percent read half or more, as compared with 39.7 percent and 16.9 percent for the closest contender. Most importantly, the overall effectiveness index for the ad rated 332 points; the closest competitor only received 210 points!

The American Bank Note Company recently placed an ad that included a hologram in *Food and Drug Packaging* magazine. A Readership Research Company survey revealed that 87 percent of the readers remembered seeing the ad, and 58 percent found the ad of interest. These scores were from three to almost seven times above average. The 58 percent "found of interest" score is the highest score ever recorded in the magazine, and the 87 percent "remembered seeing" is the highest percentage score for all three-page ads, and is the third highest ever recorded.

During the American Heart Association Convention, Syntex Laboratories sent out an invitation displaying a heart hologram for doctors to visit their hospitality suite. The theme for the event was Holography and Hospitality. Six medically related holograms provided by Holaxis Corporation were on display for doctors to view in the hospitality suite.

Twofold was Umphrey Beefmasters' use of holography. One piece was a direct mail brochure featuring the Phantom Warrior bull for breeding syndication. The brochure displayed a hologram of this magnificent specimen. The second use was an insert in a publication, *Beefmaster Cowman*.

Recently, in one of its brochures, General Electric Apparatus and Engineering Services used a hologram of a cocoon changing into a butterfly.

Formats

Five formats are commonly used in holography. The following is a short guide to each.

- Embossed holograms
- Transmission holograms
- Reflection holograms
- Integral stereograms
- Pulsed holograms

EMBOSSED HOLOGRAMS

Embossed holograms are geared to low-cost mass production. Direct mail pieces, catalogs, brochures, packaging, and security and anti-counterfeiting applications have all used this format. Embossed holograms are essentially white light transmission holograms that are backed with a mirror (metallized Mylar) so they can be seen similar to a reflection hologram.

The Polaroid Mirage Film is the newest type of hologram available for mass production. This film falls into the reflection format. It is easy to see in most light. The mirage film is the closest representation of the real object available in holography today.

TRANSMISSION HOLOGRAMS

Images that are hit with light from behind are known as *transmission holograms.* They are illuminated with a laser or arc lamp specifically designed for holographic viewing, or a high-intensity white light. They can range in size from a few square inches to four-by-eight feet, and can produce images of great depth and projection.

In a recent breakthrough, the larger sizes offer the best type of holography for display applications (especially when exhibiting large pieces of equipment). They can be either monochrome when illuminated with a laser or arc lamp, or multicolored when lit with a white light source.

REFLECTION HOLOGRAMS

Another type of display holography lit from the front is known as *reflection holograms.* They can be hung on walls and illuminated simply with a clear high-intensity bulb. Reflection holograms can range from jewelry size, which can be made into pendants, key chains, and paper weights, up to very large ones of four to five feet, which are ideal for educational or display purposes. These have less depth and projection than transmission types, but can be layered with more than one image and color to achieve dramatic effects.

INTEGRAL STEREOGRAMS

A form of holography that incorporates conventional motion picture technology and holographic technique to produce the only type of hologram that can represent motion is known as *integral stereograms.* The integral holograms are usually adhered to a drum or an acrylic arc rear lit, but can also be produced flat. They can range from 9 to 11 inches in a flat format, up to 33-by-3 feet for cylindrical displays. These exhibits can be revolved automatically with a small motor, or remain stationary, requiring the viewer to walk around them. One of the most exciting aspects of the integral stereogram is that subject matter can now be computer generated, thereby opening the process to an infinite number of images.

PULSED HOLOGRAMS

A laser that pulses a burst of laser light in 5-billionths of a second is known as *pulsed holograms*. Portraits of live people and items can be captured in a hologram, thereby opening a whole range of applications.

Key Is Subject

Most people think in two dimensions rather than in three; therefore, their concept of what will and will not make a good hologram is usually incorrect. This is one of the major problems with designing a hologram. A hologram should be thought of as a volume of space or as a window. The holographic plate can be considered a window glass, and the volume of space in front of and behind the window glass is the design area.

Hologram Production

Three areas must be addressed when producing a hologram for a printed embossed piece:

- **Origination**: Creation of the hologram.
- **Production**: Running of a hologram.
- **Application**: Putting the holograms on a substrate.

ORIGINATION

The process of taking creative ideas and giving them form in holography is known as *origination*. This is comparable to the pre-press process in printing.

First, there should be a consultation between the customer and someone skilled in holography. Second, camera-ready artwork should be prepared.

Since the object and image relationship is one to one and there is no reduction or enlargement as in other graphic arts processes, only certain types of artwork are acceptable. These include sculpted models, multiplaned flat art, a combination of these two techniques, and multiplex (film generated).

The elements going into a hologram must be stable; in taking a shot, nothing can move even one micron. Black objects must be lightened to enhance reflectivity, and highly reflective items have to be subdued so their reflections don't overexpose the holographic film.

Storyboards should be created to represent the exact arrangement of the elements of holograms. The drawings should incorporate three views:

- A side view to illustrate the items to be projected forward and those that will be behind the plate (some can straddle the entire design volume area).
- A straight-on view (as one would view a picture).
- A top view.

To aid in executing even the most difficult designs, the combination of storyboards will give the holographer a *map*. A series of photographs of the actual objects might prove useful in combination with the storyboards.

Used as a proof for the customer to view before production, the next step is the actual shooting of the hologram, followed by the production of a glass master. After the proof has been approved, holographic stamping dies are prepared for mass production. The cost for the origination process ranges from $4,000 to $9,000 depending on the complexity of the design. The cost can be reduced if the customer furnishes suitable approved artwork.

PRODUCTION

A *production* master is created, once the glass master has been approved. This production master is a microfine relief replica of the original holographic pattern not unlike the relief surface of a phonograph record, but with a frequency of 10,000 to 16,000 grooves per inch. Copies of the production master are made and are mounted to a narrow web rotary embossing press. Under considerable heat and pressure, the dies transfer their relief image to the surface of metallized polyester, creating a precise replica of the hologram.

Embossed holograms can vary in size from a minimum of a half-inch square to a maximum four-by-five inches. Production costs vary according to volume. Three cents an inch in low volumes (11,000 four-by-five-inch holograms) to .5 cents an inch in high volumes (600,000 four-by-five-inch holograms) is a good rule of thumb. The timeframe for a new hologram is approximately 80 days from the time a customer provides conceptual artwork.

APPLICATION

Embossed holograms can be produced for two different modes of *application:* hot-stamping foil or pressure-sensitive labels. Hologram labels can be applied either by hand or by machine. They can also be provided as a piggyback label or temper-evident pressure security label. Application costs are approximately .5 to 1.5 cents per label.

Hot stamping is similar to conventional roll-leaf stamping. Where an integral bond with the substrate is desired, hot stamping is the preferred method of application. Application costs are in the range of 2 to 5 cents per impression.

POLAROID MIRAGE PHOTOPOLYMER HOLOGRAM

The *Polaroid Mirage photopolymer hologram* is dissimilar to its embossed counterpart, but similar in many respects to others. While the embossed hologram is a reproduction of an original using a die, heat, and pressure, the photopolymer is produced in a manner similar to a film reproduction system. Exposing an emulsion on plastic film and then developing it creates the finished product.

The costs for setup are similar to embossing ($4,000 to $9,000). The image can be as small as one-half-inch square or as large as 10-by-15 inches. Mirage holograms can be made pressure sensitive, but not hot stamped. These holograms are extremely realistic in color tone and dimension. They do not have a shiny mirror back; they can be produced clear, with a black background or a color of the customer's choice. The cost of this material is about three to four times more than embossed holograms.

Both types work well for different applications. Design, budget, and other specifications will determine which process to choose. Now, let's look at pulsed laser holography.

Pulsed Laser Holography

The images that can be used to make holographic displays are produced by bursts from a pulsed laser. The rapid pulse, or burst, of laser energy lasts only 20 billionths of a second, yet is powerful enough to expose the sensitive holographic plate. Holograms can be made of people, scenes, and very large objects using this technology. In addition, the *stop action* nature of the pulsed laser allows the 3D rendition of objects in flight or in motion. The imagination of trade show exhibit designers and marketers is the only limit to the possibilities for holography (see Figures 1-1 to 1-5)[1].

As early as 1970, the pulsed laser hologram proved its effectiveness as an attention-getter. Cartier, the world famous jewelry store, exhibited a hand wearing a diamond ring and holding a diamond bracelet in the window of its New York shop. It was positioned and illuminated so that it projected out and over the sidewalk. The display attracted thousands of passersby, and drew major television and network news media. One viewer went so

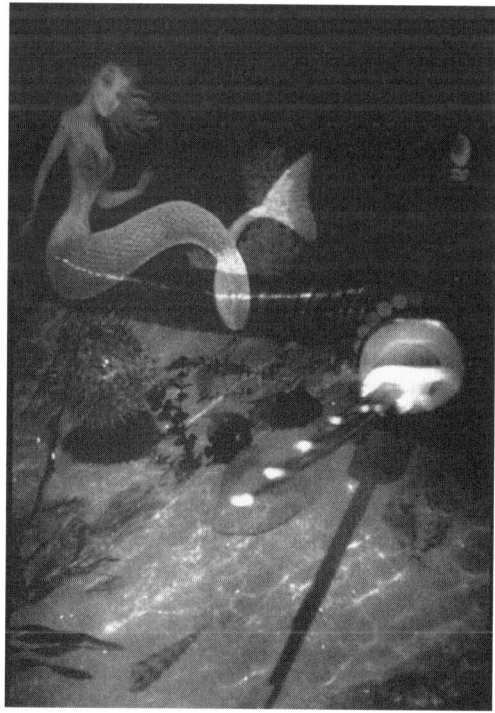

FIGURE 1-1: The AT&T "Catch Fish, Not Cables" hologram.

12 CHAPTER 1

FIGURE 1-2: The AT&T "Catch Fish, Not Cables" trade show display.

FIGURE 1-3: The Miles Laboratories holography theatre.

PRACTICAL HOLOGRAPHY **13**

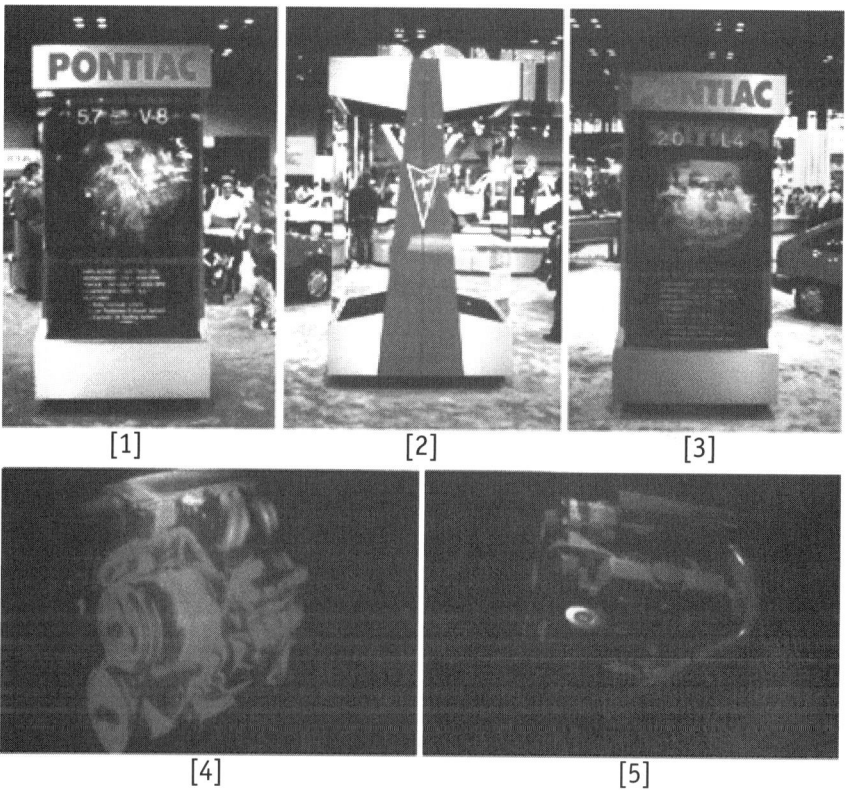

FIGURE 1-4: The Pontiac Engine holograms: [1] Pontiac 5.7L tuned port V8 engine; [2] side view, Pontiac display; [3] 2.0L overhead cam turbo L-4; [4] detail, Pontiac 5.7L tuned port V8 engine; and [5] detail, 2.0L overhead cam turbo L-4.

far as to attack the image with her umbrella, claiming it was the "work of Satan."

With the more familiar practices of photography, holography is a medium for communication of visual information that combines the new technology ushered in by the development of the laser beam. Light from an object is recorded and played back at will, so that the image appears in true three dimensions.

No strangers to trade shows are pulsed holography displays like the one shown in Figure 1-5. The effective use of holography in exhibitions by companies such as the Litton Corporation dates back to the early 1970s. One display entitled "Girl with Electronic Parts" was produced for their

14 CHAPTER 1

FIGURE 1-5: The Pontiac trade show display.

Advanced Circuitry division. This life-sized, mid shot of a young woman holding complicated electric boards was repeatedly utilized in shows from coast to coast. Part of the great success of the exhibit was attributed to the display environment specifically developed to maximize the effectiveness of the hologram. The actual hologram was only about 50 percent of the system, while the display was the other 50 percent.

A free-standing box positioned in a portable cove formed by panels was the proper environment for a Litton display. The display was usually nine feet in height and covered with black velour. This cove guarded the hologram against glare and provided a dramatic setting for its viewing. The transmission hologram was positioned at normal height and back lit by a mercury arc lamp that emitted a soft green glow. Many viewers paused as if transfixed by the display. Some thought the young woman actually stood motionless in the box, silently displaying the circuit boards for all to see.

For their industry's very important Toy Fair, the Tonka Toy Company recently developed a holographic display. As was the case with Litton, Tonka chose to go with a large transmission hologram. In transmission work, the holographic plate is lit from behind. This ensures that viewers do not block the light necessary to play back the images; the light is projected toward them. The Litton piece was a laser-viewable one, which means that a laser beam, or special light source such as a mercury arc lamp, is needed for viewing. Tonka decided to go one extra step and make their piece *white light* viewable, thereby enabling their hologram to be played back with a more conventional spotlight or high-intensity white light source. This type of lighting is less costly and easier to set up. They also opted for *rainbow* holography, adding the dimension of color to the work.

Existing technology allows the colors of a *rainbow* hologram to be controlled to a certain degree. The colors shift and change in and across the work as a viewer changes viewing angle by walking in front of the piece. The Tonka piece was a rendition of characters in battle. The models were actual people dressed in costume to represent *Lionheart* and *Burnheart*, two of the SuperNatural line.

A feat that could only be accomplished through the medium of holography, perhaps the most startling aspect of this exhibit was the depth and projection of the images. One figure appeared to be life-sized, but set back at least 10 feet from the viewer.

At a recent conference of the American Heart Association, depth was a factor taken into consideration when Intermedics Corporation incorporated a hologram into their exhibit. Their image was that of a surgeon as shown in Figure 1-6, complete with sterile mask and cap [HyperMedia Technologies Inc.]. In his hand he held a spoon on which, instead of medicine, was placed their high-tech product—the smallest pacemaker. The spoon and product projected out of the holographic plate toward the viewer, as if the doctor were actually offering the lifesaving device to a patient (see Figure 1-7 [HyperMedia Technologies Inc.]).

For their annual National Owner/Operators convention, McDonald's Corporation used pulsed holographic portraits of their Ronald McDonald in an exhibition. Graphics and design work were rendered holographically by McDonald's and placed on their inhouse folder and covers.

A holographic portrait of an individual can only be produced by the rapid burst of a pulsed laser. As mentioned previously, these portraits are realistic and play back in 3D. The themes for these portraits need not only be of clowns or models in costume. They can be, as was the case of the Litton work, of a real person. Pulsed lasers in the hands of experienced technicians are actually quite safe.

For example, for reasons of hospitality, Mike Ditka, the former coach of the Chicago Bears, had his holographic portrait done. Mr. Ditka placed his portrait in his restaurant, Ditka's City Lights. Being a local Chicago

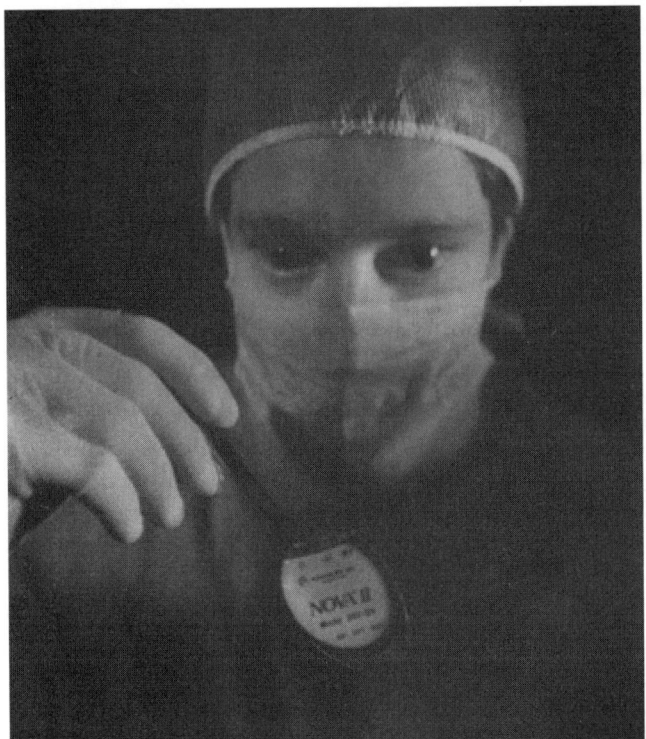

FIGURE 1-6: The Intermedics "Surgeon with Titanium Pacemaker."

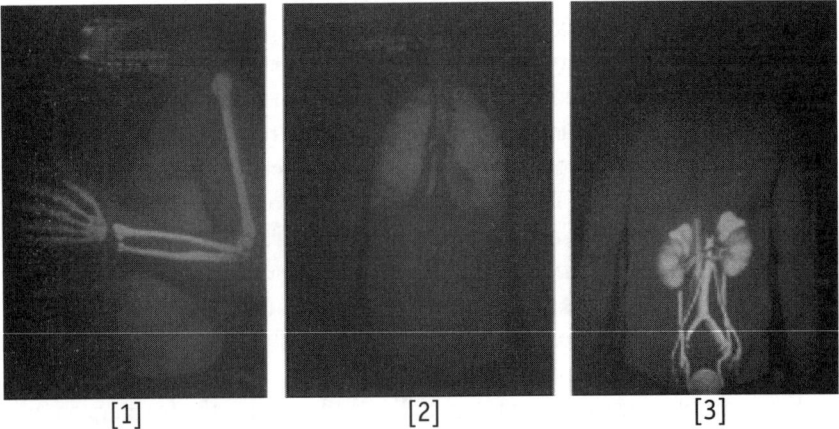

FIGURE 1-7: The Miles Laboratories holograms: [1] bone and joint; [2] respiratory system; and [3] urinary tract.

celebrity, he found that since he could not be there all the time to greet his customers, the holographic portrait was the next best thing.

In the future, more and more holography will be done with the pulsed laser. The holographic portrait business is already starting to take off in the United States, and for trade shows, the possibilities are endless.

Holographic Development

Because holography is now everywhere (on *National Geographic* covers, credit cards, baseball and football trading cards, Microsoft software packages, etc.), we are all interested observers. One can't help but wonder how they are made and what goes into making them (see Chapter 25, "How to Make Your Own Holograms"). Before we go there, let's first look at how a hologram works.

BACKGROUND AND THEORY

How does a hologram work? First, a hologram can be thought of as being a modulated grating or a modulated interference filter. When a grating is illuminated by a collimated wave, waves exit from the grating at different angles.

The amplitude of the wave influences the amplitude of the grating, and the position of the rulings of the grating depends on the phase of the wave. Therefore, the waves from the grating form images rather than plane waves.

Since a hologram is like a grating, the illumination of the hologram must be done with a single wavelength or a narrow band of wavelengths. Otherwise, the grating effect causes multiple images to be formed, resulting in a color blur instead of an image.

The first wave can be ignored, since it propagates along the same axis as the illuminating wave. The second wave is the same as the wave used in making the hologram, and is usually a virtual image. The third wave is the complex conjugate of the recorded wave (usually a real image). The second and third waves propagate along axes at angles to the axis of the illuminating wave. The type of images formed depends on the curvature of the reference illuminating waves.

Many types of holograms can be made, but they are all based on two types: *transmission* and *reflection* holograms. In the transmission type (see Figure 1-8), the film is exposed on only one side (the object and reference beams are on the same side of the film plate), and the original laser light is used as the reconstructing beam.[2] When a reconstructing light source (laser for transmission hologram) illuminates the film, the wavefront of the original object (or scene) is reconstructed, and to our eyes it looks as if the object is really there. However, what you see is a virtual image, there is a real image also, but it is usually very difficult to find. Transmission holograms have a small depth of scene, which is associated with a certain property of the laser beam. A laser beam has a limited coherence length that limits the depth of scene that can be obtained. It is for this reason that the transmission hologram needs a highly monochromatic light to reconstruct the scene, which is the light from the laser beam. If a transmission hologram is viewed in ordinary white light, the image seen is a smear.

It is possible to record holograms in (approximately monochromatic) coherent light and obtain images under *white light* illumination by a suitable modification of the technique of coherent holography. In this case, the reconstructed wavefronts are obtained by reflection from the hologram, rather than by transmission through it. For reflection holograms, the film is exposed on both sides (the object and reference beams are on opposite sides of the film).

To record a reflection hologram, the coherent object and reference waves are introduced from opposite sides of the photographic emulsion as shown in Figure 1-9 [Amberg and Hecox, 2]. After the development process, the interference of the two waves yields a transparency containing stratified layers of metallic silver that act as reflecting planes. Since the angle is near 180 degrees, the reflecting planes run nearly parallel to the surface of the emulsion.

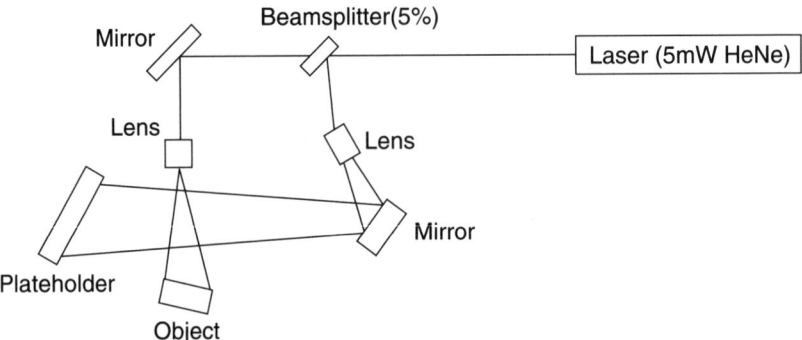

FIGURE 1-8: This diagram shows the basic setup for a transmission hologram.

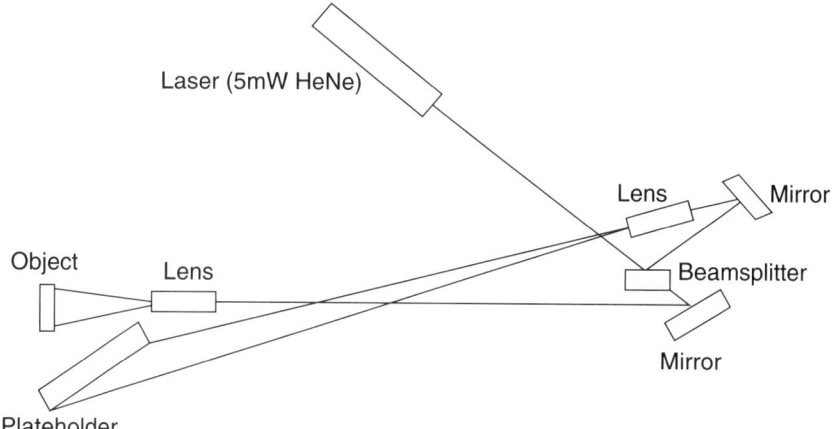

FIGURE 1-9: This diagram shows the basic setup for a reflection hologram.

The developed transparency must be illuminated with a duplication of the reference wave in order to obtain a virtual image of the original object. The original object wavefront is then reproduced, which in this case means that the image is formed from reflected light.

 Note: For a given illumination and observation geometry, there is only one reconstructing wavelength that will satisfy the Bragg's Law condition. Bragg's Law refers to the simple equation (as shown in Figure 1-10) where n is the Index of Refraction, lambda is the Wavelength, d is the thickness, and theta is the angle from the normal. This law was derived by the English physicists Sir W. H. Bragg and his son Sir W. L. Bragg in 1913 to explain why the cleavage faces of crystals appear to reflect X-ray beams at certain angles of incidence (theta). The variable d is the distance between atomic layers in a crystal, and the variable lambda is the wavelength of the incident X-ray beam (see Figure 1-10); n is an integer. This observation is an example of X-ray wave interference (Roentgenstrahlinterferenzen), commonly known as X-ray diffraction (XRD), and was direct evidence for the periodic atomic structure of crystals postulated for several centuries. Although Bragg's Law was used to explain the interference pattern of X-rays scattered by crystals, diffraction has been developed to study the structure of all states of matter with any beam (ions, electrons, neutrons, and protons) with a wavelength similar to the distance between the atomic or molecular structures of interest.[3]

20 CHAPTER 1

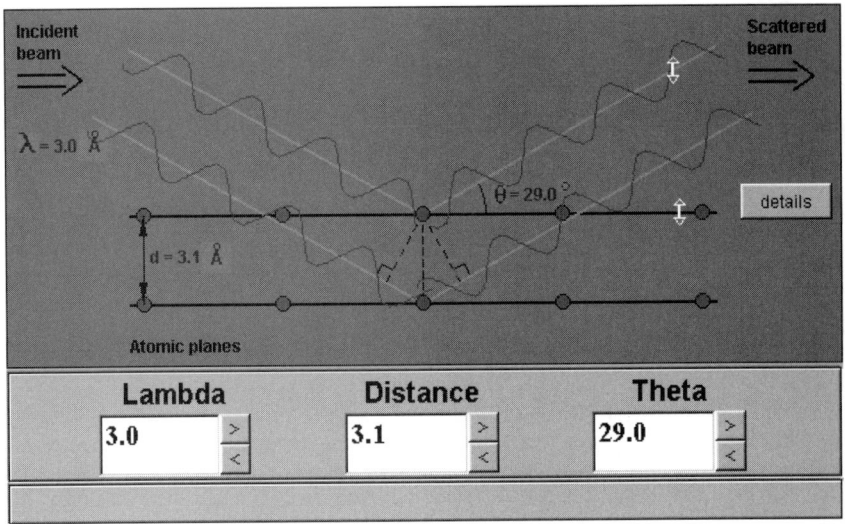

FIGURE 1-10: How Bragg's Law works.

Furthermore, the light of the wavelength required by Bragg's Law is reflected, while the remainder passes through the emulsion or is partially absorbed if white illumination is used. To obtain a real image, the transparency is illuminated from the opposite side, and again the image forms from single-color reflected light.

One would expect the color images to be about the same as the color of the light used for recording the hologram when viewed in white light. Other colors are submersed under the red reflected light, since the intensity of the Bragg reflected light is much less for colors other than red. In addition, the range of angles over which the red light is reflected overlaps the Bragg angles for other colors.

Finally, a greenish image may result instead of a red one if the emulsion shrinks during chemical processing. The shrinkage reduces the spacing between the Bragg planes, so that the hologram appears to have been made with light of a shorter wavelength than was actually used.

In Summary

A handful of companies are beginning to use holography for a multiplicity of applications: direct mail, catalogs, brochures, and collateral material.

These companies are running tests that are turning into rollouts. When asked about the success of the programs, few companies would respond, but many are using holography a second and third time.

- Because of the rapid burst of laser light, composite holograms can be made of machinery or other products in motion.
- Because of the great depths that can be hologrammed, entire rooms and scenes can be brought to the viewer.
- Scenes too dangerous for anyone to visit, even as dangerous as the inside of a nuclear reactor, can be hologrammed and studied or displayed.
- The pulsed laser, although costly and requiring special knowledge for handling, is fast becoming an important tool with which holographers can aid marketers and designers in the realization of their visions.
- Many types of holograms can be made but they are all based on two types: transmission and reflection holograms.

The next chapter discusses the basic hologram principles. This includes, but is not limited to, vibrations, light, and interference.

End Notes

1. Martin Berson, "Holographic Design and Production for Commercial Displays," HyperMedia Technologies Inc., P.O. Box 1282, Avon, CT 06001, USA, 2001.
2. Matt Amberg and Tim Hecox, "Holography," University of Washington, College of Engineering, P.O. Box 352180, Seattle, WA 98195-2180, USA, 1999.
3. Paul J. Schields, "Bragg's Law and Diffraction: How Waves Reveal the Atomic Structure of Crystals," Center for High Pressure Research, Department of Earth & Space Sciences, State University of New York at Stony Brook, Stony Brook, NY 11794-2100, 2000.

CHAPTER 2

HOLOGRAPHIC PRINCIPLES

In This Chapter

- Holographic vibrations
- Holographic light waves
- Holographic interference
- How to make single-beam reflection holograms

In short, holography is a science of recording the light waves of a laser reflected off an object. The hologram will *replay* the image of the original object when re-illuminated (placed under an ordinary light). The observer sees the recreated light waves, perceiving the object as if it were still there!

So, how does a hologram differ from a photograph? The holographic film records an infinite number of views of the object, whereas a photograph records only one view. Thus, when viewing a hologram, our left eye sees a different apparent view of the object than our right eye, and the image appears three-dimensional.

With that in mind, can holograms be made from flat artwork or photographs? Yes, but the laser recording will recreate only what it sees. A hologram created from flat art looks quite different from the astounding, fully dimensional image created when a hologram is made from an object or sculpture.

24 CHAPTER 2

Can moving holograms be made? Yes. Specifically filmed motion pictures from 4 to 30 seconds in length can be used to create animated holographic images that are astounding! This format is usually referred to as *integral* holography.

Can holograms be reduced or enlarged? Most holographic techniques produce images that are identical in size to the original object. New developments allow some enlargement and reduction if necessary.

Can you control the color of a hologram? Yes, to some extent. Certain types of holograms allow *individual color positions* for varied objects in your *scene*, while other types allow partial color *blends* at particular viewing angles.

Finally, can holograms be made from living subjects? Yes. Using a special *pulse* laser, moving subjects can be frozen in time!

Are the preceding questions answered to your satisfaction? Probably not! This chapter answers all of these questions and more, for those "inquiring minds who need to know," by covering the following basic holographic principles:

- Vibrations
- Light
- Interference
- Reflection

Vibrations

In itself, *vibration* is not a problem in holography. A holographic camera can be shaking violently without disturbing the hologram. You see, vibration is not the problem. To understand the problem, you have to think about how holograms are made. Basically, two waves of light meet and interfere on a film plane. The interference of the light waves forms a pattern. If one wave moves, it smears this pattern, comparable to taking your hands and smearing a fingerprint. It loses its information as a blur. The same happens with the hologram. The more one wave moves, the more the overall wave interference pattern is destroyed, until you dim totally out. However, no smearing occurs if both light waves vibrate in unison.

There is much less chance of one wave moving if your table is stiff. The stiffness prevents the table from bending. It is the bending that allows one of the light paths to change, thereby moving that light wave in the interference pattern. Therefore, stiffness is the real consideration in building an isolation table. The stiffer the table, the brighter the hologram.

RESONANCE

Resonance is another interesting related topic. Too many people build isolation tables based on some drawing they saw in a poorly written book. Usually, they end up making layers of unsuitable material to absorb vibration, while totally failing to deal with stiffness! In doing so, they create a whole new problem: *resonance.* Do you remember seeing an old film clip about a bridge in the Seattle-Tacoma, Washington region called *Galloping Gertie*?[1] The winds in the canyon hit the bridge at a resonant frequency and caused it to wave like a flag before collapsing.

Another example of resonance is when you sing in the shower and hit a note that makes the whole room seem to amplify that note. It happens when waves combine *very* constructively, adding as they match in size with the container in which they are placed.

Having too many layers is like putting a spring on a spring in table construction—you can bet it will resonate. In addition, instead of canceling the harmful vibrations, you are now amplifying them—not a good idea.

That being said, you should be very careful not to build a system that only makes your work more difficult. Make it a stiff one! Practice the KISS principle: Keep it simple, stupid!

LIGHT

There seems to be great deal of confusion about the nature of light; actually, it is pretty simple. Think of a basic atom. You have a nucleus, and electrons spinning around it. Electrons are the stuff that makes electricity. They have an electrical charge. Normally, the atom is a fairly balanced system. However, if you put energy into this system, you can pump it up. For instance, if you send in electricity, the electrons of the electricity bump up against the electrons of the atom. As they collide, it's like a billiard ball hitting another billiard ball. The first ball (electron) transfers some of its energy to the second ball (electron) and sends it off. In the case of an atom, the electron is sent temporarily into a higher orbit. It's something like blowing air into a balloon; the sphere the balloon occupies gets larger. Now, if you don't tie the end of the balloon, the air you just blew into it will come back out. The same thing happens in the atom. In addition, as the electron returns to a lower orbit, it releases the energy that originally sent it flying into the higher orbit.

A good way to envision the release of energy is to picture yourself in a pool of water chest high. If you sweep your arm just below the surface of the water, you make little whirlpools. This is the energy transferring from the movement of your arm to the water. The energy swirls like a vortex or a *black hole*. A wave of light is like a black hole or whirlpool of electromagnetic energy released as an electron returns from an excited orbit. The sweeping motion of the electron back to its normal or ground state is like the sweeping motion of your arm in the water. The electrical charge of the electron is transferred to a whirlpool of electromagnetic energy spinning off the atom—and it looks like a black hole. Think of how a typical swirling black hole with an accretion disk looks as it's gobbling up stars and debris. Now, look at it on its side from a cut-away view. It's like a sine wave, but that's only if you look at it sideways. Most drawings in books show light waves in this way.

Nevertheless, light waves are three-dimensional; hence, the black-hole model. You can think of the electron as being sheathed in an electromagnetic field. When it is energized by a collision, it gains a bit too much of this electromagnetic jacket. As the electron returns, a little bit of it twists and swirls free and tears off, thus becoming a free electromagnetic field swirling like a black hole. It swirls because of the spinning motion of the electron. This is a photon—a single wave of light.

Light

As discussed earlier, the best way to think of holograms is to envision them as impressions on light waves. In other words, light is a wave. All waves behave more or less the same. For one thing, they tend to echo. They reflect off many surfaces, similar to sound waves echoing to make *sonar*, or microwaves in *radar*. The wave is sent out, hits an object, and bounces back—pretty simple idea. However, what you don't think about is that when a wave bounces up against an object, it takes its shape. It's like pressing a piece of clay against a key. The key leaves a three-dimensional impression in the clay. With that in mind, if you imagine the clay as a light wave, basically the idea of holography is throwing the clay up against the key, having the key make an impression on the front of the clay, letting it bounce off, and finally storing the shape of the clay permanently.

However, you are dealing with waves that are not visible with sonar or radar. You can't see sound waves or microwaves. Nevertheless, with light waves, you are working in the visible spectrum; consequently, things that are visible tend to record on photographic film. Therefore, in effect, a hologram is a photograph of the impression left on the surface of a light wave after it has bounced off an object.

FILM EMULSION

Now, let's look at a hologram recorded on silver halide film. What is film? Well, first there is a base material of clear plastic or glass. Then there is a very important layer that contains the photoreactive chemistry. This is called the *emulsion*. It's a very special composition, and there's always room for it. It's like Jello™; plain old gelatin without any flavor or color, of course. Inside the gelatin there are two chemicals joined in a molecule. They are suspended like fruit in Jello. In an emulsion, each chemical retains its own identity, akin to what each piece of fruit floating in Jello does. First, there's silver. As we all know, silver has a unique property: it tarnishes when it combines with oxygen, and it turns black. Next, there's iodine, the stuff you use to kill germs in a cut. It's a very reactive chemical, so reactive that when it mixes with the silver to make silver iodide, it results in a silver that tarnishes very quickly.

A light wave's energy is transferred to the silver iodide molecule when it goes into this layer of Jello. Remember how a light wave looks like a swirling black hole? Well, try to imagine this black hole swirling its energy into the silver iodide molecule just like a wind-up toy. You give it a good twist and the energy goes into making the toy run. In the case of the silver iodide molecule, you give it a good wind-up of light energy. It's sort of like setting a mousetrap: You put your energy into pulling the trap open, and now it is set to snap shut. The same thing is happening in the silver iodide molecule. Light gives it energy to be ready to snap onto another atom. When you put it in a bath of photographic developer, it grabs the oxygen in the bath and tarnishes. That's why black-and-white negatives are black—so are holograms before they are bleached. Therefore, you can think of photography or, by extension, holography, as the art of selectively tarnishing silver in Jello where light has energized it.

In the case of a hologram, the patterns of light wave impressions are what is photographed in the layer of emulsion. Generally, film emulsion in holography runs about 10-microns thick; a micron being a millionth of a meter (a meter is approximately a yard) in size. That's pretty small, but a photon measures about a half of a single micron in size. That's smaller than an ant's eye. By comparison, the emulsion seems large to a photon. That's how people are able to photograph this microscopic wave impression in film and make holograms. Holograms are photographs of the three-dimensional impressions stored on light waves—sort of like fossils.

When you pour plaster into a fossil, you let it harden, and then remove it. You should have a three-dimensional sculpture of the impression that was left in the stone. Similarly, when you pour Jello into a mold, you let it set, and then remove it. You should then have a three-dimensional sculpture of the shape of the mold. Finally, when you put light into a hologram, you get a three-dimensional sculpture in light of the object that left its impressions on a photon and was captured within the thickness of a photographic emulsion.

Interference

As previously mentioned, holograms are photographs of three-dimensional impressions on the surface of light waves. Therefore, in order to make a hologram, you need to photograph light waves—this presents something of a dilemma.

As we all know, it can be problematic to take a photograph of a quickly moving object. If you've ever had a picture come back blurred from the film lab, you know the problem all too well. When a person moves too quickly in a photograph, the image blurs; and he or she is only moving at about 20 miles an hour. Try to imagine the problems associated with trying to photograph a photon. First, a light wave moves at the speed of light, which is about 186,000 miles per second. It's also more than half way to the moon in a second. It's considerably faster than someone's hand waving. In fact, it's so fast that the very idea of even capturing it on film would seem impossible. What we need is a way to stop the photon so it can be photographed—a technique known as *interference*.

Imagine standing on a small bridge over a pond of still water. Now, let's further imagine that you drop a pebble into the pond. As it hits the water it creates a circular wave. This wave radiates outward in an ever-growing circular path. We've all done this at some point in our lives.

Suppose you drop two pebbles in the water. Now you are creating two circular waves, each of which grows in size, eventually crosses the path of the other wave, and then continues on its individual expanding path. Where the two circular waves cross, you might say that they "interfere" with each other, and the pattern they make is called an *interference* pattern. Not too difficult to envision, is it? Interference is two waves interfering with each other as they cross paths. No permanent impact is left on either wave once it leaves the area of overlap. Each wave looks the same as it did before it crossed the path of the other wave. Maybe it's grown a bit, but that's about it. In that case, what's the big deal about interference?

Well, first, as waves cross paths and interfere, the pattern they make is called a *standing wave*. It is called a standing wave because it stands still, and since it stands still, it can be photographed.

Does this solve the problem of how to photograph something moving at the speed of light? It does answer the question in some ways, but, it doesn't answer the bigger question: Why does it stand still?

Let's envision a photon in order to understand all of this. Remember, it looks like a swirling black hole. If we view it from the side, it looks like a sine wave. Now, try to imagine a river whose streambed lies on a wavy rock formation that looks like a sine wave. This river would be full of rapids; in fact, it would be great for whitewater rafting. Although the water in the river is flowing furiously downstream, the pattern of water above the rapids is stationary. You might think of it as a standing wave. The wave energy is flowing through this standing wave without altering it, and vice versa. It is just a momentary pattern that the water takes as it passes over a bump.

When two waves pass through each other, each light wave acts like a bump to the other. Their respective swirling black hole shapes interact, the result of which is like rapids of light. The standing wave patterns are stationary, even though the light wave's energy continues to move.

Waves perform subtraction and addition when they meet. When two waves of equal size meet at their high points (called *crests*), they combine to make a wave twice as high at that point. Conversely, where two waves of equal size meet at their low points (call *troughs*), they combine to become twice as low. In addition, when one wave at its high point meets another wave at its low point, they subtract and cancel each other out, but not cancelled out in the sense of being destroyed. Actually, it's more a case of there being no light at that spot. If you follow the wave down its path just a drop further, it will meet the other wave at a different relationship and, again, be visible. It's really a situation of infinite possibilities. It's just like the patterns that are possible when the waves of two pebbles meet in a pond. At any point, you may notice that the standing wave pattern produces a place where the waves add together to get higher, or subtract to become lower, or even just go flat. A few terms are used to describe the possible encounters. If the waves add and get higher, it's called *constructive* interference. If the waves subtract or cancel each other altogether, it's called *destructive* interference.

Interference patterns could be considered the fingerprints of the encounter of two individual waves. Each object you make a hologram of creates its own interference pattern that identifies it.

There are two basic waves that come together to create the interference pattern in holography. First and foremost is the wave that bounces off the object that an individual is making a hologram of. Since it bounces off the object (thereby taking its shape), it is called the *object* wave. You can't have interference without something with which to interfere. A second wave of light that has not bounced off an object is used to perform this function, and is called the *reference* wave. Therefore, when an object wave meets a reference wave creating a standing wave pattern of interference, it is photographed and called a hologram.

Making Reflections

A single-beam reflection hologram is the simplest hologram. It can be seen in white light. It can be made with a small 5-milliwatt (mW) Helium-Neon (HeNe, pronounced *he-knee*) laser (producing a continuous red beam); a 10 to 20 power (10x – 20x) microscope objective (lens); and some holographic film. You can use a less powerful laser, but you will need to increase your exposure time. Conversely, if you have a more powerful laser, you should shorten your exposures. Polarized lasers have more useful power for making holograms. Make sure that the laser is rated TEM 00. This means that there will be no dark spots when you spread the beam out with a lens. For example, TEM 01 would produce a beam with a black hole in its center, and is referred to as the *donut mode*. TEM stands for Transverse Electromagnetic Mode, and the two numbers following are the graphic Cartesian coordinates of holes in the beam. Always exercise care when using a laser. Do not point it at anyone or look directly into it.

To start making single-beam reflected holograms, you will need a simple darkroom with a dark-green safelight with a 20-watt-or-less bulb at least eight feet from the film. You also need at least three developing trays;

holographic developer (consult with the people from whom you purchase the holographic film); a good holographic bleach; and a tray with water and a half a capful of Kodak Photo-Flo™ as a final anti-spotting bath. You also need a sink. A bathroom is okay if nothing else is available—just make sure it is light tight. You should also test the safelight by placing a small piece of your holographic film out where you do your developing for about 10 to 15 minutes. Develop it and see if it is exposed by the safelight. If it is, then you need more green on your filter or less light. A dimmer switch on the light can help.

The 5mW HeNe laser will cost around $200 to $300 from Meridith Instruments (http://www.mi-lasers.com/index1.htmlMeridith Instruments)[2] or MWK (http://www.mwkindustries.com).[3] A microscope objective is cheap. Film costs around $6 for a glass plate. Actual plastic film costs a bit less. Plastic film on a roll is the cheapest, but you have to buy a lot at once. Glass plates are considered best and are easiest to use. Developing and bleach formulas are available from the people who sell you the film. Try GEOLA Labs (http://www.geola.com/)[4] or Intergraph (http://www.intergraph.com/).[5]

This is just one of ways in which single-beam reflection holograms are made. Feel free to experiment with the geometry.

You should make an isolation table for your hologram setup if you can. While not essential for a successful exposure, it will help you obtain brighter holograms in a noisy environment. An isolation table can be as simple as floating a slab of granite, marble, or half-inch-thick steel on an inner tube. This acts as a shock absorber. Some people make a sandbox and float that on the inner tube. The table top should be rectangular, at least 4 feet by 5 feet. Longer is better than wider. You can also work on a hard-top table or a concrete floor.

The laser should be aimed through the lens. This will spread the beam out like a flashlight. This beam of light is called the *reference beam*. Mount the holographic film at a distance of about a meter or so from the lens. Now, mount the film at a 30- to 45-degree angle to the beam. This will be the *reference angle* you will use when you are lighting the completed hologram and are viewing it. Once you have aligned these elements, you should secure them in place so they don't move. You can sandwich the film between two pieces of quarter-inch-thick glass using carpenter's spring clamps. Glue the bottom rear back edge of the back plate of glass to the table top with hot glue available from a crafts or hardware store. Get a hot glue gun with a trigger feed. Make sure that the glue is only on the back edge. Do not touch the hot glue after it comes out of the gun; it is very hot and can cause serious burns. Wait about a minute or so until it cools and hardens. To remove it after it cools, you can wet it with a little acetone to help break its bond.

You will need to make a glass plate holder if you are shooting glass plates. A pair of U Channel metal rods can be glued to a bottom and top support bar so that the plate can slide down into it to the table. This holder should be hot-glued to the table with buttress supports from the top of both U channels to the table top. The plate holder can also be made of hard wood that has grooves routed out. You can use some modeling clay in each corner of the plate to prevent it from moving after you slide it into the channels. The plate and object are mounted sideways. You should paint all surfaces and plate holders flat black to prevent any unwanted light from degrading the quality of your hologram.

An *object beam* is the laser light reflecting off the object. It then bounces back through the holographic film where it meets the *reference beam*. The two beams pass through each other, creating an *interference pattern*. This interference pattern is what you are photographing on the holographic film.

Directly behind the glass (the opposite side of the glass than the laser), you should now mount your object sideways with the top of the object nearest to the laser. It helps to actually have it touch the glass. The laser beam will *reflect* off the object. You should try to use objects that are not too deep—a good beginner's image is coins. The choice of your object is critical to your success. The object must be very solid. Metal, ceramic, seashells, plaster, or stone are good. Paper, plastic, feathers, and string are very temperamental and are not good for your first attempts because they are more prone to vibration problems that might ruin your exposures. The object should be colored white, red, gold, or another light color. Black, green, or blue will disappear in a hologram, so do not use objects made exclusively of them. However, these colors can be used as long as they are within light colors. Remember, the laser is shining red light on your object. The color red next to the color white on your object will all look like a single color. Red print on a white background will disappear. Try viewing your object through a red filter to get a good idea of how it will look in laser light. You can use hot glue to adhere them to another plate of glass that you press up against the back plate (the far side) of the glass being used as your film holder. You can also use more than one layer of glass plate model holders for more depth. If you do, you must glue the bottom of each model holding plate to the table surface. In addition, take wooden or metal rods and glue them at an angle from the top of each plate of glass to the table top as a flying buttress. This helps to eliminate vibration problems that would cause the hologram to be dim or not work at all. The sturdier you secure the model and plate holder, the more likely it is that your hologram will be successful.

Now take a piece of black cardboard and lean it up against the front of the laser as a *shutter* to stop the beam of laser light from reaching the film holder before you load holographic film. Next, take a piece of film, and in the dark, put a corner of it between your lips. The side that sticks to your lips is the emulsion. Place that side toward the object (the coins), and place the front plate of glass over the film and clamp it to the back plate with the spring clamps. You should use three clamps: one on each side, and one on the top. Now, walk out of the room for at least 30 minutes. This allows the *camera* you have just made to *settle* and prevent vibration problems. Remember, light waves are incredibly small, and it takes very little to move them.

Now, come back in the room and walk next to the shutter. Wait one minute, and then lift the *shutter* for about 16 seconds. Close the shutter. You might want to do a bracket exposure. In this process, you cover three quarters of the film with a black card. After you have exposed the first quarter of the film, you slide the black card back to allow the next quarter of the film to be exposed. Wait a minute for the system to settle, and expose the film for an additional 6 seconds. Then, you should replace the shutter and move the black card back once again to allow another quarter of the film to be exposed. After you wait another minute to allow the film to settle again, open the shutter for another 6 seconds. Finally, you need to remove the black card completely, wait another minute, and expose a final 6 seconds.

 Tip: After you develop the film, you will be able to see which amount of exposure time worked best. You can then adjust future exposures successfully.

Next, take the film out from between the glass plates and develop it in your darkroom. If the darkroom is far from the film, you might want to transport it there in a lightproof box. Follow the film manufacturer's procedure of development. Generally, you will develop the film until it is black, and then bleach it clear and wash it. You wear latex gloves and an apron or lab jacket. Make sure the darkroom is well ventilated. It is now safe to turn on the room lights. You will not see an image until the film is dry. Nevertheless, you can see rainbows through the film if you hold it up to a light. This is the diffraction from the interference pattern you have just photographed on the film. It is a good sign that your hologram has worked. While it is wet, the emulsion is swollen and thereby visible only to those who can see infrared light. You should now hang your hologram to dry, but don't blow-dry the film in front of a fan. Also, don't use a hot blower; it can distort the film or start a fire.

By using a bright spotlight (halogen is best), you can now view a dry hologram. You can also use a clear light bulb, not a frosted bulb. Do not use florescent lighting to view holograms; it will make the image look blurry. Sunlight will work, but it also will bombard the hologram with UV rays that tend to darken the film over time, thus causing it to *print out*. It will still be a good hologram, but the film will no longer be clear. Do not get the hologram wet once it has dried; it can erase the image or spot it permanently. If you do get it wet, quickly soak it in water and carefully re-dry it.

Finally, you may have some trouble finding the image the first time. Remember, it must be lit with light coming from the same angle that it was created with the *reference angle*. Try slowly rotating the film and turning it over until you see the holographic image. It should appear behind the film as a *virtual* image, and should look exactly the way your model did when you shot it (*orthoscopic*). If the object appears to be sideways, you forgot to mount it sideways when you mounted the model. In that case, it's time to go back to your setup, rotate the object in your holographic camera 90 degrees, and re-shoot. Now, if you flip the film over, the holographic image will appear to jump out in front of the film as a *real* image. This image is inside out, or *pseudoscopic*. It helps to have a piece of black card stock behind the film to make it easier to see.

In Summary

Think of a hologram as a window. Anywhere you look through a window, you see what's on the other side. If you were to paint the window black and scratch a hole in the paint on the left side of that window just big enough to look through, you would see everything on the other side of the window. It's like looking through a peephole. If you then scratch another viewing peephole somewhere on the right side of the window, you still can see through, but from a different perspective. This is the same effect that each broken piece of a hologram would display. Just remember that if you have two broken pieces taken from opposite sides of the hologram (and you are looking at an object that looks differently from each side), one piece may let you see just one of those sides, while the other piece will let you view the other side. Therefore, you might say that each piece of a hologram stores information about the whole image, but from its own viewing angle. No two pieces will give you exactly the same view.

- The more one wave moves, the more the overall wave interference pattern is destroyed, until you dim totally out. However, no smearing occurs if both light waves vibrate in unison.
- An example of resonance is when you sing in the shower and hit a note that makes the whole room seem to amplify that note.
- A wave of light is like a black hole or whirlpool of electromagnetic energy released as an electron returns from an excited orbit.
- After it has bounced off an object, a hologram is a photograph of the impression left on the surface of a light wave.
- Generally, film emulsion in holography runs about 10-microns thick.
- Interference patterns could be considered the fingerprints of the encounter of two individual waves.
- Reflection holograms are made by interfering the reference beam with the object beam from opposite sides of the holographic film. It is visible in white light.

The next chapter discusses how to shape empty space with light. This includes, but is not limited to, the following: experiments, spectral harmonics, holographic optical elements (HOES), techniques, dreamscapes, and reflections and projections.

End Notes

1. Jason Sapan, HOLOGRAPHIC STUDIOS, 240 East 26th Street, New York, NY 10010-2436, USA, 2001.

2. Meredith Instruments, P.O. Box 1724 / 5420 W. Camelback Rd., #4, Glendale, AZ 85301, USA, 1999.
3. MWK Industries, 1269 W. Pomona, Corona, CA 91720, USA, 1999.
4. GEOLA, P.O. Box 343, Vilnius 2006, Lithuania, 1999.
5. Intergraph Corporation, Huntsville, AL 35894-0001, USA, 1999.

CHAPTER 3

USING LIGHT TO SHAPE EMPTY SPACE

In This Chapter

- Holographic experiments
- Spectral harmonics
- Holographic optical elements
- Holographic landscapes
- Holographic imagery
- Holographic art
- Holographic display
- White-light-transmission holograms
- Full-color holograms
- Laser diodes

In this chapter, we begin by looking at some early work in holography, and work our way to the present. Early experiments in holography included working with black-and-white and Benton white-light transmission. Pioneering of holography also took place with the use of Holographic Optical Elements (HOEs) in the image-making process to produce animated imagery. All of the preceding and more is discussed in this chapter, along with the technological and conceptual considerations of holographic art and high-tech art in general.

Early Experiments

In early holographic experiments (early 1970s), the initial goal was to make a black-and-white transmission hologram that could be seen in white light. Attempts were made to do this by using a large lens system, and then focusing the image on the holographic plate. This is best described as a white-light one-step process. It did produce a black-and-white image. However, the small depth of field and the color separation that occurred at the edges were problematic.

The next approach involved using the Benton white-light-transmission technique,[1] plus an extra reference beam to arrange and line up the different colors that produce white. The image comprised three small light bulbs in three porcelain sockets, with a wire connecting the electrical terminals. Each element lent itself perfectly to this technique. Real results showed the paradox of one real light bulb activating three holographic ones. At the same time, the experimenters were frustrated by the fact that diffraction made each color image a different size; this meant that when the piece was viewed from different sides, the colors separated, especially where the image extended in front of or behind the plate. This limited depth of field and color misalignment made dealing with realistic imagery a cumbersome experience.

In the late 1970s, experiments with holograms involved creating required 8" × 10" glass plates. The use of glass, as opposed to acetate or Mylar, was necessary to achieve the quality the experimenters wanted. They also found that using glass when creating the master eliminated the *wobble* seen in previous experiments, and made the positioning of the master more flexible.

The experimenters then decided to start new transmission experiments on a basic level by exploring a space defined by lines and dots. They created the lines by illuminating the sanded edges of plate glass, and the dots by making stationery air bubbles along the sides of three water-filled tanks measuring 14" × 14" × 1". The three tanks were separated by several inches to create a spatial effect. The bubbles acted like little lenses, and their random distribution gave the field an organic appearance. In recording the transfer, the experimenters combined several master holograms and placed the different masters in varying locations relative to the transfer plate to create multicolored images.

Spectral Harmonics

Holographic experimenters also discovered during the beauty that results from using several different colors that are visible simultaneously. In synthesized music, one note can sound cold, but several notes can create a chord with many harmonics, causing a richer and more beautiful sound. The same happens with spectral colors. A monochromatic image looks artificial and cold, but through the use of two or more different spectral colors, all visible at the same time, the work starts to beam *harmonics*. For the experimenters, it was important to make sure that the imagery could handle spectral color. A familiar object, with its natural subtlety of mixed colors, is often disappointing in monochromatic light. The dots and lines in the first photon studies were conducive to spectral color, and at the same time created an unfamiliar space—prerequisites for good holographic harmonics.

Now let's look at image-making techniques in holography; the use of the white light transmission setup, the specific design and use of Holographic Optical Elements (HOEs), and imperfect collimating lenses in the mastering steps.

Making Holographic Images with HOEs

The holographic image shown in Figure 3-1 was made in 1980.[2] It incorporates an image element that was created with light itself.

Figure 3-2 shows a Spatial Location Program (SLP) linear image [Berkhout, 2]. This one in particular is a linear image multiplier. It is a two-beam recording with a small angle in between, resulting in a simple holographic grating. The angle determines how many orders of diffraction can be expected around the zero order. Depending on the exposure and the development of the holographic plate, different orders of diffraction can dominate or can be balanced in relative brightness. When used in subsequent setups, it results in a set of images displaced linearly. Care needs to be taken to keep the optical noise level of the developed holograms as low as possible.

FIGURE 3-1: Event horizon.

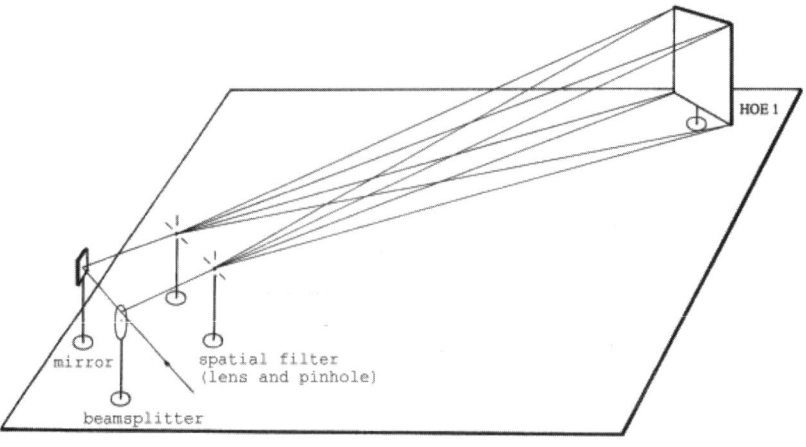

FIGURE 3-2: HOE linear image multiplier.

Figure 3-3 on the other hand, is a diagonal image extender [Berkhout, 3]. The two important distances in this setup are the 45" between the spatial filter and the plateholder, and the 40" between the plateholder and the diagonal line. This almost makes it a lensless Fourier-Transform hologram with a large angle. It was an intuitive approach to create the desired transformation from a point of light into a line of light.

The angle in between the two beams is determined by the requirements of the setup into which it has to fit. In this case, the element is used to make Figure 3-7, as shown later.

The width of the diagonal line is about 2mm, and it makes a 45-degree angle with the table. This angle is suggested by the design of the image and the nature of the white-light-transmission technique. A horizontal line would not show any perspective when viewed from side to side.

The masters for making a white-light-transmission (WLT) hologram are made from 2" glass strips cut from 8" × 10" or 12" × 16" plates as shown in Figure 3-4 [Berkhout, 4]. A 2"-wide master allows for a tight setup and more flexibility.

HOE #3 provides the collimated reference beam to record Master #1 (M1). It can be replaced by a large cylindrical lens or mirror, as long as it collimates the beam.

The sphere is a large, clear, plastic ball attached with hot glue to its support as shown in Figure 3-5 [Berkhout, 4]. The object beam is directed into

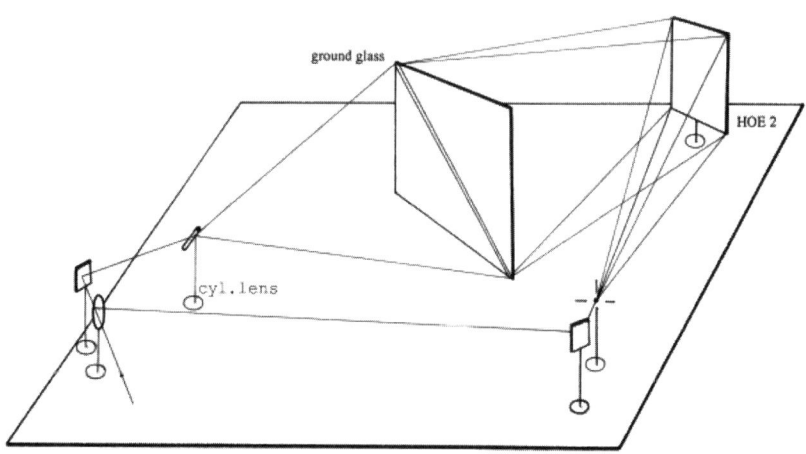

FIGURE 3-3: HOE diagonal image extender.

42 CHAPTER 3

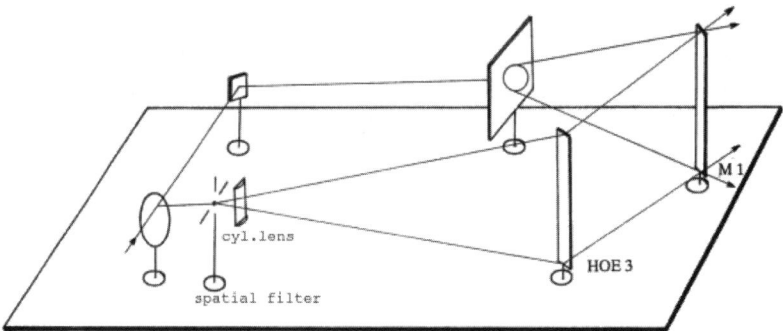

FIGURE 3-4: White-light-transmission (WLT) hologram—the sphere.

FIGURE 3-5: The collimated reference beam.

the hot glue, which acts as a scatterer and makes the sphere radiate from the inside out.

With the help of a spatial filter and a cylindrical lens, the object beam is shaped into a (1"– 2"-wide) strip of light as shown in Figure 3-6 [Berkhout, 5]. This strip is multiplied by HOE #1 and illuminates the field that is constructed from foam core, glued to 3/4" plywood and standing on edge. All

USING LIGHT TO SHAPE EMPTY SPACE   **43**

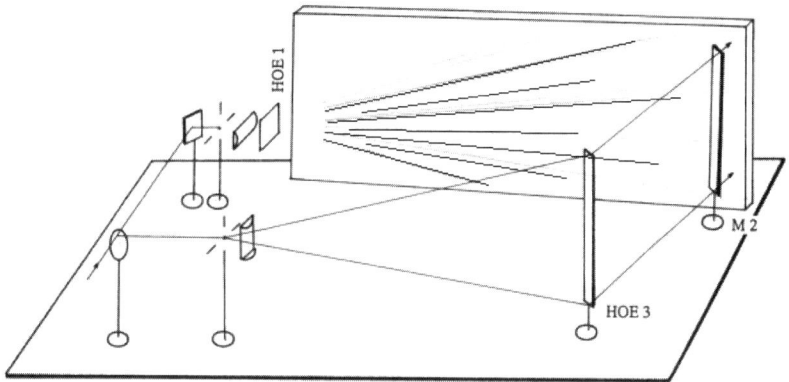

FIGURE 3-6: The object beam—the field.

the setups used to be sideways; however, today the masters are right side up—using the HOE #3 collimator positioned horizontal and overhead.

The skimming light illuminating the field from the back circumvents the coherence length limitation of most HeNe lasers. At the same time, it provides excellent object depth and an interesting illusion of perspective.

The hologram shown in Figure 3-7 makes use of the aberrations of a large glass lens (16" diameter) that was salvaged from an old projector [Berkhout, 5]. Its real focal *point* is *smeared* in 3D space. HOE #1 multiplies this focus, and HOE #2 turns this set of moving foci into a sequence of

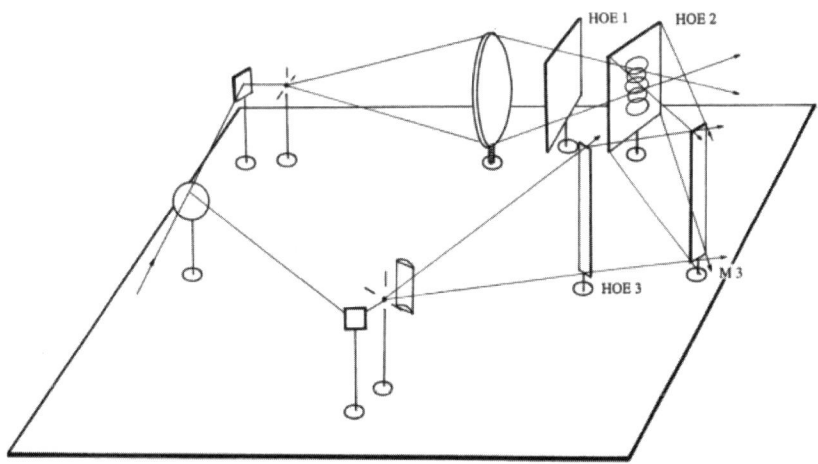

FIGURE 3-7: The aberrations of a large glass lens—moving energy.

bending and moving diagonals, visually expressing the optical characteristics of the glass lens and the two HOEs.

The animated nature of this setup requires a balance between the amount of image movement, image size, and the perception of space. Moving the real focal point of the glass lens along its optical axis, and/or changing the distance HOE #2—Master #3 (M 3), can adjust for the most comfortable binary vision as shown in Figure 3-8 [Berkhout, 6]. This optical transform together with the collimated reference beam from HOE #3 is recorded as Master #3.

The geometry of this setup determines the important angles and distances of the previous optical configuration as shown in Figure 3-9 [Berkhout, 6]. In particular, the angle between the object and reference beam and the distance between the object and master are crucial, as they determine the colors and the composition of the final work.

HOE #3 provides the collimated playback light for the three masters as shown in Figure 3-10 [Berkhout, 7]. They are placed in line, using the zero-order light of the previous master. This makes efficient use of the space and light that is available in the setup. However, the playback beam is attenuated with the optical noise from the previous master, and this can show up as noise in the final image.

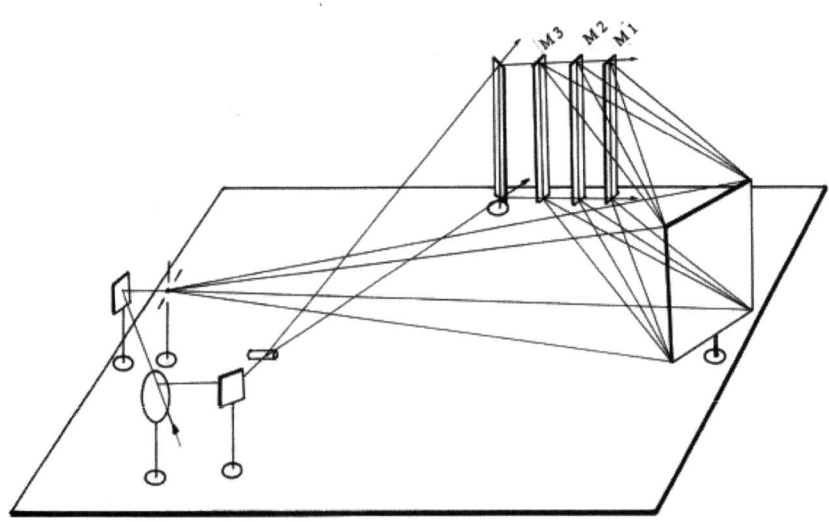

FIGURE 3-8: The transfer setup—event horizon.

FIGURE 3-9: The geometry of this hologram determines the important angles and distances.

A 2D diagram of the transfer setup is included. The three real images from the masters are recorded simultaneously, together with the reference beam, as a white-light transmission hologram. This recording is ideally accomplished with a collimated or converging reference beam. Also, all aberrations and blemishes of a collimating lens will show up in the final image. It does distort the holographic space, because the conjugate of the reference beam is used in the playback. See the sidebar, "Types of HOEs," for further information on recent HOEs technology.

■ TYPES OF HOES

Recent advances in optical pickup heads using holographic optical elements (HOEs) have brought about a practical servo-signal detection system that can realize compact integration of HOE, laser diode, photodetectors, and objective lens. This new type of holographic servo method employs a blazed HOE placed in close proximity to the objective lens so that wide allowance range of the optical system is maintained, thus allowing the lens and HOE to move simultaneously.

Hybrid singlet lenses, on the other hand, are composed of a plano-convex or a plano-concave glass lens and of a volume phase HOE in dichromated gelatin, coated on the plane surface of the glass lens. A design method that combines usual ray-tracing through a refractive lens and, for the HOE, is now available. This latter method is based on a ray-tracing procedure that uses the minimization of the mean-square difference of the propagation vector components between the actual

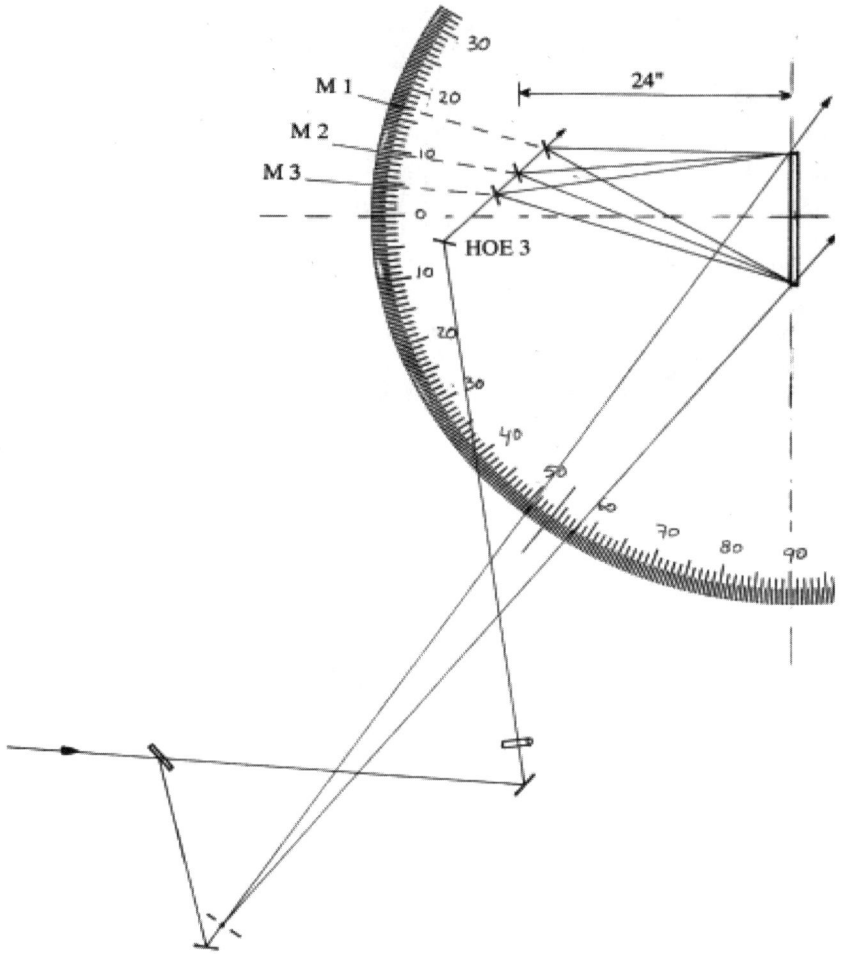

FIGURE 3-10: The playback beam.

output wave fronts and the desired output wavefronts. Considering the one-dimensional case and a given spectral bandwidth, one can determine by simulation, the image spot sizes, the distortions at the image plane, and the mean-square difference of the propagation vectors. At the end, the results can be compared with the performance of the optimal HOE alone, the spherical HOE, the refractive lens alone, and a diffractive system made of two HOEs.

Many experimental investigations are performed to produce new optical recording materials for holography. They intend to improve applications such as the construction of aberration HOEs, filters, aberration correctors, and nondestructive testing, as well as optical data storage. All of these applications require materials with high optical performances. Because of their excellent ability to record high spatial frequency gratings, photopolymers are seen as promising holographic recording media. Indeed, one of the greatest advantages of photopolymers is their high theoretical molecular resolving power. The addition of dichromate salts to a variety of water-soluble polymers such as gelatin, albumin, and poly (-vinyl alcohol) (PVA) generally results in photosensitive materials that become insoluble in water after exposure to the action of light.

Furthermore, a novel copying technique specially developed for the batch reproduction and manufacturing of large format HOEs that are used in the manufacturing of solar concentrators is now available. The HOEs are designed by means of computer programs that facilitate the optimization of the recording geometry for on-axis or off-axis operation. The HOEs are recorded in dichromated gelatin layers (DCG) and are subsequently subjected to chemical and thermal after-treatment processes in order to promote the desired characteristics and suppress the undesired properties. These *master* holograms are used for the manufacturing of the copies. The technique developed for the reproduction of transmissive holograms (holographic lenses) requires the copying procedure that the master hologram generates, an object wave that carries the information, and a reference wave needed for the recording of this information. Since the master is placed in front of the copy during the reproduction process, the master must have a 50-percent diffraction efficiency across the entire aperture. This requirement restrains the permissible variation of the exposure energy during the reproduction process to very narrow bounds. Usually, dichromated gelatin films exhibit a steep dependence of the diffraction efficiency with the exposure energy. Hence, a small variation of the intensity at a given point of the hologram may produce a very large change in the diffraction efficiency. A new method has been developed that is based on film properties control that facilitates the manufacturing of master holograms with 50-percent diffraction efficiency. The industrial fabrication of large-format holographic solar concentrators requires simple and inexpensive reproduction technology. A novel, dry copying technique was developed and tested in the laboratory.

In addition, a holographic system formed by two holographic lenses that, when working with white light, allows selection of a particular wavelength is also available. If adequate spatial and temporal coherence conditions are obtained, it is possible to copy holographic optical elements (HOEs) with partially coherent light by using optical systems formed by holographic optical elements.

Novel photoreplication technology for deep groove surface relief holographic optical elements (HOE) was developed recently at Du Pont. Dry photopolymer embossing (DPE) technology uses Du Pont proprietary materials and processes to replicate precisely almost any type of surface relief or embossed holograms. Very sophisticated surface relief holograms with the high width/depth aspect ratio of 1:20 can be faithfully replicated by this technology. Dimensions of the replicated grooves or other surface relief structures vary from 0.1 μm to 3.5 μm. Such HOEs can be produced in different geometrical configurations and sizes that are actually dictated by the master hologram. Embossed HOEs can be fabricated on a plastic film or sheet substrate of different types, thickness, and shapes. Glass and other inorganic materials can also be used as substrates in some applications. To replicate such deep groove surface relief HOEs, the master hologram should be recorded in metal, glass, or other hard-surface material. DPE technology may provide substantial technological and economical advantages over existing conventional replication processes, such as thermo-embossing, injection molding, wet photopolymerization (2P-process), and injection-reaction molding, in replication of different types of surface-relief HOEs.

Techniques specifically developed for the design and manufacturing of specialized holographic optical elements (HOEs) that are used in the fabrication of integrated optical systems for laser-Doppler velocimetry (LDV) applications are now available. The specialized HOEs needed for the construction of an integrated LDV-optics are beam-splitters, lenses and/or lens arrays, phase correction plates, and the corresponding waveguide structures. The HOEs are designed by means of computer programs that facilitate the optimization of the recording geometry and provide information for the correction procedures needed for optimized performance. The HOEs are recorded in dichromated gelatin films and are subsequently subjected to chemical and thermal after-treatment. These procedures guarantee the realization of HOEs with high diffraction efficiency and low scattering losses. The optimized holographic process has been previously described.

Finally, the design of the testing canal of adaptive optical systems based on the hologram optical element (HOE) coated on the active element is now considered. The paraxial analysis of the HOE recording and the testing canal are now available, as well as the analysis of holographic testing canal with respect to the fourth-order wave aberrations. Based on this analysis, the tolerances for the hologram construction optical parameters are determined. The results of modeling of the aberration arising both in the main canal and in the testing canal for the various deformations and the displacements of the active surface are also available.

Now, let's look at a compact hologram display and the design for a full-color version using diode lasers; in other words, white-light-transmission (WLT) holograms.

White-Light-Transmission Holograms

Unlike reflection holograms that can hang on the wall, a WLT hologram requires placement at eye level, several feet in front of the wall on which the light fixture is positioned. To accomplish this, the image model should be positioned on a tripod that stands on the floor, or suspended from the ceiling using monofilament. Both arrangements work successfully for a gallery or a museum, but it takes up a lot of space at home.

The need for space in the display of WLT holograms results from the holographic recording geometry—a two-step process of making a hologram of a hologram. This requires the use of a conjugate reference beam, the beam that is traveling in the exact opposite direction. In the recording step, this can be accomplished with a large collimating lens or parabolic mirror. A long reference beam should be used in the recording. In the display, you should use a long distance between the light source and the hologram.

Therefore, a compact display can be designed for white-light-transmission (WLT) holograms (30cm × 40cm). The unit, measuring 40cm × 40cm × 15cm, contains a tungsten halogen bulb and a folded playback beam, providing an aesthetically pleasing alternative to the usual large space needs of WLT holograms. The inevitable availability in the near future of red, green, and blue diode lasers will bring the possibility of their use in the playback of full-color transmission holograms. This part of the chapter explores the recording and playback requirements, and several approaches for future display designs.

The best illumination for a WLT hologram is a point source of white light. A close approximation to that is a clear glass light bulb with a single, straight filament, positioned vertically (as viewed from the hologram surface). This arrangement will ensure a single, clean, and sharp image with good depth of field.

Various Display Options

Now, let's look at how to do away with the light bulb approach previously discussed and incorporate it into the hologram. The first approach is to design a laser-viewable transmission hologram using a very short reference beam and two or possibly three diode lasers for playback illumination. In this way, you should have duplicated the original recording beam, allowing a deep space experience without distortions.

Another solution would be the edge-lit holographic technique that was developed at MIT.[3,4] Unfortunately, the size of the work (80cm × 30cm) would require more research and development.

The next available option is to shorten the playback beam and to fold it up into the smallest possible package, while making sure that the image distortions are acceptable as shown in Figure 3-11.[5] The outside dimensions of the final design ended up measuring 40cm × 40cm × 15cm deep, the hologram being 30cm (height) × 40cm (width) as shown in Figure 3-12 [Berkhout, 3]. The center axis of the playback beam makes a 60-degree angle with the axis perpendicular to center of the hologram (the normal).

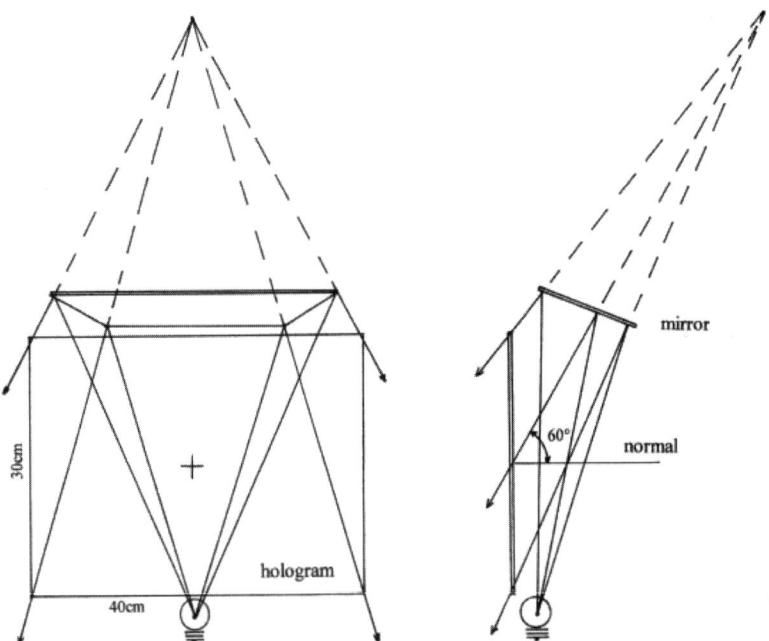

FIGURE 3-11: Playback beam—front and side views.

USING LIGHT TO SHAPE EMPTY SPACE   **51**

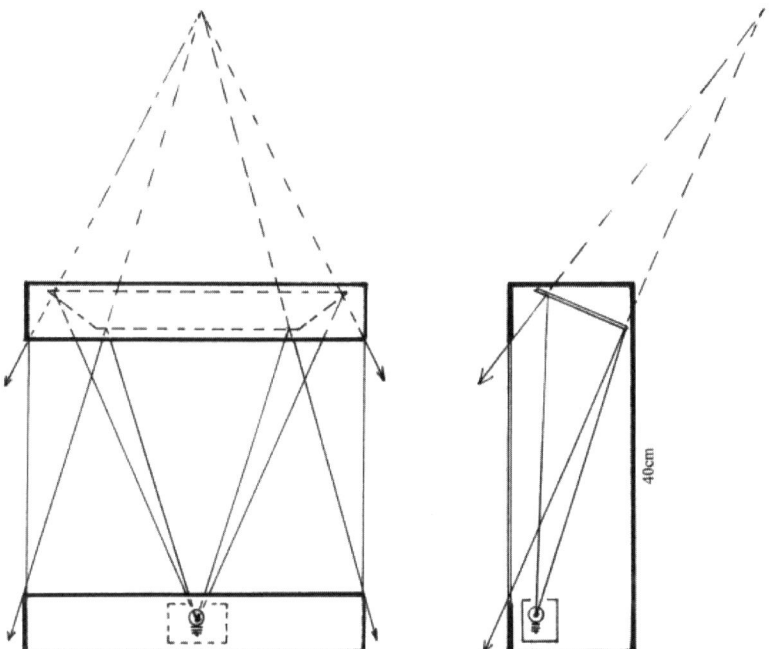

FIGURE 3 12: Framing for the hologram and light source—front and side views

The distance from the bulb's filament to the center of the hologram is 57.5cm. To accomplish this, a 55cm diameter, 225cm focal length parabolic mirror should be used to shape the reference beam for the recording of the hologram into a converging beam as shown in Figure 3-13 [Berkhout, 4].

For the light source, you should use a 12-volt, 35-watt, tungsten halogen bulb—small compared to the usual 150-watt halogen bulb that is normally used. The bulb should have no reflector or focusing lens. The lamp housing should allow only light coming directly from the filament (positioned vertically) to reach the hologram. The rest of the light is blocked and absorbed by the housing.

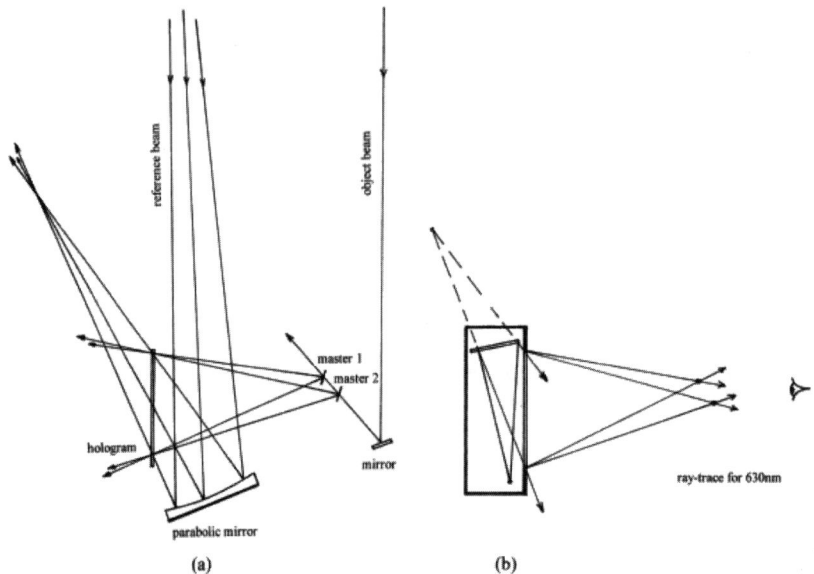

FIGURE 3-13: The optical setup to record (a) and display (b) a WLT hologram with two masters.

Playing Back Holograms with Diode Lasers

You can also use laser diodes to view holographic images. Laser diodes have many untapped possibilities. For one, using laser light for playback will keep both the horizontal and vertical parallax intact, with no astigmatism, and no distortions from the two-step process. By using three colors from three diode lasers, theoretically holography could open up to natural color work. Beyond that, the compact size, low cost, long life time, and low energy needs of the diode laser are key features that could make them very useful in holography.

Looking at a laser-viewable transmission hologram is always exhilarating. The *solid* experience of virtual space without glasses and the exquisite detail possible makes this holographic technique stand out. However, up to the present day, the need for expensive lasers has confined this experience to the laboratory. New solid-state laser development could change that.

COLOR-RECORDING A TRANSMISSION HOLOGRAM

Colors can be recorded simultaneously in a holographic emulsion that is panchromatic, but to prevent cross-talk between the colors during play-

USING LIGHT TO SHAPE EMPTY SPACE **53**

back, each reference beam has to be spatially separated from the other colors by 90 degrees. In playback with three colors, cross-talk would result in a red, green, and blue image for each recorded color—nine misaligned colors in all.

The playback geometry has to duplicate the recording geometry. Accommodating three or four separate reference beams makes this recording setup truly three-dimensional (see Figure 3-16), as opposed to the usual two-dimensional structure of a holographic setup with one reference beam. For the illumination of the subject matter, the beams need to be combined and aligned with dichroic mirrors before reaching the object.

Unlike a photograph, a hologram cannot record fluorescent colors, and it can be difficult to record colors of high saturation. For instance, when the object's reflected narrow band of wavelengths falls outside the region of wavelengths of the illuminating lasers, those colors are not recorded.

PLAYING-BACK LASERS

Figure 3-14, drawn to scale, shows the display for a laser-viewable hologram, measuring 60cm × 60cm [Berkhout, 5]. To keep the undiffracted

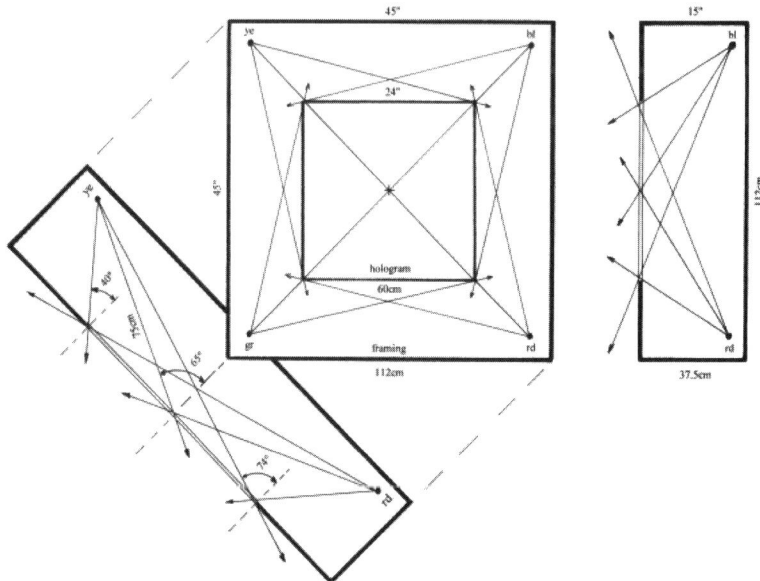

FIGURE 3-14: Three-color laser display with mirrors—front and side views.

54 CHAPTER 3

(zero order) light out of the viewing area, you should position two reference beams off center, in the top two corners. The third reference beam should be centered and comes from below.

To illuminate the hologram evenly, you should fold and direct each beam with a front surface mirror. The lasers with expanding optics should be placed at the foci of the separate color holograms (marked by dots in the drawing, rd, gr, and bl). Those foci are recorded in the hologram as the locations in space where the reference beams originated (usually a pinhole in front of a lens). The alignment is done by positioning the mirrors, the laser, or both. A fiber optic, delivering laser light and fitted with an expanding lens, could be very effective

A simpler version, one without mirrors, requires fast expanding optics. In a four-color version (see Figure 3-15), the distance from each laser to the center of the hologram is 75cm as compared to 142cm in the mirrored version as shown in Figure 3-14 [Berkhout, 5 and 6]. In addition, because of the extreme off-axis nature of the recording and playback, the intensity

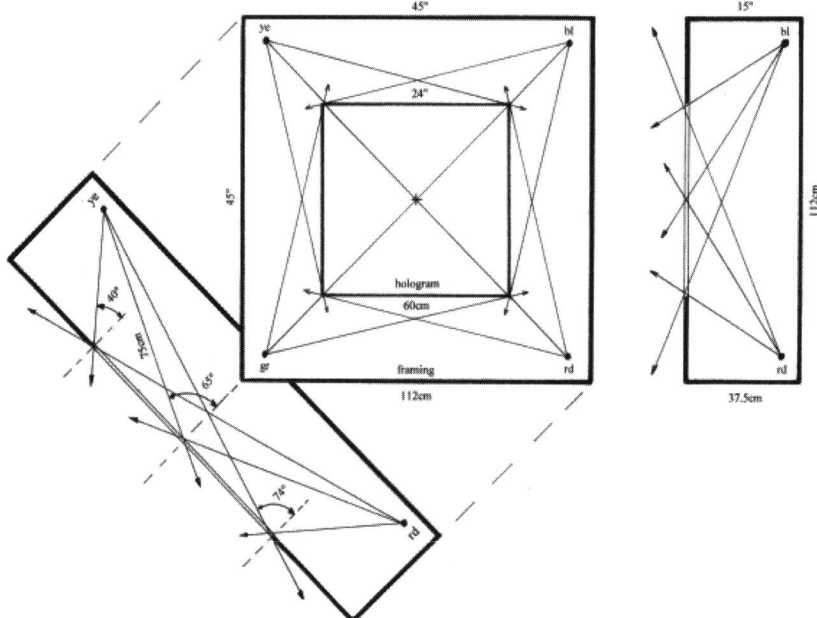

FIGURE 3-15: Four-color laser display—diagonal, front, and side views.

USING LIGHT TO SHAPE EMPTY SPACE 55

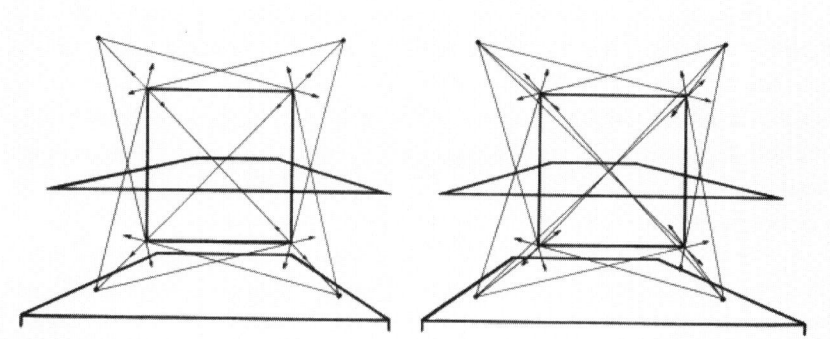

FIGURE 3-16: Location of four reference beams on optical bench (in free-view cross-eyed stereo).

distribution of the light over the hologram surface becomes more uneven. Positioning of the beams for the recording is a challenge. Hopefully, fiber-optics will make this more manageable. Figure 3-16 helps to visualize the position of four reference beams on an optical bench, and the object space that is available in the hologram [Berkhout, 7].

PLAYING BACK A USER-FRIENDLY LASER

Finally, the laser (diode) housing incorporates the adjustable expanding lens that illuminates the hologram evenly at the intended distance. At the same time, this lens should minimize astigmatism of the lazing source itself. Diode lasers usually have a linear aperture emitting light. Astigmatism could be a problem when the recording and playback lasers are different. Using the same lasers to record and play back the hologram ensures the most accurate spatial image reproduction (no size change from wavelength difference). This leaves the optical alignment and the shrinkage of the emulsion as the two remaining sources of distortion in the system.

Smearing (blurring) the laser speckle (while maintaining spatial coherence) would improve the visual appearance of the holographic image and erase the imperfections of the expanding optics at the same time. This makes spatial filtering, to *clean* the beam, unnecessary. It could be accomplished opto-electronically or electrically by shifting (moving) the phase of the emitted light randomly.

In the best of scenarios, a fiber optic, coupled on one side to the laser diode and at the other end fitted with a built-in expanding lens to illuminate the hologram, would make the display lightweight, easy to adjust, and aesthetically exciting. Depending on the efficiency of the hologram and the ambient light, a 10mW laser at 635nm is adequate to play back a hologram measuring 20cm × 25cm. When fiber optics are guiding the light, the power requirements might double due to coupling losses.

In Summary

- HOEs can be used very successfully in the creation of holographic artwork. Specifically, the image-making possibilities are greatly enhanced.
- When illuminated with white light, HOEs exhibit spectral lines or fields in a unique and pure optical space.
- The pure optical space is often referred to as hyperspace because an object, multiplied along the Z-axis, appears to get larger and larger the further it recedes from the emulsion.
- The pure optical space is the opposite of what happens in familiar space, and not an easy one to make visible due to the contrary clues about space. However, this property is one of many that make this medium so fascinating and eye-opening.
- The prospect for a successful color display depends a lot on new photonic tools that will become available in the future.

The next chapter answers many questions about holographic interferometry and the procedures that are used in holographic photography.

End Notes

1. Stephen A. Benton, Spatial Imaging Group, MIT Media Laboratory, 20 Ames Street E15-416, Cambridge, MA 02139-4307, USA, 1999.
2. Rudie Berkhout, "Using HOEs in the Holographic Image Making Process," 223 West 21 Street, New York, NY 10011, USA, 1999, p. 1.
3. S. A. Benton, S. M. Birner, and A. Shirakura, "Edge-Lit Rainbow Holograms," Proceedings of the SPIE, Vol.1461, 1991, pp. 149–157.

4. J. Upatnieks, "Edge-Illuminated Holograms," Applied Optics, Vol. 31, 1992, pp. 1048–52.
5. Rudie Berkhout, "A Compact Hologram Display, and the Design for a Full-Color Version, Using Diode Lasers," 223 West 21st Street, New York, NY 10011, USA, 2001, p. 2.

CHAPTER 4

HOLOGRAPHIC INTERFEROMETRY

In This Chapter

- Optics of holographic interferometry
- Conventional 2D photography
- Reflection holography
- Transmission holography
- Double-exposure interference hologram method
- Real-time interference hologram method
- Spatial filtering
- Michelson interferometry setup
- Photographic processes for holographic interferometry
- Step-by-step photoprocessing

In this chapter, I continue to examine the practical uses of hologram technology by looking at some basic research work in holographic interferometry. Initially, the researchers[1] proposed to differentiate between mono and bimetallic pennies in interference holograms showing thermal expansion within the penny. Their proposal tells more about this, but the basic idea is that older, pre-1983 pennies were an alloy of 95-percent copper and 5-percent zinc, and so would thermally expand differently from newer, post-1983 pennies, which were more than 97-percent zinc with just a thin electroplated coating of copper. In fact, the researchers wanted to do this with three different types of holographic interference, including reflection, transmission, and real time. This turned out to be too eager a proposition. Here are the steps they took:

1. Start with a Michelson interferomet~ setup to test coherence length.
2. Learn a few things about using the equipment.
3. Learn about reflection hologram setups.
4. Hold a penny.
5. Apply spatial filtering.
6. Make a test strip.
7. Use darkroom.
8. Make a one-beam reflection hologram.
9. Make one-beam reflection interference holograms.
10. Cooperate in switching to transmission holography.
11. Produce magnified penny interference holograms.
12. Some of the equipment the researchers used.
13. Epilog.

Michelson Interferometer

The researchers began by generating the basic concentric circles interference pattern of Michelson to check the coherence length of the laser and become familiar with aligning optical elements. Michelson split a laser beam at right angles, had the two resulting beams travel equal distances, and then recombined them to shine on a screen. Their result was an attractive pattern of concentric circles of light that was used to prove that ether, the supposed substance through which light waves travel, either did not exist, or if it did, wasn't important to consider.

Note: The Michelson Interferometer consists of a half-silvered mirror placed at a 45° angle to the incoming beam. Half the light is reflected perpendicularly and bounces off a beamsplitter; half passes through and is reflected from a second beamsplitter. The light passing through the mirror

HOLOGRAPHIC INTERFEROMETRY

must also pass through an inclined compensator plate to compensate for the fact that the other ray passes through the mirror glass three times instead of one. Because this light is parallel, it must be focused with a lens. A net phase shift of radians must also be included, since the parallel components reflect off the front of the first and second mirrors, while the perpendicular ray reflects off the back of the first mirror and then the front of the third. When the mirrors are not parallel, Fizeau Fringes result (Fringes that result when the mirrors of a Michelson Interferometer are not parallel. Changing the orientation of the mirrors can produce straight, circular, elliptical, parabolic, or hyperbolic fringes).

As shown in Figure 4-1, this involves a laser beam, a beamsplitter, two mirrors, a beam-expanding microscope lens, and a vertical sheet of paper for a projection screen [Adama, Chinowsky, and Wang, 2]. The researchers set this up and fiddled for a few hours with the alignment and were never able to produce the nice pattern that Michelson did. What they were able to produce, with the table properly floated, was concentric arc segments of approximately 10 to 15 degrees of a circle. They called this success and went on to the next step.

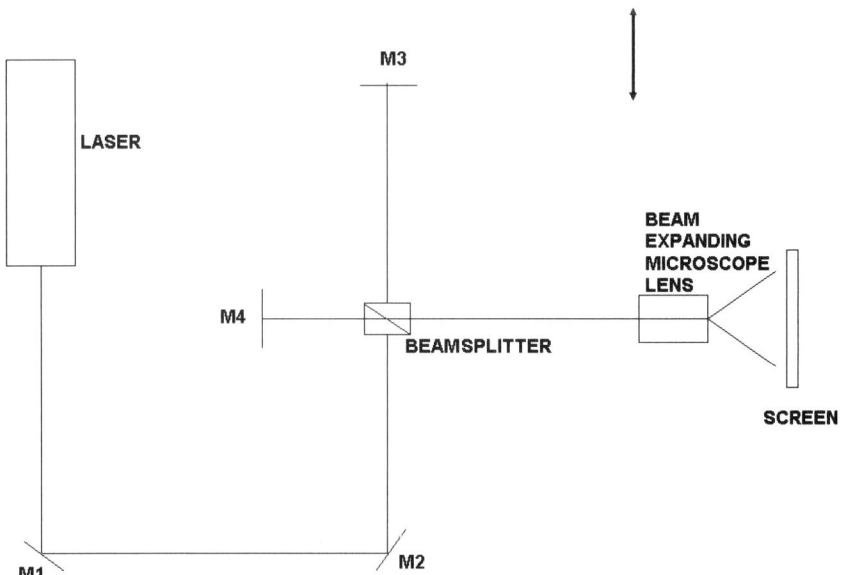

FIGURE 4-1: Michelson configuration.

COHERENCE LENGTH

The researchers then used a Melles Griot 5mW Helium Neon (HeNe) laser. This is labeled as a 10mW laser, and only outputs 5mW. The alternative was to use a 1mW (labeled as 2mW) HeNe. The advantage of using the 5mW was that it would permit the researchers to have shorter exposure times. These shorter exposure times would allow them to *filter* out vibration that could wash out their interference patterns. The problem with using the 5mW laser is that it has a shorter coherence length. Among other things, the coherence length dictates the difference between the lengths of the object and reference beams. The researchers changed the length of one leg of the setup and crudely judged the quality of the interference pattern. They definitely saw no interference for a difference of 9 inches. Using the 5mW HeNe, they saw interference patterns for a path length difference of about 2 inches. It is supposedly much longer for the 1mW. In the case of one-beam reflection holograms (explained in the next section), this affects the depth of the object one can use. Since the researchers were dealing with pennies, this didn't turn out to be much of a problem. They determined the coherence length by adjusting the path length of one of the beams in the Michelson setup.

Goggles

The other thing about 5mW HeNe lasers is that they are potentially damaging to the eye. Supposedly, after the beam is expanded and shown on the holographic plate, there is little enough power reflected that it is safe to look at with the naked eye. Truly cowardly hologram viewers will want to wear goggles at all times. However, this makes it more difficult to view a hologram that is illuminated by a laser.

TABLE FLOATING

The table upon which the researchers had their optical setup was a large Melles Griot optical table with isolation mounts connected to a tank of compressed nitrogen. The metal happily accepted magnet clamps. The holes received screw-threaded holders. When the gas was turned on, the four legs filled with nitrogen and pushed up on the table to create a perfectly level and vibration-immune surface for experiments. At least, that was how it was supposed to happen. What actually transpired was that the researchers would squint at the dials and turn the knobs on the tank until they heard a hissing sound, and the table would promptly go into some off-kilter configuration. The table was the lightest model, but the mounts were for heavier models. The researchers pushed and pulled on the surface of the table, readjusted the knobs, waited, and would get it off-kilter in a different direction. For some reason, one leg always wanted to be pushed to the upper or lower limits of its range where it had no immunity to vibration from the building. The researchers did learn that they should turn the gas up slowly and do a lot of waiting for the table to equalize. This requires much patience.

Configuration

Early planning included bleaching and an indication that single-beam reflection holograms were very easy to make. Single-beam reflection holograms are also viewable in white light.

Figure 4-2 shows a two-beam reflection hologram setup [Adams, Chinowsky, and Wang, 2]. In it, a laser beam is split into two beams: object and reference. The reference beam bounces off a mirror, goes through a

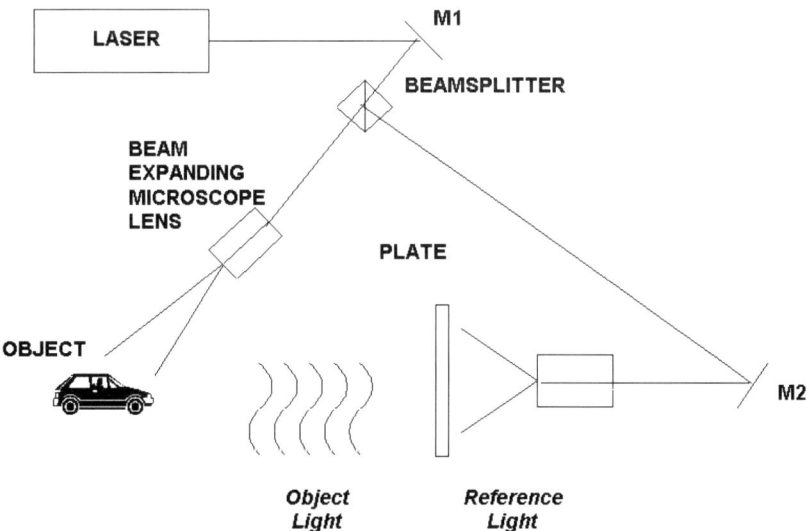

FIGURE 4-2: Two-beam reflection hologram setup.

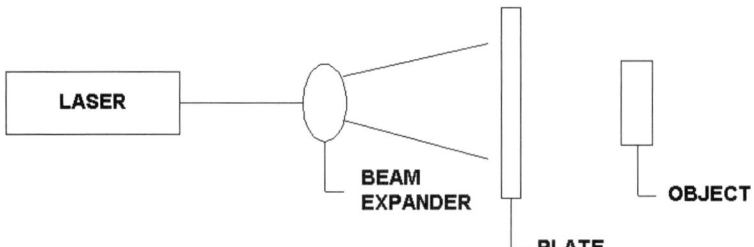

FIGURE 4-3: One-beam reflection hologram setup.

beam expander, and is displayed on the holographic plate. The object beam bounces off a mirror, goes through a beam expander, and is reflected off the object (penny) onto the holographic plate. Importantly, the object and reference beams are displayed on opposite sides of the plate.

Figure 4-3 shows a one-beam reflection hologram setup [Adams, Chinowsky, and Wang, 2]. In it, a laser beam passes through a beam expander, hits the plate (this is the reference beam), and reflects off an object behind the plate on to the plate's back side (this is the object beam).

Because the one-beam system looked a lot simpler, the researchers decided to go with it, possibly graduating to two-beam reflection systems after they succeeded with one beam. However, one-beam systems have some problems. All the light of the reference beam hits the plate, but much less light is reflected off the object. Because holograms record both intensity and phase information from the combination of object and reference beams, if either the reference or object beam is much stronger than the other is, the interference will be messed up. As previously noted, the researchers used relatively flat pennies as objects, but because of coherence length limitations of the 5mW HeNe, one-beam setups may not be appropriate for thicker objects.

Penny Holder

Related to the coherence length issue was the issue of how to hold the penny. Early before-the-researchers-reached-the-lab ideas on how to hold the penny included a three-prong style test tube holder, a lens holder, and super glue. Because they intended to change the temperature of the penny by means of liquid nitrogen, they wanted a holder that would be liquid-nitrogen robust. Moreover, because of the coherence length issue, the researchers had to have something that would get the penny as close as possible to the holographic plate.

For the test strip exposure (described later), the researchers tried to use a three-pronged lens holder that seemed to combine the best aspects of their before-they-reached-the-lab ideas. Unfortunately, it held the penny back a couple of centimeters from the plate while its own frame was very near the plate. This created a penny-in-a-dark-hole effect that they didn't like. They switched to mounting pennies on a flat black card with double-sided sticky foam tape. Eventually, the researchers realized that they could use more than one penny on the card. They soon had a row of five pennies, with an extra one above the row on the right-hand side.

MAKING PENNIES REFLECT LIGHT

Since the researchers wanted to have as much light reflect off the penny onto the plate as possible, they wanted their pennies to be highly reflective. Previous holography experiments had used white objects and objects that had been spray-painted white. They initially tried both white painted pennies and BRASSO™[2] polished pennies; the benefits and drawbacks are explained later.

Spatial Filtering

Dust motes in the laser beam's path, tiny as they are, cause diffraction of the beam, and result in funny-looking splotches on the screen after the beam is expanded. One way to get rid of these is by spatial filtering. For their proposal, the researchers dug up a pair of photos that showed dramatic difference in screen-projected, expanded laser beams with and without spatial filtering.

Naturally, the researchers wanted the perfectly smooth spatially filtered projection. Filtering light typically involves passing it through a lens, a screen with carefully positioned holes cut out of it, and a second lens akin to the setup as shown in Figure 4-4 [Adams, Chinowsky, and Wang, 2].

To get rid of the high frequency noise caused by mote diffraction, the researchers wanted just a pinhole in the center of their filter. However, because they didn't really know what they were doing, they tried to get away without using any lenses. This didn't work.

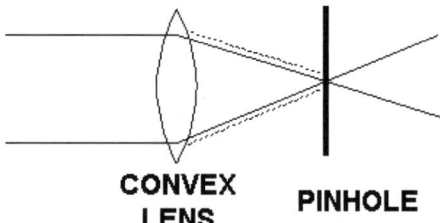

CONVEX LENS **PINHOLE**

FIGURE 4-4: Spatial filtering setup.

The researchers still had monstrous splotches in their projected, expanded beam. They thought that ideally, they would place their microscope lens directly in the beam path with the cross section of the lens perpendicular to the beam. But this didn't work. In order to get an expanded output beam, the researchers found it necessary to put the lens off center and skewed. This was no big deal as long as it worked. The problem arose in those strange blobs and splotches that came out of the lens. They drowned the lens with cleaning alcohol and rubbed it gently (in one direction only) with optical quality Kodak lens cleaning sheets, but it didn't make the situation any better. Eventually, when it came to expanding the beam, the researchers simply fiddled with the positioning of the lens until they found a portion of the pattern that was mostly devoid of the anomalies.

Test Strip

The researchers used a bleaching/fixing holographic plate development process that converted the black, light-absorbing, opaque, unexposed areas of emulsion to have an index of refraction different from the exposed areas. This came from a company called Photographer's Formulary.[3] Bleaching has a couple of effects. It allows far more light to reach the eye when viewing, which results in a brighter, clearer holographic image. It also permits us much greater latitude in exposing plates, since the process makes the overexposed plates substantially brighter. Bleaching turns out to be far more important in reflection holograms, because far less light strikes the plate than with transmission types.

All the same, because the researchers were using the 5mW HeNe rather than the 1mW, and because they were using the new and untested bleaching process, they decided to burn a plate with strips of different amounts of light in order to find the ideal exposure time. Instructions that came with the chemicals suggested that 1 second at 2mW worked well. They created a ridiculously elaborate curtain of opaque strips of black cardboard taped to a ruler suspended in front of the holographic plate on two mirror holders. It was slow, clumsy, and didn't work at all. What did work was a handheld sheet of cardboard that was slid off to one side at regular intervals of time while the laser was turned down to one-tenth of its power with a polarizer. Later, the researchers learned that because of the bleaching process, they could largely ignore the issue altogether. In the short term, they determined that they could get a pretty decent exposure by using about Y2 to 1 second overall exposure time at the 5mW HeNe's full power.

Darkroom Technique

The researchers exposed a holographic plate for 1 second with a single expanded beam to make their first real reflection hologram, and then hurried, ridiculously eager, to the darkroom to develop it. One nifty innovation that occurred in the laser lab rather than the darkroom involved a 35mm camera with its back off in the path of the beam just as it came out of the laser. The neat thing about this was that it allowed the researchers to precisely time the length of exposure with the camera's built-in shutter speed adjustments. Most important of the darkroom equipment was a glow-in-the-dark darkroom timer.

The researchers also arranged the chemical trays and donned plastic gloves (probably not necessary). The process required 2 minutes in the developer, 10 seconds in distilled water, 3 minutes in running water, 2 minutes in the bleaching fixer, however long as necessary in the vitamin-C post treatment under bright light to turn the plate from pink to light brown, and then another 3 minutes of washing in running water.

Later, when the researchers began to feel wild and crazy, they tossed out the 10 seconds in distilled mountain spring water. They also added a dunk and slosh in isopropyl alcohol step at the very end. The idea behind this was that the fast evaporation of the alcohol would dry the plate faster and result in fewer marks due to streaking of the water during drying. This method worked well and provided immediate viewing gratification, but over several uses, the alcohol became saturated with water and needed replacing. Incidentally, the darkroom is completely unventilated and the isopropyl has a rather heady bouquet. The even moderately intelligent hologram developers will want to keep the alcohol Tupperware™ dish sealed until the time of dunking. The researchers failed to follow this rule several times, and left rather dizzy from the experience.

Except for the frustration of trying to open the plateholder boxes in the dark, the researchers had rather uneventful 3-minute intervals in the dark listening to the sloshing of liquids.

Hologram

The hologram involved the penny in the three-pronged penny holder exposed for 1 second in the single-beam reflection setup. As already mentioned, this created a penny-in-a-dark-hole effect. Although it looked pretty decent under white light, it wasn't very impressive when viewed under laser light, and the image wasn't all that clear. When the researchers looked at the image under laser light some days later, perhaps due to additional drying and shrinking of the emulsion, the image was much better. They also discovered that as is true of reflection holograms in general, the image was viewable under white light.

VIEWING REFLECTION HOLOGRAMS

The researchers held the plate under a big bulb and found an amazingly realistic-looking penny. The only problem was that in order to get the best image, the plate must be tilted so that the bulb reflects directly into your eyes. One method that works to remove the glare is to tilt the plate so that the bulb's reflection is right in the middle of the penny and shining right into one eye, and then close that eye and view the penny with the other eye. It sounds weird, but it works surprisingly well. Also, they found a spotlight source designed to work with a low-power microscope in the laser lab, which produced even better images (until they broke it on the final day)!

The ideal viewing arrangement for reflection holograms is a bright point source at infinite distance. In the rare moments during which Seattle's weather cooperated, the researchers were able to take the plates outside and view them under the sun. I liked the microscope light better. But either under a bulb or under the sun, that source of illumination turned out to be a real nuisance when photographing the holograms. The result is a big white spot in the middle of the image.

Interference Hologram

After successfully creating a one-beam reflection hologram, the researchers next tried an interference hologram. This would involve two ~/2 second exposures that were easily timed with the camera's automatic shutter. It also involved changing the penny's temperature. They initially proposed to apply liquid nitrogen to the penny to cool it for what they hoped would be at least a 100-degree overall change in temperature. However, they also thought that liquid nitrogen might be overkill, and that condensation during the processing might become an issue. In the end, the researchers decided to use a heat gun.

This particular heat gun has a heating element, a fan, and a shape similar to a hair dryer—but you would never use this on your hair. It gets really, really hot! The researchers were able to heat pennies to over 125 degrees Celsius and nearly incinerate their penny holder in a matter of 20 to 30 seconds. They heated the penny to about 120 degrees Celsius and timed how long it took to cool to room temperature (about 21 degrees Celsius). It took about 10 minutes. However, it occurred in a somewhat exponential decay shape; most of the heat was dissipated from the beginning, and the rest of time was spent waiting for the penny to finally return to room temperature. Because the researchers had the pennies mounted on a sheet of cardboard positioned so close to the plate as to nearly touch it, they found it necessary to completely remove the penny holder to heat the pennies after the unexposed plate had been positioned. They also used an electronic thermometer so that they knew how hot they were getting the pennies.

One trick that came into play was to put the plate in the holder, and then put a cardboard box on top of it while the researchers waited for the pennies to cool to room temperature. Lights were still off here, but the ambient light leaking in through the window covers from the construction site made them worry about fogging or pseudo-solarizing the plate. Another trick involved putting the stupid plate in the stupid plateholder. This is amazingly easy in the light. But in the dark, they fumbled with it for what seemed like an eternity. The trick is to place one edge of the plate on the stationary peg on the right-hand side (as viewed from the back), and then rotate the plate until you find the sliding restraining peg on the left-hand side. Pull out the peg, slide the plate in, and it's doable even in the dark. Eventually, the researchers burned a double-exposure (V2 sec + V2 sec) interference hologram onto the plate and got it developed.

INTERFERENCE PATTERNS

There are several ways to get interference lines in a hologram. What the researchers wanted were interference lines related to the thermal expansion/contraction of the metal and of the bimetallic junction in the newer penny. What would really make them happy would be something looking like that cool acoustic penny hologram that got them interested in this project in the first place. The researchers didn't see this. What they did see is what looks like a fingerprint on the center penny on their penny mount.

The researchers tried additional single-beam reflection holograms after they carefully removed all vestiges of fingerprints, and one that involved nudging the penny holder rather than heating/cooling the pennies. They got similar fingerprint interference on the thermally stressed pennies, and dramatic Lincoln-in-jail interference on the nudged interferogram. The researchers are confused at this point.

Why aren't the researchers getting better interference patterns? Perhaps the penny holder itself is expanding, and that motion is washing out the interference patterns of the thermal expansion. Because the center penny is the one closest to the penny holder stand, maybe it receives less lateral translation due to its own thermal expansion. This would explain why the center penny is the one that always gets the best fingerprints. The researchers suggest supergluing a penny to the side of a post. In this way, when it expands, the penny holder will not be translating the penny.

The Penny Holder Dispute

The researchers have been doing the one-beam reflection holograms for a while now. They are starting to get bored with it. Also, those images are so small (actual penny size) that they have a hard time getting a really good look at them. How about something new and different?

After a heated debate, the researchers decided to go to a transmission setup using a magnifying lens to get a bigger image of the penny as shown in Figure 4-5 [Adams, Chinowsky, and Wang, 3]. Transmission setups must have two beams.

The beam from the laser is split 50/50 by a beamsplitter. The reference beam bounces off a mirror, is expanded by a skewed microscope lens, and hits the front of a holographic plate. The object beam bounces off a mirror,

HOLOGRAPHIC INTERFEROMETRY

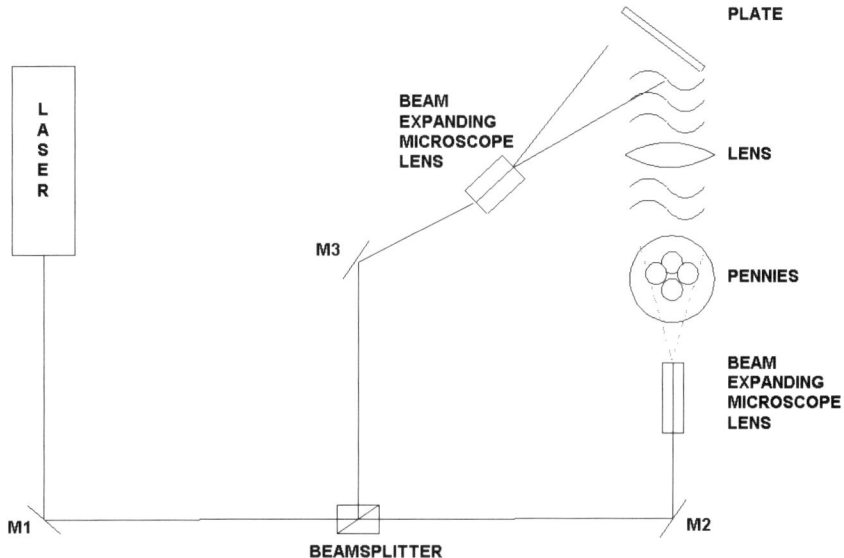

FIGURE 4-5: Transmission setup.

is expanded by a microscope lens, and reflects off the object onto the same front side of the holographic plate. This will require the beamsplitter, two mirrors, two beam expanders, the object, and the photographic plate.

There is also concern about the penny holder. Some of the researchers imagine that they have too much expansion on the penny holder and that they are heating the pennies too much. An earlier temperature test strip produced inclusive results when it came to determining how much or too little they should be heating the pennies to produce the best interference patterns. From here on in, the researchers take a pretty lackadaisical approach to heating the pennies.

But, what about that penny holder? The researchers glued the penny to the side of a post. However, they thought that there was too much heat transferred to the post during the heating of the penny. The researchers wanted to heat the pennies and then place them on an upside-down petri dish. The idea is that the glass will conduct less heat and be less likely to move. Yeah, sure, but that means that the pennies will have to be lying flat on a horizontal surface. How are they going to get the light onto the pennies and back to the plate? And remember, even though the researchers are doing transmission, they still have to have the difference in length between the object and reference beams be less than the coherence length. Also, it was confusing for them to decide which approach to take, since they were simultaneously attempting to figure in a penny image magnifying lens to boot.

Finally, the researchers started setting up as if they had a penny on a post. Meanwhile, they also constructed a stand with about 50 arms attached to it. One for a beam expander. One for the lens. One for a rectangular mirror. About 47 others as counterweights. They figured that if the post wasn't sufficiently thermally nonconductive, they could glue the penny to the side of a little Pyrex beaker that appeared on their lab bench the other day. Or, better yet, to the side of an upside-down test tube on top of a post. This would, however, still be subject to the thermal expansion of superglue.

Pretty soon, the researchers started setting up their horizontal Pyrex petri dish configuration on the other side of their half of the table. Pretty soon, their table starts to get crowded. They haven't got all the details worked out yet, but they want the light to go through the lens, bounce off the mirror, hit the pennies, pass back through the lens, and fall on the photographic plate. At the moment, a beam entering the lens is hitting the mirror and bouncing off toward the ceiling. The researchers think they know how to fix this. After a while, they go to help and they all futz with it for a while. The beam isn't doing what it's supposed to do. Aha! This proves the superiority of the vertical penny!

The researchers now think they have an idea for fixing this setup. This confocal sort of in through the lens, off the penny, and back out through the same lens sort of thing isn't going to work. Also, they're trying to expand the beam before it bounces off the mirror. The researchers start positioning spare posts as laser beams and simulate what is going to happen. They disassemble the octopus and put each arm on its own post. The researchers even draw out the *ideal* setup diagram on the whiteboard. The researchers finally come up with a setup that has a darn good chance of working, and which turns out to be a surprisingly orthodox transmission configuration.

The light is split at a 90-degree angle by a 50-percent beamsplitter. The reference beam bounces off a mirror, is expanded by a microscope lens, and hits the plate. The object beam bounces off a rectangular mirror held at a *downward* angle, passes through a mind-expanding microscope lens, bounces off the pennies on the horizontal petri dish, passes through the image expanding lens, and hits the holographic plate. Well, it seems impressive enough, at least from the sheer number of parts used. Also, the researchers can look through the big lens from the position of the plate and see the enlarged pennies.

The researchers use the light power meter to make sure that the object and reference beams are providing a similar number of lumens to the plate. These started out as equal halves of the original beam from the laser. However, due to specular reflection off the pennies, the object beam is somewhat spotty when compared to the reference.

Magnified Penny Transmission Holographic Interference

The researchers burn a plate. When the plate is developed, they get a pretty good image. Their favorite fingerprint interference is there.

Because this is a transmission hologram, the researchers can't view it outside the laser setup. They start taking photographs of the holograms. These too turn out surprisingly well. The researchers now decide they want to try to scan a reflection hologram. The sidebar, "Scanning a Reflection Hologram," covers that effort in greater detail. In the meantime, let's look at the optics of holographic interferometry itself.

■ SCANNING A REFLECTION HOLOGRAM

They all think it won't work; nevertheless, they scan it anyway. It doesn't work. What they get is a black slide with a white label. But by turning to their maximum, the contrast, intensity, and brightness of the image, a circular disk appears—something faintly resembling a penny.

Time is approaching the project deadline. The researchers make two more interference holograms. These also turn out well and they come to realize that the best fringes appear with the least temperature span. Now, the researchers want to use a bigger lens to focus the image on the plate and further minimize blurring and movement. They have procured a big kahuna lens.

This was the researchers last opportunity schedule-wise. Besides, there are only three more plates remaining. The researchers decide that the plates likely won't last for two years. In a fit of practicality and frugality, they decide to burn all three plates using the big kahuna lens. They make one interferogram using two 4-second exposures, and one with two one-eighth-second exposures. Feeling totally wild and crazy, for the last plate, the researchers use a quarter, a dime, and a nickel in addition to the penny.

> These come out surprisingly good. One problem, however, is that with the big kahuna, the researchers have a much smaller field of view. Definitely only one person can see the image at a time. It turns out that two one-eighth-second exposures do not give as good an image as do two one-quarter-second exposures. The multiple-coin image came out the best of all.
>
> In all of the transmission holograms, the researchers have peripheral visibility of the other equipment sitting nearby. Also, in the holograms with the smaller lens, it is possible to look around the lens and see the pennies unmagnified.

Optics of Holographic Interferometry

What the researchers can see in a photograph is the recorded image of reflected light from the object. However, a photograph only records the intensities of the light as shown in Figure 4-6.[4] Therefore, the wavefronts from a photograph are incomplete, as compared with the original light. This causes a viewer to see the object only from a certain perspective.

A hologram is created by recording the interference patterns produced by the object light and the reference light (see Figures 4-2 and 4-7). In this way, the plate not only records the intensities of the light, but also the phase differences. After the hologram plate is processed, the researchers used the same reference light to recreate wavefronts of light just as if they were coming from the original object.

Holograms can be divided into two groups by their different image recording and reconstruction methods. In the reflection hologram shown in Figure 4-2, the reference light and the object light shine on opposite

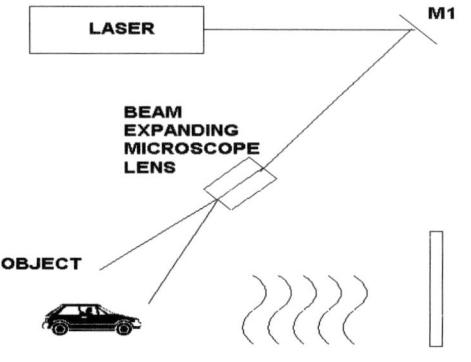

FIGURE 4-6: Conventional 2D photography.

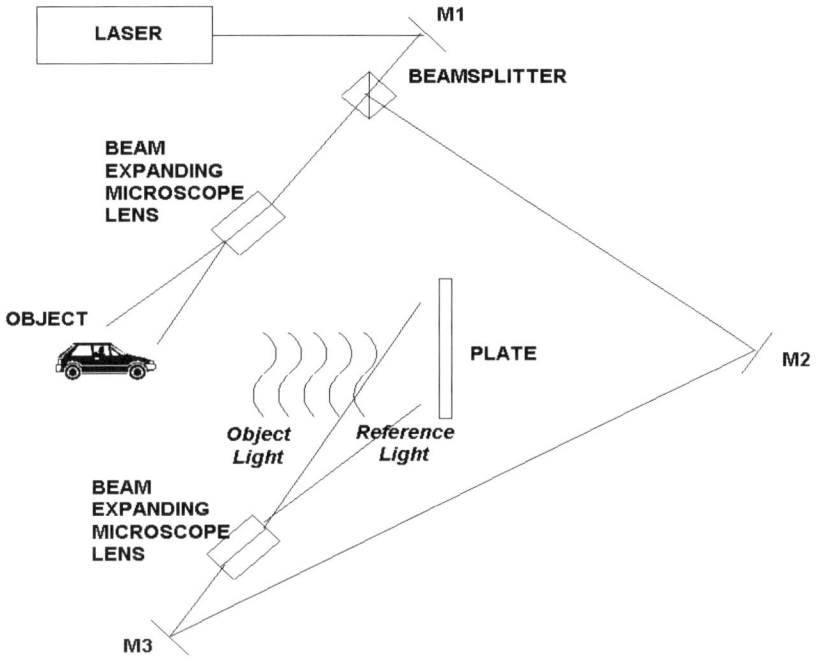

FIGURE 4-7: Transmission holography.

faces of the hologram plate, so the image is reconstructed by the reflection of the reference light. In the transmission hologram (see Figure 4-7), the reference light and the object light shine on the same face of the plate, and the viewer sees the image reconstructed by transmission of the reference light through the plate [Chinowsky, Adams, and Wang, 4].

WHAT IS AN INTERFERENCE HOLOGRAM?

Interference holograms result from the superposition of two or more holograms on the same photographic plate. The simplest type of holographic interferometry is the double-exposure method. The researchers made a hologram by exposing the object to the light twice: first when the object is undisturbed, then while the object is distorted. These two holograms interfere with each other and generate interference patterns when the image is reconstructed. Therefore, they could measure micro deformation by counting the fringes and calculating the optical path difference. The researchers used two different types of methods as listed in the sidebar, "Measuring Micro Deformation."

MEASURING MICRO DEFORMATION

Double-Exposure Method

1. Make a hologram of the undisturbed object.
2. Distort the object.
3. Re-expose the hologram.
4. Process the hologram plate.

Real-Time Method

1. Make a hologram of the undisturbed object.
2. Process the hologram plate.
3. Put the processed hologram back in place.
4. Distort the object, and observe interference fringes.

The second method has the advantages that in real time, different types of distortion can be applied and motion pictures can be taken.

Important Optical Issues

1. The direction of the reference wave should be well separated from that of the object waves, so that the viewer can see the reconstructed image without being in the direct path of the reference laser beam.
2. Since the hologram plate records the relative intensities rather than absolute intensities, the reference light should shine uniformly on the plate.
3. The object wave and the reference wave exposing the plate must be mutually coherent to result in the interference pattern; that is, the optical path length of both light beams should be the same.
4. The interference pattern must remain stationary during the exposure period. Therefore, the stress on the penny must be constant throughout the exposure.

SPATIAL FILTERING

Spatial filtering is a method used to eliminate the dark swirls in the illumination patch of the expanded laser beam. These swirls are caused by the interference of the parts of the beam that have been diffracted by dust and imperfections in the laser or optical components.

A spatial filter is composed of a convex lens and a pinhole as shown in Figure 4-4. The lens works as a Fourier transformer to convert the beam to the spatial frequency domain on the focal plane, and the pinhole works as a low-pass filter that removes the undesired high-frequency components.

Use of a spatial filter can result in a cleaner holographic image. However, it also reduces the amount of light that strikes the plate. Spatial filters were likely not necessary for the researchers' purposes, but they kept them in mind.

Photographic Processes for Holographic Interferometry

The photographic process used in holography is very similar to the conventional process used for black-and-white photography (see the sidebar, "Step-by-Step Photoprocessing" for more information). The main difference is that a hologram must record images of much higher resolution: it must be able to resolve features as small as the wavelength of light that is used. For example, an electron microscope image of a cross section of a holographic emulsion has fringes that are stored in a hologram and are spaced about 1 micron apart. The recording of such detail requires special emulsions (about 10 times the resolution, and 1000 times slower than conventional films) and great stability of the holographic setup to prevent any blurring of detail.

■ STEP-BY-STEP PHOTOPROCESSING

1. Prepare plate. The researchers used Agfa 8075HD 4" x 5" glass plates.
2. Hold plate. Positioning of the plate must be stable and repeatable. The researchers used a commercial holographic plateholder.
3. Expose plate. Typical exposure times are 10 to 60 seconds. The researchers initially determined the correct exposure by test strip, and verified successive exposures using a light meter. The researchers kept vibration as low as possible during the exposure. For double-exposure interferograms, two exposures, probably of half normal exposure times, must be made.
4. Process plate. Temperature of all solutions should be kept stable, around 20 C.
   1. Develop in D-19, 4–7 minutes.
   2. Rinse/Stop Bath, 30 seconds.
   3. Fix, Kodak Rapid Fixer, 5 minutes.
   4. Wash/Hypo Clearing Agent, 15 minutes or so.
   5. Bleach. This in an optional step, not used in conventional photography, in which the dark areas of the hologram are turned into clear areas with a high refractive index. This gives brighter, more easily visible holograms.
   6. Dry.
5. Hold plate. For live interferometry, repositioning must be exact.
6. View plate

Possible Difficulties

Double-exposure holographic interferometry should be no more difficult than normal holography, with its usual concerns of vibration control, beam coherence, and exposure time. The researchers' particular experiment had the complication that they were trying to heat or cool the subject of their hologram without moving or otherwise disturbing it (with liquid nitrogen, for example, frost could be a problem). Let's explore different ways of holding the penny and applying thermal stress.

The researchers expected live holographic interferometry to be much more problematic, as the hologram must be moved and chemically processed between *exposures*. Even if the plate itself can be repositioned exactly, changes in the emulsion during processing may have changed the stored hologram dramatically. For instance, the emulsion can absorb water during processing, causing a permanent swelling. Or, photographic processing may remove chemicals from the emulsion, causing a permanent shrinkage. As information in the hologram is stored not just on the surface, but throughout the thickness of the emulsion, a change in this thickness may change the information stored in the hologram.

This tends to be a bigger problem for reflection than for transmission holography. For example, let's look at two point sources of light, representing the object beam and the reference beam, and the interference fringes that result from interference between the two. For a transmission hologram, both beams are on the same side of the plate, and the fringes are mostly perpendicular to the emulsion surface. If the thickness of the emulsion was to change, it would not change the spacing of the fringes very much. For a reflection hologram, however, the beams are on opposite sides of the plate, and the fringes run parallel to the surface of the emulsion. A change in emulsion thickness would directly affect the spacing of the fringes, and have a large effect on the stored image.

Therefore, the researchers focused on transmission instead of reflection holography while attempting to do live interferometry. They may still have problems with emulsion changes during processing, however. One way to try to minimize these changes is to pretreat the emulsion before it is exposed. For instance, one might soak the plate in water to pre-swell the emulsion, in the hopes of eliminating any further swelling during processing. There seem to be many subtleties involved, which the researchers will address if they become an issue.

HOLOGRAPHIC INTERFEROMETRY

Finally, what kind of equipment is used in this type of project? How much does this sort of holography cost? And, what are the required materials, with purchase/rental costs?

Holography Equipment Used and Its Costs

Tables 4-1, 4-2, and 4-3 help to answer the preceding questions [Adams, Chinowsky, and Wang]. Nevertheless, be aware that these tables are not all

TABLE 4-1: Some of the equipment used in the project.

Label	Item	Description
1	Plate box	This box is unnecessarily difficult to get into in the dark!
2	Laser power supply	This supplies power to the 5mW HeNe.
3	Laser	This is a 5mW HeNe (labeled as a 10mW). It can be rented from the stockroom for $5. Beside it is its power supply.
4	Beamsplitter	This was used only for the two-beam setups. It is a 50-percent beamsplitter and costs $1 to rent from the stockroom. The researchers had it in a fancy multiple-angle adjustable holder that sat on a solid block of aluminum for height.
5	Mirror	This mirror has two angle adjustable screws. The researchers used multiples of these. For the horizontal penny holography, the researchers used a rectangular mirror in a clamp.
6	Microscope lens	This is used as a beam expander. The researchers sometimes used two of these.
7	Polarizer	This was in a precisely adjustable rotating holder. The researchers used it to dim the 5mW laser's output so that they could work with it without worrying about eye damage.
8	Plateholder	This is used to hold the holographic plate.
9	Penny mount	A piece of black cardboard with pennies attached with double-sided foam sticky tape.
not shown	Magnifying lens	This lens was placed between the pennies and the plate to enlarge the image of the pennies.

TABLE 4-2: What this sort of holography costs.

Item Description	Cost
Laser rental	$5
Beamsplitter rental	$5
Bleaching kit	$15
Photo development	$25
Isopropyl Alcohol	$2
Pure mountain spring water	$2
Various Tupperware dishes	$8
Supplies for a nifty poster	$21
Total	$83

TABLE 4-3: List of required materials, with purchase/rental costs

Item	Quantity	Cost
5mW Helium Neon Laser	1	$5.00
Laser mounts	1	–
Mirror holders	3	–
Pinhole spatial filters	2	–
Convex lenses	2	–
Microscope lenses	2	–
Beamsplitter, 50%/50%	1	$1.00
Light baffles	2	–
Holographic plateholder	1	–
Lens mounts	2	–
Mirrors	2	–
Pennies	2	$0.02
Penny holder	1	–
Holographic plates	7–10+	–
D-19 developing solution	1	–
Bleaching kit	1	$15.00
Fixing solution	1	–
Polarizing sheet (2":x2")	1	–
815 Photodetector	1	$2.00
HeNe safety glasses	3	–

inclusive. There are many hidden costs involved; and the cost of time and material fluctuates with the ever-changing economy.

In Summary

Though the researchers felt that they were successful in their project, there are some parts of it at which they did not succeed. Some of these may be ideas for future holography groups, such as:

- Attempt real-time interferograms.
- Determine the best change in temperature to see thermal expansion-related interference fringes.
- Differentiate between mono and bimetallic pennies based on thermal expansion interference fringes.

The researchers however, were successful with some parts of their project. Again, some of these may be ideas for future holography groups, such as:

- Generated single-beam reflection holograms and interference holograms.
- Generated transmission holograms and interference holograms.
- Explored new hologram development chemistry.
- Learned many procedural tricks.
- Made some nifty photos.
- Had fun and learned about holography.

The following chapter describes laser electronics.

End Notes

1. Brian D. Adams, Tim Chinowsky, and Chi-Yen Wang, "The Penny Project," Department of Electrical Engineering , EE488 Laser Engineering, University of Washington, P.O. Box 352500, Seattle, WA 98195-2500, USA, 2000, pp. 1–3.
2. Keysan.comCatalog, P.O. Box 5146, Pittsburgh, PA 15206, USA, 1999.
3. Photographer's Formulary, Inc., P.O. Box 950, Condon, MT 59826, USA, 1999.

4. Brian D. Adams, Tim Chinowsky, and Chi-Yen Wang, "Holographic Interferometry of Mono- and Bimetallic Pennies," Department of Electrical Engineering, EE488 Laser Engineering, University of Washington, P.O. Box 352500, Seattle, WA 98195-2500, USA, 2000, pp. 1–11.

CHAPTER 5

LASER ELECTRONICS

In This Chapter

- Reflection hologram
- Transmission hologram
- Polarizer
- Beamsplitter
- Object beam
- Reference beam

This chapter continues the theme of Chapter 4 by showing you how to set up a transmission and reflection hologram using laser electronics. It presents a different approach to setting up transmission and reflection holograms than what was discussed in Chapter 4. This chapter also continues to examine the practical uses of hologram technology. Before you start the setup process, you need the following revised equipment list:

- HeNe Laser: 5mW and 1mW
- Polarizer: to cut down the laser output
- Mirrors
- Lenses
- Beamsplitter
- Plateholder

84 CHAPTER 5

- 8E75HD film (unbacked)
- D-19 film-processing chemicals
- Objects: The best objects to use are those that reflect a lot of light. Mostly white objects seem to be the best.

Transmission Hologram

Transmission holograms are formed by two coherent beams hitting the same side of a plate and interfering with one another. The setup that the researchers used for making transmission holograms is shown in Figure 5-1.[1] For the 96/4-percent beamsplitter, a microscope slide works very well. It is important that the reference beam and object beam lengths are the same (within the depth of the object). Of the two types of holograms that the researchers made, this type had the best results; that is, the best image resolution. To view this hologram, the plate must be held in an expanding beam from a coherent light source. Bleaching the plates is optional for these holograms and can often improve the image quality, especially for overexposed plates.

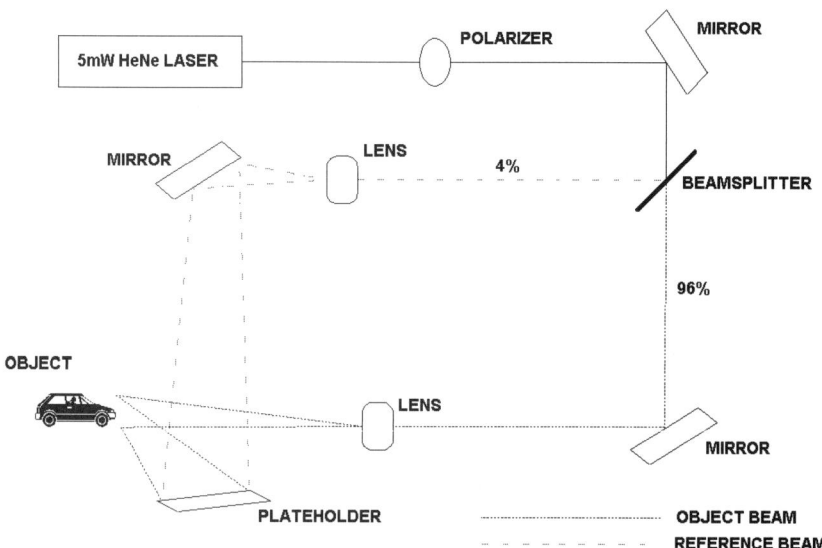

FIGURE 5-1: Transmission hologram setup.

Reflection Hologram

The second types of holograms that the researchers made was reflection holograms. This type is different from the transmission type in two ways. The first is that when viewing this hologram, you do not need to use a coherent light source; all that is needed is a point source, such as a lamp or the sun. The second difference is the way in which the hologram is recorded. In this case, the light reflecting off the object hits the front side as before, but the reference beam hits the backside of the plate. It is better for this type of hologram to use a 5mw laser due to its longer coherence length over the 5mw laser used in the transmission type of hologram.

Figure 5-2 shows a diagram of the researchers' setup [Hecox and Amberg]. Again, it is important to keep the two beam lengths the same (to a degree of accuracy consistent with the depth of the object). However, care should be taken to keep unwanted light from the reference beam from hitting the object or the side of the plate. For this type of hologram, it is necessary to bleach the plate after developing to get the phase information off the plate. Without bleaching, all that is viewable is the amplitude information, which is okay for the transmission holograms since the researchers were using a coherent light source to view them. One thing that they noticed was that on the holograms they bleached and used the *fixer* in the developing process, a whitish-colored splotch appeared on their plates,

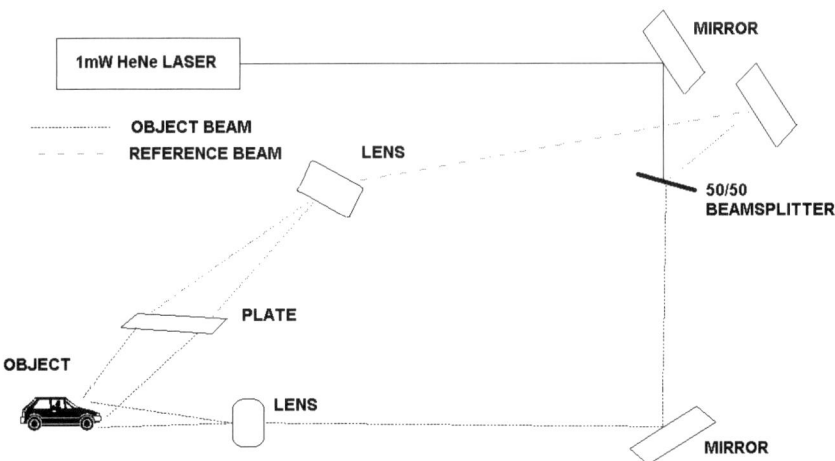

FIGURE 5-2: Reflection hologram setup.

which turned out to be smaller for lower exposure times. The researchers did not receive the same white splotch on the transmission hologram that they developed and bleached without using the *fixer*.

 Since researchers work with potentially dangerous photographic chemicals, they should be carefully instructed in the safety precautions for mixing and storing chemicals. The developer should be mixed well in advance and diluted to a working strength just before using. The developer is chemically a base and is caustic. The stop bath is diluted acetic acid, and the *fixer* or *hypo* contains sulfuric acid and other potentially harmful chemicals. Adequate ventilation should be provided, and protective clothing, gloves, and goggles should be worn while mixing.

Project Results

Overall, the researchers' results were pretty good. Several of their transmission holograms came out fairly well. None of them were particularly exceptional, but they did have a three-dimensional image of the object in them. Unfortunately, the researchers didn't have enough time to check the effect of the object shape and color on the outcome of the hologram. However, they did find that objects that are less reflective (painted flat white) do not make very good holograms unless exposure time is increased and reference beam strength is decreased.

The first transmission hologram that the researchers made was overexposed, but by bleaching it, they managed to obtain a fairly good hologram in spite of the overexposure. The next few transmission holograms that they tried did not come out due to problems such as underexposure and washout (described later). The last two transmission holograms that the researchers made came out pretty well. One of them was an object that did not reflect light as well as their test object, but they were still able to get a discernable image.

After making a couple of transmission holograms and getting a feel for the problems that arise and how to deal with them, the researchers moved on and started making reflection holograms. Reflection holograms presented a whole new set of problems. They first had to rearrange the optical elements on the table so there would be a beam on each side of the plate. In spite of pretty bad overexposure problems, the reflection holograms all turned out fairly good, with images clearly visible when the plate was held sufficiently far away from the light source. The researchers had gotten pretty discouraged with their reflection holograms since they couldn't find the image when they were holding the plate close to the light source. As it turns out, the plate needs to be far enough away from the white light source so that the light source is approximated by point source.

Project Trials and Tribulations

This section is a summary of the major problems that the researchers had in making the holograms. Their first major problem was a *washout*. A washout occurs when the hologram looks like it has been exposed properly, but there is no image on the plate. Problems that could cause an image washout include bumping the table during exposure, talking during the exposure, and even having the reference beam noticeably brighter than the object beam. In order to avoid image washout, the table must be properly floated, so that any vibration in the building won't affect the image. Also, avoid touching the table or talking during the exposure of the plate.

Overexposure is simply exposing the plate too long. An indication of this is having a quite black plate after developing. The plate will often not be uniformly black; there will sometimes be some areas along the side of the plate where you can see the reference beam's speckle pattern. Bleaching the plates can sometimes save overexposed holograms. Underexposure is not exposing the plate long enough to get an image. An indication of this is getting a clear plate of glass after developing the plate. To correct this problem, increase the time of the exposure, increase the power output of the laser, or both. Both underexposure and overexposure can be avoided by first running a test exposure, exposing sections of the plate for increasingly longer times and choosing the best exposure time based on which band looks best.

Viewing problems are when you cannot find the image and think that it is not even present. When viewing transmission holograms, it is best to try all orientations of the plate, and tilt the plate forward and backward. For viewing reflection holograms, the researchers found that the hologram needed to be held about a meter or so underneath a point source of light to start seeing the image. This is because the light source needs to approximate a point source in order for the image to appear. Originally, they were viewing the hologram a couple of inches below the light and were not seeing any image at all.

The researchers also had trouble with inconsistencies in developing processes. They used D-19 developing solutions for most of the out development process. However, the bleach solution that they used was part of another kit. This kit did not use the *fixer* in its developing process. The two methods were not completely compatible, because using the bleach made spots and other problems appear on the holograms. This problem is easily remedied by simply being consistent between developing and bleaching.

Finally, the researchers had a little trouble with the beam lengths. On one of the reflection holograms, they forgot to check the object beam and the reference beam to see if they were of the same length. It is difficult to say what the effects of this were, since the hologram didn't work very well due to overexposure. Those who do a project like this in the future should make sure to have the beam lengths fairly close to equal.

In Summary

- Two coherent beams hitting the same side of a plate and interfering with one another are known as transmission holograms.
- Reflection holograms were the second types of holograms that the researchers made. They are different from transmission holograms.
- The researchers' results were pretty good overall. Several transmission holograms came out fairly well.
- The researchers managed to obtain a fairly good hologram in spite of the overexposure. Due to problems such as underexposure and washout, the next few transmission holograms that they tried did not come out.
- The researchers moved on and started making reflection holograms after making a couple of transmission holograms and getting a feel for the problems that arose and how to deal with them. A whole new set of problems presented itself in the creation of reflection holograms.
- *Washout* was their first major problem. *Washout* occurs when the hologram looks like it has been exposed properly, but there is no image on the plate.
- Exposing the plate too long is known as *overexposure*. Having a black plate after developing is an indication of this.
- Viewing problems are when you cannot find the image and think that it is not even present. It is best to try all orientations of the plate, and tilt the plate forward and backward when viewing transmission holograms.

- In developing processes, the researchers also had trouble with inconsistencies. For most of the out development process, they used D-19 developing solutions.
- Finally, the beam lengths gave the researchers a little trouble. They forgot to check the object beam and the reference beam to see if they were of the same length on one of the reflection holograms.

The next chapter discusses the practical uses of European holography.

End Notes

1. Tim Hecox and Matt Amberg, "Holography," Department of Electrical Engineering, EE488 Laser Engineering, University of Washington, P.O. Box 352500, Seattle, WA 98195-2500, USA, 2000, pp. 1–4.

CHAPTER 6

HOLOGRAPHY MARKETING IMPLICATIONS

In This Chapter

- Lasers that produce large-format holograms
- Creating large-format studios
- Creating new pulse lasers
- General Optics Laboratory
- Large-format holography
- Australian Holographics
- Arts in the electronic landscape

It seems that a hologram has become a generic concept, simply meaning anything that is sufficiently futuristic and impressive. Unfortunately, somewhere between the hype associated with the technology and the true nature of holography, something has become lost in the general understanding of the wonderful and real potential of the medium.

This chapter concludes the discussion on the practical uses of hologram technology. It examines the holography marketplace through the eyes of the father of Australian Holographics (AH), Dr. David Ratcliffe. This chapter also examines the practical uses of holographic art in the electronics marketplace (commerce).

Large-format Studio Creation

Dr. David Ratcliffe[1] founded Australian Holographics (AH) with the deliberate intention to develop large-format holograms as its main priority. Dr. Ratcliffe appreciated the artistic possibilities that large-format holography seemed to promise, but he could not find a studio that had the type of technology he wanted to develop.

The market for large holograms was then almost nonexistent, and the financial risks were great. These risks were compounded by the fact that Dr. Ratcliffe used many unconventional and untested techniques in his method of producing large holograms; an aspect that would, however, later bear fruit.

The AH project necessitated building a large, climate-controlled studio incorporating a 6 × 5 meter (19.5' × 16.3') optical table weighing some 25 tons. The system had to allow for the creation of large-depth scenes for mastering, and afford the space required for the effective production of ultra-large-format rainbow transmission and reflection hologram copies. The resulting studio now produces transmission holograms up to 1.1 × 2.2m (3.6' × 7.2') and reflection holograms up to 1 × 1m. (3.3' × 3.3'). A heavy, sand-filled cavity steel construction was used for the table. The suspension system was constructed around nine Firestone air bags connected to a standard pneumatic setup with needle-valves, ballast tank, and compressor. Overhead towers were designed to carry large transfer mirrors at heights of over three meters above the table. These towers were constructed from hollow steel tubes filled with sand. Over the years, lifting systems for the large glass filmholders evolved from hand operated, to mechanical, and finally to pneumatic.

The first AH laser, which is still being used faithfully today, was a Coherent small-frame 6W (all-lines) Argon laser that produced at 514.5 normal mode (nm) service factor (SF) at around 2WATT (W). For reflection work, this was supplemented with a large-frame Russian-Krypton laser from the Plasma Research Laboratory (PRL).[2] This gives around 1W SF at 647nm as fairly reliable, if you don't mind changing the seals on the water circulator from time to time. Different recording materials and chemistries have been used over the years. Dr. Ratcliffe started with Ilford materials, but changed to Agfa[3] when Ilford[4] stopped production. The Agfa now seems to be superior to the older Ilford emulsions. Since Dr. Ratcliffe's work is almost exclusively large format, he tends to always work with film. For Rainbows, he uses 8E56 green material, and for reflections and transmissions, either the 8E56 or the 8E75, depending on what color laser light he is using. He has also had cause from time to time to use large glass plates made in various places in the former Soviet Union. Chemistry varies depending on the application. For Rainbows, Dr. Ratcliffe consistently uses an ascorbic acid developer and a simple potassium dichromate bleach. Tests by Adelaide University[5] have shown excellent diffraction efficiencies with these chemicals (reaching 70 percent with optimized drying, and in some cases higher); and, for the standard Rainbow hologram, he hasn't found anything better. For reflections, it is difficult to cite a single formula. For Argon transfer holograms, a Pyrogallol developer is often used, as this allows a uniform image in the yellow with appropriate choice of bleach and presoaking in water. For Krypton or dye transfers, he'll usually use another developer, sometimes with shrinking bleach, sometimes without. Of course, for uniform color control, he frequently uses pretreatment with TEA.

Note: *Normal mode (*nm): The term refers to electrical interference that is measurable between line and neutral (current carrying conductors). Normal mode interference is readily generated by the operation of lights, switches, and motors.

New Pulse Laser Creation

In 1992, Dr. Ratcliffe relocated to Europe to establish a base in Lithuania and to develop a network of contacts with laser and optics manufacturers in the former Soviet Union. Having worked with many Russian scientists in his time, as a mathematical physicist he understood very well the strong potential synergy that could, in principle, be realized by finding partners in Eastern Europe. After years of often frustrating work experimenting with a variety of technologies and after collaborating with widely separated companies and institutions, he has finally established a strong presence in the region.

Dr. Ratcliffe soon realized that in order to maximize the potential of mastering for large format, it would be necessary to start working with *pulsed* holography. This is because the nanosecond (ns) exposure time (typically 25ns) of a pulsed system allows one to shoot otherwise unstable images, and also does away with the need for a vibration isolation table. This frees the holographer from some of the limitations of a vibration isolation system, and has the added benefit of theoretically allowing the laser/camera system to be taken to the subject, rather than always the other way around. This allows the possibility, for instance, of making holograms of scenes like the excavation of China's Entombed Warriors in situ, or perhaps a hologram of the inside of Tutankhamen's tomb. Because of the speed of the flash, it is possible, for example, to generate a hologram that freezes the motion of a bullet as it travels through a glass of water, long before the water has even thought of leaving the glass. The implications of this innovation allow for a whole new world of unique holographic images to be recorded: a beehive colony in flight, or a person in a shower showing every droplet frozen in time and floating in suspended three dimensionality.

The problems with available pulsed laser systems at the time were that they were not designed specifically, and hence not completely appropriate for artistic and commercial holography, or that the systems were inadequate technologically. All systems, whatever their construction, were extremely expensive. Dr. Ratcliffe realized that by building a base in Lithuania, he could bring together the superior Soviet optical technology with Western design requirements and reliable electronics to produce the laser that many others in the holography community needed.

The design of a pulsed laser for holography is restricted by many issues. There are methods of obtaining just about any frequency in the visible spectrum, but there is clear evidence that solid-state systems are the most appropriate. This leads to a choice between different laser crystals. Ruby was the first laser to be invented, and although historically the favorite laser for pulse holography, it has many problems that actually make it rather inappropriate. Neodymium (Nd), as it turns out, is much more suitable. This crystal is a four-level laser and so is intrinsically far more efficient than ruby, which is a three-level laser. In practical terms, for this factor, this translates into much smaller power supplies.

There are many varieties of Neodymium. The Nd:YAG laser is perhaps the best known. It has an emission at 1064nm in the infrared, and can be frequency doubled with a nonlinear crystal such as DKDP or KTP to produce an emission in the green at 532nm. For large-energy applications, an ND:YAG oscillator is traditionally paired with silicate glass amplifiers. Despite the fact that ND:YAG/silicate glass can produce a potentially good holography laser, there are significant reasons for choosing the lesser-known material Nd: Yttrium Lithium Fluoride (YLF). Nd:YLF lasers at 1053 and can be paired with phosphate glass amplifiers for high energy; hence, its doubled emission is at 526.5nm. Phosphate glass has significantly more gain than silicate glass, and this is a great advantage. In addition, YLF, with lower gain than YAG, allows more energy to be stored in the oscillator without risk of superluminescence, an effect that can be a problem when designing a YAG system. In addition, YLF has weaker thermal lensing than YAG; hence, even though it has higher thermal conductivity, YLF can tolerate more optical pumping without propagation modification. This is extremely important when you want to design a laser that has a high-frequency (with high thermal stress) low-energy alignment mode in addition to the normal high-energy, high-frequency mode.

Note: The power source for the Nd:YAG laser is a synthetic single-crystal *Neodymium–YAG Laser*. The crystal is composed of the elements Yttrium, Aluminum, and Garnet; hence, Nd:YAG.

After taking into account these and many other constraints, Dr. Ratcliffe proceeded to develop a laser along these lines. Initially, the idea was to build a 5J system based on an ND: YLF ring cavity oscillator with two phosphate glass amplifiers. The oscillator design incorporated a Lithium Fluoride (LiF) passive Q switch and etalon. Matching the oscillator to the first amplifier was done by magnification of the far-field. This assured a perfect seed distribution. The first amplifier was two pass and used SBS correction. This was vitally important in obtaining a constant beam divergence and propagation direction. The second amplifier was single pass and used focal plane translation incorporating vacuum spatial filtering between the two amplifiers. Frequency doubling was done by a DKDP crystal, which gave approximately 60-percent conversion efficiency. Such a system, when matched with Western quality electronics, represented the best marriage of Western design philosophy and Soviet optical expertise. This system, now marketed as the G5J, can actually give up to 8J in the green. Its clear advantages are its almost perfect beam parameters, its long coherence, (typically 3–5 meters), and its almost perfect shot-to-shot reliability. Also important are its capacity to produce a 1-Hz low-energy mode for alignment with exactly the same beam parameters as in a normal high-energy pulse, compact size, and near perfect output color at 526.5nm, which in practical terms means great portraits without the need for skin make-up usually required when using a ruby laser. More recently, G5J has been supplemented by 1 and 2 Joule systems. Currently a 1J system goes for around $65,000, which is comparatively cheap for a new holography laser.

Tip: Every time the computer scans up and down in frequency, a data set in point numbers is acquired rather than in frequency. That is, a transmission signal is collected through the oven for each step in frequency, but since the laser frequency's ramp is not perfectly linear, one cannot assume that equal steps in PZT voltage correspond to equal steps in frequency. The etalon provides users with a way of quantifying this frequency nonlinearity. As the laser's frequency changes, it periodically comes into resonance with the etalon's mirror cavity: for every gigahertz (roughly), one sees a peak in transmission through the cavity. If the user records the transmission through the etalon while ramping the laser, the user will get a trace that can be described by an Airy function (AIRY (X)) that returns the Airy function Ai of real argument X. The file SHARE1;AIRY FASL contains routines to evaluate the Airy functions Ai(X), Bi(X), and their derivatives dAi(X), dBi(X). Ai and Bi satisfy the AIRY eqn diff(y(x),x,2)-x*y(x)=0.)).

Most frequently for pulsed holography, Dr. Ratcliffe used the Agfa 8E56 material, and when using an Sm6 developer, good results can be obtained. Dr. Ratcliffe has also used the green-sensitive Russian plates, but while they are cheaper, he still preferred the higher-resolution, higher-sensitivity Agfa emulsions. He is currently testing plates from Holographic Recording Technologies (HRT), GmbH in Germany that he hopes will provide an alternative to Agfa. He is particularly interested in these plates, as HRT is offering sizes up to 50 × 60 cm. Specializing in large format, this is really the minimum master size for a meter-square transfer reflection hologram, and although everything is possible using film, glass is easier.

The Laser and the Business: GEOLA

Dr. Ratcliffe's endeavors in Lithuania have resulted in the formation of GEOLA, short for *General Optics Laboratory*,[6] a business specializing in the manufacture of specialized Nd/YLF/phosphate glass pulse lasers for holography and also offering a state-of-the-art pulsed holographic mastering facility in Lithuania's historic capital, Vilnius. GEOLA has recently built new premises in Vilnius that house modern stylish offices, a studio environment, and laser fabrication laboratories. The image room of the pulsed laser studio has a ceiling height of six meters and allows for the creation of holograms with image volumes of up to 100 cubic meters and hologram formats of more than two meters square. This, coupled with the fact that virtually everything of importance in the lab is made inhouse especially for this application, has given the lab the potential to make wonderful, large-depth holographic images. The studio is available for rent to holographers for artistic work.

Large-Format Holography Advances Over Time

Although pulse holography has added considerably to the capacity of Australian Holographics to generate new kinds of holographic images, there are still many occasions in which Continuous Wave (CW) mastering is desirable, if not essential. The requirement for stability in the CW mastering process has a surprisingly beneficial aspect in that it allows for the utilization of unstable curtained areas to effectively render invisible unwanted elements in the field of vision. This trick is unique to CW and is sorely missed during pulse mastering, in which the problem is that often too many things are visible and there are limited methods available to conceal them. Thus, if a large object is required to apparently float unsupported in space, CW mastering, rather than pulsed, provides the means to easily achieve this illusion. Many important elements involved in producing high-quality large-format holograms rest not so much with the traditional concerns of holography, but rather with aesthetic concerns that relate to table layout; lighting techniques that endeavor to feature the subject without visual distractions; and, to control glare and reflections that lead to nonlinear noise. Over time, a vocabulary of devices is built up to deal with the changing demands of each project.

It is worth noting that it took 50 years from the appearance of the first photograph in 1826 to invent the roll-film camera, and another 48 years to put perforations in that film to arrive at the 35mm camera in 1924. In the 49 years since the invention of holography, we have made great advances in the technology. However, like photographs taken in 1924, at which time most photographers were still coming to terms with the medium, clever use of holographic technology has been limited to the small number of individuals who strive to push the envelope. Progress, however, is definitely being made. Consider, for example, that Dr. Dennis Gábor's first hologram achieved a depth of only a couple of millimeters. Today, in Dr. Ratcliffe's studios, although limited by available film size, he makes 2.2 × 1.1 meter holograms with depths of up to 6 meters. This allows images from full-size holograms of cars to a 4-meter model of a white pointer shark or a 20-square-meter model of the Earth with the MIR space station hovering above.

Note: Dr. Dennis Gábor (b. 1900, Budapest—d. 1979, London) won the Nobel Prize in 1971 for his investigation and development of holography (father of holography). Holography dates from 1947, when British/Hungarian scientist Dennis Gábor developed the theory of holography while working to improve the resolution of an electron microscope. Gábor, who characterized his work as "an experiment in serendipity," coined the term *hologram* from the Greek words *holos*, meaning *whole*, and *gramma*, meaning *message*. Further development in the field was stymied during the next decade, because light sources available at the time were not truly *coherent* (monochromatic or one-color, from a single point, and of a single wavelength).

Current Technical Work in Progress: Australian Holographics and GEOLA

Dr. Ratcliffe's recent exploratory work with simultaneous red and green flash pulse laser research (raman scattering) has given rise to some important results with a direct relevance to color holography. He has been able to achieve very significant nonlinear raman conversion efficiencies that approach the quantum limit while preserving the spatial and temporal coherence in the two-color beam. With experimental results showing such promise in this field, this project will hopefully lead him to large (>1m2), two-color image-planed pulsed reflection holograms. This is a technological solution that seriously addresses comments from the advertising industry that large-format holography needs to be able to offer realistic product colors for advertisers. Incidentally, the new three-color PFG plates from Slavich[7] similarly address this problem for small-format CW work and have been producing excellent results in the Denisyuk format.

Tip: For practical applications, Denisyuk holograms offer important advantages. This recording geometry is utilized to develop a compact holographic double-exposure interferometer based on a photorefractive SBN:Ce crystal as storage material. A preliminary report on the realized interferometer is given, present limitations are determined, and proposals for further improvements are outlined.

Dr. Ratcliffe has also been collaborating with the optics department of Adelaide University to build a ring-dye laser for color transfer holography. This laser, which is being pumped by a 4W multimode multiline Russian (POLARON) Argon laser, is a potentially great laser source and is of significant interest for color holography. To Dr. Ratcliffe's knowledge, such lasers, although used from time to time in holographic applications, have not been studied sufficiently. He is presently running the laser with DCM, but plans soon to experiment with other dyes including R6G, which has a higher yield but with the disadvantage of falling in a low-sensitivity region of the Agfa 8E emulsions.

Other current research projects include experiments in underwater holography and high-speed shutters for pulsed rainbow cameras for full daylight use. Dr. Ratcliffe is confident that making masters of large holograms under water with a frequency-doubled Nd/YLF pulsed laser at 526.5nm is quite feasible, and will lead to the development of a new camera system that could be used to make large holograms in oceanariums, and ultimately on the ocean floor. For daylight holography, high-speed shutters will synchronize to the nanosecond flash of the pulse laser to allow the film to be exposed effectively only by the laser flash, and then close before the film is affected by the sunlight. The resulting portable, if bulky holographic camera could be used to capture events like the split-second timing on the finish line in the men's 100 meters final at the 2004 Olympics. What an appropriate debut for a 21st-century technology that would be! So, Dr. Ratcliffe has innovative current technology and promising new technology on the drawing board, but it is all in the end just technology, that will hopefully one day seem commonplace and ordinary in the 21st century. The challenge, Dr. Ratcliffe believes, is to continue to use that technology as a creative toolbox, and not see it as an end in itself.

The marriage between art and science over the last few years has come to see Dr. Ratcliffe's lab as something resembling a large photo-processing machine, like the ones down at the local shopping mall. The difference is that he is not so automatic, and he actually creates pictures as well as develops and prints them. The similarity lies in the fact that the lab is, after all, a fairly finite technological concern. Human creativity, on the other hand, the source of the ideas behind each hologram, is infinite. A hologram, like a photograph, is a product of technology, but in the end, is judged more on its creative content than its degree of chromatic balance or diffraction efficiency. In the final analysis, a hologram will be appreciated for the concept behind it (people make concepts, not machines).

In recognition of this, Dr. Ratcliffe welcomes collaboration with artists working in the field who can assist his growth by bringing unusual projects to his studios. The expansion of the potential of the holographic medium is in his interests.

The necessity for high-quality 3D models led Dr. Ratcliffe to develop a business joint venture with the South Australian Museum. This arrangement gives him legal access to the museum's vast collection of exhibits. As a natural progression from this activity, he has become involved with all of the Australian Natural History Museums and various science museums and businesses in Asia and Europe to develop the concept for a thematic touring holographic exhibition called Prehistoric Lives. This exhibition will document key milestones in the development of evolution on Earth. This will be done with 30 multichannel rainbow transmission holograms and a series of five hexagonal laser transmission modules with a hologram on each face. The specific elements of the story and the choice of the actual prehistoric animals will be made by the international community of paleontologists and archaeologists via an interactive homepage called Prehistoric Lives at MCM's Web site (http://www.mcm.com.au) as shown in Figure 6-1. Projects like this will bring state-of-the-art holographic tech-

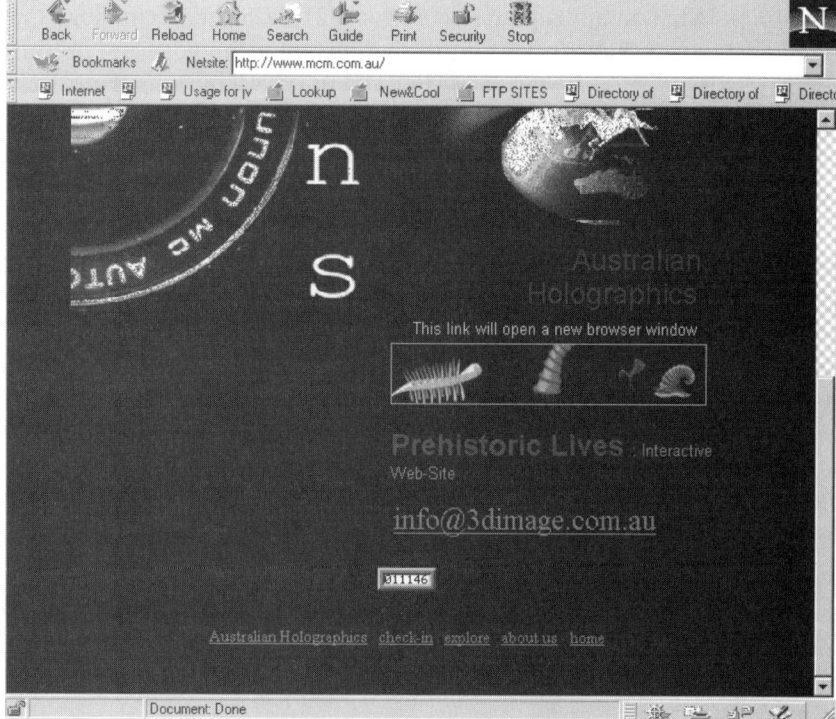

FIGURE 6-1: Prehistoric Lives interactive Web site.

nology to a world audience, providing a new benchmark exemplifying the power and new aestheticism of today's holographic technology.

In Summary

Within the cultural psyche, holography has come to occupy a special place. Far from being feared as something foreign and unapproachable, it has been embraced and almost eulogized as being synonymous with a kind of futuristic view of the world. Thanks to the scriptwriters of TV shows like *Quantum Leap* and *Star Trek*, and films such as *Star Wars* and *Back to the Future*, the public seems to have a perception of holography that actually has more in common with cinematic special effects. Contrast this acceptance to the very slow integration of holography into either advertising (with the exception of the ubiquitous embossed security holograms on credit cards) or the visual arts, and it is clear that holography has probably been disadvantaged by the unrealistic publicity it has received. Somewhere between the hype associated with this new and intriguing technology and the true nature of holography, something has become lost, and that seems to be an understanding of the wonderful and real potential of the medium:

- In essence, a laser is no more than a light source; and holography, rather than being a futuristic novelty, is in fact a very close cousin of photography. Like photography, the subject is illuminated by light, and the exposure is recorded on a piece of film. After the exposure is made, the film must be bathed in developer and fixer, just like a photograph. The key difference is that holography records spatial information in a way that makes photography look woefully inadequate.

- A photograph of an object is somewhat like viewing that object through a pinhole, giving a fixed perspective. A hologram, on the other hand, is a truly three-dimensional image, recorded on a fixed substrate, which appears to either float behind or in front of the transparent substrate, allowing the viewer to see it from a multitude of angles.

- A holographic master is only viewable under laser light. However, when a copy is made from that holographic master, the image can become viewable under ordinary white light, and in the process, the object can be *pushed* through the image plane to appear to float in mid-air (a virtual image of the object)—visible yet untouchable. And this is just the beginning.

- Because a hologram is created by the cross referencing or superposition of light waves creating what is known as an interference pattern on the film, the recorded information is actually encoded at a resolution close to a wavelength of light, or about 2,000th of a millimeter—in other words, microscopic.
- With holography, it is in fact theoretically possible to put a microscope lens in front of a hologram and be able to see microscopic organisms on the surface of whatever model was the subject of the hologram. Although this degree of resolution is imperceptible, and therefore irrelevant to most applications of holography, it has in part enabled the new field of holographic computer memory to be developed, which will produce crystals of lithium niobate that it is theorized will allow storage capacities of a trillion bytes of data in a crystal smaller than a sugar cube.
- In practical terms, two types of lasers produce holograms: continuous wave lasers and pulse lasers. The analogous concepts in photography are the long exposure and the flash bulb, respectively.
- The continuous wave laser produces the linear beam of colored light usually associated with the concept of a laser. This beam is split into two parts to illuminate the model and the film. The beams are then recombined at the film plane, thereby creating the interference pattern, and the hologram is recorded.
- Holograms made with a continuous wave laser require the model and laser beam and optics to not move above about 5,000th of a millimeter to ensure that the superposition of the light waves forms a perfect interference pattern. This precision is achieved by using a vibration isolation platform on which the holographic recording environment is created.
- In his main studio, Dr. Ratcliffe has a 6 × 5 meter platform weighing 25 tons that floats on a series of interconnected air cushions, thus reducing vibrations to an absolute minimum. Using this platform, it is possible to hold flat a 2 × 1-meter piece of film and maintain its stability and that of the related optics, beneath the 5,000th of a millimeter tolerance for the 4- to 8-second exposure. This procedure allows bright, white-light-viewable hologram copies of this large size to be recorded.

- A pulse laser generates an enormously bright flash that occurs in about one 40-millionth of a second. This pulse of light is one of the fastest manmade events, creating a superfast flash that effectively freezes the subject, registering the interference pattern on the holographic film before any vibration or movement is possible.
- Pulse hologram masters do not require vibration isolation platforms to record their images, and because of this, it is now possible to take the pulse laser/camera setup to the subject, rather than the other way around. This theoretically allows the possibility of making holograms of scenes like the excavation of China's Entombed Warriors in situ, or perhaps a hologram of the inside of Tutankhamen's tomb.
- Because of the speed of the flash, it is possible to generate a hologram that freezes the motion of a bullet as it travels through a glass of water, long before the water has even thought of leaving the glass. The implications of this innovation allow for a whole new world of unique holographic images to be recorded; a beehive colony in flight, a person in a shower, showing every droplet frozen in time and floating in suspended three dimensionality.
- The necessity for high-quality 3D models led Dr. Ratcliffe to develop a business joint venture with the South Australian Museum. This arrangement allows his company access to the museum's vast collection of exhibits. As a natural progression from this activity, Dr. Ratcliffe has become involved with all of the Australian Natural History Museums and various science museums and businesses in Asia and Europe to develop the concept for a themed touring holographic exhibition called Prehistoric Lives. This exhibition will document key milestones in the development of evolution on Earth. Projects like this will bring state-of-the-art holographic technology to a world audience, providing a new benchmark exemplifying the power and new aestheticism of current holographic technology.
- Clever use of holographic technology has been limited to the small number of individuals who strive to push the envelope.
- It seems unlikely that the future of holography will follow a path similar to photography. We won't all be carrying around miniature holographic cameras and taking our 3D holiday snapshots.

- Holography and photography are closely related, but in true evolutionary fashion, they will grow apart, and holography will adapt and change as it makes new friends along the way.
- One of holographic memory's first commercial incarnations will be a fusion of the CD and holographic emulsions. In other words, it is called the DVD, and looks like a CD, (only slightly thicker) holding 200 times more information.
- Holograms will become more commonly seen in the form of 3D billboards on special light boxes, and perhaps in automatic teller machines popping up to greet us when we make a transaction.
- The ability of a hologram to contain a number of images that change as the viewer walks past will facilitate new ways of portraying concepts of transition for visual merchandising. The key to the integration of visual holography in the marketplace is the concept of interactivity; involving the viewer in the control of the experience.
- One piece of holographic film can contain a multitude of different holographic images. These images can be triggered electronically, and thereby invite a myriad of possibilities for new kinds of multimedia interactivity such as biofeedback sensors reading your body and projecting an appropriate holographic image. This is not the future; it's starting to happen now!
- As to the relationship between holography and art: Art, of course, being thoroughly indefinable, will refuse to make its potential nexus with holography either easily predictable or straightforward. There is equal potential for a fusion of holography with computer art via stereographic techniques, as there is for continued development with traditional and avant-garde sculpture or portraiture. The key surely rests with the ability of artists in any genre to acquaint themselves with the intricacies and idiosyncrasies of holography. Once cognizant of the potential of the medium to portray new ideas in a new way, the employment of the technology by a greater number of practitioners as *art* will be inevitable.

Chapter 7, "The Hologram Universe," covers holographic production, conversion, and commercial applications, as well as a number of related topics.

End Notes

1. Simon Edhouse, 3DIMAGE, P.O. Box 95, Sydney, Australia 2001, 1999.
2. Plasma Research Laboratory (PRL), Research School of Physical Sciences and Engineering, The Australian National University, Canberra, ACT 0200, Australia, 1999.
3. Agfa-Gevaert N.V., Septestraat 27, B-2640 Mortsel, Belgium, 1999.
4. Ilford Imaging Australia Pty Limited, P.O. Box 144, Mt. Waverley, Victoria 31494511, Australia, 1999.
5. The University of Adelaide, North Terrace, Adelaide, Australia, 5005, 1999.
6. General Optics Laboratory, P.O. Box 343, Vilnius LT-2006, Lithuania, 1999.
7. Slavich International Wholesale Office, uab GEOLA, P.O. Box 343, Vilnius 2006, Lithuania, 1999.

PART **II**

COMMERCIAL APPLICATIONS

The second part of this book identifies intranet security trends on the Web: client and server, procedures and tools, and system and intranet administration currently in place within most organizations. It also covers holographic production, conversion, and commercial applications, as well as a number of other related topics.

Part Two also addresses the design and anatomy of a hologram by looking at computer graphics in general. It explains how to master holograms by showing you how to create and combine the necessary photographic and holographic elements to produce a *master* hologram on special photosensitive emulsions. You'll also learn how the *master* hologram can be copied repeatedly in a variety of formats, depending on your final application. Also discussed is how to affix the metal shims to a high-speed embossing machine that stamps the holographic pattern into rolls of very thin plastic or foil. In addition, you'll learn how metallized films are routinely used for holographic applications. The converting stage of the holographic process that manufacturers go through to create commercial holograms is examined, along with the *products* stage of the commercial holograms process. Finally, you'll learn about the *final* stage of the commercial holograms process: *patterns*.

CHAPTER 7

A UNIVERSE OF COMMERCIAL HOLOGRAMS

In This Chapter

- Master holograms
- Production proofs
- Converting
- Integration/finishing
- Custom-made holograms
- Holographic applications

This chapter begins Part Two, "Commercial Applications," with a brief look at the types of custom-made holograms. First, there will be a brief discussion about companies in general (no specific company) that specialize in making, designing, developing, and applying commercial holographic technology to a wide spectrum of products and services.

A need exists today for companies that produce commercial holograms to satisfy the growing demand for holographic materials and the need for quality project design and management. The hologram producers of quality products and services today are many, and the technology is evolving fast. New techniques and better materials are being developed constantly.

No one company can have all of the skills and produce all the material needed to keep pace with the speed at which the high-tech world of holography is moving. Even the largest companies in the world depend on the materials and services supplied by smaller, more specialized companies.

Today, the mission of these companies is to offer their clients *state-of-the-art* holographic products and services, and supply customers with the newest and best technology available for their application.

The combination of 25 years in the holographic industry, a sound business structure, artistic creative design team, and a state-of-the-art *worldwide network,* is the formula for success for these companies. Understanding the fine balance between *when* to offer a technology that can be delivered per specification and still be innovative is derived from experience. Today's hologram-producing companies are striving to bring your products to that *cutting edge* with the newest innovations *only when proven reliable* in production.

These companies are also dedicated to staying on the forefront of the newest developments in the holographic industry, and making those advancements available to their customers. They are dedicated to delivering you the newest and best possible technology at the most competitive prices.

From sourcing the best suppliers, designing the concept, monitoring and reporting the status of your job and delivering the finished products, these holographic-producing companies can make the difference. The success of a holographic project is the combination of all steps coming together to form the completed product. *The whole is only as good as the sum of its parts.* This philosophy guides these companies to success with your holographic project!

Let's look at the steps that are necessary for these companies to follow to make your holographic project a real success.

Successful Steps

Once contact and preliminary meetings are made and the scope of the holographic project is clear, the previsualization of the product is formalized by the following steps:

1. The brief
2. Artwork
3. Master hologram

4. Production proof
5. Production
6. Converting
7. Integration/finishing[1]

THE BRIEF

With the combined effort of the holographic producing company's artistic and design team, and the information the client supplies, the project is previsualized from beginning to end. Communication between the client and the holographic vendor is critical for the success of the project. The goal is that the client's only surprise is that the project goes beyond the projected results.

Depending on the application and environment in which the hologram is to be seen, the best design, materials, and lighting are recommenced. Sketches, mockups, and storyboarding are created at this stage. Once the perimeters are agreed on, time flow charting and budgeting considerations can enter into the determination of the final design. Once prototypes and final design are determined and approved, final artwork is produced.

ARTWORK

Artwork is carefully prepared from approved design concepts. This is usually in the form of 3D models and 2D graphic color separations, film, or computer graphics.

MASTER HOLOGRAM

The artwork is translated into holographic masters by lasers and holographic technology. The artwork is also illuminated by the laser and the holographic film that records the interference patterns of light.

Making the master hologram is like producing the printer's stone; if changes are required, the holographer must start over. Since the master hologram is technically demanding, artwork should be thoroughly approved before the mastering step.

PRODUCTION PROOF

The production proof is a hand-pulled hologram that looks very close to what the final hologram will be like. It is used to produce the working prototype that shows the client the hologram integrated into the finished product. Any final changes to the overall design integration are done at this step before the client gives final approval to run the production.

PRODUCTION

Production is the step during which the specified quantity of holograms is produced. Tooling is performed, embossed holograms are stamped, photopolymer holograms are run, and silver halide holograms are shot. The paper goes to press. All other production materials are produced for the converting step.

CONVERTING

Converting is the step during which holograms are die or kiss cut. This also includes gluing, backing, and preparing for the final integration and finishing steps.

INTEGRATION/FINISHING

Integration is the step during which all holograms and other materials are assembled into the finished product. Hot-stamped holograms are applied; overprinting, cutting, folding, or other final assembling is performed.

From beginning to delivery, your project will be monitored to keep on budget and schedule with a critical eye on quality control. Most holograms have a written guarantee of satisfaction that specifications are met as agreed in the terms of sale agreement.

Something about Holograms

The use of holograms represents a new visual language in communication. As with any media, the successful implementation is a combination of many talents and disciplines. Like all new media, usage will increase while price decreases over time. Since the embossed holograms were first introduced commercially in the late 1970s, large-volume orders are nearly 50 percent less in cost.

Holography, like photography, can be used for a broad range of applications. However, the best holograms use direct light to replay their message, which makes them different from photographs. We are moving into the Age of Light as the media of the future. Holograms use light as a form of inkless printing. Fiber optics will carry communications at the speed of light, and optical computers will soon be a reality. Light is the future. Holography is part of the Light Age of information and communications.

TYPES OF CUSTOM-MADE HOLOGRAMS

Although there are many types of holograms with many variations, only a few types have proven to be effective in each application. Unless a display is to incorporate a light (and some applications require one), surface image holograms are the most effective for in-store use where no direct light is provided, such as:

- **2D/3D**: No direct light required. Surface image.
- **3D**: Direct light required. Deep image.
- **Dot matrix**: No direct light required. Surface image.
- **LCD stereogram**: Direct light required. Deep image.[2]

Commercial Applications

The following are types of custom-made commercial holographic applications:

- Art
- Card authentication
- Corporate displays
- Display
- Entertainment
- Executive gifts
- Fashion
- Illustration
- Incentives
- Jewelry
- Labels
- Ornamental
- Pack decoration
- Point-of-purchase displays

- Premiums
- Rigid box packaging
- Security documents
- Stationery
- Stock images
- Trade show displays

Next, let's examine some of the preceding applications and see how they are used.

Art
Art holography is a very special area of holography that is divided into fine art and commercial art holography. All holography could be considered art, since it all deals with making some sort of visual. We usually think of fine art when we refer to original pieces or limited editions, and commercial art when referring to unlimited or mass-produced reproductions.

Display
Display holography includes *point-of-purchase, trade show,* and *corporate display*. Point-of-purchase displays can use all types of holograms, from decorative foils to eye-catching stereograms showing products or fashion models. Trade show and corporate display applications usually involve large-format holograms for their dramatic effect. Commissioned artistic holograms have been used effectively in corporate headquarters, adding a special high-tech statement. Continental Tire[3] used a hologram in its in-store display and received the highest-ever dealer response.

Entertainment
Holograms have proven to be effective sales promotional tools on books, records, and CD covers in the entertainment industry. The simulation of holograms in movie special effects has been a common occurrence in science fiction films for many years. *Star Wars'* Princess Leia, *Star Trek's* Holodeck, *Hologram Man,* and *Escape from LA* are just a few of the hundreds of simulated holo references in movies.

Labels
Label-type holograms are those that are applied to previously produced items. They are usually provided with self-adhesive backing or as hot-

stamping foil. Often for low-volume applications, a preexisting stock image will serve for cost-effective label application with the additional overprinting step for customization. Label applications include advertising pieces, direct mail, promotional products, and point-of-purchase displays. Sales took a leap when Wardley[4] fish food used holographic labels on its products.

Packaging
Packaging includes flexible packaging, rigid box, board packaging, and pack packaging. The eye-catching visual impact, added value, authentication, and collectible nature of holograms have been used effectively in many packaging applications. Hologram premiums and incentives have been included in packaging and back-of-box panel promotions. Undoubtedly, the creative use of holograms in the packaging industry has unlimited possibilities. Brach & Brock Candy Company[5] realized a threefold increase in sales using holographic packaging.

Interactive Graphics
Interactive graphics is the newest and maybe the most exciting area for printing. By using holograms and holographic diffraction effects for the background of over-printing lithography, some of the in-store lighting problems of pure holograms can be overcome. Color printing with the magic of holography combines to form stunning interactive visual effects that go beyond what either can do alone in the marketplace. Trading cards using the holo/litho combo are trading at two to three times the regular price.

Gifts
Ornamental objects; jewelry, including watches, earrings, pendants; and executive gifts have always been a staple in many holographic product lines. Executive gifts such as desk lights, and wall decor such as clocks and calendars using holograms make ideal functional office products. Blanton Whiskey added a hologram to its bottle, and sales surpassed targets.

Security
Probably the fastest-growing area for the use of holograms is security and product authentication. Added to documents, tamper-proofing, anti-counterfeiting, customizing ticket protection, identification documents including drivers licenses, credit and telephone cards holograms have proved

to be unsurpassed. Drivers Licenses Bureaus across the United States have documented elimination of counterfeit vehicle registration cards. The trend that most all credit cards carry a hologram for anti-counterfeiting purposes is a good indication that security holography has proven to be very effective.

In Summary

- There exists a need for companies that produce commercial holograms to satisfy the growing demand for holographic materials, and for quality project design and management. The hologram producers of quality products and services today are many, and the technology is evolving fast. Better materials and new techniques are being developed constantly.
- The best design, materials, and lighting are recommended, depending on the application and environment in which the hologram is to be seen. At this stage, sketches, mockups, and storyboarding are created. Time flow charting and budgeting considerations can help to determine the final design, once the perimeters are agreed on. Final artwork is produced once prototypes and final design are determined and approved.
- Artwork is carefully prepared from approved design concepts, usually in the form of 3D models and 2D graphic color separations, film, or computer graphics.
- Producing the printer's stone is like making the master hologram. The holographer must start over if changes are required. Artwork should be thoroughly approved before the mastering step, since the master hologram is technically demanding.
- Your project will be monitored to keep on budget and schedule, with a critical eye on quality control from beginning to delivery. Most holograms have a written guarantee of satisfaction that specifications are met as agreed upon in the terms of sale agreement.

The next chapter discusses the design and anatomy of a hologram by looking at computer graphics in general.

End Notes

1. Larry Lieberman, The Hologram Universe, 1642 Euclid Ave., Miami Beach, FL 33139, USA, 2001.
2. Ibid.
3. Continental Tire, 136 Summit Avenue, Montvale, NJ 07645, USA, 1999.
4. Wardley, 283 Wilson St. E, Suite 163, Ancaster, ON, L9G 2B8, 1999.
5. Brach & Brock Confections, Inc., 401 N. Cicero, Chicago, IL 60644, USA, 1999.

CHAPTER 8

DESIGN: ANATOMY OF A HOLOGRAM

In This Chapter

- Computer graphics
- Animation
- Reduced or enlarged images
- Onsite recording
- Artwork
- Scanner
- Graphic software
- Graphic effects
- Color palettes
- Holomagic

Holographic applications designers have fun working with holographic papers and boards. As a designer, you will want to know a few basic things about the creation of custom commercial holography. You will want to understand the use of opaque and transparent inks. You may want to create your design using opaque white underneath the color printing process to alter the visual result of your project. It is also important to look at the integration of lithography before you begin your design in order to achieve your desired effect.

This chapter examines the design stage of the holographic process that designers go through to create commercial holograms. As with all holograms, computer graphics plays a major role in their development.

Computer Graphics

Computer graphics is confined chiefly to flat images. Images may look three-dimensional, and sometimes create the illusion of 3D when displayed; for instance, on a stereoscopic display. Nevertheless, when viewing an image on most display systems, the human visual system (HVS) sees a flat plane of pixels. Volumetric displays can create a 3D computer graphics image, but fail to provide many visual depth cues (shading, texture gradients), and cannot provide the powerful depth cue of overlap (occlusion). Discrete parallax displays (such as lenticular displays) promise to create 3D images with all of the depth cues, but are limited by achievable resolution. Only a real-time electronic holographic *holovideo* display can create a truly 3D computer graphics image with all of the depth cues (motion parallax, ocular accommodation, occlusion, etc.) and resolution sufficient to provide extreme realism. Holovideo displays promise to enhance numerous applications in the creation and manipulation of information, including telepresence, education, medical imaging, interactive design, and scientific visualization.

The technology of electronic interactive 3D holographic displays is in its second decade. Though fancied in popular science fiction, only recently have researchers created the first real holovideo systems by confronting the two basic requirements of electronic holography: (1) computational speed, and (2) high-bandwidth modulation of visible light. This part of the chapter describes the approaches used to address these problems, as well as emerging technologies and techniques that provide firm footing for the development of practical holovideo.

ELECTRO-HOLOGRAPHY BASICS

Optical holography, used to create 3D images, begins by using coherent light to record an interference pattern. Illumination light is modulated by the recorded holographic fringe pattern (called a *fringe*), subsequently diffracting to form a 3D image. As illustrated in Figure 8-1,[1] a fringe region that contains a low spatial frequency component diffracts light by a small angle. A region that contains a high spatial frequency component diffracts light by a large angle. In general, a region of a fringe contains a variety of spatial frequency components, and therefore diffracts light in a variety of directions.

An electro-holographic display generates a 3D holographic image from a 3D description of a scene. This process involves many steps, grouped into two main processes: (1) computational, in which the 3D description is converted into a holographic fringe; and (2) optical, in which light is modulated by the fringe. Figure 8-2 shows a map of the many techniques used in these two processes [Lucente, 3].

The difficulties in both fringe computation and optical modulation result from the enormous amount of information (or *bandwidth*) required by holography. Instead of treating an image as a pixel array with a sample spacing of approximately 100 microns as is common in a 2D display, a holographic display must compute a holographic fringe with a sample spacing of approximately 0.5 microns to cause modulated light to diffract and form a 3D image.

A typical palm-sized full-parallax (light diffracts vertically as well as horizontally) hologram has a sample count (*space-bandwidth produc*t or simply *bandwidth*) of over 100 gigasamples. Horizontal-parallax-only (HPO) imaging eliminates vertical parallax, resulting in a bandwidth savings of over 100 times without greatly compromising display performance.

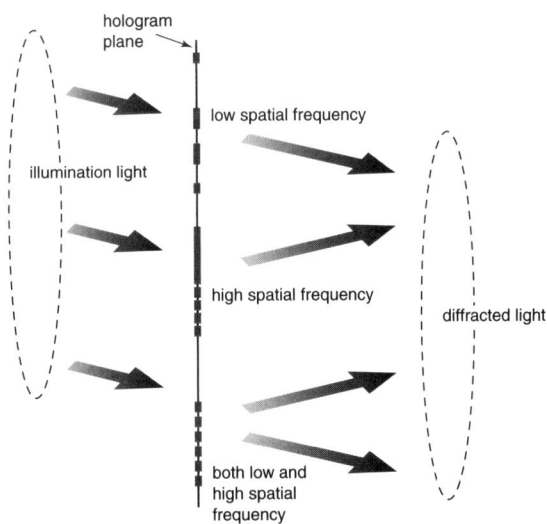

FIGURE 8-1: Diffraction of illumination light by holographic fringe patterns. Fringes with higher spatial frequencies cause light to diffract at larger angles. Fringes containing many spatial frequencies diffract light in many directions.

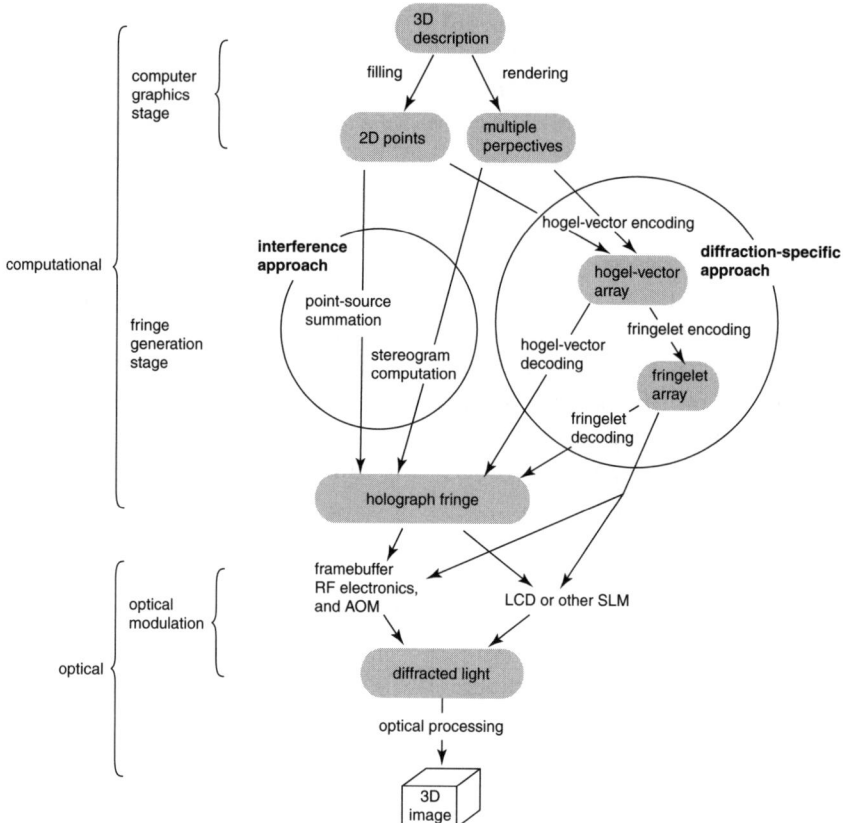

FIGURE 8-2: Information flow in interactive 3D holographic imaging. Each path traces the steps required for a particular method. Computation is generally faster for the methods that are more to the right-hand side.

Holovideo is more difficult than 2D displays by a factor of about 40,000, or about 400 for an HPO system. The first holovideo display created small (50 ml) images that required minutes of computation for each update. New approaches, such as holographic bandwidth compression and faster digital hardware, enable computation at interactive rates and promise to continue to increase the speed and complexity of displayed holovideo images. At present, the largest holovideo system creates an image that is as large as a human hand (about one liter). Figure 8-3 shows typical images displayed on the MIT[2] holovideo system [Lucente, 4].

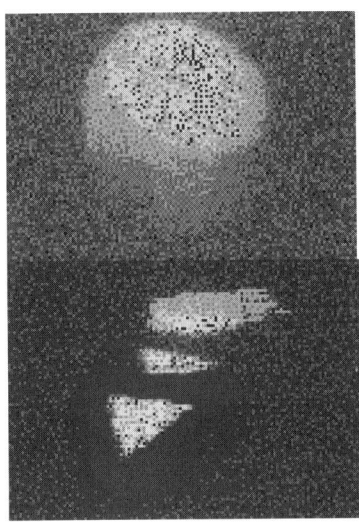

FIGURE 8-3: 6MB holovideo images on the MIT full-color display. Top: If this were a color image, there would be a reddish apple with multicolor specular highlights, computed using hogel-vector bandwidth compression. Bottom: If this was a color image, there would be red, blue, and green cut cubes, computed using a stereogram approach.

HOLOGRAPHIC FRINGE COMPUTATION

The computational process in electro-holography converts a 3D description of an object or scene into a fringe pattern. Holovideo computation comprises two stages: (1) a computer graphics rendering-like stage, and (2) a holographic fringe generation stage in which 3D image information is encoded in terms of the physics of optical diffraction (see Figure 8-2).

The computer graphics stage often involves spatially transforming polygons (or other primitives), lighting, occlusion processing, shading, and (in some cases) rendering to 2D images. In some applications, this stage may be trivial. For example, Magnetic Resonance Imaging (MRI) data may already exist as 3D voxels, each with a color or other characteristic.

The fringe generation stage uses the results of the computer graphics stage to compute a huge 2D holographic fringe. This stage is generally more computationally intensive, and often dictates the functions performed in the computer graphics stage. Furthermore, linking these two computing stages has prompted a variety of techniques. Holovideo computation can be classed into two basic approaches: interference based and diffraction specific.

The Interference-Based Approach

The conventional approach to computing fringes is to simulate optical interference, the physical process used to record optical holograms. Typically, the computer graphics stage is a 3D filling operation that generates a list of 3D points (or other primitives), including information about color, lighting, shading, and occlusion.

Following basic laws of optical propagation, complex wavefronts from object elements are summed with a reference wavefront to calculate the interference fringe. This summation is required at the many millions of fringe samples and for each image point, resulting in billions of computational steps for small, simple holographic images. Furthermore, these are complex arithmetic operations involving trigonometric functions and square roots, necessitating expensive floating-point calculations. Researchers using the interference approach generally employ supercomputers and use simple images to achieve interactive display. This approach produces an image with resolution that is finer than can be utilized by the human visual system.

STEREOGRAMS

A stereogram is a type of hologram that is composed of a series of discrete 2D perspective views of the object scene. An HPO stereogram produces a view-dependent image that presents in each horizontally displaced direction the corresponding perspective view of the object scene, much like a lenticular display or a parallax barrier display. The computer graphics stage first generates a sequence of view images by moving the camera laterally in steps. These images are combined to generate a fringe for display.

The stereogram approach allows for computation at nearly interactive rates when implemented on specialized hardware. One disadvantage of the stereogram approach is the need for a large number of perspective views to create a high-quality image free from sampling artifacts, limiting the computation speed. New techniques may improve image quality and computational ease of stereograms.

The Diffraction-Specific Approach

The diffraction-specific approach breaks from the traditional simulation of optical holographic interference by working backward from the 3D image. The fringe is treated as being subsampled spatially (into functional holographic elements or *hogels*) and spectrally (into an array of *hogel vectors*). One way to generate a hogel-vector array begins by rendering a series of orthographic projections, each corresponding to a spectral sample of the hogels. The orthographic projections provide a discrete sampling of space (pixels) and spectrum (projection direction). They are easily converted into a hogel-vector array. A usable fringe is recovered from the hogel-vector representation during a decoding step employing a set of precomputed *basis fringes*.

The multiple-projection technique employs standard 3D computer graphics rendering (similar to the stereogram approach). The diffraction-specific approach increases overall computation speed and achieves bandwidth compression. A reduction in bandwidth is accompanied by a loss in image sharpness—an added blur that can be matched to the acuity of the HVS simply by choosing an appropriate compression ratio and sampling parameters. For a compression ratio (CR, the ratio between the size of the fringe and the hogel-vector array) of 8:1 or lower, the added blur is invisible to the HVS. For CR of 16:1 or 32:1, good images are still achieved with acceptable image degradation.

SPECIALIZED HARDWARE

Diffraction-specific fringe computation is fast enough for interactive holographic displays. Decoding is the slower step, requiring many multiplication-accumulation calculations (MACs). Specialized hardware can be utilized for these simple and regular calculations, resulting in tremendous speed improvements. Researchers using a small digital signal processing (DSP) card achieved a computation time of about 1 second for a 6MB fringe with CR=32:1. In another demonstration, the decoding MACs are performed on the same Silicon Graphics RealityEngine2 (RE2) used to render the series of orthographic projections. The orthographic projections rendered on the RE2 are converted into a hogel-vector array using filtering. The array is then decoded on the RE2, as shown in Figure 8-4 [Lucente, 6]. The texture-mapping function rapidly multiplies a component from each hogel vector by a replicated array of a single basis fringe. This operation is repeated several times, once for each hogel-vector component, accumulating the result in the accumulation buffer. A computation time of 0.9 seconds was achieved for fringes of 6MB with CR=32:1.

FRINGELETS

Fringelet bandwidth compression (Figure 8-2) further subsamples in the spatial domain. Each hogel is encoded as a spatially smaller *fringelet*. Using a simple sample-replication decoding scheme, fringelets provide the fastest method (to date) of fringe computation. Complex images have been generated in less than 1 second for 6MB fringes. Furthermore, a *fringelet display* can optically decode fringelets to produce a CR-times greater image volume without increased electronic bandwidth.

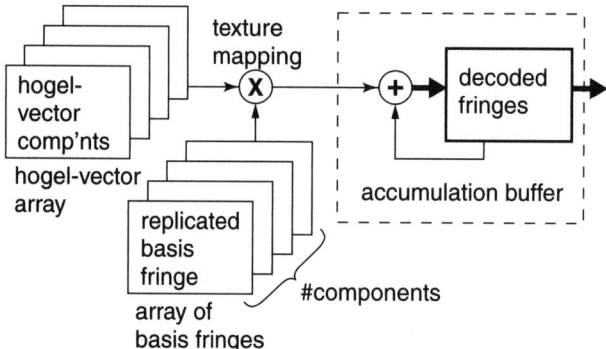

FIGURE 8-4: Hogel-vector decoding on the graphics subsystem. The inner product between an array of hogel vectors and the precomputed basis fringes is performed rapidly by exploiting the texture-mapping function and the accumulation buffer.

OPTICAL MODULATION AND PROCESSING

The second process of a holographic display is optical modulation and processing as shown in Figure 8-5 [Lucente, 6]. Information about the desired 3D scene passes from electronic bits to photons by modulating light with a computed holographic fringe using spatial light modulators (SLMs). The challenge in a holographic display arises from the many millions of samples in a fringe. Successful approaches to holographic optical modulation exploit parallelism and/or the time-response of the HVS.

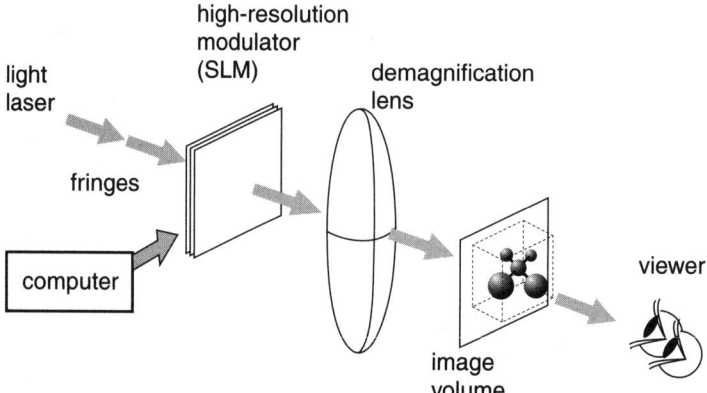

FIGURE 8-5: Holographic optical modulation using a typical high-resolution modulator (SLM). A minimum of 2 million modulation elements is required to produce even a small image the size of a thumb.

Liquid-Crystal and Related SLMs

A liquid crystal display (LCD) is a common electro-optic SLM used to modulate light for projection of 2D images. A typical LCD contains about 1 million elements (*pixels*). A 1-million-sample fringe can produce only a small flat image. A magneto-optic SLM that uses the magneto-optic effect to electronically modulate light, often contains less than 1 million elements. Early researchers using LCD SLMs or magneto-optic SLMs created small planar images. The low pixel count of typical LCDs is overcome by tiling together several such modulators.

For any modulation technique, several issues must be addressed. Modulation elements are too big—typically, 50 microns wide (in an LCD) compared to the fringe sampling pitch of about 0.5 microns. Demagnification is employed to reduce the effective sample size, with the necessary but unattractive effect of proportionally reducing the lateral dimensions of the image (see Figure 8-5). Holographic imaging may employ either amplitude or phase modulation. LCDs are basically phase modulators when used without polarizing optics. Phase modulation can be more optically efficient, and so is most often used. Finally, it is desirable to employ modulators possessing many levels of modulation (grayscale). Common LCDs have nominally 256 grayscale levels, sufficient for producing reasonably complex images.

Deformable micro-mirror devices (DMDs) are micromechanical SLMs fabricated on a semiconductor chip as an electronically addressed array of tiny mirror elements. Electrostatically depressing or tilting each element modulates the phase or amplitude of a reflected beam of light. A phase-modulating device was used to create a small, flat holographic image, and a binary amplitude-modulating DMD was used to create a small, interactive 3D holographic image.

Scanned Acousto-Optic Modulator (AOM)

The time-multiplexing of a very fast AOM SLM has been used in holovideo. A wide-aperture AOM phase modulates only about 1,000 samples at any one instant in time, using a rapidly propagating acoustic wave within a crystal2E. By scanning the image of modulated light with a rapidly moving mirror, a much larger apparent fringe can be modulated. The latency of the HVS is typically 20ms, and the eye time-integrates to see the entire fringe displayed during this time interval. This technique was invented and exploited by researchers at the MIT Media Laboratory to produce the world's first real-time 3D holographic display in 1989. A generalized schematic of this approach is shown in Figure 8-6 [Lucente, 8].

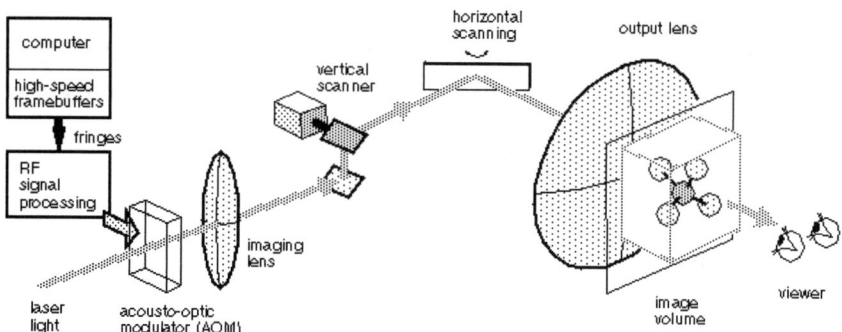

FIGURE 8-6: Schematic of the scanned-AOM architecture used in the MIT holovideo displays.

After Radio Frequency (RF) signal processing, computed fringes traverse the aperture of an AOM (as acoustic waves), which phase-modulate a beam of laser light. Two lenses image and demagnify the diffracted light at a plane in front of the viewer. The horizontal scanning system angularly multiplexes the image of the modulated light. A vertical scanning mirror reflects diffracted light to the correct vertical position in the hologram plane.

One advantage of the scanned-AOM system is that it can be scaled up to produce larger images. The first images produced in this way were 50 ml, generated from 2MB fringes. More recently, by building a scanned-AOM system with 18 parallel modulation channels, images created from a 36MB fringe occupy a volume greater than one liter.

One disadvantage of the scanned-AOM approach is the need to convert digitally computed fringes into high-frequency analog signals. The 18-channel synchronized high-speed framebuffer system used at MIT was made for this application, and was a major practical obstacle in this approach. LCDs, DMDs, and other SLMs are more readily interfaced to digital electronics. Indeed, LCD SLMs are commonly constructed to plug directly in to a digital computer, or are built on an integrated circuit chip. Another disadvantage of the scanned-AOM approach is the need for optical processing. Typical LCD-based holographic displays require only demagnification and the optical concatenation of multiple devices. The time-multiplexing of the scanned-AOM system requires state-of-the-art scanning mirrors that must be synchronized to the fringe data stream. Despite these obstacles, the scanned-AOM approach has produced the largest holovideo images.

Other Techniques
Now, let's look at some other techniques: Color and SAW AOM.

COLOR
Full-color holovideo images are produced by computing three separate fringes. Each represents one of the additive primary colors (red, green, and blue), taking into account the three different wavelengths used in a color holovideo display. The three fringes are used to modulate three separate beams of light (one for each primary color).

SAW AOM
Recently, researchers have used an AOM device with multiple ultrasonic transducers. These multiple electrodes are fed a complex pattern, and launch surface acoustic waves (SAWs) across the device aperture. Diffracted light forms a holographic image. Preliminary results show that this approach may eliminate the need for time-multiplexing and, consequently, scanning mirrors. However, the large number of electrodes may be prohibitively expensive. Also, the array of SAW electrodes necessitates an additional numerical inversion transformation, making rapid computation difficult.

PRESENT AND FUTURE

Currently there are no off-the-shelf holographic displays. Holographic display technology is in a research stage, analogous to the state of 2D display technology in the 1920s. What, then, does the future hold? The future promises exactly what holovideo needs: more computing power, higher-bandwidth optical modulation, and improvements in holographic information processing.

Computing power continues to increase. A doubling of computing power at a constant cost (a trend that continues at a rate of every 18 months) effectively doubles the interactive image volume of a holographic display. Inexpensive computation (around $200 per gigaMAC) is the most crucial enabling technology for practical holovideo, and should be available in 2004.

Although optical modulation has borrowed from existing technologies (transmissive LCDs, AOMs), new technologies will fuel the development of larger, more practical holovideo displays. Because bandwidth is most important, one should use as a figure of merit the number of bits that can be modulated in the latency time of the HVS (typically 20 ms). An AOM can modulate about 16Mb in this time interval, at a cost of about $3000 (or

$188 per Mb), including the associated electronics. The DMD, a new technology for high-end 2D video projection technology, delivers approximately 100Mb in 20 ms, for a cost of about $4,000, or $40 per Mb. Future mass-production could reduce the cost further. Reflective LCDs are another possible technology. Several researchers create small reflective LCDs directly on a semiconductor chip using VLSI technology.

The bandwidths of computation and modulation are likely to increase steadily. Improvements in holographic information processing will likely provide occasional dramatic improvements in both of these areas. Already, holographic bandwidth compression increases fringe computation speed by 3,000 times for same-hardware implementation. Standard MPEG algorithms can be used to encode and decode computed fringes. Nonuniformly sampled fringes provide lossless bandwidth compression and promise further advances.

User demand may be the one additional key to the development of holovideo2E. As other types of 3D display technologies (autostereoscopic displays) acquaint users with the advantages of spatial imaging, these users will grow hungry for holovideo, a display technology that can produce truly 3D images that look as good as (or better than) actual 3D objects and scenes.

Next, as part of the design stage of the commercial holographic process, let's look at animation. The techniques available for the animation of 3D images by computer are discussed, and exemplified by still and animated images. Monocular techniques discussed are hidden lines and surfaces, depth cueing, background blurring, perspective, surface shading, and temporal parallax. Binocular techniques covered are parallel stereo pairs, crossed stereo pairs, Anaglyphs, image switching, polarized images, dual displays (virtual reality), Lenticular film, Pulfrich effect, Multiplex holograms, and Autostereograms. Polyocular techniques discussed are volume scanned displays, vibrating mirrors, intersecting beams, and full-computed holograms. The last may require a processor capable of around 1,019 operations per second.

Animation

A number of techniques are at last providing computer professionals and users with 3D spatial images that can be animated. Such animated sequences can be thought of as four-dimensional entities. Other methods are available for presenting 4D entities (using combinations of RGB color space plus geometric space), but none have the impact of spatial 3D + time, as may be judged by the popularity of virtual reality and Autostereograms with the general public for recreational purposes. The combination of 3D imaging and animation is the best way to present complex holographic images.

MONOCULAR TECHNIQUES

We obtain much information about the 3D nature of the world that we see by simply using the information in an image at one eye. This usually involves the following monocular techniques:

- Hidden lines and surfaces
- Depth cueing
- Background blurring
- Perspective
- Surface shading
- Temporal parallax

Hidden Surfaces
The opacity of many objects in our world means that a nearer object can obscure parts of a more distant object. This led to the *hidden line* and *hidden surface* (see the sidebar, "Hidden Surface Algorithm" for more information) problems in computer graphics. The many algorithms for the rapid solution of this problem have been a mainstay of computer graphics academic examinations for two decades. Simple examples for comparison are shown in Figures 8-7 and 8-8.[3] Cyclic looping images of the holographic animations for Figures 8-7 and 8-8 are included on the companion CD-ROM.

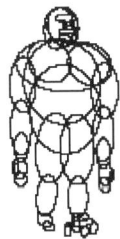

FIGURE 8-7: Example of a hidden line and surface.

FIGURE 8-8: Another example of a hidden line and surface.

■ HIDDEN SURFACE ALGORITHM

When one renders a scene in the *real world*, some objects are often hidden by other objects. A Hidden Surface algorithm attempts to identify these surfaces, and display them logically as shown in Figure 8-9 [Herbison-Evans, 1].

Many different approaches have been proposed for determining how to display polygons in the proper *depth* order. Literally hundreds of algorithms have been written, and many schools of thought exist on how *best* to do this. As in most computing applications, the primary concerns are speed, flexibility, and efficiency.

There Are Essentially Four Approaches to Hidden Surface Algorithms

It is difficult to generalize these algorithms, as each approach is fundamentally different, but a few things they must all have in common:

- They must use coherence over a given polygon.
- The basic algorithm cannot be hardware specific, in order to maintain flexibility and portability.
- In one way or another, they must support transparent surfaces.
- They must work in color to be ultimately useful on modern machines.

The basic approaches are:

- z-Buffer algorithm
- Scan-Line algorithm
- Painters algorithm
- Warnock's algorithm [Herbison-Evans, 1–2]

FIGURE 8-9: On the left, Polygon 2 is obscured by Polygon 1, and clipped appropriately. This would be a different scene if displayed as it is on the right. Here, it is apparent that one polygon pierces the other.

Chiaroscuro

Most animals have an upper surface that is darker than the underside. This has been interpreted as an attempt at camouflage, because monochromatic objects are brighter on the side closest to a source of illumination: outdoors, averaging over all normal upright orientations of an animal, the sun is overhead. Thus a simple shading of a computed image of a 3D object from lighter on top to darker underneath is a good way to present a solid look to the object. Artists have used this process, known as Chiaroscuro, for many years (see the sidebar, "Chiaroscuro Defined," for more information). Many computer graphics programmers in the past have put an implicit light source at the observer's eye, rather than at infinity above the scene. This has the advantage of requiring no computation of shadows. Except for the shadows, the difference is computationally trivial: just a question of referring the shading to the y component rather than the z component of the surface normal. However, visually, the impacts are very different. Vertical light gives a simple natural feel to the solidity of objects in a scene, even if no shadows are computed. Light from the observer washes out any shading due to curvature of the objects in a scene. It also gives the feeling of being down in a mine, with a light on the hat illuminating the scene. For many observers, this is associated with a claustrophobic sensation, which is usually not what the programmer had in mind. Examples can be seen in Figures 8-10 and 8-11 [Herbison-Evans, 3]. Cyclic looping images of the holographic animations for Figures 8-10 and 8-11 are also included on the companion CD-ROM.

FIGURE 8-10: Example of a claustrophobic sensation.

FIGURE 8-11: Another example of a claustrophobic sensation.

■ CHIAROSCURO DEFINED

Chiaroscuro is a method of applying value to a two-dimensional piece of artwork to create the illusion of a three-dimensional solid form as shown in Figure 8-12.[4] This process was devised during the Italian Renaissance and used by artists such as Leonardo da Vinci and Raphael. In this system, if light is coming in from one predetermined direction, then light and shadow will conform to a set of rules.

A highlight will mark the point where the light is being reflected most directly. This is most often bright white, although in my illustration it is 10-percent black. As one's eye moves away from this highlight, light hits the object less directly, and therefore registers a darker value of gray. In Figure 8-12, there is a regular transition until you reach the point where the shadowed area of the form meets the lighted side.

In addition, in Figure 8-12, there is a more sudden transition to darker values because no light is hitting that side. Some indirect light is available because the dark side does not turn solid black. This is the result of reflected and refracted light that naturally occurs. As you look at the extreme edge of the form, you will notice that it is markedly lighter than the shadowed area of the object—light in the environment is illuminating the back edge.

The cast shadows are usually divided into separate values as well. The area closest to the object is usually the darkest area that is being portrayed. Then, as light

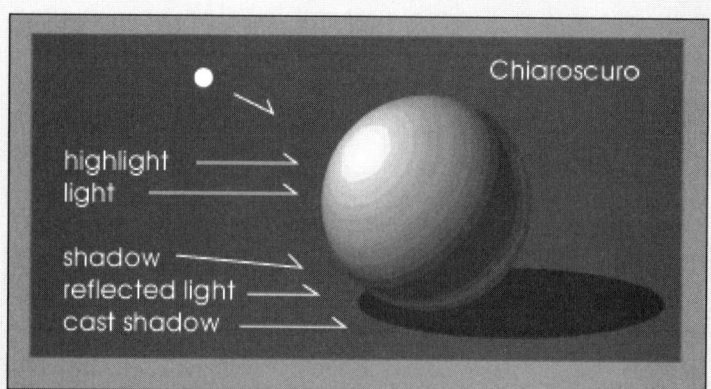

FIGURE 8-12: Example of a Chiaroscuro.

becomes more available, the same cast shadow lightens in increments until it reaches the shadow's edge.

Tip: Often times, a drawing does not have this exact transition of grays. One can control and manipulate this formula to create interesting moods and character in a piece of work. A High Key drawing is one that has mostly light values, probably with no value of more than 60 percent at the darkest points. A Low Key drawing would be one that has mostly dark values. In both Low and High Key pieces, this system of Chiaroscuro can be used to create the illusion of three-dimensional space in a drawing [Southern Arkansas University, 1–2].

Background Blurring

An early observation was that backgrounds appear to be more blurred (see the sidebar, "Blurred Objects") than the objects in front of them that have our visual attention. This is probably due to both the limited focusing depth of field of the ocular lens, and the reduced resolution available outside of the 1/2 degree field of the fovea centralis, as well as the effects of mist and dust in the intervening atmosphere (*aerial perspective* (see the sidebar, "The Aerial Perspective Effect")). Examples are shown in Figures 8-13, 8-14, and 8-15 [Herbison-Evans, 3–4]. The effect has been widely used by painters, photographers, and cinematographers to give a 3D feel to their images. Cyclic looping images of the holographic animations for Figures 8-13, 8-14, and 8-15 are included on the companion CD-ROM.

FIGURE 8-13: Example one of background blurring and aerial perspective effect.

FIGURE 8-14: Example two of background blurring and aerial perspective effect.

DESIGN: ANATOMY OF A HOLOGRAM 137

FIGURE 8-15: Example three of background blurring and aerial perspective effect.

■ BLURRED OBJECTS

When a photograph is taken with a camera, the lens is focused at a particular distance. Objects nearer or farther than this focal distance will appear blurred. By changing the focus of the lens, near or distant objects can be made to appear in sharp focus. If you want to create an image in which both distant and close objects are in focus, two or more images can be merged together to make an image with increased depth of field. This is done using a simplification of a pyramid-based technique.

The Technique

Figure 8-16 shows two images of the same scene, one focused close and the other focused at a distance.[5] Next, let's combine the in-focus parts of both photographs using the following procedure. First, each input image is blurred as shown in Figure 8-17 [Haeberli, 2].

Next, you should subtract the blurred image in Figure 8-17 from the original in Figure 8-16, and create an image that shows the magnitude of the difference as shown in Figure 8-18 [Haeberli, 2]. This image will be dark where the original image is smooth, and will be bright where the original image has edges. The strength of the edge information maps directly into the brightness of these edge images.

FIGURE 8-16: Multifocus images of the same scene.

FIGURE 8-17: The in-focus parts of each photograph.

Now, let's compare the two edge images, and make an image that is black where the left image has more edge information, and is white where the right image has more edge information as shown in Figure 8-19 [Haeberli, 3]. Finally, this is used to create an image with the best parts of each original image, as shown in Figure 8-20 [Haeberli, 3]. Where the image above is black, pixels from the left image are used. Where the image above is white, pixels from the right image are used. A simple extension of this technique can be used to combine the in-focus parts of any number of photographs [Haeberli, 1–3].

DESIGN: ANATOMY OF A HOLOGRAM   139

FIGURE 8-18: The magnitude of the difference in the image.

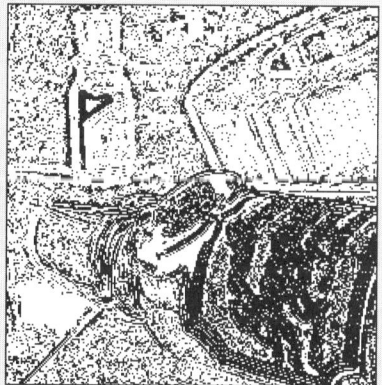

FIGURE 8-19: Comparison of the two edge images.

FIGURE 8-20: Creation of an image with the best parts of each original image.

■ THE AERIAL PERSPECTIVE EFFECT

The atmosphere scatters light. You have probably heard that that is what causes the sky to be blue. Well, it is true, but the fact that the atmosphere scatters light also plays a role in our depth perception. The short or blue wavelengths of light are most easily scattered every which way by the particles in the atmosphere, which is why the sky is blue, although scattering does occur to some extent for other wavelengths of light. In addition, the scattering occurs for all light regardless of the direction that the light comes from (there is nothing special about sunlight that causes it to be scattered more). Thus, light coming from a distant object should have some of its light scattered. That will have two effects on the light reaching our eyes: (1) since blue is scattered more, more distant objects should appear bluish, and (2) since not all of the light is traveling in a straight line to us, then more distant objects should appear a bit fuzzy and not as in focus. Examine the picture of the Grand Canyon shown in Figure 8-21; especially compare the appearance of the foreground to the portion of the rim of the canyon in the distance.[6] The texture on the near side is clear, sharp, and tinged with reddish hues (if this were a color image). The far rim is less distinct, and less distinct than can be accounted

FIGURE 8-21: Multifocus images of the same scene.

for just because it is farther away. The walls, made of the same material as the near side, have a decidedly bluish cast.

The animation in Figure 8-22 attempts to illustrate a real perspective in a dynamic way by adding and removing the depth cue [Krantz, 2]. The oval in the upper right-hand part of the image starts out identical to the other oval. Then the edges are blurred and the color is tinged with blue (by simply letting some of the blue background seep through). When these changes are made, the oval seems to be slightly farther away than before. There is no relative size change that helps the depth perception. In fact, when the oval is blurred at the edges, it is slightly larger to compensate for the fact that it is harder to see the edges. Therefore, the apparent depth change has to be due to the change in the coloring and clarity of the oval.

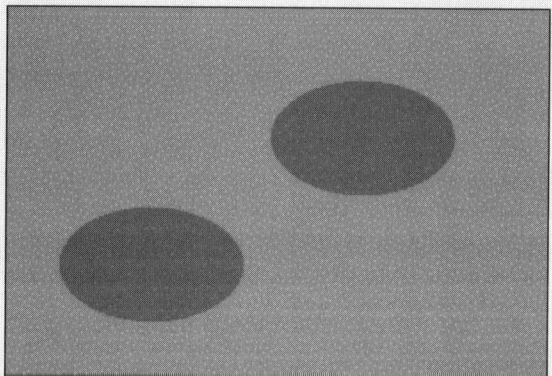

FIGURE 8-22: The in-focus parts of each photograph.

As is clear in the photograph shown in Figure 8-21, distance affects the clarity and color tinting of an object [Krantz, 2]. The bluing and slight blurring is called aerial perspective. These techniques have been used in painting and do help give the impression of depth. Look at the painting Near Salt Lake City by Albert Bierstadt as shown in Figure 8-23 [Krantz, 2]. The mountains are less distinct in the distance. In addition, the artist uses a clearly bluish tint (if this were a color image) on some of the distant peaks. In many ways, these effects mimic those seen in the photograph of the Grand Canyon as shown in Figure 8-21.

Two paintings that have been used as earlier examples additionally help to illustrate how aerial perspective can add the impression of depth along with other depth cues. The first is "Madonna of the Magnificat" by Botticelli, as shown in Figure 8-24 [Krantz, 2]. Now, compare the foreground figures to the background. Note the detail and sharpness with which the close figures are painted compared to the blurring of the more distant trees and buildings.

The second painting is "Paris Street" by Caillebotte, as shown in Figure 8-25 [Krantz, 2]. Here, compare the sharpness of the building just to the right of the closest figures at the right of the painting, to the sharpness of the building at the back between where the streets split off from each other [Krantz, 1–3].

142 CHAPTER 8

FIGURE 8-23: Near Salt Lake City by Albert Bierstadt.

FIGURE 8-24: Madonna of the Magnificat by Sandro Botticelli.

FIGURE 8-25: Paris Street: A Rainy Day by Gustave Caillebotte.

Depth Cueing

A version of this effect, called *depth cueing* (see the sidebar, "Motion Cue," for more information), has been used for many years in chemistry and in computer animation. In this technique, items that are close to the observer are made more intense or thicker than items farther away. This has been common practice for many years in diagrams used in stereochemistry as shown in Figure 8-26.[7]

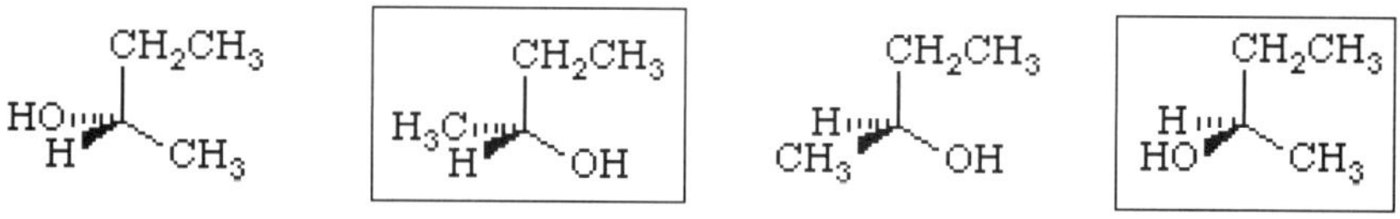

FIGURE 8-26: Stereochemistry diagrams.

Motion Parallax

Motion parallax (see the sidebar, "Motion Cue," for more information) can be obtained using only one eye by means of animation: near objects move faster on the visual field than do distant objects. And while this is true in general, more specifically the brain infers 3D character most effectively from the comparison of two views of a scene, taken with the scene rotated a couple of degrees about a vertical axis between the views. This is the basis of binocular 3D imaging, which is discussed later. Nevertheless, it works even with one eye. The brain can apparently use the same algorithms for depth perception, whether the two images arise from different eyes at the same time or from one eye at different times. This effect has been used in computer graphics for many years by having the default state of the display of a 3D entity to be a continuous, slow rotation about a vertical axis. The reader may care to compare the apparent solidity of the molecules rotating about the vertical and the horizontal axes as shown in Figure 8-27: if the theory is correct, the ones rotating about the vertical axis should appear more solid than the ones rotating about the horizontal axis [Herbison-Evans, 4]. Cyclic looping images of the holographic animation for Figure 8-27 are included on the companion CD-ROM.

FIGURE 8-27: Comparison of apparent solidity of the molecules rotating about the vertical and horizontal axes.

■ MOTION CUE

The following illustrates how depth can be portrayed in two-dimensional art. Many representational paintings give a very strong feeling of depth, despite the fact that they are painted on flat surfaces that lack any depth.

There are several restrictions on the techniques that an artist can use when trying to depict depth. First, paintings and drawings are two-dimensional. There is no actual depth in the artwork, so the artist must understand, at least intuitively, what information is in the environment that allows us to perceive depth. These sources of information are commonly called *depth* or *distance cues*. A consequence of the two-dimensional nature of painting and drawing is that we lose all the depth information that comes from the fact that we have two eyes. These binocular, or two-eye, depth cues require true depth, and thus we will not discuss them in context with paintings or drawings. For instance, there is the binocular depth cue called *disparity*. Disparity arises from the fact that our two eyes have a slightly different view of the world. To allow you to see disparity requires either real depth or two images developed as if from different positions like our eyes. This latter approach is illustrated in the photograph from the California Museum of Art as shown in Figure 8-28 [Krantz, 1]. The person is looking at a slide with two images, much

FIGURE 8-28: Example of disparity.

like a viewmaster, with each image being taken from a slightly different position to allow the brain to see disparity and, thus, see a very realistic three-dimensional picture.

The next two pictures (see Figures 8-29 and 8-30) are single-image random dot stereograms, or SIRDS, that illustrate how depth is generated using binocular disparity.[8] There are actually two sets of random dots overlaid on these images, one for each eye. Stare at the image in Figure 8-29. Next, try to relax your eyes, and try to *look through* the image. What often helps is to place your nose near the image in the book and slowly move backward. Most people report seeing a three-dimensional face. Some people cannot relax their eyes enough (though it is easier on the screen than on the printed SIRDS); so, there's a second image, that of a cone as shown in Figure 8-30. For this image, instead of relaxing your eyes, cross them slowly. Eventually, you will see a cone that is pointed toward you. Try one or the other or both.

All of the other cues to depth are called *monocular* or *one-eye*. One such cue, the relative motion of objects at different distances, can be a powerful cue to depth, but is unavailable to the painter. Perception of motion requires only one eye and is thus monocular. The artist is even more limited than, say, television or

FIGURE 8-29: The face.

FIGURE 8-30: The cone.

movies that use moving pictures. So that you can get some idea of how motion can give a powerful sense of depth, see the computer-generated movie *Flight through the Grand Canyon* on the companion CD-ROM. Even though the picture is low resolution and small, most viewers will report that the depth of the canyon is vivid. Nearby portions of the canyon are faster moving than more distant parts (a motion cue called *motion parallax*), and the direction of motion can be told by where the picture seems to expand from.

The artist, in trying to paint or draw, is, therefore, limited to depth cues that (1) need no more than one eye to work, and (2) do not require a moving world. Fortunately, there is a collection of such depth cues, a subset of monocular cues called *pictorial cues* by some. Before continuing, let's correct the impression that some readers may have that the depth generated on paintings is inferior. Actually, if you look at a detailed painting using many of the depth cues discussed here with a tube (as found in the center of a roll of paper towels), the impression of depth can be very realistic. Seeing a painting hanging on a wall probably has much to do with weakening the impression of depth [Krantz, 1–2].

Perspective
Another early observation was that images of nearer objects appear larger than images of more distant objects. This led to the science of *perspective*, epitomized by the projective transformation learned by all computer graphics students. This transformation does have two minor problems, however. The first is that the perspective is for viewing from only one point in space. If the observer's eye is placed anywhere else, the image is incorrect, and the 3D effect is strained. The second problem is that the perspective image is projected mathematically onto a flat plane, whereas the retina of the eye is approximately hemispherical. This again leads to a distortion around the periphery of images generated by the flat perspective projection. Examples can be seen in Figures 8-31 and 8-32 [Herbison-Evans, 5]. Cyclic looping images of the holographic animation for Figure 8-32 are included on the companion CD-ROM.

Texture Gradient
A similar effect is called *texture gradient*, in which the pattern of some texture such as grass varies with distance. Statistically: the autocorrelation function of the texture contracts as distance increases.

In other words, related in a sense to relative size, but a depth cue in its own right is what has been termed *texture gradient*. Most surfaces, such as walls and roads and a field of flowers in bloom, have a texture. As the sur-

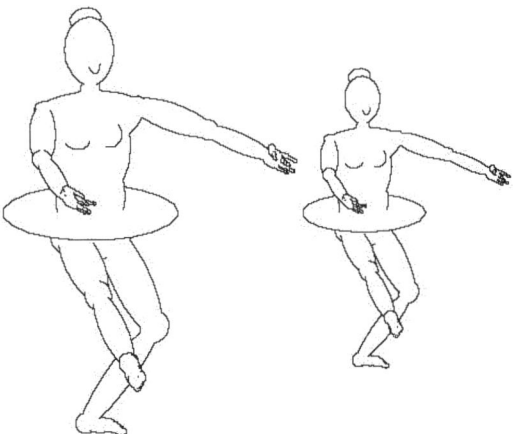

FIGURE 8-31: Example one of a distortion around the periphery of images generated by the flat perspective projection.

FIGURE 8-32: Example two of a distortion around the periphery of images generated by the flat perspective projection.

face gets farther away from us, this texture becomes finer and appears smoother. For example, the animation as shown in Figure 8-33 makes the image look like a wall rising up before us [Krantz, 1–2]. When the texture units stay the same size and no depth is indicated, this outcome can be seen both at the beginning of the animation and for the top part of the pattern at the end of the animation. However, when the spaces are shaped so that the top is smaller than the bottom, and the sides tilt in toward the middle, then the pattern appears like a floor receding in depth. So, when the texture units change in size, then depth is indicated; in this case, the depth of a surface like a floor spread out before you.

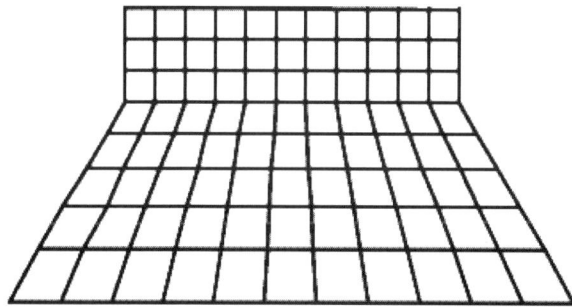

FIGURE 8-33: Texture gradient.

Texture gradient is carefully used in the painting previously shown in Figure 8-25, *Paris Street: A Rainy Day by Gustave Caillebotte*, to show that the street recedes in depth. Here, square cobblestones are used. These cobblestones get progressively smaller as the road recedes in depth until the stones are not clearly distinguishable from each other. In the distance, only the general roughness of the street is noticeable; and then this roughness becomes less easily noticed as well.

A surface or field that recedes in depth has a texture that gets finer. That is very different from a wall in which the surface is approximately the same distance from a person at all points. For example, imagine yourself standing and staring at a brick wall that, instead of receding in depth like a cobblestone road, rises up in front of you. Here the texture (in this case, the brick alternating with the mortar) will have about the same roughness all over the surface and provide a clue that the surface does not recede in depth. In addition, texture may play a role in helping us determine the size of an object. Regardless of how far an object is from us, it covers roughly the same amount of surface, and thus texture, which can help us determine the actual size of an object.

BINOCULAR TECHNIQUES

The illusion of solidity is particularly strong if the left and right eye images can be separately computed, and each presented to the appropriate eye. Doing this by computer has problems. The eyes can only appreciate the solidity of a scene if the ratio of the nearest to farthest distance is greater than about 2:3, and the left and right scenes differ by a rotation of about 2 or 3 degrees about a vertical axis (clockwise as viewed from above if deriving the right eye view from the left). Given a screen resolution of about 4 pixels/millimeter (for most binocular techniques), this quantifies depths to about 16 possible values, which can give an unfortunate layering effect. Printed stereo images, having an order of magnitude better resolution, can give nearly continuous depth perception, as the eye/brain has itself only a resolution of about 250 values.

Freeview
Also called *parallel viewing*, this simple technique places the stereo images side by side, left on the left, and right on the right. The viewer must then keep the axes of the eyes parallel as if looking at infinity, while focusing on

the display. Initially, this is not easy, and some practice may be necessary. This display technique is called *freeview*. A restriction is that the images must be placed no farther apart than the spacing of the eyes (and hence they can each be no wider than this); and each is limited to using only half the display width. Examples are shown in Figure 8-34, and the cyclic looping images of the holographic animation for Figure 8-34 are included on the companion CD-ROM [Herbison-Evans, 6].

FIGURE 8-34: The restriction is that the images must be placed no farther apart than the spacing of the eyes.

Crossview

Alternatively, the images can be placed side by side, left on the right and right on the left. This is commonly called the *crossview* or *transverse view* technique (see the sidebar, "Stereoscopic Photographs," for more information). To view such images, the viewer must keep the eyes crossed, as if looking at something closer than the screen, while focusing on the screen. This is not initially easy either. The images, however, are now no longer restricted in their spacing, and can be of arbitrary width. However, again, only half the area of the display can be used for each image, effectively halving the display resolution. Examples are shown in Figure 8-35, and the cyclic looping images of the holographic animation for Figure 8-35 are included on the companion CD-ROM [Herbison-Evans, 6].

152 CHAPTER 8

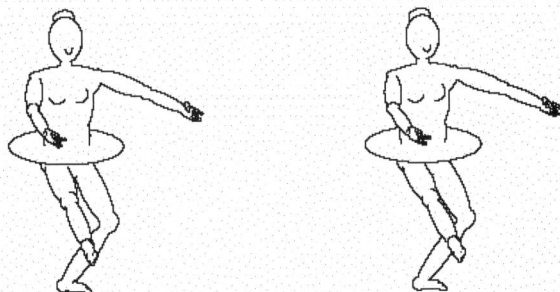

FIGURE 8-35: The images are not restricted in their spacing, and can be of arbitrary width.

■ STEREOSCOPIC PHOTOGRAPHS

There are two ways to view stereo photographs in a 3D art gallery as shown in Figure 8-36.[9] If you already know how to *cross-view* and *parallel-view*, continue on and enjoy! Otherwise, you might want to read the rest of this sidebar.

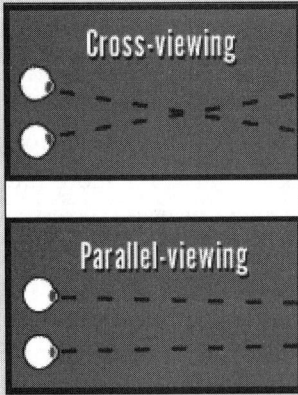

FIGURE 8-36: Parallel-viewing and cross-viewing.

Parallel-Viewing versus Cross-Viewing

Many people find one method easier than the other. Parallel-viewing requires the ability to relax the eye muscles, *let go*, and look beyond the plane of the image. This can appeal to those who like to gaze off into the distance and scan the horizon. Cross-viewing requires that the eyes be aimed at a point between the viewer's nose and the image, so it can be comfortable for people who are good at focusing in on details and close objects.

It's good exercise for your eyes to alternate between cross-viewing and parallel-viewing. The cross-viewing method offers a design advantage over parallel-viewing when setting up stereo pairs for free-viewing. With the parallel-viewing method, the two images can be set only so far apart. That's because there's a physical limit as to how much the human eyes can angle outward. This limitation forces the parallel images to stay small (this will become evident as you move through a 3D gallery). With cross-viewing, the images can be larger and farther apart. When you cross-view a stereo pair, the virtual 3D image in the middle appears smaller than the actual image [Grossman and Cooper, 1–2].

Anaglyphs

A 3D scene can be displayed in Anaglyph form. The image for one eye is displayed in red, and the other in green. The observer must wear glasses with one red lens and one green lens. The major disadvantage of this technique is that only monochrome images can be displayed. In addition, the observer needs to use special eyewear, which is troublesome for people who normally wear spectacles already. Examples are shown in Figure 8-36, and the cyclic looping images of the holographic animation for Figure 8-37 are included on the companion CD-ROM [Herbison-Evans, 6].

FIGURE 8-37: 3D scene displayed in Anaglyph form.

Shuttered Glasses

The left and right images can be displayed alternately at a rate faster than the critical flicker frequency of the eye (about 20 per second). The observer again needs to wear special shuttered glasses (see the sidebar, "Stereoscopic 3D Images" for more information), this time having lenses that alternately blacken out each eye in synchronism with the display. The images with this technique can now use the whole screen and be in full color, but the eyewear problem remains.

> ### ■ STEREOSCOPIC 3D IMAGES
>
> CrystalEyes is a lightweight, wireless eyewear system that delivers high-definition, stereoscopic 3D images in conjunction with compatible software and standard workstation displays as shown in Figure 8-38.[10] For engineers and scientists who build 3D computer graphics models, stereoscopic 3D delivers realistic representation of graphical information. Many common software applications used in mechanical CAD, molecular modeling, GIS/mapping, and medical imaging support StereoGraphics' CrystalEyes on all major Unix platforms and Windows NT workstations.
>
>
>
> **FIGURE 8-38:** Lightweight, wireless eyewear system.

Cross Polarized Images

Two normal displays can be used and their images *polarized* at right angles to each other, and then combined with a half-silvered mirror. Basically, projection of polarized 3D images involves:

- Use of a dual-lens projector, or two single-lens projectors.
- Polarized filters over each projection lens, aligned at right angles.
- Use an aluminum (silver) or lenticular (with vertical lens ridges) projection screen. Normal beaded or white projection screens will de-polarize the light projected, destroying the stereo effect. The silver or lenticular screens will preserve the polarization of the projected light, essential for maintaining a left and right image.
- Polarized glasses are worn that match the alignment of the projector lens polarizing filters. In most projectors and glasses, the orientation of the polarization is in a *V* shape.[11]

The observer must now wear glasses with cross-polarized lenses. This technique is the one used most often for *3D movies*. The complexity of the display and the glasses are problems.

Artificial Reality

Two tiny displays, one for each image, can be mounted on a headset, and mirrors used to present the images to each eye. This is the technique normally used in virtual reality systems Again, the headset is a problem.

Lenticular Film

A *lenticular* film composed of a set of thin, vertical cylindrical lenses (each typically with a width of 1/4 millimeter) can be placed over the screen, and the two stereo images dissected into vertical stripes and interlaced so that the lenses present each stripe to the appropriate observer's eye. This technique has been used on picture postcards for many years (see Figure 8-39).

It has the advantage of not needing the observer to train the eyes or wear special gear. Its main disadvantages are the reduced horizontal resolution, and the difficulty of aligning the images' stripes with the lenses.

Pulfrich Effect

The Pulfrich effect can be used to give some 3D illusion. This depends on the fact that the eye requires longer to process dim images than bright ones. People look at Pulfrich glasses and think something is wrong. Most feel 3D

FIGURE 8-39: A pair of stereo images with the left and right images interlaced. A suitable lenticular film must be placed over the image to see the 3D effect.

glasses should be blue/red (see Figure 8-40).[12] Others think both lenses should look alike, such as with polarized glasses.

Watch out for glasses with colored lenses. The Pulfrich illusion or effect is most effective when viewed with glasses having *one dark lens*. Colored lenses alter the color perception of the picture. True Pulfrich glasses do not require *colored* lenses, therefore allowing perfect, unaltered program color.

FIGURE 8-40: Pulfrich glasses.

Furthermore, if animation is created of a scene rotating about a vertical axis, and the appropriate eye is covered with a dark filter, the scene will appear in 3D. If the left eye is the one darkened, then the scene must be rotating clockwise at about 40 degrees per second for the effect to occur. An interesting variant is the possibility that if one eye is dominant, it may process an image faster than the other eye, in which case no special glasses need be used. Unfortunately, people having a dominant left eye would require the scene to be rotating in the opposite direction to those with a dominant right eye. Examples may be compared in the two molecules as shown in Figure 8-27, and the cyclic looping images of the holographic animation for Figure 8-27 are included on the companion CD-ROM [Herbison-Evans, 8].

Multiplex Hologram

A multiplex or *rainbow* hologram (see the sidebar, "Computer Generated Rainbow Hologram," for more information) can offer a 3D monochrome image, albeit multicolored! It is composed of a series of thin, vertical holograms of flat images of the scene, each generated with the scene rotated a degree or so from the last one about a vertical axis. Each eye of an observer sees a different hologram stripe, each reconstructing the scene at the appropriate angle for that eye as if the scene were solid. The generation of a multiplex hologram takes some time, so no interaction with the image is possible, but limited animation is possible in the precomputation of the set of flat images. The horizontal resolution is restricted, as well as the scene having to be monochromatic. However, observers need no special eyewear, and multiple simultaneous observers are easily accommodated. A number of commercial firms offer the service of custom making multiplex holograms of virtual objects.

■ **COMPUTER-GENERATED RAINBOW HOLOGRAM**

A simplified model is proposed here to calculate the computer-generated rainbow hologram fast as shown in Figure 8-41.[13] The rainbow hologram is very practical to display three-dimensional images, because the images can be reconstructed by white light. Since the rainbow hologram has horizontal parallax only, less computation is required compared with full parallax transmission holograms. In the proposed method, one can simply generate the final hologram from intermediate data,

FIGURE 8-41: Computer-generated rainbow hologram.

whose total number of samples can be less than one-tenth of the final hologram. This intermediate format makes fast computation and effective storage/transmission possible. It can be done only with multiple and additional operations to convert the intermediate data to the final data. Therefore, a hardware implementation to the electro-holographic display or printer is possible.

Autostereograms

The latest technique is to overlap the left and right images, and indeed to overlap multiple copies of each. The result is called an *autostereogram*. The left and right images can use random dots, or indeed any arbitrary pattern, for displaying the depth, with no coloring information (making a random dot stereogram or single-image stereogram), or be in full shaded color (resulting in a wallpaper stereogram). Possibly a better technique is to combine the shaded color at each pixel 50/50 with a random value, giving a sort of speckled autostereogram. This combines the proper color shading of the wallpaper stereogram with the depth precision of the random-dot technique. Random-dot stereograms pose a problem for some systems because they are impossible to compress. When using random numbers, the same set of random numbers can be used for every frame, which gives peculiar effects for the observer, because some of the lines change numbers and some do not, so patches of the image appear not to animate. Alternatively, different random numbers can be used for each frame, which gives a sparkling effect to the images, as though the objects were covered in sequins. Examples of wallpaper stereograms can be seen in Figure 8-42, and the cyclic looping images of the holographic animation for Figure 8-42 are included on the companion CD-ROM [Herbison-Evans, 9].

Examples of random-dot stereograms can be seen in Figures 8-43 and 8-44, and the cyclic looping images of the holographic animation for Figures

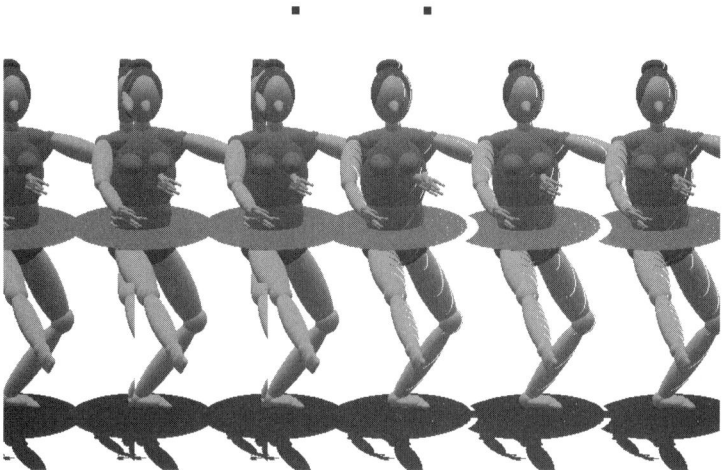

FIGURE 8-42: Wallpaper stereograms.

8-43 and 8-44 are included on the companion CD-ROM [Herbison-Evans, 10]. Random-dot stereograms are very hard to see, so some simpler ones with the same figure but a plain background are shown in Figure 8-44.

Examples of speckled stereograms can be seen in Figure 8-45, and the cyclic looping images of the holographic animation for Figure 8-45 are included on the companion CD-ROM [Herbison-Evans, 11].

FIGURE 8-43: Random-dot stereograms.

FIGURE 8-44: Another example of a random-dot stereogram.

FIGURE 8-45: Speckled stereograms.

DESIGN: ANATOMY OF A HOLOGRAM   **161**

POLYOCULAR TECHNIQUES

Polyocular techniques create an image of solid appearance that is viewable by any number of eyes, and so, for example, are viewable by several people at once. Various methods have been tried to create this effect.

VOLUME SCANNING

A display screen can be driven mechanically to *scan a volume*, preferably faster than the critical flicker frequency of the eye. For example, a cathode ray tube can be constructed so that the screen oscillates or rotates inside the vacuum.

Alternatively, a semitransparent screen on which a scannable laser beam shines can be arranged to scan out a volume. A *helically shaped screen* has been suggested for this. For example, a 3D volumetric display system allows true three-dimensional visualization of images as shown in Figure 8-46.[14] The three color beams from the system are guided by the acoutsto-optic deflectors to a display medium. The display medium allows

FIGURE 8-46:   3D volumetric display—the wave of the future in air traffic control!

the laser beam to create discrete visible dots of light, called *voxels*, at any point within its imaging volume. Arrays of voxels are used to create images that are perceived by an observer from the perspective of his or her position. The volumetric display is comprised of three major subsystems of technical information on the 3D display:

- Laser optics system
- Computer controller
- Helical display system

In addition, a *rotating array of LEDs* has been tried. With any of these techniques, the points on the image are painted in appropriate synchronism with movement of the screen so that they appear at the required point in the 3D space scanned by the screen.

Vibrating Mirror
Another technique is to view a normal CRT via a *vibrating mirror*. This technique can give a large, virtual scanned volume by making the mirror curved, so magnifying the CRT screen image. If the mirror is driven by a loudspeaker mechanism, its focal length can be oscillated, and so a virtual image of considerable depth created.

Double-Beam Display
An optically nonlinear material can be used to create *double quantum* transitions. Two scannable invisible infra-red beams are shone into the material at an angle to each other. Where they intersect, visible light is emitted, so that the beams can be scanned to draw 3D shapes in the medium.

Holograms
The ultimate 3D display may be the computed projected full *hologram*. The computational problems, however, are immense. For example, a 10-cm square hologram would require approximately 10^{12} pixels, the value at each being a Fresnel transform of the light coming from the scene. For animation, these 10^{12} transforms would need to be computed faster than the eye's critical flicker frequency. Suitable materials exist that change their optical properties when an appropriate voltage is placed across them, and this can be done using transparent electrodes. Thus, a hologram could be made from such a material. In addition, the electrodes can be striped and painted

at right angles across opposing faces, so that a tiny volume of material can be addressed where the stripes cross. Thus, suitable displays seem quite possible. The problem is that each output pixel is an integral (approximated presumably as a finite sum) over all the input pixels in a depth map of the scene. Therefore, the problem appears to require a processor capable of approximately 10^{25} multiplications and additions per second. One of the best hopes for reducing this might be to compose the scene from flat triangles, and to approximate the Fresnel transform of each by its Fourier transform multiplied by a quadratic phase factor representing the depth of its center. If the Fourier transforms are precalculated and held to appropriate accuracy in tables, a scene of, say, 10,000 triangles (enough for a human figure given current screen resolutions) would only need 10^{19} operations per second. Further research may lead to improved ways of cheating.

Nevertheless, an animated holographic image of solid appearance viewable by any number of unencumbered observers is still a dream popularized by such images as that of Princess Leia in the movie *Star Wars*, and Selma in the TV series *Timetrax*. A number of techniques are closing in on this dream, many requiring rather specialized types of hardware. With hardware speeds currently reaching 1 Teraflop (10^{12} floating operations per second) and improving at approximately a factor of 2 every 18 months, it may take only 37 years for this dream to become reality.

Next, as part of the design stage of the commercial holographic process, let's look at reduced and enlarged images via computational holographic bandwidth compression. Here, hogel-vector holographic bandwidth compression is a novel technique to compute holographic fringe patterns for real-time display. This diffraction-specific approach treats a fringe as discretized in space and spatial frequency. By undersampling fringe spectra, hogel-vector encoding achieves a compression ratio of 16:1 with an acceptably small loss in image resolution. Hogel-vector bandwidth compression attains interactive rates of holographic computation for real-time, three-dimensional electro-holographic (holovideo) displays. Total computation time for typical 3D images is reduced by a factor of over 70 to 4.0 s per 36MB holographic fringe, and under 1.0 s for a 6MB full-color image. Analysis focuses on the trade-offs among compression ratio, image fidelity, and image depth. Hogel-vector bandwidth compression matches information content to the human visual system, achieving *visual-bandwidth holography*.

Reduced or Enlarged Images

Electro-holography (also called *holovideo*) is a new visual medium that electronically produces three-dimensional (3D) holographic images in real time. As previously discussed, holovideo is the first visual medium to produce dynamic images that exhibit all of the visual depth cues and realism found in physical scenes. It has numerous potential applications in visualization, entertainment, and information, including education, telepresence, medical imaging, interactive design, and scientific visualization. Electro-holography combines holography and digital computational techniques. Holography is used to create 3D images using a two-step coherent optical process. An interference pattern (*fringe pattern* or simply fringe) is recorded in a high-resolution, light-sensitive medium. Once developed, this recorded fringe diffracts an illuminating light beam to form a 3D image.

As early as 1964, researchers computed holographic fringes to create images that were synthetic and potentially dynamic. Both the computation and display of holographic images are difficult due to huge fringe bandwidths. A computed (discrete) fringe must contain roughly 10 million samples per square millimeter to effectively diffract visible light. Interactive-rate computation (about one frame per second or faster) was impossible. In 1989, researchers at the MIT Media Laboratory Spatial Imaging Group created the first display system that produced real-time 3D holographic images. Computation of the 2MB fringe required several minutes for small, simple images using conventional computation methods.

A new diffraction-specific computation technique named *hogel-vector bandwidth compression* achieves interactive-rate holographic computation as shown in Figure 8-47.[15] The main features of this technique reported here are:

- Its architecture, based on the discretization of space and spatial frequency.
- The use of hogel-vector bandwidth compression to reduce bandwidth by 16:1 and higher, allowing for easier display, transmission, and storage.
- Fringe computation that is over 70 times faster than conventional computation.
- The trade-offs among the system parameters of image resolution, image depth, and bandwidth [Lucente, 349–350].

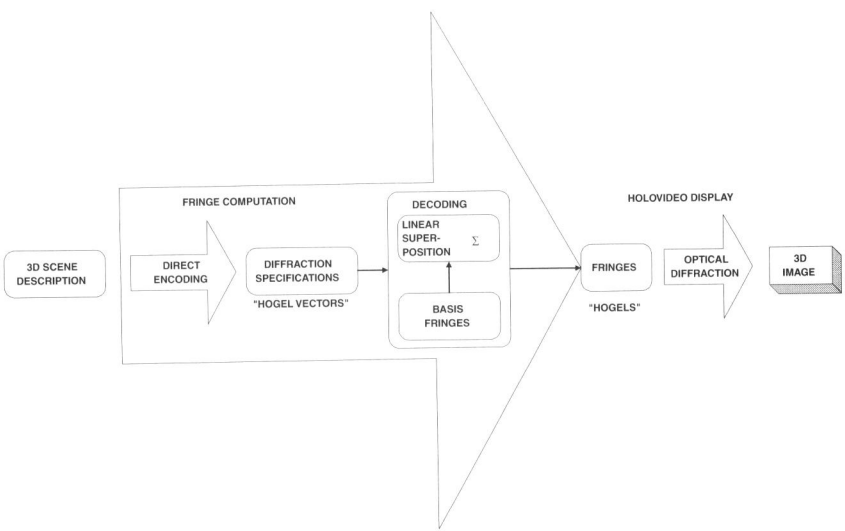

FIGURE 8-47: Hogel-vector bandwidth compression: direct encoding and decoding using superposition of precomputed basis fringes.

Now, let's continue with a look at computational holography, holographic displays, and past work in holographic information reduction.

COMPUTATIONAL HOLOGRAPHY

Computational holography begins with a 3D numerical description of the object or scene to be imaged. Traditional, conventional holographic computation imitated the interference of optical holographic recording. Speed was limited by two fundamental properties of fringes: (1) the myriad samples required to represent microscopic features (>1000 line-pairs per millimeter [lp/mm]), and (2) the computational complexity associated with the physical simulation of light propagation and interference.

In a computer-generated hologram, the number of samples per unit length (in one dimension) is defined as the pitch, p. To satisfy fringe sampling requirements, a minimum of two samples per cycle of the highest spatial frequency is needed. A typical full-parallax 100mm × 100mm hologram has a sample count (also called *space-bandwidth product*, or simply *bandwidth*) of over 100 gigasamples. The elimination of vertical parallax provides savings in display complexity and computational requirements without greatly compromising display performance. This part of the chapter deals with horizontally off-axis transmission horizontal-parallax-only

(HPO) holograms. Such an HPO fringe is commonly treated as a vertically stacked array of one-dimensional holographic lines.

A straightforward approach to the computation of holographic fringes resembled 3D computer graphics ray-tracing. The complex wavefront from each object element was summed, with a reference wavefront, to calculate the interference fringe. Interference-based computation requires many complex arithmetic operations (including trigonometric functions and square roots), making rapid computation impossible even on modern supercomputers. Furthermore, interference-based computation does not provide a flexible framework for the development of holographic bandwidth compression techniques.

HOLOGRAPHIC DISPLAYS

Holographic displays modulate light with electronically generated holographic fringes. Early researchers employed a magneto-optic spatial light modulator (SLM) or a liquid-crystal display (LCD) to produce tiny planar images. The time-multiplexing of a very fast SLM provides a suitable substitute for an ideal holographic SLM. The research presented in this part of the chapter employs the 6MB color holovideo display and the 36MB holovideo display developed by the Spatial Imaging Group at the MIT Media Laboratory. These displays used the combination of an acousto-optic modulator (AOM) and a series of lenses and scanning mirrors to assemble a 3D holographic image in real time. The 6MB display generated a full-color 3D image with a $35 \times 30 \times 50$mm width by height by depth. The 36MB (monochromatic) display generated a $150 \times 75 \times 160$mm image. By incorporating the proper physical parameters (wavelengths, sampling pitch), the fringe computation described in this part of the chapter can be used for other holographic displays.

INFORMATION REDUCTION IN HOLOGRAPHY

Several researchers have attempted to reduce bandwidth in *optical* holographic imaging. Holographic fringes contain more information than can be utilized by the human visual system. By employing a dispersion plate, a full-size image can be generated using a reduced-size hologram. However, image quality suffers. Image resolution or signal-to-noise ratio is reduced. Good images can be reconstructed with information reduction factors of six in each lateral dimension.

Recent experiments have exploited the redundancy inherent to holographic fringes. Essentially, researchers subsampled (spatially or spectrally) to reduce bandwidth. Image quality suffered (dispersion plates caused graininess and noise). Such artifacts were inevitable because researchers could not directly manipulate the recorded fringes. *Computed* fringes can be directly manipulated. This part of the chapter discusses the translation of optical information-reduction concepts into computational holography, where they are more useful and realizable.

HOGEL-VECTOR BANDWIDTH COMPRESSION

Hogel-vector bandwidth compression is a diffraction-specific fringe computation technique. Stated simply, the diffraction-specific approach is to consider only the reconstruction step in holography. In practical terms, it is the spatially and spectrally sampled treatment of a holographic fringe. Although numerical methods can compute diffraction backwards, they are far too slow for interactive-rate computation. Diffraction-specific fringe computation provides a fast means for generating useful fringes through calculations that relate the fringes to the image through diffraction in reverse. It has the following features (see Figure 8-47):

- **Spatial discretization**: The hologram plane is treated as a regular array of functional holographic elements named *hogels*. In HPO holograms, a horizontal line of the hologram is treated as regular line-segment hogels of width, each comprising roughly 100 to 2000 samples.
- **Spectral discretization**: A *hogel vector* is a spectrally sampled representation of a hogel. Each component represents the amount of spectral energy near a particular spatial frequency. A hogel vector is the diffraction specification of a hogel. Three-dimensional object scene information is encoded as an array of hogel vectors.
- **Basis fringes**: A set of precomputed basis fringes combine to decode each hogel vector into one hogel-sized fringe. Each basis fringe represents an independent part of the hogel spectrum and is precomputed with appropriate width and component.
- **Rapid linear superposition**: In the decoding step, hogel vectors specify the linear real-valued super-position of the precomputed basis fringes to generate physically usable fringes [Lucente, 352].

By encoding the 3D scene description as diffraction specifications (an array of hogel vectors), this technique reduces required bandwidth. Speed results from the simplicity, efficiency, and directness of basis-fringe summation in the decoding step.

Sampling and Recovery
Encoding hogel vectors and decoding them into fringes is based on a spatially and spectrally sampled treatment of the fringe. The spatial and spectral sample spacings are selected to allow the fringe to be recovered from the hogel-vector array and used to diffract light to form the desired image. The first-order diffracted wavefront is the physical entity being represented by a fringe. Diffraction is linear, and the wavefront immediately following modulation by a fringe can be expressed as a summation of plane waves, each diffracted by a spatial frequency component.

Diffraction-specific computation treats a one-dimensional HPO fringe (at some vertical location y) as a two-dimensional (2D) localized spectrum $S(x,f)$, where x is the spatial position on the hologram and f is the spatial frequency. This is a Wigner distribution, possessing a continuously varying amplitude as a function of space and spatial frequency. The discrete representation of $S(x,f)$ is an array of hogel vectors (the diffraction specifications for the one horizontal line of the fringe). When sampled and recovered correctly, $S(x,f)$ causes light to diffract and to form image points throughout one plane of the image volume as shown in Figure 8-48 [Lucente, 352]. To effectively sample $S(x,f)$, sufficiently small spatial and spectral sample spacing (width and component) must be chosen. As discussed later, these spacings are selected using an empirically verified model that relates these parameters to the quality of the reconstructed image.

For the spectral dimension f, the convolution is performed in the spatial domain by the weighted summation of basis fringes, in which each basis fringe represents one of the spectral regions indexed by j. These convolutions are equivalent to performing a low-pass filtering. For the spatial dimension x, the sinc function is approximated by the rectangular envelope of the basis fringe combined with the low-pass process of diffraction. In practice, no ideal low-pass filter exists. In this part of the chapter, the basis fringes had Gaussian spectral shapes rather than sinc-function shapes. The resulting spectral cross-talk theoretically added some noise to the image, though little additional noise was observable. Spectrally, Gaussian basis fringes were used because they produced superior image results. A properly decoded (recovered) fringe has a smooth, continuous spectrum.

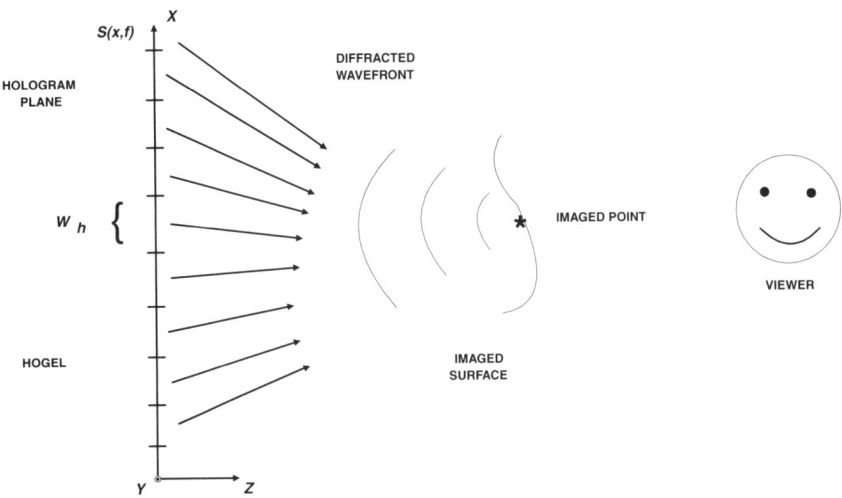

FIGURE 8-48: Computed fringe diffracts light to form an image.

Bandwidth Compression: Spectral Subsampling

Hogel-vector bandwidth compression reduces the number of encoded symbols through subsampling of hogel spectra. The information content of an encoded fringe (the number of symbols in the hogel-vector array) is equal to the product of the number of hogels times the number of components (N) in each hogel vector. Therefore, the sample spacings *(width* in space and component in spatial frequency) determine the information content of the encoded fringe. Hogel-vector encoding reduces the number of symbols by sampling hogel spectra in large frequency steps. The amount of bandwidth compression in an encoded fringe is measured by the compression ratio (CR), in which the spectrum is assumed to range from 0.0 to the sampling limit of 0.5 cycles/sample. A CR > 1 means a reduction in required bandwidth; fewer symbols are required to represent a fringe because each symbol represents a larger portion of a hogel spectrum. The spectrum of a hogel comprising N_h samples is encoded as $N=N_h/CR$ symbols.

 Note: Although the uncompressed spectrum has N_h samples for magnitude and N_h for phase, there are only N_h independent samples, because the spectrum of the real-valued fringe has even conjugate symmetry, $S(f) = S^*(-f)$.

Conventional image and data compression starts with the desired data and then encodes the data into a *compressed* format. This compressed format is subsequently decoded into a (sometimes approximate) replica of the desired data. This approach can be applied to holographic fringes. However, total (model-to-fringe) computation speed is increased by computing the encoded format directly. This *direct-encoding* approach involves only two computation steps: direct encoding and decoding. Hogel-vector encoding is a *direct-encoding* technique, giving it the speed necessary for holovideo interactivity.

Basis Fringes

Each basis fringe is used to contribute spectral energy with a Gaussian profile. The spectral phase is uncorrelated among the basis fringes to make effective use of dynamic range. Spatially, a basis fringe has a uniform magnitude of 1.0 within the hogel width, and zero elsewhere. The spatial phase contains the diffractive information. The inter-hogel phase continuity is assured by constraining the endpoints of each basis fringe. The many constraints on basis fringes make their computation intractable using analytical approaches. For the present research, nonlinear optimization was used to design each basis fringe to have the desired spectral characteristics. The synthetic basis fringes for given sampling spacings (width and component) and a given display (pitch, p) were precomputed and stored for use in hogel-vector decoding.

Direct Encoding

Hogel-vector encoding converts a given 3D object scene into a hogel-vector array. Direct encoding is composed of the following:

- Diffraction tables
- Use of 3D computer graphics rendering
- Color

DIFFRACTION TABLES

The mapping from image element (x, z) to hogel position and vector component is precalculated and stored in a *diffraction table*. Indexing by the horizontal and depth locations (x, z) of each element, the diffraction table contains a spatially and spectrally sampled representation of the fringe that diffracts light to form the image element. The table lists which nonzero components of which hogel vectors are needed to generate the correct fringe. In an HPO hologram, each line of the fringe pattern is treated independently, and indexed by the vertical (y) location of the image element.

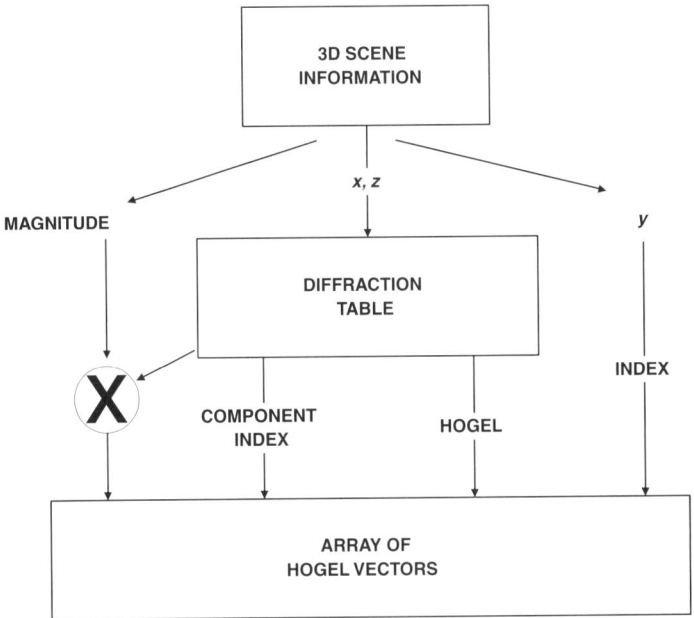

FIGURE 8-49: Generating hogel vectors using a diffraction table.

During hogel-vector encoding as shown in Figure 8-49, the diffraction table rapidly maps a given (x, z) location of a desired image element (a point) to components in the hogel-vector array [Lucente, 355]. The diffraction table includes an amplitude factor at each entry. This factor is multiplied by the desired magnitude (taken from the 3D scene and lighting information) to determine the amounts of contributions to each hogel-vector component. The magnitude of an image element is determined from its desired brightness. This brightness is represented as an intensity that is equal to the square of its magnitude. Therefore, in theory, the square roots of desired brightness values are used when calculating hogel vectors. In practice, however, nonlinearities in the MIT holovideo display systems necessitated a brightness correction approximately equivalent to squaring magnitudes—canceling out the need to calculate square roots.

During computation of hogel vectors for a particular 3D object scene, each image point is used to index the diffraction table. The contents of the table—for each indexed hogel-vector component—are summed to calculate the total hogel-vector array for this object. This direct-encoding step is fast because it involves only simple calculations.

A diffraction table is computed by spatially and spectrally sampling (with spacings *width* and component) the continuous spectral (Wigner) distribution at the plane of the hologram ($z = 0$). The spectrum is related to an imaged element through optical propagation (diffraction) The Wigner distribution of the desired image element is back-propagated from z to $z = 0$ using the transport equation of free space. For point image elements (the most common case), the spectral distribution of the fringe is a uniform distribution. The x-location of this spectral energy (which hogels are to contain this spectral energy) is a function of the x-position of the image point, and its spread is a function of image point depth. The continuous spectrum is sampled to calculate the value at each discretized location in the spatial and spectral dimensions (each hogel-vector component). A sampled value is calculated using a bi-Gaussian kernel with full widths of and components in the spatial and the spectral dimensions.

Hogel-vector encoding provides higher-level image elements. For example, if line segments of various sizes are useful for assembling the image scene, then a diffraction table is used to map location, size, and orientation of the desired segment to the proper hogel-vector contributions. Furthermore, the amplitude factors in the diffraction table allow for directionally dependent qualities (specular highlights) when a diffraction table is used to represent more complex image elements.

USE OF 3D COMPUTER GRAPHICS RENDERING

Another approach to performing direct encoding employs 3D computer graphics rendering software. This facilitates advanced image properties such as specular reflections, texture-mapping, advanced lighting models, and scene dynamics. A series of views are tendered, each in the direction corresponding to the center of one of the spectrally sampled regions. The view window (the plane upon which the scene is projected) must be coincident with the plane of the hologram ($z = 0$). Each rendered view is an orthographic projection of the scene from a particular view direction. The rendered views provide a discrete sampling of space and spectrum. These views are converted into a hogel-vector array using either a modified diffraction table or filtering, depending on hardware. In the modified-table method, the picture element (pixel) spacing in the 2D rendering of the scene is half the hogel spacing. This allows for subsampling. A special diffraction table uses view direction and rendered pixel location to select hogel-vector components, providing a sampling of the spatial-spatial-frequency space that matches width and component: The second method uses the features of advanced rendering hardware. Anti-alias filtering combined with nearest-neighbor spatial linear interpolation (as part of a texture-mapping) gives a roughly Gaussian sampling.

COLOR

Full-color holovideo images are produced by computing three separate fringes, each representing one of the additive primary colors (red, green, and blue) taking into account the three different wavelengths used in a color holovideo display. Three separate hogel-vector arrays are generated, and each is decoded using the linear summation of one of three sets of precomputed basis fringes. Basis-fringe selection via hogel-vector components proceeds using a single diffraction table, with the shorter wavelengths limited to a smaller range of diffraction directions.

Decoding Hogel Vectors to Hogels

Hogel-vector decoding is the conversion of each hogel vector into a useful fringe in a hogel region. Decoding performs the convolutions through a linear summation of basis fringes using the hogel vector as weighting coefficients. To compute a given hogel, each component of its hogel vector is used to multiply the corresponding basis fringe. The decoded hogel is the accumulation of all the weighted basis fringes. Looking at the array of precomputed basis fringes as a two-dimensional matrix, hogel decoding is an inner product between the basis-fringe matrix and a hogel vector.

Decoding is the more time-consuming step in hogel-vector bandwidth compression. However, the simplicity and consistency of this step means that it can be implemented on specialized hardware and performed rapidly. Various specialized hardware exists to perform multiplication-accumulation (MAC) operations at high speeds.

IMPLEMENTATION

Direct encoding of the hogel-vectors was implemented on a Silicon Graphics, Inc., Onyx workstation—a high-end serial computer. Encoding involved a wide range of calculations, but was relatively fast. The second computation step, the decoding of hogel vectors into hogels, was implemented on three systems: the Cheops framebuffer system used to drive the MIT 36MB holovideo display; the Onyx workstation; and a Silicon Graphics, Inc., RealityEngine2 graphics subsystem.

Implementation on Cheops
Hogel-vector encoding begins with a 3D image scene description generally consisting of about 0.5MB of information or less. After the appropriate transformations (rotations, translations) and lighting, it is direct-encoded as a hogel-vector array. For a compression ratio (CR) of 1:1 (no bandwidth compression), the hogel-vector array comprises 36MB. For larger compression ratios, this number is proportionally smaller (a CR: 16:1 gives a 2.2MB hogel-vector array).

The Cheops image processing system as shown in Figure 8-50 is a compact, block data-flow parallel processor designed and built for research in scalable digital television. The P2 processor card communicates to the host via a small computer standard interface (SCSI) link with 1MB/s bandwidth. Six Cheops output cards provide 18 parallel analog-converted channels of computed fringes to the MIT 36MB holovideo display. The P2 communicates data to the output cards using the fast Nile Bus. The P2 also supports a type of stream-processing superposition daughter card (the *Splotch Engine*) that performs weighted summations of arbitrary one-dimensional basis functions. The Splotch Engines (two were used in this research) perform the many MAC operations required for the decoding of hogel vectors into hogels. The hogel-vector array was downloaded to the Cheops P2 card, where it was decoded using two Splotch Engines.

In addition, the Cheops output cards store each fringe sample as an 8-bit unsigned integer value. Computed fringes are normalized to fit within these 256 values. Normalization generally involves adding an offset and multiplying by a scaling factor. In the hogel-vector technique, normalization is built into the computational pipeline. For example, when using

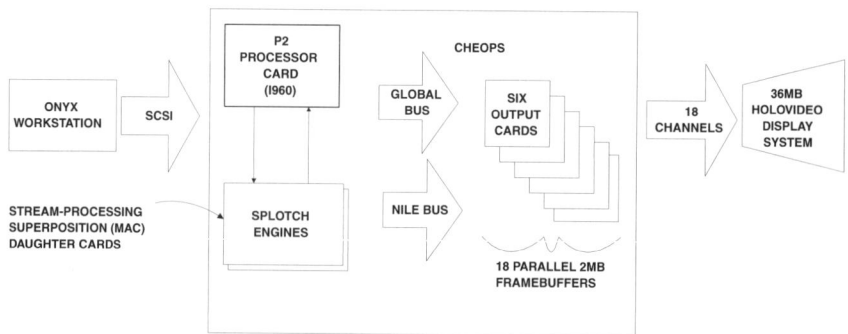

FIGURE 8-50: Hogel-vector decoding on the Cheops modular framebuffer and image processing system.

Cheops, the hogel-vector components are prescaled to produce useful fringes in the higher 8 bits of the 16-bit result field. Only this high byte is sent to the output cards.

The decoded 36MB fringe was transferred to the Cheops output cards and used by the MIT 36MB holovideo display to generate images. After radio frequency (RF) processing, computed fringes (in the form of acoustic waves) traversed the aperture of an acousto-optical modulator (AOM), which phase-modulated a beam of laser light. Two lenses imaged the diffracted light at a plane in front of the viewer. The horizontal scanning system angularly multiplexed the image of the modulated light. The vertical scanning mirror positioned diffracted light to the correct vertical position in the hologram plane. Electronic control circuits synchronized the scanners to the incoming holographic signal.

Implementation on a Serial Workstation
The entire diffraction-specific computation pipeline was also implemented on the SGI Onyx serial workstation. The process for generating a 36MB fringe was the same as above, except that a simple linear loop performed the decoding step. The computed fringe was downloaded to Cheops to generate images on the 36MB holovideo display.

Implementation on a Graphics Subsystem
The SGI RealityEngine2 (RE2) is a computer graphics subsystem generally used to render images. The rapid texture-mapping function and the accumulation buffer were used to perform rapid multiply-accumulate operations. The RE2 rendered directionally dependent 2D views of the object scene. These rendered views were converted into a hogel-vector array that was then decoded in the RE2. The texture-mapping function rapidly multiplied a component from each hogel vector by a replicated array of a single basis fringe. This operation is repeated, once for each hogel-vector component, accumulating the result in the accumulation buffer. Transfer times were negligible because all computations occurred inside the graphics subsystem that included the framebuffer to drive the display. For CR = 32:1, a 2MB fringe was decoded from a 64KB hogel-vector array for each of the three colors.

MODEL OF POINT SPREAD

A fringe generated using hogel-vector bandwidth compression generally loses some of its ability to produce a sharp image. A given image point appears slightly broadened or blurred. The increased point-spread results from several processes:

- Aperturing due to spatial sampling, $blur_{spatial}$
- Spectral sampling blur due to sparsely sampled hogel spectra, $blur_{spectral}$
- Aberrations in the display, $blur_{displ}$
- Quantization and other noise [Lucente, 359]

As the number of symbols per hogel decreases, spectral sample spacing increases. Each hogel-vector component carries information about a wider region of the hogel spectrum, limiting the achievable image resolution.

The spectral sampling blur and aperture blur add geometrically with other sources of blur. Blur caused by the display was measured for various z locations in the image volume. Additional contributions to point spread are for now neglected.

Experimental Verification of Model

To verify the point-spread model, a series of experiments was performed. Pictures of individual imaged points were used to measure point spread as a function of the spatial and spectral sampling parameters (width and components), and to compare these data to the model for point spread. For the deepest points at $z = 80$ mm and for smaller depths, the point-spread model fit very well to the measured data.

Model-Based Selection of System Parameters

The model for point spread provides an analytical expression that relates the various parameters of the holovideo system. This expression is used to select certain parameters (such as hogel width and compression ratio) given other parameters, such as desired image resolution and image depth. For practical imaging, blur must be below the amount perceivable to humans—about 0.18 mm at a typical viewing distance of 600mm.

IMAGING RESULTS

Fringes were computed using hogel-vector bandwidth compression and used to generate a variety of 3D images on the MIT holovideo displays. Speeds were measured. Digital photographs were taken of the images.

The MIT 6MB color holovideo display was used to generate images computed using hogel-vector bandwidth compression. Hogel-vector direct encoding and decoding were performed by the RE2, using three sets of basis fringes, each precomputed for the specific wavelength used in the full-color display. Three 2MB fringes were decoded from three 64KB hogel-vector arrays (one per color). The resulting images (computed at interactive rates) showed good quality and possessed the full range of lighting features (specular highlights and transparency).

Earlier work reported the use of the RE2 to compute stereogram-style holograms. The stereogram images had noticeable blur and artifacts, especially when computed rapidly. In comparison, hogel-vector bandwidth compression maintained image fidelity, even at the high compression ratios (CR = 32:1) necessary to achieve computation at interactive rates. These images did not exhibit the artifacts of vertical dark stripes or jumps. Moreover, images computed using the diffraction-specific technique produced real 3D images.

Spatial Coherence
Hogel-vector bandwidth compression added a noticeable speckle-like appearance to the image. These brightness variations at infinity likely resulted from the use of coherent light in the display. Diffraction-specific fringe computation assumes that light is quasi-monochromatic with a coherence length. This ensures that the diffracted light adds linearly with intensity. In practice, the effective coherence length of light in the holovideo displays was approximately 2.0mm. To reduce the speckle effect, a random set of phases was introduced into each hogel via the basis fringes. This reduced interhogel correlation. Implementation employed a set of 16 different but spectrally equivalent precomputed basis fringes. To decode a given hogel vector, each basis fringe was selected at random from the set of equivalent basis fringes.

SPEED

Hogel-vector bandwidth compression achieved an increase in speed by a factor of over 70 compared to conventional interference-based methods. Computing times were measured on three platforms: the Onyx workstation (alone), the Onyx (for hogel-vector generation) and the Cheops with two Splotch Engines (for decoding), and the Onyx/RE2. The results from three computational benchmarks are described: a conventional interference-based technique, and two diffraction-specific cases in which hogel-vector compression ratios are CR = 1:1 and CR = 32:1.

Using the hogel-vector technique, total computation time consists of the initial direct-encoding step, the time to transfer the hogel-vector array to the decoding system, and the hogel-vector decoding step. The first step (generation of the hogel-vector array on the Onyx workstation) was very fast. For most objects, typical times were 10 seconds for CR = 1:1, and 0.5 seconds for CR: 32:1. The downloading of the hogel-vector array over the SCSI link was slow. However, the use of hogel-vector bandwidth compression (CR = 32:1) reduced data transfer of the 1.1MB hogel-vector array to only 1.0 second.

Appropriate scene complexities were chosen to ensure equivalent benchmarks. For hogel-vector bandwidth compression, speed is basically independent of image scene complexity, whereas the computing time for interference-based ray-tracing computations varies roughly linearly.

Speed on a Serial Workstation
The conventional interference-based method used was to sum the complex wavefronts from all object points (plus the reference beam) to generate the fringe. Because it involved complex-valued, floating-point precision calculations, it was not implemented on either of the specialized hardware platforms (Cheops/Splotch, or the RE2). A fairly complex image of 20,000 discrete points (roughly 128 imaged points for each line of the fringe) was used. Implemented completely on the Onyx workstation, a 36MB fringe required 23,000 seconds (over 6 hours). This timing was extrapolated by computing a representative 2MB fringe.

For comparison, hogel-vector bandwidth compression was implemented on the Onyx workstation. For a 36MB fringe computed using CR = 1:1, total time was 9,600 seconds. For CR = 32: 1, the time was reduced to only 300 seconds. Including the time for hogel-vector generation (0.5 seconds), this represents a speed increase of 74 times compared to the conventional computing method. Another advantage of hogel-vector bandwidth compression is that the simplicity of the slower decoding step allows for its implementation on very fast specialized hardware.

Speed on Cheops
When implemented on the Cheops system containing two Splotch Engines, hogel-vector decoding time was 190 seconds for CR = 1:1, and only 6 seconds for CR = 32:1. These timings are worst case (most complex

image), measured using a fully nonzero hogel-vector array. In practice, typical image scenes produced many zero-valued hogel-vector components. Skipping zero-valued components resulted in faster decoding times. Typical test images (CR = 32:1) were closer to 3 seconds.

The total hogel-vector encoding and decoding time in the case of CR = 32:1 was 6.5 seconds, worst case. Although it is not quite fair to compare this to the conventional method implemented only on the workstation, the relative speed increase of over 3,500 times is made possible by the simplicity and efficiency of the hogel-vector decoding algorithm.

Speed on a Graphics Subsystem
When implemented on the RE2 graphics subsystem, for a 6MB (2MB per color) fringe, hogel-vector decoding time was 28 seconds for CR = 1:1, and only 0.9 seconds for CR: 32:1. The texture-mapping function of the RE2 graphics subsystem and its accumulation buffer performed rapid multiplications and additions. The 0.9 seconds is a total time, from model to image, since the MIT color holovideo display was driven directly by the RE2. Conventional methods cannot use this specialized hardware or achieve interactive-rate fringe computation.

Analysis of Speed
In hogel-vector bandwidth compression, the decoding step required the great majority of computing time. To decode a *CR*-component hogel vector to a sample hogel requires calculating an inner product: *CR* multiplication-accumulation operations (MACs). For example, for a 36MB fringe, a hogel width of 1,024, and a CR =32:1, the decoding step requires 1.2 GMACs (over 1 billion multiplies and adds).

Note: As shown in Table 8-1, the speed increase from CR = 1:1 to CR = 32:1 is about 32 times. This was due to the reduction by a factor of 1 / CR in the number of time-consuming MAC calculations required. Because each fringe sample requires *CR* MACs, the speed of hogel-vector decoding increases linearly with CR. Faster speeds can be achieved by sacrificing image quality.

Finally, further speed increase during interactivity was achieved by exploiting the scalability of a hogel-vector array (its ability to supply variable degrees of precision as required). For instance, in one interactive demonstration of the 6MB display, a subset of the hogel-vector array was decoded

TABLE 8-1: Computation times for different hardware and techniques

Platform Fringe Size	Conventional (seconds)	Hogel-Vector Bandwidth Compression	
		CR = 1:1 (seconds)	CR = 32:1) (seconds)
Workstation, 36MB	23,000	9,600	300.5
Cheops/2-Splotch,36MB		200	6.5
RE2, 6MB		28	0.9

Note: Transfer times are not included. Hogel vector times are worst case, and include encoding and decoding.

to produce a quick-and-dirty image that was subsequently replaced by the full-fidelity image when interactivity ceased.

Next, as part of the continuing design stage of the commercial holographic process, let's very briefly look at how scientists have used simple in-line onsite recording and off-axis viewing for single-exposure holographic flow visualization of a drop of fluid falling into water, and have extended the technique to combustion flow studies.

Onsite Recording

Researchers at the Wright Patterson Air Force Base in Dayton, Ohio are using holography and laser technology in an effort to develop a practical 3D diagnostic technique for evaluating aircraft combustion flow as shown in Figure 8-51.[16] In a program jointly funded by the Air Force base's Aero Propulsion and Power Directorate and its Office of Scientific Research, the team is developing two techniques (holographic flow visualization and holographic particle image velocimetry) for combustion studies.

Flow visualization techniques currently are limited to a 2D slice of a flow illuminated by a laser sheet or a 2D field that results from the integration of a 3D density field along the path of a laser beam. These techniques do not provide full visualization of 3D information.

Holographic imaging, which provides 3D representation of spatial objects instantaneously, holds great promise as a qualitative and quantitative 3D diagnostic tool for spatially and temporally evolving complex flow structures. Under this program, S. Gogineni, J. Estevadeordal, and L. Goss of Innovative Scientific Solutions Inc., and Professor H. Meng of Kansas

DESIGN: ANATOMY OF A HOLOGRAM 181

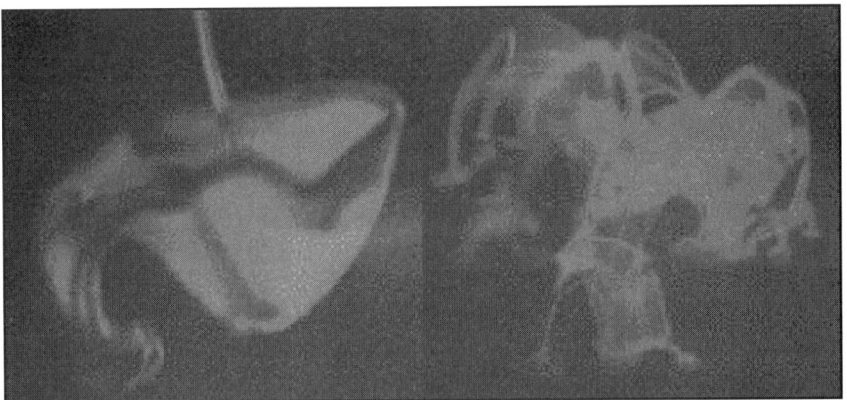

FIGURE 8-51: Inline onsite recording and off-axis viewing for single-exposure holographic flow visualization.

State University in Manhattan, Kansas have demonstrated a holographic recording and reconstruction system that can provide 3D flow visualization and velocity measurements in simple combustion flows.

MATCHING WAVELENGTHS

To record the holograms, the researchers use two 532-nm Spectra-Physics Nd:YAG lasers that are injection-seeded to provide a coherence length greater than 1m. They also use a single Spectra-Physics Millennia Nd:YVO$_4$ continuous-wave laser at 532 nm to reconstruct holograms. They use the Millennia not only because of its convenience, but also because its wavelength matches that of the onsite recording lasers.

The team has employed simple inline onsite recording and off-axis viewing techniques for single-exposure holographic flow visualization of a drop of fluid falling into water. The results demonstrate the effectiveness of holographic visualization in flows under unstable conditions (see Figure 8-51). They have extended this technique to combustion flows and have demonstrated the limitations that result from distortion of the reference beam by temperature and density gradients.

As an alternative to this technique, the group has developed an off-axis-based system that can perform holographic flow visualization and holographic particle image velocimetry. This system was applied to reacting and nonreacting propane jet flows in the presence and absence of a coflowing air stream. The flow field seeded with TO$_2$ particles provided images suitable for 3D flow visualization, and the same flow seeded with Al_2O_3 provided images suitable for qualitative and quantitative 3D velocity measurements. With this research well under way, 3D imaging should allow scientists to study complex flow phenomena.

Finally, there are a few additional things to consider when designing a commercial hologram. Let's take a look.

The Design Itself

The creation of a hologram is only as good as its designer. With that in mind, and as part of the continuing design stage of the commercial holographic process: computer graphics, animation, reduced or enlarged images, and onsite recording are but a small part of this process.

There are a few things to consider when designing a commercial hologram. First, in what application will the hologram be used? The following are commercial application examples that answer the preceding question:

- For packaging, with over-printing, *Dotz!* (see Figures 8-52 and 8-53 and Table 8-2) patterns work best for this application.
- Where true color is important, a Hi-View 3D hologram (see Figures 8-54 and 8-55 and Table 8-2) displays full color.
- Where dimensional representation is important, a 3D stereogram (see Figures 8-56, 8-57, and 8-58, and Table 8-2) provides dimension, color, and animation.
- As a security product, a combination hologram (see Table 8-2) is best.
- For an unusual premium, a holodisk (see Figures 8-59 and Table 8-2) is a unique collector's item.[17]

Nevertheless, there are many other commercial applications for holograms; this is just a starting point. When creating an actual commercial hologram, the following holographic design specifications are required:

- Artwork
- Scanner
- Graphic software
- Graphic effects

DESIGN: ANATOMY OF A HOLOGRAM 183

FIGURE 8-52: Viper. Medium: embossed hot-stamping foil. Special Features: This hologram of the Viper car was made using CFC Applied Holographics' patented Dotz! process. As the viewer changes position, the image changes from a daytime scene to a nighttime scene.

FIGURE 8-53: Spiral. Medium: embossed, metallized foil, with pressure sensitive liner. Special Features: Spiral is a Dotz! image, made using CFC Applied Holographics' patented dot-matrix process for creating holograms. The pretty spiral pattern, called a fractal, is made from a computer graphics image. In the transfer process, each pixel of the computer image is represented as a dot on the hologram. When light shines on the hologram, each dot plays back in a designated color.

TABLE 8-2: Hologram commercial application types and art requirements

Type of Hologram	Description	Art Requirements
Dotz!	• Images are made from thousands of holographic dots. • Animation effects result from light playing through the dots, diffracting light through a range of angles that form kinetic patterns. • Performs excellently under any lighting conditions, including fluorescent. • Images are 2D rather than 3D. • Up to four images can be overlapped (providing a *jewelry-like* look). See Figures 8-52 and 8-53.	Storyboard, line art, grayscale images, continuous tone grayscale photography: • For over-lapped images, elements must be congruent. • Specify animation preferences, if any.
Hi-View 3D	• Full-color, dimensionally layered hologram (typically referred to as 2D3D). • Designed especially to be viewed in fluorescent lighting or ordinary lighting. • Made from continuous tone flat art. • Elements are separated and placed on different layers providing a sense of depth. • Can have 2-channels for *animated effect*. See Figures 8-54 and 8-55.	Storyboard, photography, line art (black & white or color): • Specify color requirements, if any (remember, bright, saturated colors work best). • Specify preferred dimensional layout (what elements will be in the foreground, midground, and background?).

TABLE 8-2: *(Continued)*

Type of Hologram	Description	Art Requirements
3D Stereogram	• Full-color, 3D hologram from live subject or computer-generated 3D image. • Can include animation. See Figures 8-56, 8-57, and 8-58.	Storyboard, computer 3-D model files, actual model (live person, small or real-size object) for filming: • Specify animation (motion should be limited to slow, short movements to prevent blurring the image).
Combination hologram	Any of the preceding types of holograms can be combined providing visual effect and added security.	Depends on what types of holograms are combined: • Specify which type of hologram for each component.
Holodisk	Unique monochromatic, 360°, 3D, standup hologram. Image appears to stand up from flat disk surface when lit with point-lights from above. As flat disk is rotated, 360° view of image is seen. Mono-color hologram. Can include animation. At present, limited to 6" disk. See Figure 8-59.	Actual model (live person or object): • Supply storyboard for animation (motion should be limited to slow, short movements to prevent blurring the image). • Details should be high contrast to provide a clear image (remember, this is a mono-color hologram).

Note: Proofs and art approval: At what stage is art approval required? Is it normal for a client to approve art after it is readied for holographic production? Who should receive a proof (or proofs) of the finished hologram? Who will sign off on art approval?

FIGURE 8-54: Legends. Medium: embossed hot-stamp foil, attached to cardboard stock. Special Features: This collector's edition card shows baseball legends Babe Ruth and Roger Maris. The image is a computer-generated composite of flat art (photographs) and of objects modeled on the computer (the baseball bats).

- CMYK values
- Color palettes
- Comparison
- Registration
- Viewing software

DESIGN: ANATOMY OF A HOLOGRAM   **187**

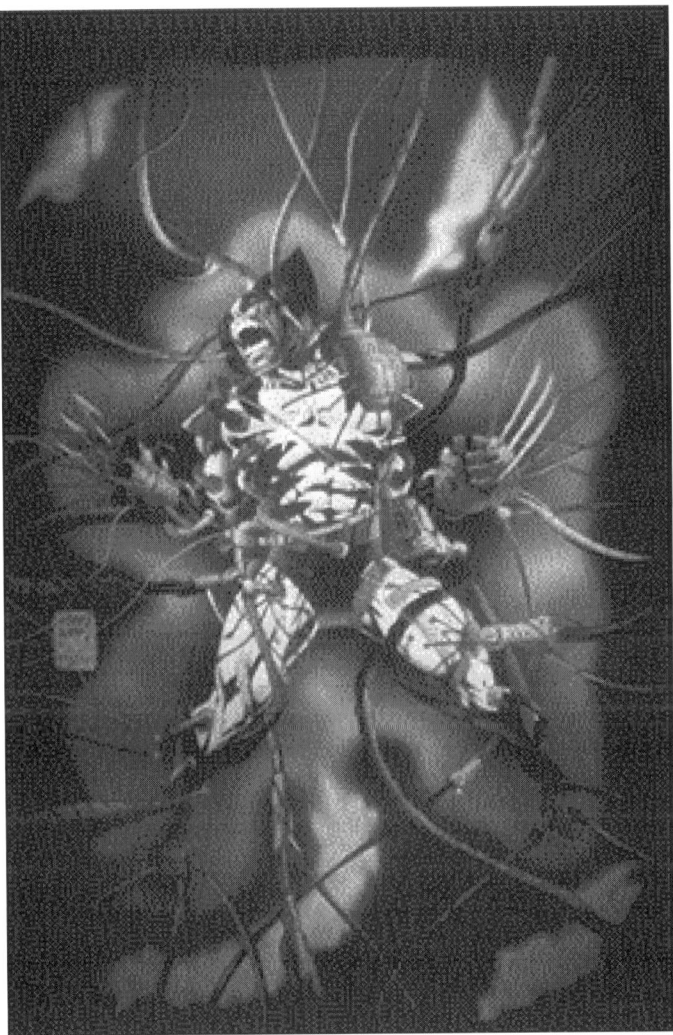

FIGURE 8-55: Wolverine. Medium: embossed metalized foil with a unique die cut, applied to a paper comic book cover. Client: Marvel Comics. Special Features: Wolverine was produced for a collector's edition comic book. This two-channel hologram was made from two pieces of flat artwork by the illustrator Adam Kubert. The first illustration, depicted in the preceding photograph, is of the Wolverine about to receive a powerful electric shock. Note the richness of the saturated color. (Holograms are best at producing saturated color, rather than pastels.) In the second illustration (the second channel of the hologram), we see the agonized character as a skeleton, reeling from the shock. The original artwork was separated into five layers that were each printed individually on the hologram, so as to give the illusion of depth. This technique of using layered flat artwork to produce depth is known as 2D3D. As the viewer moves the hologram, the image switches from the Wolverine about to receive the shock, to the Wolverine skeleton. The saturated colors and the vivid imagery make this a striking hologram indeed.

188   CHAPTER 8

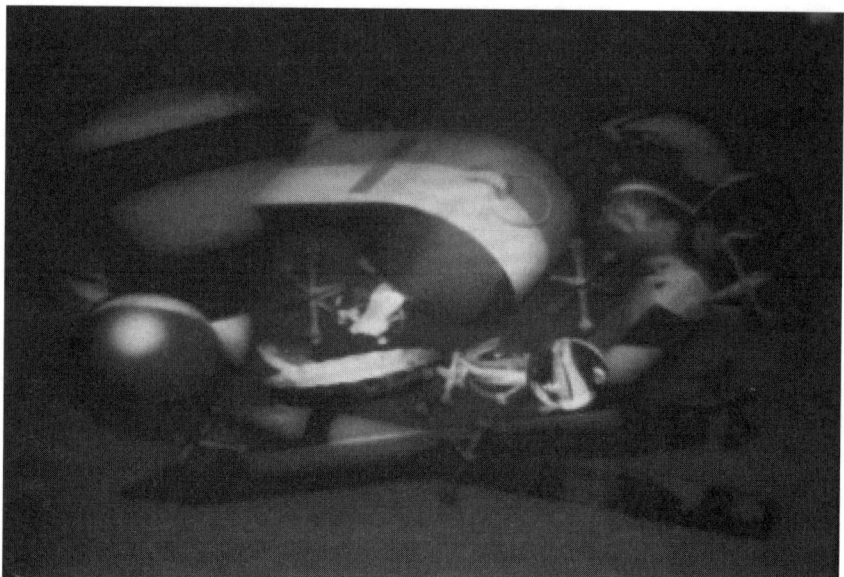

FIGURE 8-56: From My Pocket. Medium: embossed, metalized foil, with pressure-sensitive liner. Special Features: From My Pocket looks like a hologram of toys taken from a child's treasure box, but this is just an illusion. In fact, all of the objects depicted in the hologram are computer-generated images, created by artist Barbara Roman. The computer program, Alias Animator, uses a technique called ray-tracing to provide the realistic reflections. This hologram illustrates the double-illusion CFC Applied Holographics can achieve: the illusion of 3D, and the illusion of seeing an image of a "real" object. Like the previous gallery image, EV-1, From My Pocket is a full-color stereogram.

ARTWORK

You should contact your holographic design vendor for tips in designing artwork. Most storage formats are acceptable; Adobe Photoshop is preferred. Gradients work well to create movement in the hologram.

So, what kind of art is necessary to make a hologram? Table 8-2 answers that question.

SCANNER

Traditional artwork is first digitized on a flatbed scanner; then the image is cleaned up using graphic design software. Special effects can be added in Photoshop.

FIGURE 8-57: Flowers. Medium: embossed, metalized foil, with pressure-sensitive liner. Special Features: This is a hologram of a real flower arrangement. It is remarkable for the realistic color it achieves. It is a type of hologram known as a full-color stereogram. It was made using film of the flower arrangement that was shot with a movie camera that moved slowly past the flowers. Since the camera was moving, each image in the film strip shows the flowers from a slightly different point of view. The different images on the strip were recorded in the hologram so as to give the illusion of depth when viewing the flowers. This technique of combining a series of 2D filmed images to obtain 3D is known as a stereogram.

GRAPHIC SOFTWARE

Use software that allows you to manipulate colors and shapes. Adobe Photoshop will give the most support in graphic design. The use of *layers* in Photoshop is preferred.

GRAPHIC EFFECTS

Experienced holographic designers can make the most of the effects and treatments available with software. This software will enhance your graphic design for holographic output. Radial and linear gradients produce *movement* effects when made into a hologram.

190 CHAPTER 8

FIGURE 8-58: EV-1. Medium: glass coated with holographic emulsion. Special Features: This is a hologram of the chassis of GM's electric car, the EV-1. The surprising thing about this hologram is that it was made before the car chassis was manufactured! The hologram was produced directly from output of a CAD (computer-automated design) program. Like Wildflowers, EV-1 is a full-color stereogram. That means that it was made by recording, in a hologram, a strip of movie film that pans across the object. In this case, since the object was a computer-generated image, the computer generated a series of different views of the car shell, and then recorded the series of views onto movie film.

COLOR PALETTES

The RGB color palette and its values are converted to the holographic color palette ready for output. Designing in the *Spectrum* color palette works well.

COMPARISON

A comparison is made between the four-color output and the holographic color output. You need to step and repeat the actual-sized image to meet embossing requirements.

REGISTRATION

You should use crop marks to optically convert the holographic image for printing and cutting. The use of eye marks as a guide for cutting and racing stripes to control web weave is also important to registration.

FIGURE 8-59: Rita. Special Features: Rita is a kind of hologram called a holodisk. Rita was made by making a 360-degree film of a model. When the hologram is illuminated from above by a point source of light, the image of Rita will appear to project out 2 to 3 inches in space above the holodisk. Rotating the disk produces a 360-degree view of the model.

VIEWING SOFTWARE

Holo Mac or Holo PC are programs that can be used to view most of the holographic images. They are proprietary software packages that simulate the animated movement and colors of the holographic image.

Commercial Holographic Design Tips

Finally, the first tip in designing a custom commercial hologram is to have fun. Anything you can imagine and anything you can print can basically be designed into a custom hologram design or pattern. Looking at different examples of your finished projects will help you imagine what can be created for your custom design. Communicating with holographic designers is also a good idea. They will send you a sample kit of finished pieces and sample sheets.

Once you have an idea, you can print a mockup. To see your artwork with a holographic background, print out separate layers on a clear film overlay. Print different sections of your design in white as well. You probably don't want to see a person's face look holographic, or your corporate logo may not suit holographic qualities—that's fine and easily rectified. It is recommended that you print out those sections in white ink or an opaque color on the clear acetate. Now, take a sheet of your holographic paper from your sample kit and lay the clear *printed* overlays on top. It's easy, fun, and a great way to sample your holographic design.

Custom Pattern Design Tips

There are thousands of designs from stock pattern examples such as Lots-O-Dots and Shear Patterns, to hearts, teddy bears, coins, spaghetti, squares, and more. It's easy and fun. If you want the design to be a repeating corporate logo wallpapered across the entire roll, or just a rose bud in the upper-left corner of each sheet, it can be done. If you design it, it can be turned into a hologram. If you want help, you need to get designers on staff who will gladly work with you to design a custom pattern. When designing for a custom pattern, a full roll of holographic paper must be purchased as required in the manufacturing process.

Inks

Considering inks and the printing process will always benefit you as a holographic designer. Opaque white inks are often used in various degrees to alter the visual effect of your design. The use of opaque white from 10 percent to 200 percent underneath the printing process will give you different visual outcomes. If you desire to block out the holographic qualities in certain areas of your design (for example, a person's face or corporate logo), a double hit of opaque white over the area is necessary before you print your colors. If you want to create subtleties that show through the holographic paper and board, you can choose to use opaque whites in various degrees. The design is up to your imagination.

The varied uses of opaque inks and translucent inks will also help you create the design effect you desire. Translucent inks allow the holographic qualities of prismatic illusions papers and boards to show through at a higher visual percentage versus the use of opaque inks. Opaque inks will create a heavier cover over the holographic qualities of the prismatic paper. Some projects desire the use of both inks and opaque whites to create the final design effect. Again, speak to your designers, or let your imagination have fun—it's holography!

In Summary

- Real-time 3D holographic displays are expensive, new, and rare. Although they alone among 3D display technologies provide extremely realistic imagery, their cost must be justified. Each specific computer graphics application dictates whether holovideo is a necessity or an extravagant expense.
- Interactive computer graphics applications are normally divided into two extreme modes of interaction: the *arm's-reach* mode, and the *far away* mode. An arm's-reach application involves interacting with scenes in a space directly in front of the user, where the user constantly interacts, moving around it to gain understanding. In this mode, all of the visual depth cues are employed, particularly motion parallax, binocular disparity, convergence, and ocular accommodation. These applications warrant the expense of holovideo and the extreme realism and three-dimensionality of its images: computer-aided design, multidimensional data visualization, virtual surgery, teleoperation, training, and education (holographic virtual textbooks on anatomy, molecules, or engines).
- At the other extreme, a far away application involves scenes that are beyond arm's reach and are generally larger. The imagery of such applications (flight simulation, virtual walk-throughs) makes adequate use of the kinetic depth cue, pictorial depth cues, and other depth cues associated with flat display systems. A high-resolution 2D display may be a more cost-effective solution for far away applications.
- Hogel-vector bandwidth compression (a diffraction-specific fringe computation technique) has been implemented and used to generate complex 3D holographic images for interactive real-time display. The application of well-known sampling concepts to the localized fringe spectrum has streamlined, generalized, and greatly accelerated computation.
- Hogel-vector bandwidth compression makes efficient use of computing resources. The slower decoding step is essentially independent of image content and complexity, and simple enough to be implemented in specialized hardware.
- Hogel-vector bandwidth compression is the first reported technique for computational holographic bandwidth compression. It achieves

a bandwidth compression ratio of 16:1 without conspicuous degradation of the image, thereby eliminating transmission bottlenecks.
- Fringe generation was 74 times faster than conventional computation. This technique provides a superior means for generating holographic fringes—even if bandwidth compression is not needed.
- Hogel-vector encoding provides a foundation for *fringelet encoding*, which achieves further speed increases. It can also be applied to full-parallax holographic imaging.
- Hogel-vector bandwidth compression can be applied to other tasks. For example, holographic movies can be hogel-vector-encoded for digital recording and transmission over networks or television cable. As another example, diffraction-specific fringes have been recorded onto film to produce static holographic images. Hogel-vector bandwidth compression provides the speed and portability required to generate large fringes for a holographic *fringe printer*.
- The analysis of hogel-vector bandwidth compression has revealed a simple expression relating the fundamental system parameters of bandwidth, image resolution, and maximum image depth. Hogel-vector encoding attains *visual-bandwidth holography*—it frees holographic fringes from the enormous bandwidths required by the physics of optical diffraction. Bandwidth is instead matched to the abilities of the human visual system.

The next chapter explains how to master holograms, by showing you how to create and combine the necessary photographic and holographic elements to produce a *master* hologram on special photosensitive emulsions. It also shows you how a multidimensional image is recorded as a unique *interference pattern* that is created using laser light and precision optical techniques.

End Notes

1. Mark Lucente, "Interactive Three-Dimensional Holographic Displays: Seeing the Future In Depth," IBM Research Division, Thomas J. Watson Research Center, Box 218, Yorktown Heights, NY 10598, USA, 1997, p. 2.
2. Massachusetts Institute of Technology, 77 Massachusetts Ave., Cambridge, MA 02139, USA, 1999.

3. Donald Herbison-Evans, "4D Computer Output," Faculty of Applied Science, Central Queensland University, Macleay Museum, University of Sydney, NSW 2006, Australia, 2000, p. 1.
4. Southern Arkansas University, Magnolia, Arkansas, 1999, p. 1.
5. Paul Haebarli, "A Multifocus Method for Controlling Depth of Field," GRAFICA Obscura, Silicon Graphics Incorporated, 2011 N. Shoreline Boulevard, Mountain View, CA, 94043-1389 USA, 1999, p.1.
6. John H. Kranz, Ph.D., "Aerial Perspective," Psychology Department, Hanover College, P.O. Box 108, Hanover, IN, 47243, USA, 1999, p.1.
7. Abby Parrill and Dr. Jacquelyn Gervay, "Identifying Enantiomers Exercise #1," The Department of Chemistry, The University of Arizona, Tucson, AZ 85721, USA, 1999, p. 1.
8. John H. Kranz, Ph.D., Peter Chang, Nicolas Pioch, and Gareth Richards, "Depth Cues," Psychology Department, Hanover College, P.O. Box 108, Hanover, IN, 47243, (Ecole Polytechnique, Ecole Nationale Superieiere des Telecommunications, Paris, France). USA, 1999, p.1.
9. Dr. Marc Grossman and Rachel Cooper, "Stereoscopic Photographs," Magic Eye Inc., 67 Shank Painter Rd., Provincetown, MA 02657, USA, 1999, p. 1.
10. StereoGraphics Corporation, 2171 E. Francisco Blvd., San Rafael, CA 94901, USA, 1999.
11. Rocky Mountain Memories, 6203 Avery Island Avenue, Austin, TX 78727, USA, 1999.
12. DIMENSION 3, Fulfillment Center, 5240 Medina Rd., Woodland Hills, CA 91364-1913, USA, 1999.
13. Hiroshi Yoshikawa and Hiroyuki Taniguchi, "Computer Generated Rainbow Holograms," Presented at SPIE Photonics China, (SPIE Proceedings of Holographic Displays and Optical Elements, Nov. 1996,) 7-24-1 Narashino-dai, Funabashi Chiba 274-8501, Japan, 1996.
14. NEOS Technologies, Inc., 4300-C Fortune Place, Melbourne, FL 32904, USA, 1999.
15. Mark Lucente, "Computational Holographic Bandwidth Compression," IBM Systems Journal, Vol. 35, Nos. 3 & 4, Thomas J. Watson Research Center, Box 218, Yorktown Heights, NY 10598, USA, 1996, p. 350.
16. Bruce Craig and S. Gogineni, "Lasers and Holographic Imaging Team Up for Combustion Flow Study," Spectra-Physics Lasers, Mountain View, CA; Innovative Scientific Solutions Inc., Dayton, OH; Laurin Publishing Co., Inc., Berkshire Common, P.O. Box 4949, Pittsfield, MA 01202-4949, USA, 1998, p. 1.
17. CFC International Inc., Corporate Headquarters, 500 State Street, Chicago Heights, IL 60411, USA, 1999.

CHAPTER 9

MASTERING HOLOGRAMS

In This Chapter

- Imaging software
- Photoresist
- Interference pattern
- Exposure
- Three-dimensional
- Dot matrix
- Processing
- Comparison

*M*astering, or creating a master hologram, is the first step in the process of holography. The mastering process consists of the following elements:

- Digital image
- Photoresist
- Exposure
- 3D
- Interference pattern

FIGURE 9-1: Creation of a master hologram.

- Dot matrix
- Processing
- Comparison

This chapter examines the mastering stage (see Figure 9-1) of the holographic process that researchers go through to create commercial holograms.[1] As with all holograms, imaging software plays a major role in their creation.

Digital Image

If you are using Adobe Photoshop, the digital image is converted to a bitmap file for output to the laser machine. Laser imaging software conversion would then control the exposure(s) of the laser into the photoresist plate to create the hologram.

Some researchers, for example, are imaging living subjects using a Q-switched Nd:YAG/Nd:Glass laser system of their own custom design. Some have noticed different results and differences in recording transmission masters and digital image plane reflections using different commercially available emulsions in format sizes to 14'–30'. Most researchers always finish with a summary of the markets that they perceive for commercial holographics using live subject images.[2]

Photoresist

The photoresist is a smooth glass plate that is treated with a photosensitive coating 1.5 microns thick. The photoresist is only sensitive to certain wavelengths of light; in this case, blue light at 442nm.

Through the use of fiber optics, the computer is set to start exposing the photoresist pixel by pixel. The interference pattern is created by the lasers as it is exposed into the photoresist.

Exposure

The exposure using helium cadmium laser(s) and bitmap (bmp) file creates the dot matrix hologram by exposing it pixel by pixel. On the other hand, there are other techniques that are used to create several large-format silver-halide masters for transfer into a large-format dichromated gelatin DCG hologram. The subject here is a fossilized dinosaur egg, one of the largest such artifacts ever found. In order to present all facets of the subject—the outer egg shell, the articulated embryo bones, and a model of an artist's concept of the embryo inside the shell—several masters, were produced to create in the end a double exposure multichannel transfer image. The emphasis here was placed on a practical approach to the production of the holograms.[3]

3D

Holography is a useful technology for 3D images as shown in Figure 9-2.[4] There has been a lot of research on spatial imaging based on holographic technology. There have also been very many good results with regard to 3D electronic images: 3D TV monitor, 3D fine art, 3D measuring instruments, and so on. In addition, there have also been many good results from the stereoscopic 3D display system. This system consists of a liquid crystal device and a holographic screen formed of holographic optical elements. This display can construct animated 3D images in real time by updating LCD pixels.[5]

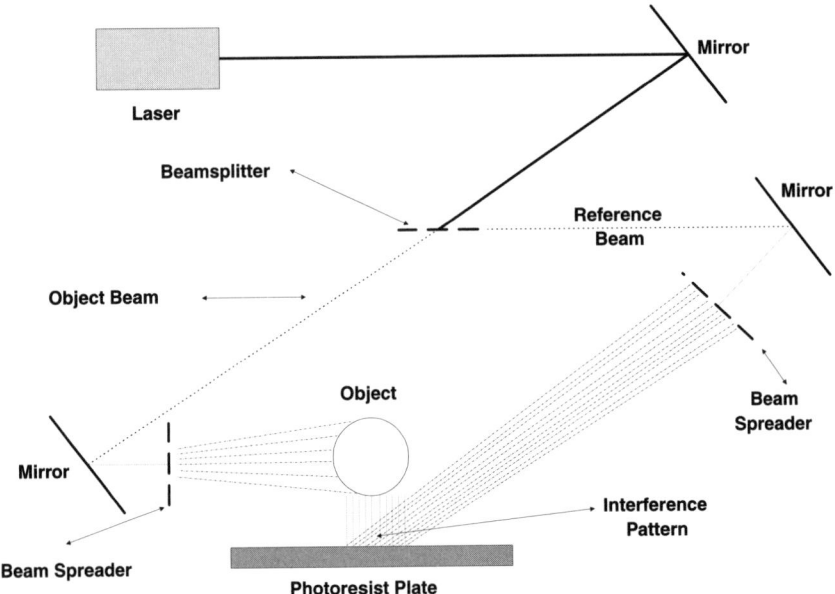

FIGURE 9-2: The 3D object's reflected interference pattern is recorded.

Interference Pattern

An *interference pattern* is created when two of the same wavelengths of light interfere with each other, producing a pattern as shown in Figure 9-3.[6] This pattern is the hologram in the size and shape of a pixel.

Researchers are also looking at a holographic system using a helium-neon laser to study the phase information in the region of space in which a plume of fuel is ejected from an impulse nail gun. A transmission hologram is then recorded in which the object is a trans-illuminated piece of round glass. While observing the virtual image of the hologram, fuel is injected into the region of space immediately upstream from the ground glass. The interference pattern between the reconstructed virtual image and the phase modulated light from the ground glass is observed. This information is recorded at a rate of 1,000 frames per second using a high-speed video system.[7]

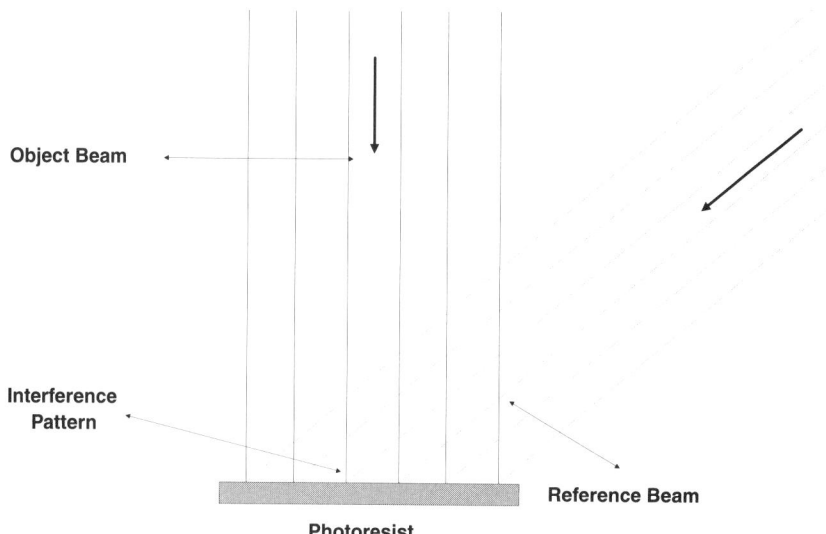

FIGURE 9-3: This pattern will produce the hologram.

Dot Matrix

Dot matrix technology allows the hologram to reflect extremely bright colors viewed from many angles as shown in Figure 9-4.[8] An example of exposure time would be approximately 48 hours for a 6" × 6" hologram that includes almost 360,000 exposures.

Processing

Processing the hologram is the next step. The hologram is developed in a chemical bath that removes all areas of the unexposed photoresist, producing the master hologram.

Nevertheless, with the withdrawal of Agfa-Gevaert[9] from the manufacturing of silver-halide recording materials, there is now a need to consider some of the superfine-grain materials such as those made by Slavich in Russia. Problems with processing have been encountered in the initial studies of appropriate chemical processing. It is considered that the most important issue is maintenance of a unique and achromatic Bragg condition over the

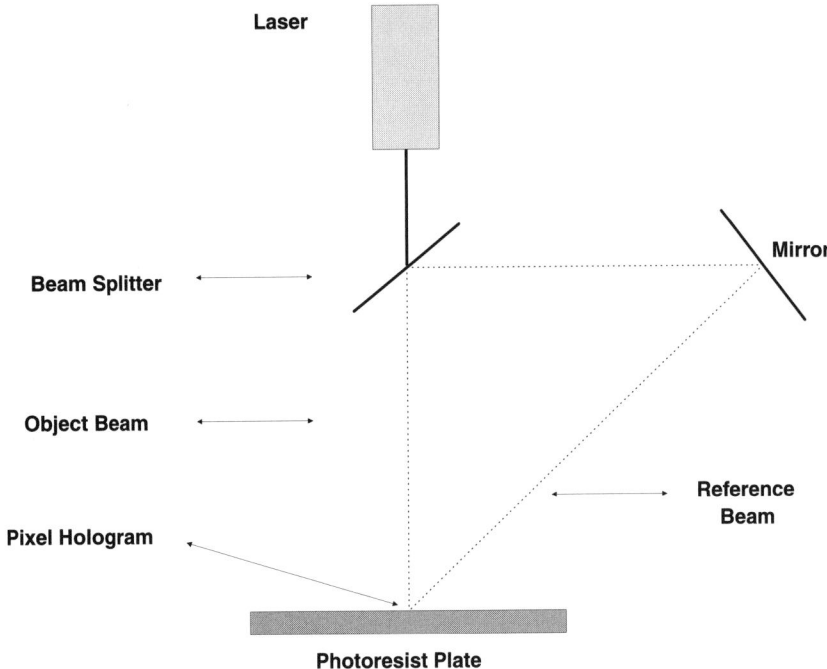

FIGURE 9-4: Each pixel is exposed individually.

entire working area of the hologram since such materials will probably have their greatest direct use in the formation of contact copies into polymer in a scan coped regime. Some of the most apparent difficulties of the process methods are being researched for positive solutions to the processing problem. A second issue of major importance appears to be associated with the current inability of these materials to create uniform developed density when exposed by precision digitally controlled exposure systems. This latter problem poses serious questions over the use of Slavich materials for lithography and digital holography.[10]

Comparison

Finally, a visual comparison is made between the photoresist and the screen output. All the pixels are inspected for brightness and quality as well.

In Summary

- Graphics file can be converted to laser-imaging software. Laser-imaging software controls the exposure of the laser.
- Smooth glass is treated with a photosensitive coating 1.5 microns thick. Photoresist is only sensitive to certain wavelengths of light.
- An interference pattern is created when two of the same wavelengths of light interfere with each other, producing a pattern. This pattern will produce the hologram.
- By using a helium cadmium laser and image file, the hologram is exposed pixel by pixel.
- A 3D object's reflected interference pattern can be recorded. It gives the perception of depth, but is limited in its viewing angle.
- Each individual pixel is exposed individually. Thus, extremely bright colors can be viewed from many angles. It takes a 6"× 6" hologram and 48 hours to expose 360,000 exposures.
- A hologram is usually developed in a chemical bath. The bath removes all areas of the unexposed photoresist, producing the hologram.
- A visual comparison is made between the photoresist and the screen output. Pixels are then inspected for brightness and quality.

Chapter 10, "Electroforming," shows you how the *master* hologram can be copied repeatedly on a variety of formats, depending on your final application. To produce embossed holograms appropriate for high-volume runs, the next chapter will show you how manufacturing facilities that have specialized in electroforming equipment are able to generate metallized shims that preserve the holographic *interference pattern* exactly.

End Notes

1. Mastering International Holographic Paper, Division of Pennsylvania Pulp & Paper Co., 300 Highpoint Drive, Chalfont, PA 18914, USA, 2001.
2. Bernadette L. Olson and Ron B. Olson, "Playmates to Primates," Laser Reflections, Inc., San Jose, CA, USA, SPIE Proceedings Vol. 3358, Sixth International Symposium on Display Holography, Lake Forest College, Lake Forest, IL, USA, Meeting Date: 07/21– 07/25/97, pp. 251–256.

3. Fred D. Unterseher, August Muth, and Rebecca E. Deem, "Silver Halide Masters for Transfer into Large Format Dichromated Gelatin (DCG) Holograms," Lasart Ltd. and Zone Holografix Studios, Glendale, CA, USA; Lasart Ltd. and Zone Holografix Studios, Santa Fe, NM, USA; SPIE Proceedings Vol. 3358, Sixth International Symposium on Display Holography, Lake Forest College, Lake Forest, IL, USA, Meeting Date: 07/21–07/25/97, pp. 328–332.
4. International Holographic Paper, Division of Pennsylvania Pulp & Paper Co., 300 Highpoint Drive, Chalfont, PA 18914, USA, 2001.
5. Kunio Sakamoto, Hideya Takahashi, Eiji Shimizu, Koji Yamasaki, and Takahisa Ando, "Three-Dimensional Display Systems with Holographic Technologies," Osaka City University, Osaka, Japan; Labs. of Image Information Science and Technology, Toyonaka Osaka, Japan; Labs. of Image Information Science and Technology, USA; Masaaki Okamoto, Labs. of Image Information Science and Technology, Toyonaka Osaka, Japan. SPIE Proceedings Vol. 3358, Sixth International Symposium on Display Holography, Lake Forest College, Lake Forest, IL, USA, Meeting Date: 07/21–07/25/97, pp. 232–238.
6. International Holographic Paper, Division of Pennsylvania Pulp & Paper Co., 300 Highpoint Drive, Chalfont, PA 18914, USA, 2001.
7. Tung H. Jeong and Louis M. Spoto, "Real-Time High-Speed Holographic Interferometric Study of a Fuel Injection System," Lake Forest College, Lake Forest, IL, USA; ITW Technology Ctr., Lake Forest, IL, USA., SPIE Proceedings Vol. 3358, Sixth International Symposium on Display Holography, Lake Forest College, Lake Forest, IL, USA, Meeting Date: 07/21–07/25/97, pp. 305–312.
8. International Holographic Paper, Division of Pennsylvania Pulp & Paper Co., 300 Highpoint Drive, Chalfont, PA 18914, USA, 2001.
9. Agfa-Gevaert, N.V., Septestraat 27, B-2640 Mortsel (Antwerp), Belgium, 1999.
10. Nicholas J. Phillips, "Slavich Recording Materials for Holography: An Individual View," De Montfort University, Leicester, United Kingdom, SPIE Proceedings Vol. 3358, Sixth International Symposium on Display Holography, Lake Forest College, Lake Forest, IL, USA, Meeting Date: 07/21–07/25/97, pp. 41–52.

CHAPTER **10**

ELECTROFORMING

In This Chapter

- Theory
- Jig
- Conductivity
- Treatments
- Tanks
- Shim

Electroforming is used to produce a copy, actually, an exact negative of the master hologram called a *shim*. The shim is a hard nickel copy of the delicate master hologram in photoresist as shown in Figure 10-1.[1] This chapter examines the electroforming stage of the holographic process that manufacturers go through to create commercial holograms.

Electroforming Holographic Shim-Making System

The holographic shim is a modular system suitable for the post-holographic origination processing of resist plates through to the production shim stage to suit all embossing methods used. The system is supplied fully operational, reliant only on local services being provided onsite as shown in Figure 10-2.[2] A basic system consists of three modules:

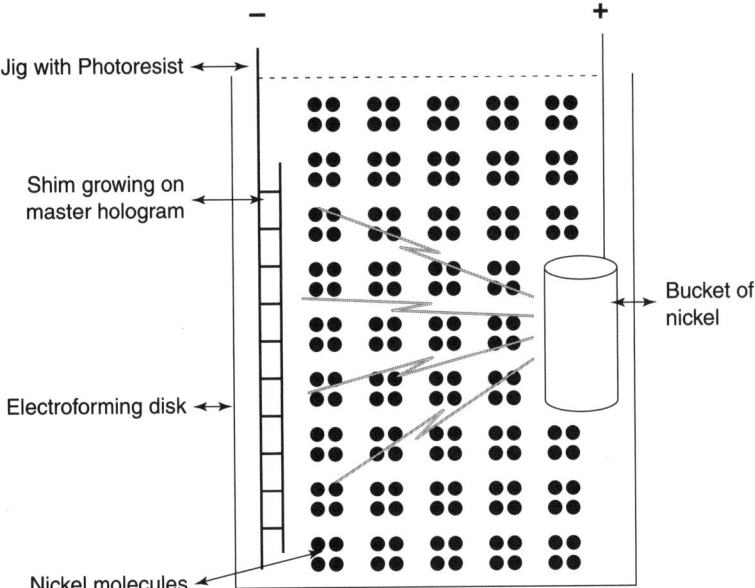

FIGURE 10-1: The electro-deposition of a metal onto a model, capable of conducting electrical current.

- Electroform module
- Processing module
- Passivation Booth/Sink module

ELECTROFORM MODULE

The Electroform module has a 50-gallon capacity. This will be used to process all nickel shim deposits such as Masters, Mothers, and embossing shims as shown in Figure 10-3.[3]

PROCESSING MODULE

Each processing module consist of tank, heater, filter, and pump as shown in Figure 10-4.[4] Modules can be added as desired for larger production requisites.

PASSIVATION BOOTH/SINK MODULE

The Passivation Booth/Sink module is a stand-alone module designed to be integral to the electroform units. Passivation and separation of

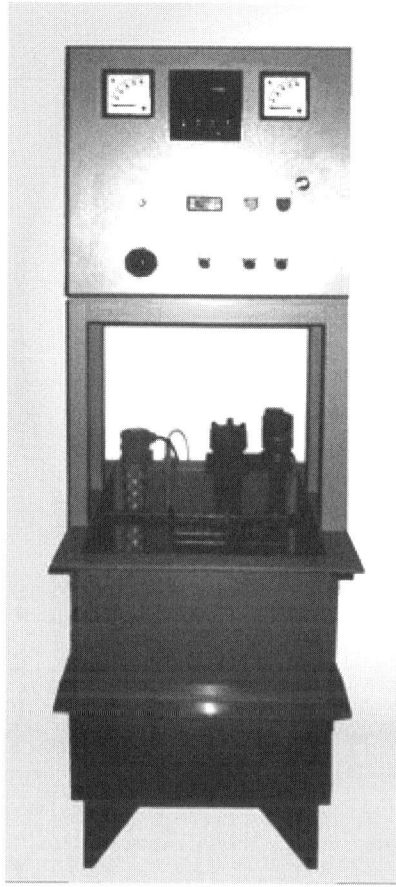

FIGURE 10-2: The basic holographic shim system.

processed shims will all be carried out in this module. Full specifications of the electroforming modules are:

- Color
- PP Beige dimensions: 24 × 24 × 24 inches
- Tank: Reinforced girth; 6-inch V legs; 2-inch flange
- Four stainless steel, PP coated shelf brackets; integral rectifier shelf
- 14-inch shelf brackets height from top of tank
- Shelf: 24 inches wide × 18 inches deep x 2 inches high
- Control Panel: 24 inches height × 24 inches wide × 12 inches deep
- 50 A 25 V air-cooled silicon rectifier rated for continuous duty at 35

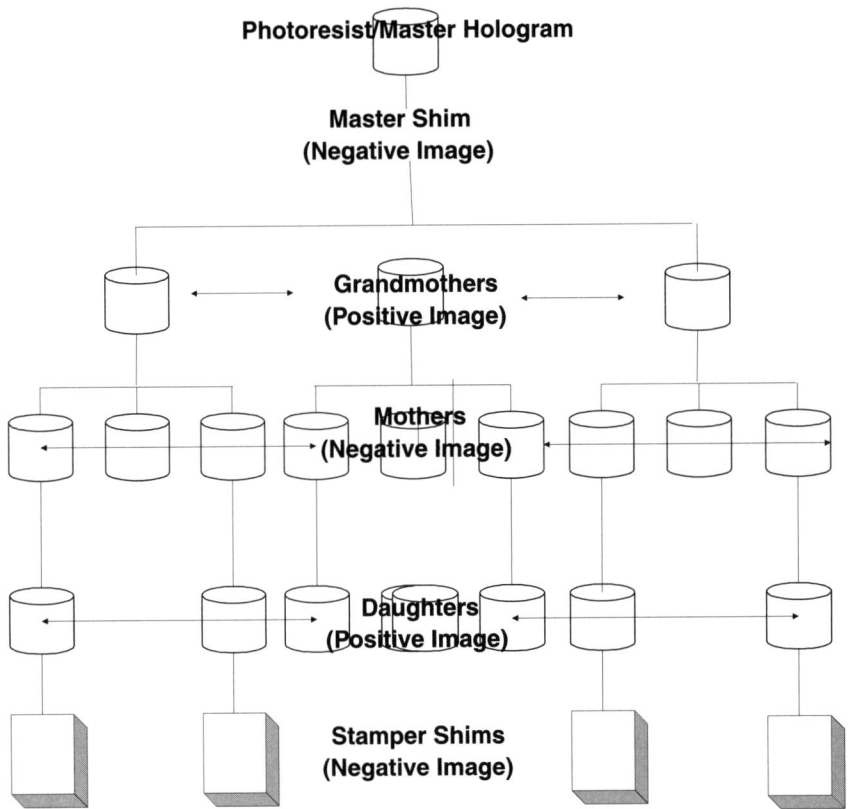

FIGURE 10-3: By repeating the electroforming process, a family of shims is created from the photoresist.

degrees; fitted with analogue ammeter and voltmeter

- On/Off push buttons, contactor primary trip, and fuse protected rectifier assembly; smoothed to less than 5% ripple; output control by hand step-less variac
- Four-digit preset ampere hour/min. meter with countdown LED display, and 8-digit LCD totalizer with reset
- Shutdown and alarm on reaching of ampere meter setting
- On/Off switches for heater, pump, and rectifier; heater controls; and thermostat readout (digital)
- Heater: 2 kW HGP heaters; immersion type with polypropylene mounting guard.[5]

ELECTROFORMING 209

FIGURE 10-4: The photoresist is cleaned and primed for painting, and the silver spray is applied. The jig is lowered into the tank and a current is introduced, causing nickel molecules to be attracted to the photoresist. The thickness of the nickel is determined by time and amperage.

Controls are incorporated in the control panel with low-level protection cut out for the solution. The controls have the following specifications:

- Filter/Pump: In-tank connected system using 10-inch filter cartridge medium.
- 250 gallons per hour solution pumping capacity. A sparge pipe work setup is used for agitation and recirculation of solution.
- On/Off is controlled by switching on the control panel.
- Power: To be denoted on order; either 220 V. 50 Hz, or 110 V. 60 Hz.
- Ancillaries: Plating Jigs are twin bus bar, acrylic framed to suit both the photoresist plate and nickel shim processing by the use of aperture 1/2 frames on an acrylic base plate.
- Anode baskets are titanium mesh with PP bags.
- All chemicals and supplies needed for the normal minimum three months of operation.[6]

PRODUCTION CAPACITY

Using advised processing instructions, 60-micron-thick shims of a size up to 12 × 12 inches can be processed normally in 2.5 hours each. Thicker shim growth can be calculated at approximately 25 microns per hour at 20 A/ft 2 growth rate. Passivation Booth/Sink and color have PP Beige dimensions of: Height 72 inches, Width 24 inches, and Depth 24 inches.

This module contains a one-piece construction of PP sink, with an aperture opening for the passivation process incorporating an extraction fan in the top part of the module. This module is coordinated in design and dimensions to suit the electroforming modules. A spray attachment is fitted to be connected to a suitable de-ionized water supply.

Now, let's look at some specifics of the holographic electroforming process.

The Basic Electroforming Process Facility

At the conclusion of *shooting* a hologram in photoresist, the result is a fragile, positive-reading resist-coated glass plate with 15,000 to 30,000 lines per inch resolution and a .3 micron depth. The electroforming process converts the hologram to metal parts and subsequent *stampers* for use on embossers.

Today, *turn key* production electroform facilities are built for precise and complete hologram conversion from glass resist plates to metal stampers. Now, with remote monitoring and controls capability as shown in Figure 10-5,[7] you can monitor the pH, temperature, amp hours, and the general operating conditions of the tanks. Control, plating speed, circulation pumps and heaters, and the status of the entire system can now be monitored from your computer or from another location via modem. You can safely monitor an unattended system for around-the-clock production. You can also diagnose problems with expert technical assistance via the Internet. A basic facility consists of three modules:

- The *work station*, which houses a stainless steel sink, acrylic hood, hot and cold plumbed water system including de-ionized water, a system for copying photoresist glass in nickel and storage shelves.
- An *unstressed 55-gallon electroform module* to form thicker nickel deposits (masters). The module consists of rectifier, automatic

FIGURE 10-5: Remote monitoring and control of the electroform module.

counter and shut off, heater and thermostat, pump, filter, fiberglass reinforced leak-proof tank, all plumbing, solutions and chemicals.

- A *stressed electroform module* for growing nickel stampers. This module is equipped as the unstressed module except for a different solution composition and inside baffles to control current distribution.[8]

Finally, most manufacturers of these basic facilities should supply you with transformers to convert almost any power to that required by their installation, all hardware, electroforming supplies, and solutions to calibrate the pH meter. Any consultations related to training will be made via telephone or fax. Manufacturers should also be able to arrange for a qualified technician to install the electroforming system at a reasonable cost and provide training at no extra charge to their customers.

In Summary

- Electroforming is used to produce a copy—really an exact negative—of the master hologram. This copy is called a shim.
- Electroforming involves the electro-deposition of a metal onto a model, capable of conducting electrical current.
- The photoresist is mounted into the jig. The jig holds the photoresist in position while in the processing tanks.
- Silver paint is applied to help increase conductivity of the jig and the photoresist.
- Photoresist is cleaned and primed for painting. Silver spray is then applied to the photoresist to make it conductive.
- The jig is lowered into the tank. A current is introduced causing nickel molecules to be attracted to the photoresist. Thickness is determined by time or amperage.
- The jig is removed from the tanks and washed. A straight edge is used to remove the shim from the master hologram. The master shim is then washed, dried, and inspected.
- Finally, by repeating the electroforming process, a family tree of shims is created from the photoresist.

The next chapter discusses holographic embossing. It will show you how to affix the metal shims to a high-speed embossing machine that stamps the holographic pattern into rolls of very thin plastic or foil. The chapter will also show you how various backings and laminations can be applied to the rolls. When properly illuminated, the embossed patterns focus light waves in specific ways to produce a multidimensional image—a hologram!

End Notes

1. "Electroform," International Holographic Paper, Division of Pennsylvania Pulp & Paper Co., 300 Highpoint Drive, Chalfont, PA 18914, USA, 2001.
2. Westmead Technology Ltd, Unit 7, St. Georges Industrial Estate, Wilton Road, Camberley GU15 2QW, England, 1999.
3. International Holographic Paper, Division of Pennsylvania Pulp & Paper Co., 2874 Limekiln Pike, Glenside, PA 19038-2234, USA, 1999.
4. Ibid.
5. Westmead Technology Ltd, Unit 7, St. Georges Industrial Estate, Wilton Road, Camberley GU15 2QW, England, 1999.
6. Ibid.
7. James River Products, Inc., 211 East German School Road, Richmond, VA 23224-1460, USA, 1999.
8. Ibid.

CHAPTER **11**

HOLOGRAPHIC EMBOSSING

In This Chapter

- Paper coating
- Polyester
- Oriented poly propylene
- Embossing machine
- Automatic precision embossing
- Precision embossing of holography and gratings

This chapter briefly examines the embossing stage of the holographic process that manufacturers go through to create commercial holograms. The holographic embossing process can simply be stated in three phases:

- Pre-origination
- Origination
- Manufacturing

Pre-origination

During the pre-origination phase, artwork, film, and the photoresist plate are prepared. Most manufacturers take advantage of the current computer technology that has allowed *them* in a short amount of time to create artwork, to separate colors, and/or to print transparencies. This artwork is eventually transferred to film with the conventional photographic method for the next step in the making of 2D/3D holograms. Furthermore, in order to create a recording of a hologram, the manufacturer would apply the glass plate with photoresist material for laser exposure by using a coating machine. This plate also is cleaned and heated for good quality recording.

Origination

The origination phase is critical for establishing a master hologram and a mask hologram. This phase requires two separate isolation table setups. Both setups will have all the necessary mounts and lenses clamped to the table, but one will use a high-power Argon laser and the other will use a high-power HeCd laser. If the procedures are followed properly, the manufacturer will have two holograms, one visible by laser beam and one visible by white light.

Manufacturing

Last, we have the manufacturing phase. This involves transferring the holographic image from the shim to the desired substrate (paper, polyester, oriented poly propylene (OPP)). Here, a nickel shim can be formed by two electroforming units. Two units are used here with four different chemicals to give an enhanced image of the 2D or 3D hologram. The nickel shim is then placed onto the embossing machine as shown in Figure 11-1.[1] Through pressure rollers, the holographic image is pressed onto the polymer foil. Standalone coating machines are also available to add adhesive to the back of the foil. In addition, an automatic die cutter is provided to allow for the mass production of labels.

HOLOGRAPHIC EMBOSSING 217

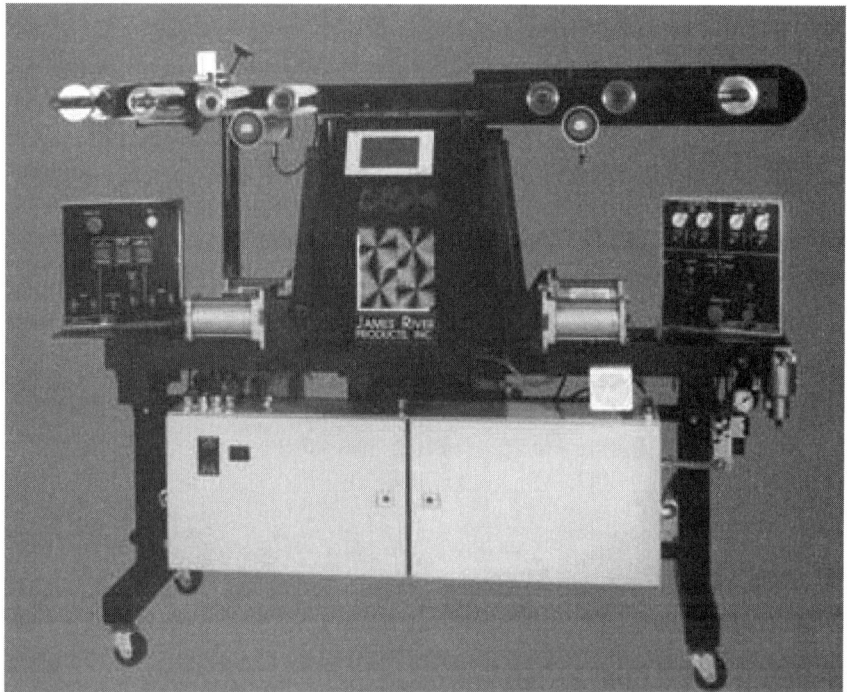

FIGURE 11-1: Automatic precision embossing machine.

PAPER

Paper is normally the substrate of choice for the embossing process. It is first treated with an acrylic coating under low heat. Coating makes the paper smooth and soft.

As the holographic paper is manufactured with a stock pattern, the same is done for a custom hologram. The hologram must be embossed into the paper as it is being manufactured and before it is metallized. The advantages of embossing into paper include low cost, availability, good image quality, and ease of converting.

POLYESTER

Polyester is another substrate used in the holographic embossing process. The advantage of using polyester is that it provides the best possible image quality for 3D holography. The disadvantages are that it is very expensive, and the production sizes are limited.

ORIENTED POLY PROPYLENE

Oriented poly propylene (OPP) is yet another substrate used in the holographic embossing process. The only advantage it has is that it has been approved by the FDA. The disadvantages are problems with converting and printing, and fair image quality.

Embossing Machines

Embossing machines use heat, pressure, and the shim to emboss the hologram into the paper. The thickness of the shim is critical in the embossing process.

The basic steps taken to emboss a pattern or custom hologram are to first unwind a roll of substrate, using paper as an example, through the impression cylinder where the hologram on the shim is embossed into the substrate, again with heat and pressure, and then rewound back into a roll. There are three types of holographic embossing machines in the industry today:

- Automatic precision embossing
- Standard precision embossing of holography and gratings
- Basic embossing machine of holography and gratings

AUTOMATIC PRECISION EMBOSSING MACHINE

The automatic precision embossing machine (see Figure 11-1) offers a constant tension for both unwind and rewind as standard features. It also acts as an edge guide to control the unwinding material's side-to-side variation.

Temperatures are controlled by multifunctioning controllers located in the operator's subpanel. All controls are accessible and conveniently located for uninterrupted production. This type of embossing machine also has an onboard computer for automatic control of most routine embossing functions.

Splice detection will automatically shut down the web for splices and the end of the roll. When this type of embossing machine automatically shuts down, embossing pressure is released by the electro/pneumatic system, avoiding any product or roll damage. Panel repeat (register) is controlled automatically and maintains +/– .0025" repeat variation. Should the repeat spacing exceed the preset control limit, a voice alarm sounds and the machine automatically shuts off.

The visual/audible *eye* also feeds data to the computer, allowing for programmed shutdowns based on image count or footage, with accurate totalizing. Add its communications port and voice communication capability, and this machine is the closest thing to foolproof precision that you will find in optical embossing.

STANDARD PRECISION EMBOSSING MACHINE FOR HOLOGRAPHY AND GRATINGS

The next type of embossing machine is a standard unit for precision embossing of holography and gratings. Automatic tension for both unwind and rewind is standard, as is an edge to control the unwinding material's side-to-side variations. Very precise manual register control is provided to maintain panel-to-panel spacing of 1/2 mm or less. Temperatures are controlled by multifunction indicating controllers located in the operator's subpanel. Everything is very accessible and neatly organized in subpanels to provide the precision control to allow the embossing of precision prototype products: holographic optical backplane and a rotation encoder disc.

Holographic Optical Backplane

In high-performance computing application, data needs to be distributed at high speed between microprocessor boards in rack-mounted systems. Existing Virtual Memory Environment (VME)-based electrical backplanes can operate at frequencies up 640 MHz; however, at higher frequencies both signal density and the distance over which the signal can be sent are limited.

One way to overcome this information bottleneck is to use light to do the signal distribution. A holographic optical backplane is a reflective glass plate that mounts onto the front or back of the processor rack as shown in Figure 11-2.[2] Laser diode transmitters and photodiode receivers are located on the edge of each processor board. Laser light hits the optical backplane and is split by submicron period diffraction gratings embossed onto

FIGURE 11-2: Photo of the embossed optical backplane.

its surface. Light is distributed within the plate by means of total internal reflection.

Another cost-effective method of manufacturing the backplanes is to use a UV embossing process. The process involves high-precision replication of the submicron period gratings (see Figure 11-3) into a thin polymer film coated on the surface of the glass.[3] Pattern transfer is achieved along the full length of a 325mm-long backplane plate.

Rotation Encoder Disc

Optical rotation encoders, as their name implies, are used to measure mechanical rotary motion. They work by detecting the patterns of light and dark reflected from the embossed surface of a spinning encoder disc (see Figure 11-4) mounted on the shaft, whose rotation is being measured.[4] Thousands of units per year are required in a wide range of industrial applications to monitor machinery spinning at high speed. Making the encoder smaller enables it to be integrated more closely with the mechanical shaft, even to the point where the encoder is located inside the head of a motor. However, smaller encoders require finer pitch marks on the encoder disc to maintain a large enough number of counts per revolution.

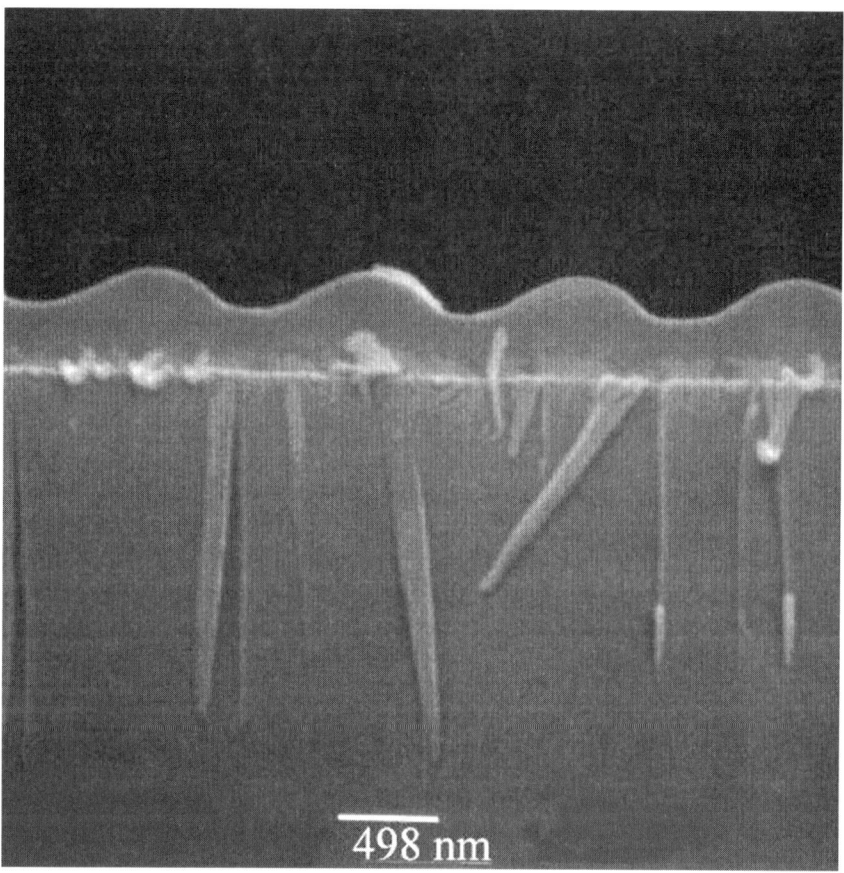

FIGURE 11-3: Scanning Electron Microscope (SEM) photo of embossed surface grating on the backplane.

Therefore, a very accurate (and expensive) master grating, with a period of less than 1m, is made by electron beam lithography and converted into a nickel mould tool by electroforming. Finally, the nickel tool is used to emboss the radial grating pattern onto a large number of encoder discs. The prototype embossed gratings have only 0.01-percent ellipticity, high diffraction efficiency, and reproduce the desired physical dimensions to within 0.03 percent as shown in Table 11-1.[5] Quartz or glass substrates can be used to give the disc improved thermal stability over similar parts made entirely of plastic.

FIGURE 11-4: Photo of embossing shim with embossed encoder discs.

TABLE 11-1: Dimension measurements

Sample	Disc Diameter (mm)	Error in Diameter (micron)	Relative Error in Diameter	Ellipticity (micron)	Relative Ellipticity
UV embossed radial grating on PET film	32.2088 (EW) 32.2053 (NS)	9 5	0.03% 0.02%	4	0.01%
Quartz radial grating made by photo-lithography and reactive ion etching	32.205 (EW) 32.205 (NS)	5 5	0.02% 0.02%	0	0

Note: All dimension measurements made by Centre Suisse d'Electronique et de Microtechnique SA.[6] Relative error in measurement of diameter is 0.015%.

BASIC EMBOSSING MACHINE OF HOLOGRAPHY AND GRATINGS

Finally, there's the basic embossing machine for holography and gratings as shown in Figure 11-5.[7] All of the controls are manual, easily accessible, and quite straightforward. Unwind and rewind tensions are controlled by an easily accessible hand wheel. The temperature of each embossing roll is

FIGURE 11-5: Basic embossing machine.

controlled by multifunction—indicating that the controllers are located in the electronic box beneath the machine. Also, 120 or 240 volts are available to the *die* roll at operator discretion.

In Summary

- The embossing process involves transferring the holographic image from the shim to the desired substrate.
- Advantages of the paper substrate include low cost, availability, and good image quality.
- Paper is first treated with an acrylic coating under low heat. The coating makes the paper smooth and soft.
- The embossing machine uses heat, pressure, and the shim to emboss the hologram.

- Embossing is the most cost-effective way to mass-produce holograms and diffraction gratings.
- Diffraction gratings that are composed of *hill-and-dale* grooves can be pressed into plastic materials, thus replicating the grating.
- Embossed gratings are typically coated with aluminum to enhance the reflectivity and the brightness of the diffraction effect.
- 3D holograms can be made with materials that produce relief structures. These are, in essence, very complex grating structures, so they can be embossed as well.
- Diffraction gratings can be made through interference of laser light or by mechanical ruling or scribing.
- Most gratings made for decorative purposes are made using laser light.

The next chapter discusses metallizing. It will show you how metallized films are routinely used for holographic applications. Some of these might be for security and anti-counterfeiting applications, while others are for graphical use.

End Notes

1. James River Products, Inc., 211 East German School Road, Richmond, VA 23224-1460, USA, 1999.
2. Dr. Tom Harvey, Dr. Tim Ryan, Epigem Limited, Wilton Research Centre, P.O. Box 90, Wilton, Middlesbrough, TS90 8JE, UK, 1999.
3. Ibid.
4. Ibid.
5. Ibid.
6. CSEM Centre Suisse d'Electronique, et de Microtechnique SA, Rue Jaquet-Droz 1, P.O. Box, CH-2007 Neuchâtel, Switzerland, 1999.
7. James River Products, Inc., 211 East German School Road, Richmond, VA 23224-1460, USA, 1999.

CHAPTER 12

METALLIZING

In This Chapter

- Loading chamber
- Vacuum metallizing
- Roll removal
- Coating
- Slit/rewind

Metallized embossing and its variants have proved to be the most cost-effective and popular means for mass replication of holograms and diffraction gratings. Alternatives such as photopolymers or silver halide films can cost eight to ten times as much as metallized embossed materials. The cost of metallized embossed materials has fallen to the point where they are being used for mainstream product packaging such as software packages, food and beverage containers, candy packages, and wrapping paper. Companies can now produce holographic images and gratings at widths of 60 inches and wider, and at speeds of hundreds of feet per minute.

This chapter briefly examines the metallized embossing stage of the holographic process that manufacturers go through to create commercial holograms. The holographic metallized embossing process can simply be stated in five primary phases:

1. **Origination**: The creation of the master plate in photoresist or other relief pattern material.
2. **Electro-forming:** The replication of the original master by electroplating to form embossing plates.
3. **Embossing:** The mounting of a nickel master on a press and making impressions into plastic film or paper.
4. **Metallizing:** The application of a thin layer of aluminum before or after embossing to enhance brightness.
5. **Application:** The conversion of the raw film into a final product by laminating, slitting, and or die cutting.

Origination

Embossed holograms and grating patterns are typically recorded in photoresist. Positive photoresist is used to record the fringe pattern in relief. Areas in the fringe pattern that receive greater energy during exposure will be removed during development to reveal a *hill-and-dale* pattern that can be mechanically reproduced. There are a number of ways to create 3D images and gratings.

WHITE-LIGHT-TRANSMISSION HOLOGRAMS

Variations on a rainbow hologram are used to create transmission holograms that can be embossed and viewed in white light. Holographic stereograms use sequences of two-dimensional images to create 3D images in color and with animation.

HOLOGRAPHIC DIFFRACTION GRATINGS

Holographic diffraction gratings are made by recording the interference pattern of two or more beams of laser light. Portions of a small grating can be replicated and combined to form larger grating patterns. These patterns are made optically, mechanically, or with electron-beam techniques. The mechanical techniques start with a simple grating and replicate it selectively. Each segment of the pattern has a slightly different grating orientation so that the final pattern flashes from all angles. A popular way to make optical patterns is to build up an image by recording thousands of minute, adjacent grating spots to create an image. Each spot can have a different grating orientation or fringe spacing so that different effects can be achieved. Some security images are made by computing fringe patterns and writing the hologram into photoresist with electron beam devices (see the sidebar, "Holographic Security Solutions and Films").

■ **HOLOGRAPHIC SECURITY SOLUTIONS AND FILMS**

Metallized films are routinely used for holographic applications. Some of these might be for security or anti-counterfeiting applications, while others are for graphical use (see Figure 12-1).[1] Using a pattern metallization process, higher security devices can be cost-effectively produced.

Substrates for Soft and Hard Embossing

Polyester and polypropylene are metallized before or after embossing. Widths range from 6–65 inches, with thickness ranging from 48 gauge to 2 mil.

Diffraction Embossed Substrates

Polyester and polypropylene films are available with 18 different patterns. Widths range up to 42 inches, and thickness ranges from 48 gauge to 2 mil.

Grid Overlaminates

Metallized 100+ lines-per-inch grid patterns are available on polyester with widths of up to 38 inches, and thickness ranging from 48 gauge to 5 mil.

FIGURE 12-1: Metallized for film holographic anti-counterfeiting applications.

Pmp **Holograms**

This is patterned metallized polyester film that is combined with a diffraction metallized layer. Low-cost security holograms are available on the market today. They are available in widths up to 38 inches, with a thickness of up to 48 gauge to 5 mil. They can be data encrypted with alphanumeric digital and/or bar codes. Low tooling costs are guaranteed for unlimited volumes.

Affirm **Holograms**

This is a retroflective patterned polyester film laminated or combined with an embossed layer to provide visual authentication of a security hologram by using a point source of light.

Heart Holograms

This is known as a Holographic Electronic Authentication Recognition Tag (HEART). It is a pattern metallized hologram. It has a unique signature when measured or interrogated by radio frequencies, complex impedance, and capacitance.

Hybrids

Hybrids are any combination of the preceding features to produce unique security, detection, authentication, or data encryption.

Electro-forming

However the master hologram is recorded, the result is a relief pattern in photoresist or plastic. The original photoresist cannot be used as an embossing tool; therefore, it is necessary to produce a durable metal replica. The most common procedure for replicating the master uses nickel electroplating. The master must first be made conductive. In one popular method, silver is applied using a dual spray gun containing a two-part solution. Alternatively, vacuum metallizing can be used for smaller parts. Electroless nickel has also been used by some in the field.

Once made conductive, the master is usually immersed in an electroplating bath where it is nickel-plated to a desired thickness. This typically takes six to eight hours. The nickel layer is then peeled from the photoresist master, which is destroyed in the process. The first nickel replica is used to make intermediate masters, which in turn yield the final embossing shims. A given job may require from 10 to 1,000 shims.

Embossing

Most embossing is done from nickel master plates called *shims*. The master shim is mounted on a heated roller in an embossing machine. Under controlled pressure, a second smooth roller presses the material to be embossed into the heated shim. The relief pattern from the shim is pressed into the film or coated paper, which retains the pattern.

FIGURE 12-2: Forty-eight-inch-wide web embossing machine.

Some materials are coated with proprietary coatings before embossing. A number of materials can be embossed directly without coatings. Figure 12-2 depicts a wide web embossing machine.[2]

Metallizing

Most hologram and grating products are vacuum-metallized with aluminum to enhance the brightness of the material. Without metallizing, the diffraction efficiency is typically limited to 30 percent or less. If the relief pattern is exposed to the air, it will be easily damaged by abrasion or contact with oils from the skin: the oils fill in the fringe pattern, thus rendering the image invisible. If the final product must be transparent, the relief pattern must be coated with a very thin layer of metal or a material of significantly higher index of refraction. This coating maintains brightness of the image while keeping it transparent. Security holograms for drivers' licenses and the like are made in this manner.

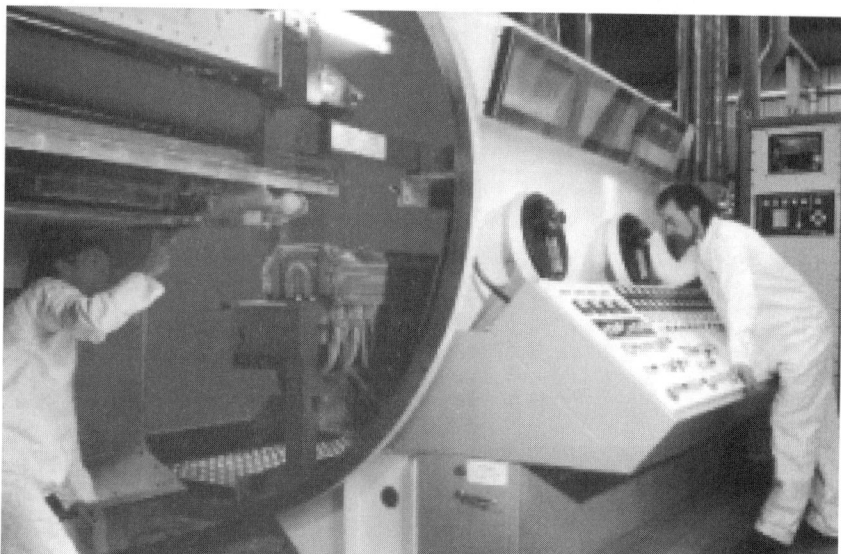

FIGURE 12-3: Vacuum metallizer for plastic film. This photo shows the chamber open for loading rolls of film.

Six- to twelve-inch narrow web embossing is often done with pre-metallized material such as polyester film. Wider film and paper is usually metallized after embossing, using a large vacuum metallizer as shown in Figure 12-3.[3] These metallizers typically cost 2.6 to 3 million dollars.

Application

Embossed gratings and holograms can be applied in myriad ways to nearly any substrate, including:

- Self-adhesive labels
- Lamination
- Hot-stamping transfer processes
- Embossed paper products
- Direct packaging of snack foods

SELF-ADHESIVE LABELS

For the self-adhesive label, there is a paper or plastic *release liner* that is coated with a rubber or acrylic-based adhesive. The coated liner is applied to the roll of embossed film between a pair of pressure rollers. The holograms are die-cut, and the waste film around each label is stripped and discarded. The labels are delivered to the customer in roll form, still attached to the release liner to be machine or hand applied to the final product. Typical uses of holographic labels include:

- Product authentication and anti-counterfeiting, particularly in the software and music industry; and for aircraft and automotive parts.
- Anti-tampering protection to stop clandestine refilling of expensive drugs, cosmetics, or motor oil.
- Marketing applications; for example, attention-getting labels on magazines or product boxes.
- Toys and decorative stickers for household use.[4]

LAMINATION

Continuous roll lamination can be used to apply plastic film onto paper or plastic. The laminated material can be cut into sheets or left in roll form for printing and subsequent conversion into cartons and boxes.

HOT-STAMPING TRANSFER PROCESSES

Hot-stamping transfer foils are made routinely for paper or plastic. The transfer film is specially prepared with multiple layers to release a thin hologram layer when heat and pressure are applied. These foils are used extensively in the printing industry to add sparkle to paperback books. Security holograms are applied to credit cards and to documents using hot-stamping. Some packaging applications using these techniques have been successful, but prove to be too expensive for anything but premium or seasonal use.

EMBOSSED PAPER PRODUCTS

Prismatic diffraction wrapping paper can be made by direct embossing into coated paper or by transferring an embossed layer of lacquer onto a paper substrate. The direct embossing technique appears to be the most cost effective for this purpose.

DIRECT PACKAGING OF SNACK FOODS

Finally, candy bars, chips, and trading cards have been packaged using overprinted embossed film that has been coated with a heat-seal layer. *Pouch* packaging machines use film from a roll and form it into a pouch, which is filled with the product and sealed at high speed.

In Summary

- Embossed holograms exposed to vaporized aluminum become reflective. Roll sizes range up to 92" wide and 72" diameter.
- The roll is then loaded into the vacuum chamber and air is removed from the chamber, creating vacuum.
- Coils are then heated to 2,000°F. A roll is then put into a rewind sequence. Aluminum wire is then vaporized by hot coils.
- Vacuum metallizing: Aluminum molecules adhere to an embossed hologram. Reflective coating allows viewing of all colors and angles.
- The metallized roll is then removed from the vacuum chamber.
- Moisture lost during metallizing is added to the paper. Top lacquer coating makes paper ink receptive.
- Finally, movable cutting blades convert larger rolls into smaller rolls for finishing.

This chapter presented a very basic overview of a complex and growing industry. After the initial popular interest in holography and diffraction grating application in the mid 1980s, the industry has grown gradually to the point where it is cost effective for holograms and diffraction gratings to be used for standard rather than seasonal or premium packaging. Packaging appears to be the application in which there is greatest potential for growth. The next chapter looks at the conversion process.

End Notes

1. "Holographic Security Solutions and Films," Advanced Deposition Technologies, Inc., Myles Standish Industrial Park, Taunton, MA 02780, USA, 2001.
2. CFC International, Inc., Chicago Heights, IL, 1999.
3, Ibid.
4. Ibid.

CHAPTER 13

CONVERTING

In This Chapter

- Lamination
- Optical sheeting
- Cutting
- Fold and glue
- PS Coating

This chapter briefly examines the converting stage of the holographic process that manufacturers go through to create commercial holograms. The converting process (see Figure 13.1)[1] includes techniques such as:

- Die cutting and foil stamping
- Folding
- Gluing
- Laminating
- Sheeting
- Printing
- Holography

FIGURE 13-1: The converting process. Source: ©2001. International Holographic Paper, Inc. All rights reserved.

Die Cutting

Metallized paper is die cut or guillotined like plain paper and board. As with plain paper, lubricating the die blade with a wax-impregnated separator will keep the blade sharp and prolong its life (see Figure 13.2).[2]

Furthermore, a die cutter should address and help solve in an efficient manner the problem of high-volume, high speed, precision die cutting of embossed holography. It is a well-known fact that in any printing/embossing operation, the problem of moving a web of material produced on one rotary piece of equipment to another piece of rotary equipment for additional processing such as *kiss cut* die cutting is formidable. The problem is solved by equipping rotary equipment with two independent die-cutting stations, one for each separate impression made by the embossing equipment. The die stations are, in fact, tuned to a discrete mark made by each impression of an embosser.

Each cutting station on the die cutter should have a dedicated, onboard computer and a precision servo drive motor. Number 1 and 2 die stations should be totally divorced from mechanical connection to the press or to each other. Each station should be completely independent of the press drive train. In operation, each station receives an independent command from the moving web by an *eye mark* signal that was *printed* on the web when it was embossed. The pulse produced by this mark is fed to the appropriate computer, where it is converted to a signal that controls the sta-

CONVERTING 235

FIGURE 13-2: Die cutter.

tion's servo-motor. The motor precisely speeds up or slows to allow the cutting die to arrive at the exact register point as the moving web passes below it.

Tension for precise register is automatically maintained throughout the die cutter by three separate, constant tension controllers for unwind, pacing, and rewind needs. Accurate tracking through the machine is maintained by a totally electric side guide. The guide receives input from the edge of the material by a totally electric side guide. The guide also receives input from the edge of the material.

Foil Stamping

Prismatic Illusions can also be foil stamped and embossed similar to conventional white paperboard. For example, a standalone converting machine like the one shown in Figure 13.3 is designed to complement the embossing equipment.[3] This machine will apply *size* (adhesive) coating to hot-stamping foils, duplex slit to 1/4 inch, apply two-side adhesive tape, apply thermal-activated textile adhesive, and apply (and dry) colored coatings using applicator rods.

This is usually available with a *hot melt extrusion system option*, including a heated slot die extrusion head, heated hose, quick recovery melting pot, variable speed pumps, interface to converting machine drive train, and quick change fittings. Also available is an *electric side guide* with a *constant tension unwind and rewind option*. Unwind and rewind tension is controlled smoothly by an electric brake and a constant tension transducer system.

Folding and Gluing

Prismatic Illusions can be folded following normal paperboard procedures. At that point in the process, a glue with a pH of 9 or lower is suggested. Metallized paper labels will perform on most standard gluing equipment.

Laminating

Paper and foil add strength to paper board. Pressure-sensitive coating is applied during lamination, which could be tinted to alter the color of the hologram.

Sheeting

Sheeting involves slitting and trimming the roll. The roll is then cut into sheets for the printing phase. This process can also involve optical sheeting

CONVERTING 237

FIGURE 13-3: Standalone converting machine.

and slitting, where eyemarks are used to sheet to a specific place on each roll. This will produce registered sheets.

For example, the holographic finishing machine shown in Figure 13-4 represents a treatment of the standard vertical ram/punch die cutter.[4] Standard features for this machine include a reinforced cutting assembly made out of precision ground; one inch thick steel plates; electric side guide; splicing platform; optical registration (with an audible signal feature) controlling the web feed and stamping foil feed; CRT computer

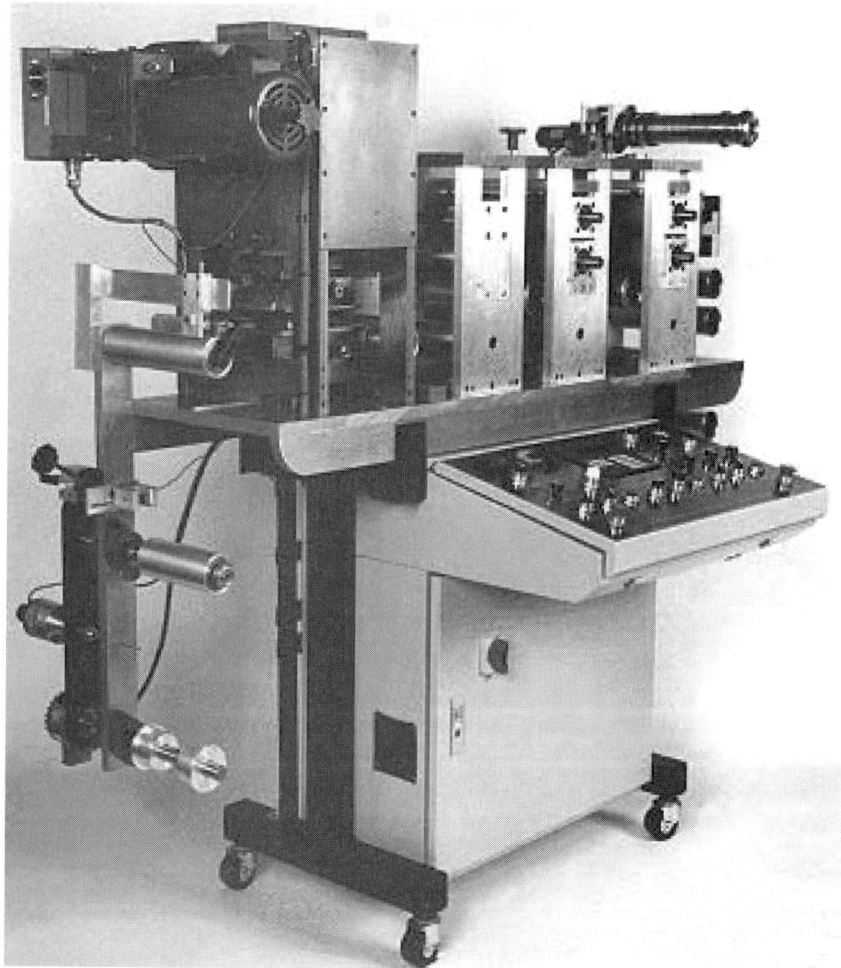

FIGURE 13-4: Holographic finishing machine.

monitor with touch entry pad for control data entry; and independent duplex rewinds. The die mount is adjustable left/right, back/forward, and rotatively for perfect alignment. The spherical cutting base allows the flat and parallel cutting surface to be infinitely adjusted for repeatable cutting or stamping. All idlers that have a material wrap around them are a minimum diameter of 3" to eliminate any curl or delamination problems.

The die-cutting head is interchangeable to a hot-stamp head in less than 10 minutes. The controls for temperature and registration of stamping foil and moving web are permanently mounted in the control panel. Stamping a preprinted die-cut web is quite easily accomplished. To make this machine simple to operate in remote areas, a universal transformer is included to handle incoming power specifications.

Printing

Printing involves the use of a high-quality four-color process, and opaque white adds to the printing quality of *Prismatic Illusions* paper and boards. Registered overprinting on a custom prismatic background is another frequently used technique.

Holography

Finally, the use of holography is not only effective and eye-catching, but also cost effective as previously stated. It achieves incredible results that justify use of holography in most projects.

In Summary

- The converting process techniques may include lamination, optical sheeting, cutting, fold and glue, and PS coating.
- With regard to the lamination stage of the converting process, paper board adds strength to paper and foil. Top coat applied during lamination could be tinted the tint color of the hologram.
- Sheeting involves slitting and trimming the roll. The roll is then cut into sheets for the printing phase.
- Printing involves a high-quality four-color process using opaque white. It also involves a registered overprint on a standard refractive or prismatic background.
- Finally, holography is not only effective and eye-catching, but also cost effective. Desired results justify the use of holography.

Chapter 14, "Products," deals with *products* stage of the commercial holograms process. We'll look at the creation of many unique holographic projects for products and customers around the globe.

End Notes

1. "Converting Process," International Holographic Paper, Division of Pennsylvania Pulp & Paper Co., 300 Highpoint Drive, Chalfont, PA 18914, USA, 2001.
2. James River Products, Inc., 211 East German School Road, Richmond, VA 23224-1460, USA, 1999.
3. Ibid.
4. Ibid.

CHAPTER 14

PRODUCTS

In This Chapter

- Holographic tags, labels, and stickers
- Folding cartons
- Specialty papers and boards
- Point of purchase (POP)
- Stock *Prismatic Illusions*® holographic paper and board
- Custom manufacturing

Many unique holographic projects have been created for products and customers around the world. This chapter examines the products stage of the holographic process that manufacturers go through to create commercial holograms. Let's look at some of the exceptional holographic projects that have been created for the following areas:

- Stock *Prismatic Illusions* holographic papers and boards
- Specialty papers and boards
- Holographic tags, labels, and stickers
- Custom manufacturing

Stock Prismatic Illusions Holographic Papers and Boards

Stock *Prismatic Illusions* holographic papers and boards consist of sheets and rolls—from cartons to skids as shown in Figure 14-1.[1] *Prismatic Illusions* paper and board quantities and sizes are listed in Tables 14-1 and 14-2. The holographic paper is specifically designed for conventional printing. See Figures 14-2 and 14-3 for further examples of stock *Prismatic Illusions* holographic papers and boards.[2]

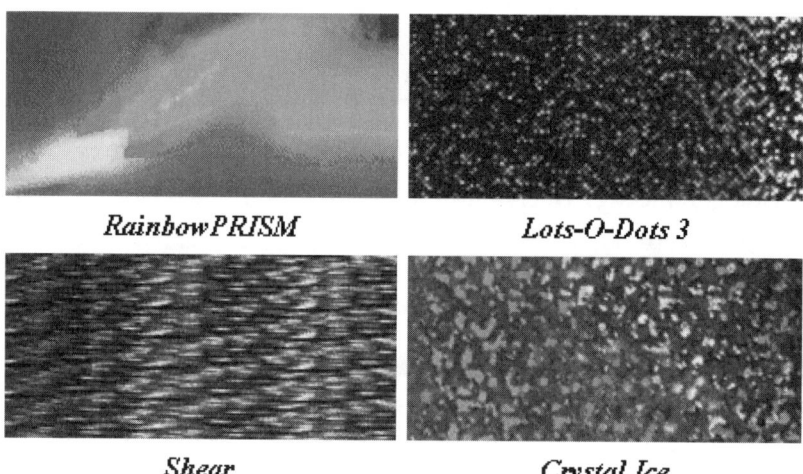

FIGURE 14-1: Stock patterns: Two-dimensional holographic paper. All items are prismatic on one side. Source: ©2001. International Holographic Paper, Inc. All rights reserved.

TABLE 14-1: Text Paper—Sheets

Nominal Bs Wt.(25 x 38) C/1/S	Approx. Caliper	Stock Size	Sheets per Carton	Patterns
85 C/2/S	.0042	28 x 40-200M	250	Rainbow
75 C/1/S	.0042	28 x 40-177M	250	Rainbow, Lots-O-Dots 3, Crystal Ice 3, Shear

Note 1: Bs 75 text uncoated back & bs 85 text coated back

Note 2: Text paper—rolls. Patterns: Rainbow, Lots-O-Dots 3, Shear & Crystal Ice 3 BS. 60 (25 x 38) uncoated back rolls 30.5" wide/ 3" core = 4,000 lin. ft. rolls 30.5" rolls available for pressure sensitizing & laminating.

TABLE 14-2: Paper board—sheets: Caliper .015 prism one side fully coated back

Stock Size	Caliper	Sheets Per Carton	Pattern
28 x 40 - 620M	.015	100	Rainbow, Shear, Lots-O-Dots 3, Crystal Ice 3

FIGURE 14-2: Coca Cola Ltd Real Summer: With the use of *Prismatic Illusions* Shear pattern and opaque whites, this 8' point-of-purchase (POP) display looks like the family is really swimming in a pool while enjoying a Coke. Source: ©2001. International Holographic Paper, Inc. All rights reserved. Image: ©2001. Coca-Cola, Ltd., Inc. All rights reserved.

FIGURE 14-3: Molson's added another sparkle to their beer and their marketing campaign years ago when they incorporated holographic paper and boards into the plan. Source: ©2001. International Holographic Paper, Inc. All rights reserved. Image: ©2001. Molson's. All rights reserved.

Specialty Papers and Boards

Stock and manufacturing of specialty papers and boards consist of such items as:

- **#1 Grade Premium Mirror Board**: Clear as glass. It looks like a mirror. You can actually see yourself using this product. It consists of a silver reflective top with a fully coated back ready for printing. It comes in 28" × 40" sheets—800M; with an approximate caliper of .016 pt; in 100 sheets per carton.

- **#3 Grade Mirror Board**: It consists of bright reflective silver top with a fully coated back. It comes in 28" × 40" sheets—500M; with an approximate caliper of .016 pt; in 100 sheets per carton.
- **Colored holographic stock**: It's easy to change the silver holographic stock to any color you choose by adding a die to the acrylic finish coating at the end of the manufacturing process (see Figure 14-4).[3] You can have rolls or sheets of colored holographic paper and board. The custom color matching4 process ensures an accurate color from beginning to end.
- **24K gold paper**: It's a 24-karat gold paper board at $96.00 a sheet. It's not cheap, but it's real gold and ready to print.
- **Litho Fuzz**: It's white, it's fuzzy, and it doesn't come off on printing blankets. You can print on it and create holographic projects.

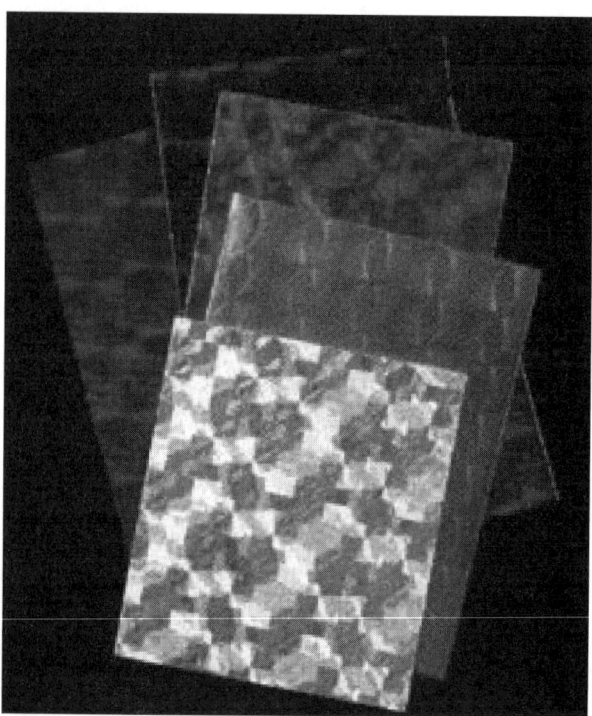

FIGURE 14-4: Colored holographic stock. Source: ©2001. International Holographic Paper, Inc. All rights reserved.

- **Litho Wood**: It's real wood and ready for printing.
- **Iridescent paper**: It looks like mother of pearl; changes pastel colors with light; has a soft opal likeness; and it's a film laminated for printing. It's also available in multiple colors, including black and satin!
- **Litho Canvas**: It's got the canvas look and feel, and it's still printable.
- **Clear holographic film**: It's a very subtle holographic touch for use in window box applications. You can see through it, but it will still refract light.

An example of specialty papers and boards are Topps trading cards as shown in Figure 14-5.[4] With years of requesting something different, something new, Topps had continued to create custom holographic trading cards for collectors of all ages. Baseball, football, hockey, basketball, and more are all incorporating the use of a multiple pattern holographic board, custom holography with registered lithography, and many different substrates to create an exciting market for trading cards.

Still another example of specialty papers and boards is the packaging for CD-ROM boxes (see Figure 14-6).[5] It is very important to use a holographic Rainbow Board for this box and matching disc holder.

FIGURE 14-5: Topps trading cards. Source: ©2001. International Holographic Paper, Inc. All rights reserved. Image: ©2001. Topps, Inc. All rights reserved.

FIGURE 14-6: X-Files unrestricted access. Source: ©2001. International Holographic Paper, Inc. All rights reserved. Image: ©2001. Fox Studios, Inc. All rights reserved.

Holographic Tags, Labels, and Stickers

Custom holographic image labels (see Figure 14-7) can be produced in a variety of configurations and patterns with narrow web holographic embossing presses.[6] You can create your own design tags and labels this way.

Almost any holographic image label, tag, or sticker you can imagine can be produced by the narrow web holographic embossing presses, such as:

- Small-volume custom labels up to 6" × 12", 6 colors
- Large-volume custom labels up to 30" × 48"
- Pressure-sensitive P/S, die cut, roll or sheet form

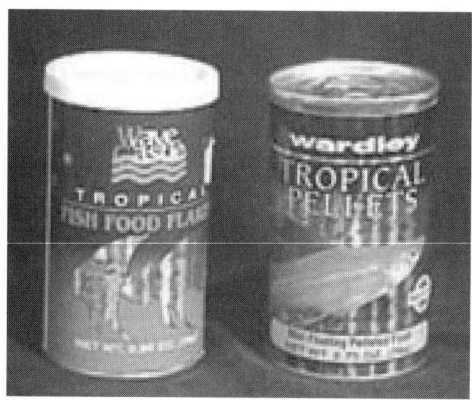

FIGURE 14-7: Wardley food labels, from fish to frog foods. Source: ©2001. International Holographic Paper, Inc. All rights reserved. Image: ©2001. Wardley Foods. All rights reserved.

- Ready for hand or machine applications
- Tamper evident- authenticate/tamper proof available
- Security tags and labels
- Permanent, removable adhesives
- Fast turnarounds, low volumes
- Multiple color registered printing available
- Sequential numbering available[7]

Diverse label products benefit from *Prismatic Illusions*, such as beer bottles, paint cans, fish food, promotional items (see Figure 14-8)[8], and product authentication.

Custom Manufacturing

Finally, custom manufacturing is specifically designed for commercial printing (see Figures 14-9 to 14-12).[9] Paper and board can be manufactured in many different sizes and basis weights to suit your needs, such as:

- Rolls up to 92" wide.
- Sheets up to 54" × 77".
- Bs. 38 lb rolls to 85 lb rolls available C/1/S coated one side and C/2/S, partial wet strength.

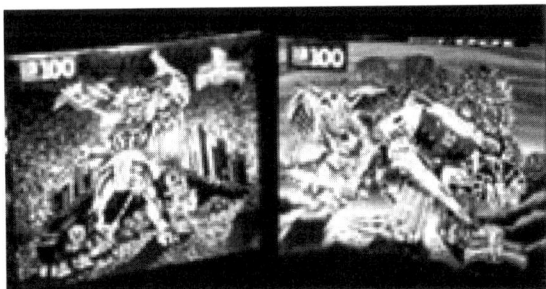

FIGURE 14-8: Milton Bradley puzzles and kids toys: prismatic puzzle pieces for Star Wars, X-Men, and Batman became hot sellers in the toy industry. Other children projects: Barbie accessories, Matchbox cars, POGs, stickers, and lots more. Source: ©2001. International Holographic Paper, Inc. All rights reserved. Image: ©2001. Milton Bradley. All rights reserved.

FIGURE 14-9: Paramount's *Star Trek Generations* video box exploded off the shelves. Designed with a custom hologram, this video box has now become a collector's item. Source: ©2001. International Holographic Paper, Inc. All rights reserved. Image: ©2001. Paramount Pictures. All rights reserved.

FIGURE 14-10: Skynet and Daggerfall were designed with the use of custom holography to create a full cool cover for their CD-ROM boxes. Source: ©2001. International Holographic Paper, Inc. All rights reserved.

- Board available from 7 pt. to 80 pt.
- Prismatic two-sided material.
- 200+ patterns, or create your own using our 2D dot matrix technology.
- Repeats: 36", 40", and 48". These repeats create a seam line. Please consider when laying out your design piece. Seamless sheets are available in different sizes and patterns, but must have sheet length of 36", 40", or 48".

FIGURE 14-11: M & M Mars Candy Cart: It's to date one of the world's largest custom 2D holograms ever created at 24" x 30" tall. This four-sided Candy Cart was shot using a 25dpi holographic Lots-O-Dots background and overlaying 100dpi custom-designed candy logos. Over 5 million holographic pixels were used to create this hologram. Source: ©2001. International Holographic Paper, Inc. All rights reserved. Image: ©2001. Mars, Inc. All rights reserved.

FIGURE 14-12: *X-MEN* Comic Books—MARVEL'S use of custom holography has added eye-catching designs that make these comic books come to life! Comic book values trading over 10 times the newsstand price are not uncommon when using holography. Source: ©2001. International Holographic Paper, Inc. All rights reserved. Image: ©2001. Marvel Comics, Inc. All rights reserved.

- Minimum quantities are required for custom orders (approx. 10,000 sheets 28" × 40" as a guideline).
- Manufacturing lead time is approximately 4–6 weeks; check with IHP for current lead times.
- Specialty substrate would include various types of paper, such as partial wet strength, synthetic, and some BGA and FDA approved grades.
- Other materials include OPP, BOPP, polyester, and PVC.
- Custom mastering of holograms: Four digital hologram labs for dot matrix holography, and 2D, 2D-3D, multichannel and stereograms.[10]

In Summary

- Diverse label products benefit from *Prismatic Illusions*: beer bottles, paint cans, and fish food.
- Holographic folding cartons acquire the necessary shelf presence. *Prismatic Illusions* holographic paper board uses traditional converting techniques.
- *Prismatic Illusions* paper board can be laminated and die cut without special requirements. It's a cost-effective substrate with visual appeal.

Chapter 15, "Patterns," deals with the *final* stage of the commercial hologram process: *patterns*. We'll look at how the defocusing of the correlation plane decreases the shift invariance of the correlators—thus increasing the number of patterns that can be stored in each correlation operation.

End Notes

1. International Holographic Paper, Division of Pennsylvania Pulp & Paper Co., 300 Highpoint Drive, Chalfont, PA 18914, USA, 2001.
2. Ibid.
3. Ibid.
4. Ibid.
5. Ibid.
6. Ibid.
7. Ibid.
8. Ibid.
9. Ibid.
10. Ibid.

CHAPTER **15**

PATTERNS

In This Chapter

- Holographic pattern recognition
- Optical holographic correlators
- Fresnel correlators
- Eyeball topographer
- Autonomous vehicle path following
- Target pattern recognition and tracking

This chapter briefly examines the patterns stage of the holographic process that manufacturers go through to create commercial holograms. Let's look at the following pattern recognition projects:

- Optical holographic correlator
- Fresnel correlator
- Real-time eyeball topographer
- Robot navigation using a peristrophic holographic memory
- Target recognition and tracking

Optical Holographic Correlators

An optical holographic correlator is very similar to a holographic memory system, except that the hologram is reconstructed with a signal beam

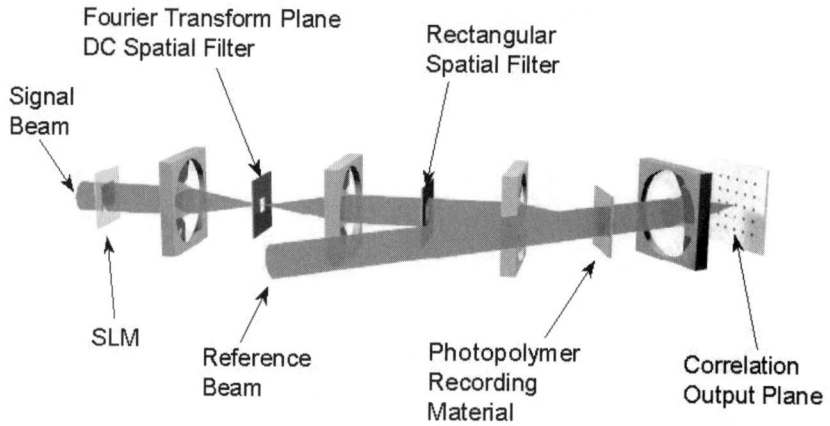

FIGURE 15-1: Holographic correlator system.

instead of with a replica of the reference beam used for recording. An example of a correlator system is shown in Figure 15-1.[1] First, an image is presented on a liquid crystal spatial light modulator. Then, a pair of lenses combined with spatial filters is used to perform edge enhancement and imaging of the Fourier transform onto the holographic recording media. A reference beam is then used to record a hologram of this template image. Multiple holograms can be recorded by varying the angle of the reference beam with respect to the recording material.

When the correlator system is used, the reference beam is shut off and new images are presented on the spatial light modulator (SLM). The image on the SLM is then correlated with all the template images that were stored as holograms (see Chapter 21, "Holographic Storage Systems: Converting Data into Light," for further information). At the correlation plane, an array of correlations is present, each at a position related to the reference beam angle used to record the particular template image. Since hundreds (or even thousands) of holograms can be recorded at a single location of material, and an off-the-shelf SLM can operate at 30Hz, this system can perform tens of thousands of correlations per second. The output of the correlator system is captured by a camera and digitized for analysis by a computer. Since the template images are known ahead of time, real-time image recognition and object tracking can be performed.

Next, we'll look at the Fresnel correlator, which is a variation of the optical correlator in which shift-selectivity can be controlled. One of the primary uses of the optical correlator system is in *target recognition and tracking*, which is discussed later in the chapter.

Fresnel Correlators

Holographic correlators can implement many correlations in parallel. For most systems, shift-invariance limits the number of correlation templates that can be stored in one correlator. This is because the output plane must be divided among the individual templates in the system. When the system is completely shift-invariant, the correlation peak from one correlator can shift into an area that has been reserved for a different template. In this case, a shifted version of one object might be mistaken for a well-centered version of a different object. This part of the chapter describes the experimental test of a technique to control the shift-invariance of a correlator system by moving the holographic material away from the Fourier plane.

The experimental setup for the correlator is shown in Figure 15-2.[2] An image of random white and black rectangles shown in Figure 15-3 was

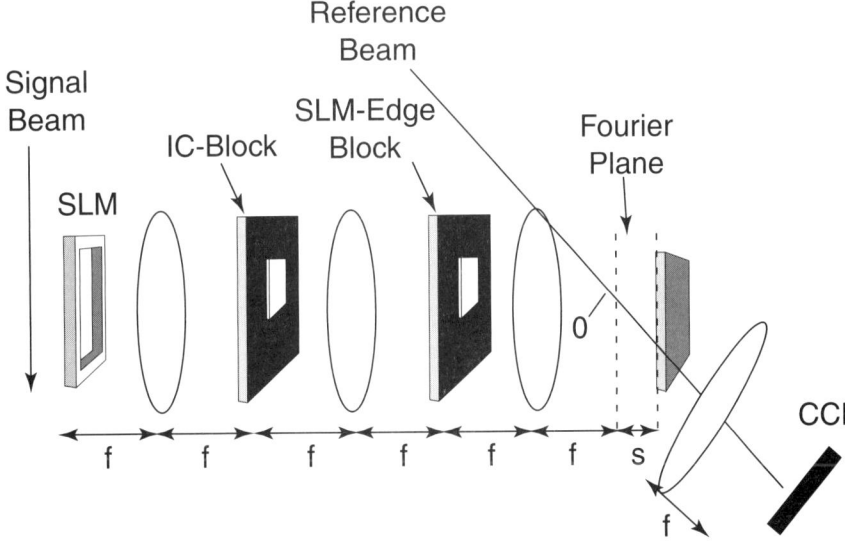

FIGURE 15-2: Experimental correlator setup.

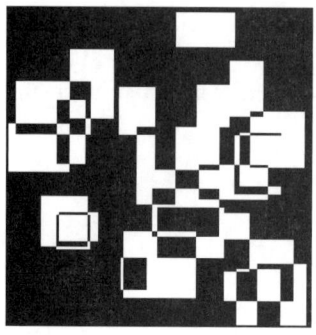

FIGURE 15-3: Randomly generated image 120 by 120 pixels in size.

displayed on a portion of the liquid crystal spatial light modulator that has a resolution of 640 by 480 pixels and a 24mm pixel pitch.[3] A device control (DC) block in the Fourier plane of the first lens edge enhances the image before correlation. The filter behind the second lens blocks the edges of the SLM, created by the edge-enhancement process of the DC block. If not blocked, the SLM creates an undesirable constant DC offset to the strength of the correlation, regardless of what image is presented on the SLM. The holographic material (a 250mm-thick LiNbO3 crystal) is mounted on a motorized translation stage to enable computerized control of the location relative to the Fourier plane. The signal beam is coincident with (and the reference beam at a 25 degree angle to) the recording material surface normal. A lens is placed along the path of the reference beam, and in its back focal plane, a Charged-Coupled Device (CCD) camera is used to capture the intensity and position of the correlation peak. The video signal from the CCD camera is digitized and analyzed by a computer.

For each hologram displacement distance Z_c, a hologram of the input pattern centered on the SLM is recorded. After recording, the reference beam is turned off and the image on the SLM is correlated with the stored hologram. The input image is shifted, electronically, on the SLM. The image is first shifted horizontally (the in-plane direction) while centered vertically. For each horizontal location, the peak of the correlation and its location on the CCD is measured. The image is then shifted vertically (the out-of-plane direction) while centered horizontally, and again, the peak intensity and position are measured. The correlation measurements are taken under a very weak illumination to prevent saturation of the CCD and erasure of the hologram.

Figure 15-4 shows typical curves of peak intensity versus image location for both horizontal and vertical displacements.[4] The shift-selectivity is

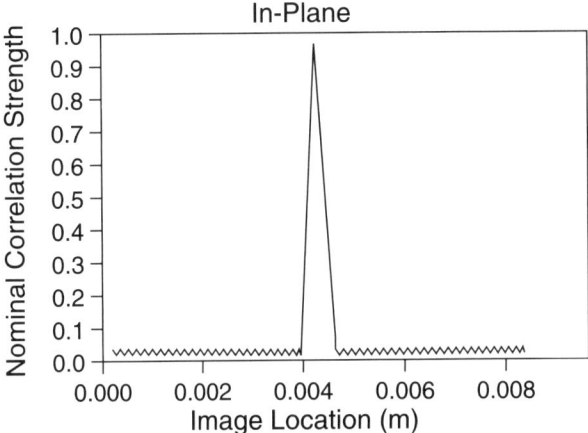

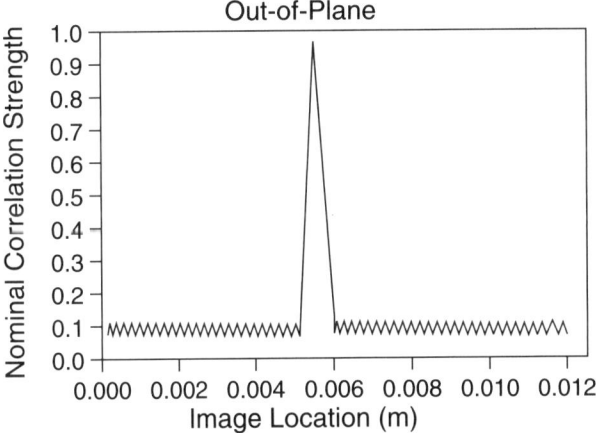

FIGURE 15-4: Correlation strength versus image displacement for both in-plane (top) and out-of-plane (bottom) shifts.

measured as the width of the curve when it attains half of its maximum value. Plots of the shift-selectivity for both the in-plane and out-of-plane directions together with the theoretical predictions are shown in Figure 15-5 as functions of the recording material location relative to the Fourier plane.[5] The correlation integral derived in the preceding section was computed with a Monte Carlo technique with experimental values for beam angle (25 degrees), material thickness (250mm), and index of refraction (2.24).

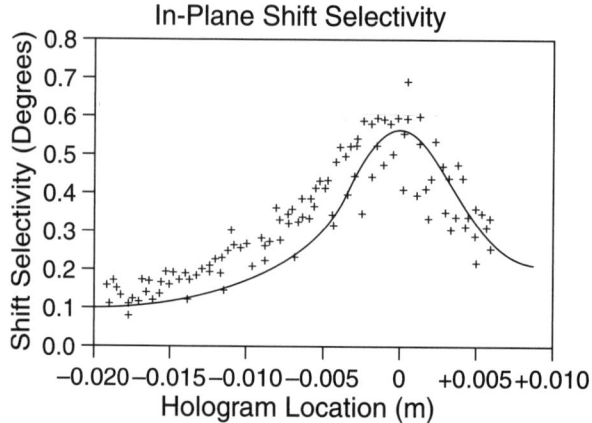

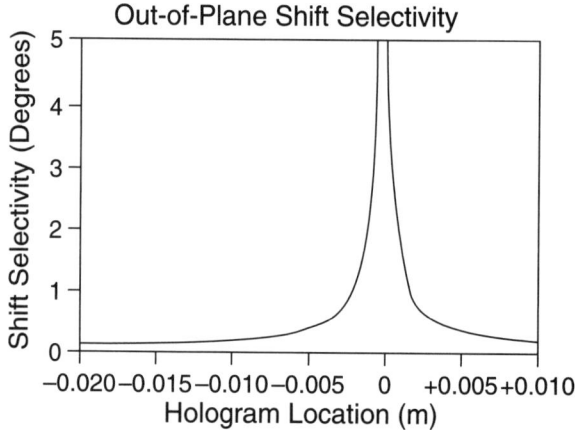

FIGURE 15-5: In-plane (top) and out-of-plane (bottom) shift selectivity with the recording material displaced from the Fourier plane.

The experiment agrees well with the theoretical calculations over a large range of material displacements. Theory and experiment deviate most for out-of-plane shifts close to and at the Fourier plane, where the predicted value of the shift-invariance shoots up to 25 degrees. Figure 15-5 does not contain the full vertical range of the theoretical curve so that the details of the wings would be evident.

Correlations *a* and *b* in Figure 15-6 show the output from an array of 81 correlators stored in 250mm LiNbO3 displaced 1 cm in front of the Fourier plane.[6] Only two different faces were used as templates, in an alternating

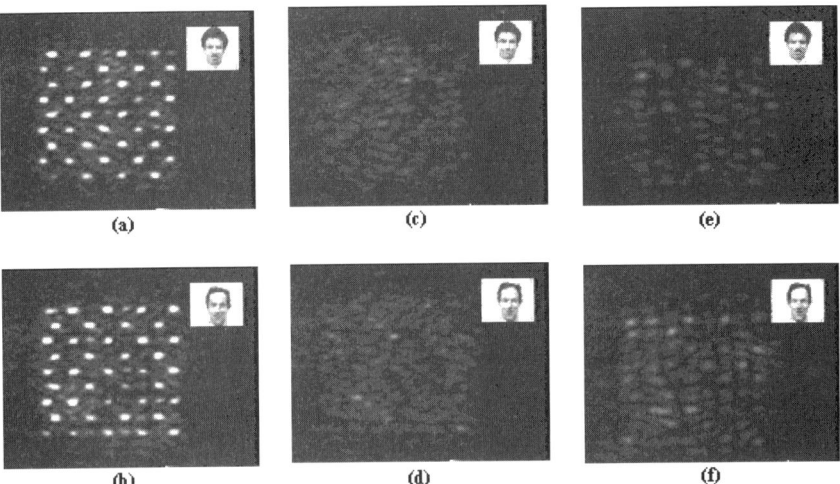

FIGURE 15-6: Array of correlations.

fashion, so that the overall array size could be easily viewed. In this experiment, the central reference beam angle was 50, and 0.08 separated each reference beam. Correlations c and d show the output when the input images are shifted just enough so that their correlation peaks would fall in the area reserved for the neighboring template. The peaks have disappeared, as intended, due to the positioning of the hologram in the Fresnel zone. Correlations e and f show the output when the holograms are stored and used in the Fourier plane. In this case, the Bragg-selectivity is not enough to prevent the sidelobes from interfering with neighboring templates. Moreover, the output of the system is noisy even for well-centered input images. This shows that by using the Fresnel correlator system, more correlators can be stored than would be possible in the Fourier plane. Figure 15-7 shows cross sections of auto-correlations for both the Fresnel and Fourier plane holograms.[7] The sidelobes of the Fresnel hologram are clearly suppressed relative to those for the Fourier hologram.

Eyeball Topographer

Conoscopic holograms are produced by the interference between the ordinary and extraordinary light modes in a birefringent crystal (calcite). The center of the eye is placed at the focal plane of the lens as shown in Figure

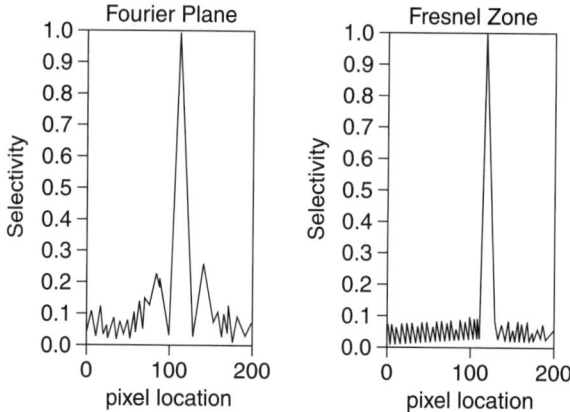

FIGURE 15-7: Auto-correlation cross sections.

15-8.[8] The light reflected from the eye is collimated using the lens. The birefringent crystal produces an interference pattern, which is the overlap of two spatially shifted wavefronts. This interferogram is called a *shear* interferogram. It contains information on the phase distribution of the wavefront, which in turn conveys information on the topography of the eye's surface. The conoscopic topographer offers advantages over other methods:

- It operates with partially coherent light, which reduces speckles due to surface roughness such as a de-epitheliated cornea.
- The depth resolution is in the order of a fraction of a wavelength.
- The alignment is relatively insensitive to mechanical movements, since the two interfering beams are produced and recombined at the calcite faces without the need for additional optics.[9]

One of the challenging tasks is to get the wavefront phase from the shear interferograms. For that, one must make use of a property of Hermite polynomials: the sheared interferogram is decomposed in Hermite polynomials up to order n = 8, since the cornea's shape is usually fitted with polynomials up to this order. A mathematical relation exists between the Hermite coefficients of the shear interferogram and the Hermite coefficient of the true wavefront. This relation is recursive, and the nicest property is that the complexity is O(n), thereby yielding a fast reconstruction.

As a first step, the phase of the sheared interferogram is measured using a phase shifting method involving four rotations of the crystal (or equiva-

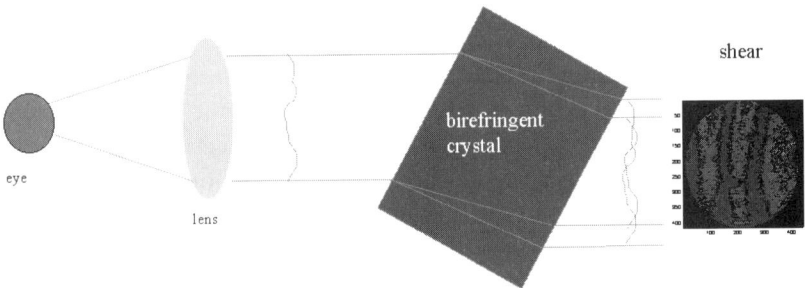

FIGURE 15-8: Conoscopic hologram.

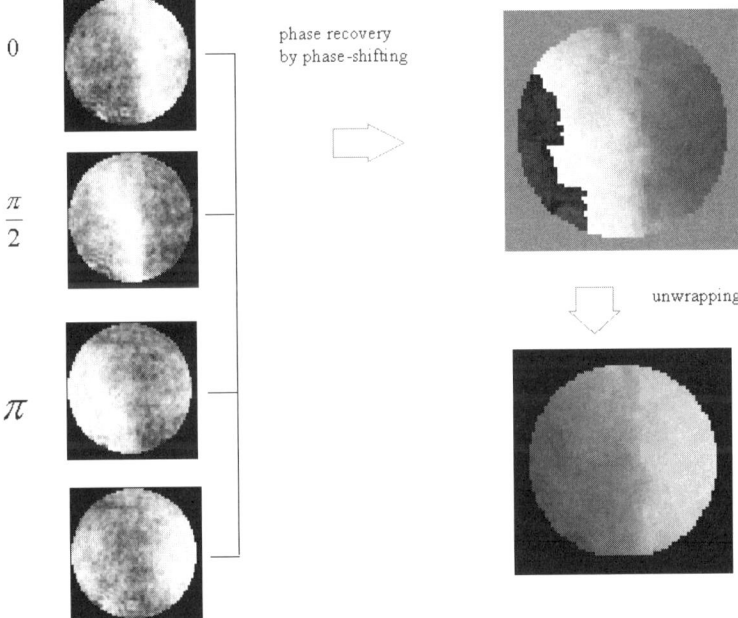

FIGURE 15-9: Holographic phase extraction.

lently the beam angle upon a fixed crystal). The phase extraction is shown in Figure 15-9.[10]

The second step involves the decomposition of the recovered sheared phase into Hermite polynomials. The true wavefront is computed using the preceding method.

MOTIVATION

Vision defect correction by corneal reshaping is becoming safer and more popular. For example, the relatively new photorefractive keratectomy (PRK) method is routinely performed on myopic and astigmatic patients. Corneal shape measurements pre-operatively and post-operatively are currently done with commercial topographers. However, these topographers cannot be used during surgery due to the computationally demanding algorithm. In this part of the chapter, a novel type of corneal topography system has been presented based on conoscopic interferometry. Z-axis resolution is on the order of half a wavelength (0.5 micrometer), and 2,500 points on the cornea are computed in half a second.

ACHIEVEMENTS

The results show that the error on the surface using conoscopic holograms is less than 0.5 micrometers in comparison to a Michelson interferometer taken as reference. The surface area was 2×2mm on a 8mm radius contact lens-like surface made out of polymethylmethacrylate (PMMA). Twenty-five hundred (2,500) points were extracted in 500 ms.

The method developed applies well for spherical reflective surfaces. The effect of eye drying (pig's eyes in vitro) is shown Figure 15-10.[11] After de-epitheliation of the eye, the surface becomes slightly rough and scattering. Therefore, the interferograms of the eye taken at different times show degradation due to drying. The eye is wet at t = 0, yielding a good interferogram, but it degrades quickly after a few seconds. The inteferogram shows no fringes after a minute, although the source used was broad band to limit the speckles. This shows that the surface is too rough to colimate the beam after the lens.

Autonomous Vehicle Path Following

In recent years, there has been a resurgence of interest in holographic memories. Most of the recent experiments in holographic storage have been in LiNbO3, in which up to 10,000 holograms have been stored in one location, or the DuPont photopolymer in which 1,000 holograms were stored. A technique called *peristrophic multiplexing* was combined with conventional angle multiplexing to store the 1,000 holograms in the polymer, which has a thickness of only 100 microns. Most of the develop-

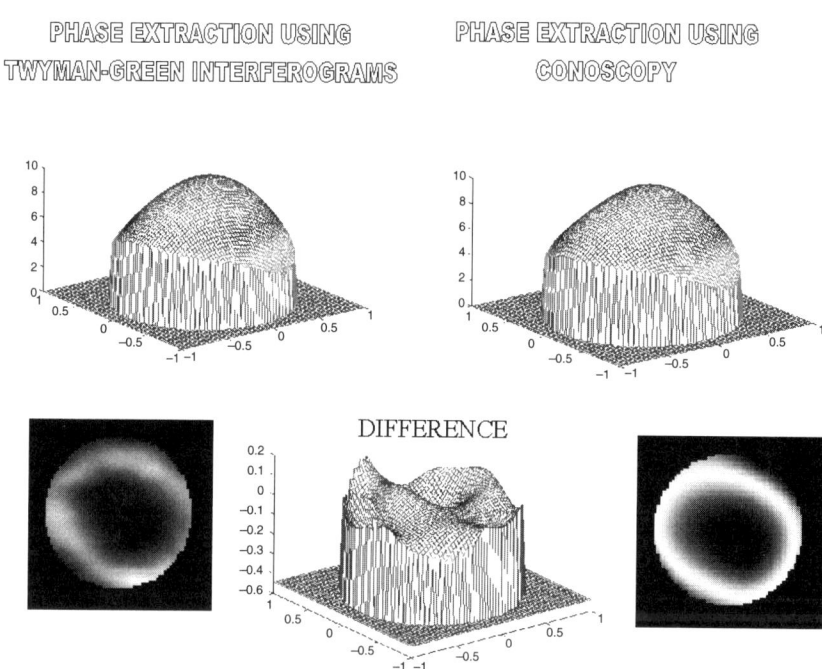

FIGURE 15-10: Interferograms.

ment of holographic memories is aimed at digital computer storage. Holographic memories can also be used for image processing applications. A peristrophic system is used as an optical database to store images to navigate a small car autonomously along specified paths.

RESEARCH

The peristrophic memory system is shown in Figure 15-11.[12] It is very similar to a conventional angle multiplexed system, with the signal beam normal to the surface of the medium, and a plane wave reference beam incident at an angle. In a peristrophic memory, holograms are multiplexed by rotating the medium around the surface normal (which is also the direction of the signal beam in this case). The film rotation causes the reconstruction of a recorded hologram to move away from the output detector array, which makes it possible to record a new hologram on the rotated film. Stored data is retrieved by illuminating the hologram with the reference plane wave and rotating the film to the appropriate peristrophic

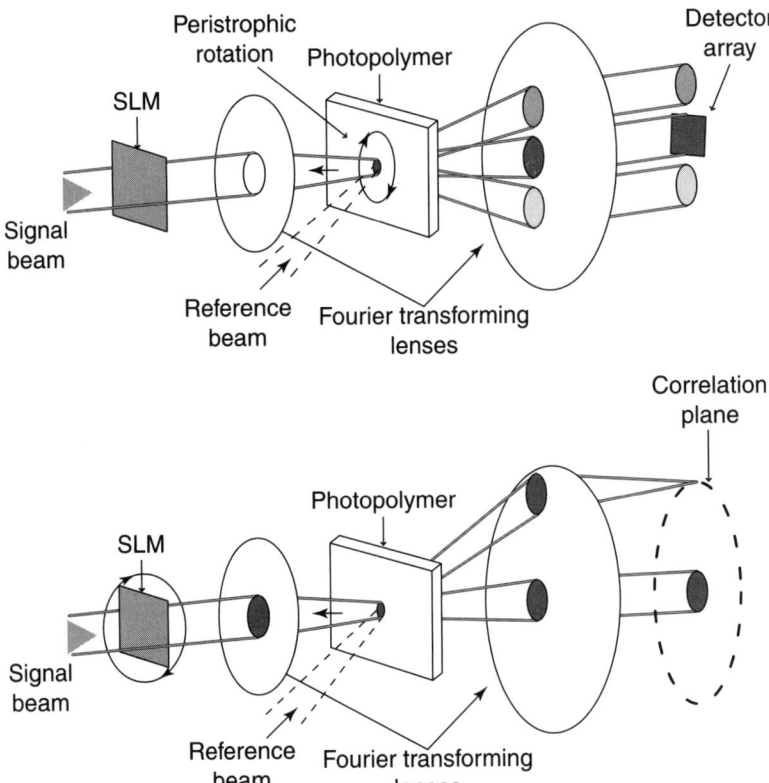

FIGURE 15-11: TOP: Holographic memory using peristrophic multiplexing. BOTTOM: Holographic optical correlation using peristrophic multiplexing.

position. Typically, 100 or more holograms can be peristrophically multiplexed independent of the hologram thickness. The same system can also be configured as an array of optical correlators (see Figure 15-11). In this case, the hologram is illuminated with the signal beam, and a *ring* of correlations is produced surrounding the image of the input SLM. If angle multiplexing is combined with peristrophic multiplexing, multiple concentric rings of correlations form at the output. Previously, up to 1,000 stored images and hence, 1,000 correlations, have been demonstrated. The correlations can be detected in parallel by multiple detector arrays. Alternatively, a single detector array can be used at one correlation position. In this case, the hologram is rotated and the memory is searched serially. This method is well suited for the car navigation problem.

The experiment was done with a small car. The car has three wheels and it carries a CCD camera, a video transmitter (to relay the video to the optical table), a remote-control receiver (to receive the control signals for turning and speed), two drive motors, and two lead-acid batteries (see Figure 15-12).[13]

First, the car is moved manually along the desired course. The images that the car-mounted camera sees are sampled periodically and recorded in DuPont's photopolymer through peristrophic multiplexing. The rotation between holograms is small enough so that three correlation peaks can fit within the detector array placed at the correlation plane. The bottom, middle, and top correlation peaks represent the previous way-point, the current position of the car, and the next way-point, respectively. Then, after the entire path has been mapped, the car is returned to the original position and the photopolymer is returned to the original angle. The video transmitted back from the car is presented on the SLM and is correlated with the stored holograms. What the car sees now is what is stored as the first hologram, so a strong correlation peak appears in the middle of the detector array. A weaker peak representing the next way-point along the path also appears above the middle correlation peak. The car is then commanded to move forward. A personal computer monitors the digitized correlation peaks as seen by the CCD at the output of the correlator. The computer extracts steering information from the lateral position of the middle correlation peak and transmits it to the car. When the intensity of the top correlation peak becomes stronger than the middle correlation peak, the car is assumed to have reached the next way-point along the path, and the computer rotates the hologram. This causes the top peak to now appear at the middle. In this way, as the car proceeds, it is steered through

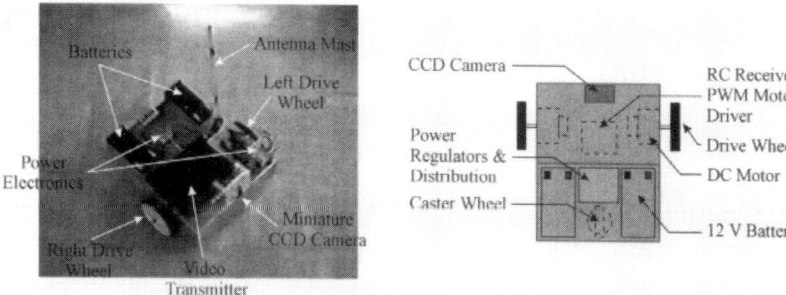

FIGURE 15-12: Car navigation experiment.

the series of way-points and it stays on the desired course. This mode of navigation is the *follow* navigation mode. The system automatically switches to other navigation modes (controlled by software in the computer) to allow the car to execute sharp turns, search for a familiar path when it is lost, or switch between two paths. In this way, the researchers were able to program the optical memory and the PC to guide the car to complete various complex trips.

An experiment was then set up to navigate the car from one lab to another. The labs are about 15 meters apart joined by a common hallway. Way-points were recorded at about 30cm intervals down the hallway. A total of 54 holograms were recorded to describe the entire path. Experimentally, the car was able to reproduce the desired path within a few inches. Furthermore, the system was very tolerant of noise, such as placing new objects in the hallway and the researchers' attempts to push the car off course.

ACHIEVEMENTS

The researchers have demonstrated a system that uses peristrophically multiplexed holograms to navigate a car in real time through their laboratory. It should be possible to build a simple system that navigates a car through the entire Caltech campus with the storage capacity of a single holographic 3D disk.[14]

Target Pattern Recognition and Tracking

Optical holographic correlators are excellent systems for target recognition and tracking tasks. First, views of targets are recorded as holograms in the template database of the correlator. Then, software is written to analyze the output of the optical correlator, which is a grid of correlations between input signal and stored images. If the input image contains a target, then the target's correlation will be very strong. If the target is not in the input scene, then correlation with that target's stored images will be weak. The software uses the results of the correlations to make decisions and take action. For a recognition and tracking system, it might simply be to control a camera by following a moving target.

The first property of the optical correlator that makes it useful is the inherent parallelism with which correlations are performed. Since multiple template images are stored at a single location of the holographic recording material, all the stored images are simultaneously correlated with the input image. The speed limit here is the video rate of the input device, and the rate of the detector at the output plane combined with the ability to digitally process the correlations. The second useful property of the optical correlator is the shift-invariant nature of the correlation function. When the target shifts, its correlation output will shift with it. This allows an object to be tracked, since its position can be determined from the position of its correlation.

With the optical holographic correlator system, multiple targets can be tracked simultaneously with limited overhead. Views of additional targets are simply stored holographically in the same location of the same material as additional holograms. When the software processes the correlation output of the system, it knows which correlations correspond to which targets, and can then determine which targets are in the input scene, and where.

REAL-TIME RECOGNITION AND TRACKING

A real-time target recognition and tracking system has been built that is capable of identifying and tracking models of four different types of vehicles. Figure 15-13 shows the four models to be tracked.[15] A single profile view of each model was recorded by the system. Software was written to monitor the optical system's output and display visual feedback to a computer screen for system performance monitoring. In Figure 15-14, for example, (at the top of the screen) is a picture of each model and a bar indicating its corresponding correlation strength.[16] When the correlation strength gets above a predetermined threshold value, the model's position is shown in the box at

FIGURE 15-13: Real-time tracking.

FIGURE 15-14: Real-time tracking at work.

the top right of the screen as shown in Figure 15-14. As more targets are brought into the scene, they are added to the position indicator box.

In Summary

- **Optical holographic correlators**: Holographic correlators store images that can be compared in real time to input images. Due to selectivity in holography, many such correlations can be carried out in parallel.
- **Fresnel correlators**: A defocusing of the correlation plane decreases the shift invariance of the correlators, thereby increasing the number of patterns that can be stored in each correlation operation.
- **Eyeball topographer**: An eyeball topographer has been developed for use in laser-assisted keratectomy. A scanning resolution set of projected lines provides near-real-time depth data on the surface of the eyeball.
- **Autonomous vehicle path following**: A small car is navigated in real time through the halls of a laboratory using a holographic memory to do way-point recognition and navigation.
- **Target recognition and tracking**: Targets (like vehicles) are stored in a holographic correlator, and their images are matched with the field of vision of some camera. Holographic correlation is well suited to this recognition problem, due to high parallelism and shift invariance.

Chapter 16, "Motion Picture Holographic Image Production," marks the beginning of Part Three, "Integral and Portrait Holography." The chapter will show you how to produce holographic motion pictures. In other words, it will show you how to transfer motion picture film, or video, directly to the holographic image.

End Notes

1. Gregg Steckman, Caltech, EE 136-93, Pasadena, CA 91125, USA, 1999.
2. Ibid.
3. Ibid.
4. Ibid.
5. Ibid.
6. Ibid.
7. Ibid.
8. "Real-Time Eye Typography," Christophe Moser, George Barbastathis, and Demetri Psaltis, Collaboration with USC Doheny Eye Institute, California Institute of Technology 136-93, Pasadena, CA 91125, USA, 2001.
9. Ibid.
10. Ibid.
11. Ibid.
12. Allen Pu and Demetri Psaltis, Collaboration with the National Science Foundation, California Institute of Technology 136-93, Pasadena, CA 91125, USA, 2001.
13. Ibid.
14. Ibid.
15. Gregg Steckman, Caltech, EE 136-93, Pasadena, CA 91125, USA, 1999.
16. Ibid.

PART III

INTEGRAL AND PORTRAIT HOLOGRAPHY

Integral holography is a two-step process. Images are first shot on 16mm or 35mm motion picture film and then transferred with a laser to hologram film. Integral holograms can be displayed flat, curved, or as a cylinder viewable from 360 degrees. Displaying the hologram on a curved or cylindrical surface makes the hologram viewable to a larger audience than normal. Integral holograms can be created from living people in motion, stop-motion animation, video, and/or computer-generated imagery.

Portrait holography, on the other hand, is producing a motion image holography image of you. For example, holograms have been created from portraits of Andy Warhol, Pierre Cardin, John Cage, Isaac Asimov, Ed Koch, John Kenneth Galbraith, Edward Heath, Phil Donahue, Sally Jessy Raphael, Oksana Baiul, Tommy Moe, Johann Olav Koss, Billy Idol, the Smothers Brothers, and Phyllis Diller, among many others.

CHAPTER 16

MOTION PICTURE HOLOGRAPHIC IMAGE PRODUCTION

In This Chapter

- Motion picture holography system
- Larger holograms
- Transferring motion picture film or video directly to the holographic image
- Medical imaging
- Data storage
- Entertainment industries
- Security products
- Credit cards
- Negotiable documents
- Event tickets
- Animation

Of all the new technologies that are changing the mediascape at the beginning of this century, holography is perhaps the medium that remains least understood by the general public and specialized art critics alike. The small number of mainstream exhibitions and published critical papers on holographic art reveals a yet uncharted territory.

One of the most common misconceptions about holography is the notion that the medium's primary visual property is that of producing *illusionistic* three-dimensional pictures—a kind of spatial photograph, with an added dimension. The *naturalistic* misconception is usually grounded on unfulfilled expectations and unproductive comparisons with other media. Furthermore, those who think of holography in these simplistic terms are just unaware of some of its most significant features and directions.

Therefore, the goal of this chapter is to dispel these delusions by demonstrating that, in fact, the holographic aesthetic experience is much more complex than it may seem at first. Holography may be thought as a perspectival system, but this approach is of no interest here. Rather, the goal here is to reveal holography as a time-based medium, and to show in what circumstances it has been explored as such by artists. Observation of a few artists' work encircles the problem and suggests manifold approaches.

It is clear that aesthetic experiences are warranted in all cases to be discussed ahead; however, by no means will it be suggested that scientific holograms, whenever mentioned, be read as artwork. If every artwork unfolds into aesthetic experiences, not all aesthetic encounters are provoked by artwork. The focus will be the issue of time, in a medium traditionally known (albeit little understood) for its spatial properties. The revelation of the range of the aesthetic and technical directions of the temporal experience in holography, it is hoped, will lead to a greater appreciation for the artistic potential of the medium.

Photonic Cinema: Experimental Holography

The uniqueness of the digital moving image as distinct from other forms of cinema is the art of organizing a stream of audiovisual events in time. It is an event stream, like music. There are at least four media through which we can practice cinema (film, video, holography, and structured digital code), just as there are many instruments with which we can practice music. Of course, each medium has distinct properties and contributes differently to the theory of cinema; each expands our knowledge of what cinema can be and can do. This observation is of particular interest because it emphasizes holography as a time-based medium, and not as a three-dimensional imaging technique. However, as we shall see, time is manifested in holographic art not only as streams of images, but also as suspended clusters and discontinuous structures.

The prospect of digital holographic movies of the future notwithstanding, the multimedia nature of the computer compels us to a redefinition or, at least, to an expanded definition of what holography is or can be (see the sidebar, "Motion Picture Holographic Images"). As practiced by a small but increasing number of artists around the world, art holography asserts time. It can be expressed as changes and transformations, and as an aesthetic feature as important as the three dimensions of space. Created with computers or not, motion-based holograms become interactive events that can be perceived in any direction—forward or backward, fast or slow; all depending on the relative position and speed of the viewer. Unlike the unidirectional *event stream* of film and music, four-dimensional holograms are *buoyant events* with no beginning or end. Furthermore, the viewer can start looking at any point. Time is suspended from its extended continuum and can flow forward or backward.

■ MOTION PICTURE HOLOGRAPHIC IMAGES

Recently, Holographic Dimensions, Inc.[1] entered into an agreement with Holographic Images, Inc., of Miami, FL to acquire key technologies related to the production of holographic motion pictures. Under the terms and conditions of the agreement, Holographic Images, Inc. (HII) will install its motion picture holography system into Holographic Dimensions, Inc.'s (HDI) R&D lab. The two companies will work together to adapt the HII system for mass replication using HDI technology.

With this new technology, HDI will be able to deliver full-color, motion picture holograms to their customer base. Additionally, the new technology will provide them with the ability to make holograms of much larger subjects than was previously possible. In the past, if HDI wanted to make a hologram of an elephant or a football stadium, for example, they were limited to models, or static two-dimensional photographs as the subject matter for the hologram. Now they will be able to transfer motion picture film, or video, directly to the holographic image.

The combination of HII's technology and that of HDI presents the consumer holography marketplace with some exciting possibilities. The next baseball card or cereal box you may buy might feature a hologram movie of Hank Aaron hitting a home run and sliding into home plate. The possibilities are endless.

Although HDI plans to market the technology through its existing distribution channels, it will also seek to develop strategic relationships with other manufacturers of holographic imagery for the distribution of the their products and technologies. They will also evaluate the possibility of introducing its new technology to the medical imaging, data storage, and entertainment industries.

Many holograms and holographic installations created today involve electronic image manipulation and digital synthesis, and draw from other artistic fields such as photography, film, and video. These works explore time in unique ways and reveal a very important aspect of the medium. Most holograms created by scientists or commercial holographers are motionless, or at best have very limited motion, because their images usually aim at reproducing a virtual environment or object with the visual stability typical of traditional holography. Since the object in most holograms is three-dimensional and stationary, many holographers use the computer to make stationary virtual objects. Holograms produced so far emphasize space instead of time, and volume instead of movement.

We need not examine in detail the technical development of automatic imaging systems, from the early nineteenth century until now, to understand the creation of historical, aesthetic, and material conditions for the current digital synthesis of holographic images. In a clear development of painting's aspiration to truth and veracity, photography first attempted to fix images as seen in nature. The camera obscura, used by painters for centuries, became the photographer's essential tool. In the next stage, photographers tried to capture different moments of an action. The analysis of motion and chronophotography paved the way for cinema. As a consequence, images representing motion could not only be recorded as stills, but set to motion themselves, allowing us to see representations of the recorded events as a temporal flux. Much later, video technology instantiated the recording, eliminating the temporal gap between the action and its playing back—thus reinforcing the congruity between the representation and the reference. More recently, personal computers seem to have demolished photography's truth ambition by allowing anyone to manipulate photographic images and to easily recombine them in any desired way. If photography forced painting to redefine its direction in the beginning and middle of the nineteenth century, computers today have a similar impact on photography. How does holography fit in this context? Holograms are already routinely synthesized from secondary sources, including silver photography, video, film, sensing devices, and computer graphics.

Mixed Memories and Media

Many artists and holographers have used computers, video, film, and photographic techniques to create complex holographic artworks of distinct beauty. For instance, some pieces reveal how the artist unites highly personal imagery to a sharp technical sensibility, layering images that define their space by an intricate kinetic articulation of light.

Other artists and holographers usually create for each piece a paradigm of photographic and cinematographic records of events clearly defined in time. For example, they then displace the time reference by associating images of events that could have taken place as many as 60 years apart. Many of the old original negatives, prints, and film footage that artists and holographers manipulate are usually in a state of decay—their silver coatings having been partially dissolved in time. These artists and holographers see in these images a way to suggest the symbolic dissolution of memory. In these types of holograms, tenuous memories, fading images, partially erased images, and dissolving environments open a gateway to the viewer's own recollections.

Drawing attention to the hologram's capability for storing information nonlocally rather than to produce three-dimensional images, some artists and holographers use more than 100 different pictures in a piece in which they combine photographic prints and stereograms made from old movies around the central images. For example, stereograms (holograms produced from many sequential two-dimensional images that usually create a three dimensional stereoscopic picture) are usually distorted and transformed beyond their common use.

Most of the holograms these artists and holographers produce should be seen from both sides because the artist wants to use the film plane—not just as a physical reference, but as a *transitional device*. The viewer should never expect to see the *other side* of the image on the *other side* of the hologram. In some artists' and holographers' work, the space is outlined by motion and the relative position of images, not by stability of forms. Also, in some holographic images, a single-pane piece can be viewed from both sides. For example, here the artist can employ several dynamic forms, such as an image of a walking figure both in negative and positive, and a spinning metal artifact that rotates more than 360 degrees around its vertical axis, disobeying conventional stereoscopy. The action can usually take place in several planes, and the structure of the space reveals itself as the viewer sees both sides and activates several other real and virtual images that shift in multiple directions simultaneously.

In some of their work, artists and holographers use computer imaging techniques either to make texture animation or to manipulate the grayscale (and other features) of old prints and movie frames. For example, take an image of a woman's face that is subjected digitally to successive tonal changes. Simultaneously, take an image of a small artifact that rotates slowly in its visual integrity. Since this time the stereoscopy was preserved, the computer imagery is double exposed with the stereogram and produces a complex animation as if decades of decomposition were compressed in a holographic event.

A typical working method for most artists and holographers starts with selecting images fixed in the past by means of light. They then manipulate these faint images with other photographic processes, sometimes with film editing or computer techniques. The resulting images are finally manipulated holographically and become propagating light again. These reminiscent images seem to show, with their chromatic delicacy and collapsing structure, that there is a whole world to be discovered when fragments of memory are brought into light.

Choreographies of Light

One of the most intriguing and fascinating styles of holographic work revolves around the modulation of luminous structures whereby light itself is controlled as a new visual medium. The art of light in motion, devoid of referential subject matter, acquires in holography yet an unexpected sense of wonder. Artists such as Rudie Berkhout (see Figure 16-1) create subtle orchestrations of light forms that are at once communicative of human spiritual qualities and revelatory of unforeseen luminescent vistas. In their works, the sense of play is extended from the dynamic manipulation of light forms to an interplay between the work and the viewer, since it is

FIGURE 16-1: Event Horizon.

through the latter's movements that the holographic images are activated. With holography, these artists explore not the action of light on surfaces, as in the photogram, but the actual manipulation, recording, and reconstruction of light phenomena.

Note: Rudie Berkhout: Born 1946, Amsterdam, The Netherlands. Lives in New York City and Cairo, New York. Rudie Berkhout was first introduced to holography in 1975 when he attended a now historic survey exhibition of holograms at New York City's International Center of Photography. After initial investigations using representational subjects, Berkhout experimented with abstract and geometric imagery. The holograms that resulted established Berkhout as one of the preeminent artists using holography. The lyrical, rolling animation of Event Horizon (1980) balances the elusiveness of Ukiyo (1981) and the crisp image and vivid colors of Deltawerk (1982).

For example, Dutch artist and holographer Rudie Berkhout has combined an interest in the spatial properties of the hologram with delicate manipulations of fluid images (not stable objects). Since the beginning, he intentionally avoided the use of immediately recognizable forms to invite the viewer, instead, to new worlds he created within the parameters of holographic space and time.

Berkhout's holograms often engage the viewer both for their new sense of composition, in which empty space defies the inevitable link between three-dimensionality and matter, and carefully controlled movement, which can oscillate between the violent fusion of images to the more delicate undulation of disembodied colors. Many of Berkhout's holograms, for example, invoke the appearance of a spacescape, or a lightscape, suggesting through receding lines or protruding visual elements (that actually come out of the picture plane) an extended vista into a subjective domain. However, these are not contemplative spacescapes. Because images are often in motion in his holograms, the viewer is invited to explore them dynamically, discovering as he or she moves in front of the piece subtle chromatic and spatial changes. These often reveal new elements not seen at first, which can be experienced simultaneously by multiple viewers. The artist sees in these simultaneous structures, which can only be perceived as such in holography, a form of revelation of dimensions of human experience that cannot be rendered visually otherwise.

Berkhout also creates other works that show self-contained light forms with very subtle colors. His *Light Flurry* (1992) series, for example, is composed of what in other media might be identified with gestural brushstrokes, but which in holography becomes more difficult to describe. These abstract forms are obtained through the reflection and refraction of light on irregular surfaces, and are further chemically manipulated to produce pastel hues. Rudie Berkhout uses the hologram not only to capture the form of this irregular diffraction of light, but also its behavior. This enables the viewer to see the modulation of light in a constant state of transformation.

While Berkhout is fascinated by natural phenomena and tries to bridge the new medium of holography with traditional painting subject matters, such as the landscape, other artists and holographers focus on the unique properties of pure light itself. In other artists' and holographers' holograms, it is usually impossible to identify specific forms or compositions, because they work with the evanescent behavior of diffracted white light to produce fleeting apparitions. These insubstantial luminous phenomena are in flux and invite the viewer to move vigorously before the work. The slightest change in point of view is enough to activate their patterns of oscillation.

Other artists and holographers also identify visual clarity, light intensity, and color saturation as holography's unique qualities, and set out to combine them in infinite possibilities. For instance, their images are usually not premeditated; they result from an intense experimentation in the laboratory. Other artists and holographers usually make coherent light (laser) pass through arrangements of lenses, blown, cut, and broken glass, opaque masks, and other materials with an eye for the ways in which this coherent light will break up when the final piece is at last illuminated and reconstructed under white light. This exploration requires a certain level of previsualization, but by no means can the artist fully determine beforehand what the behavior of the work will ultimately be. The discoveries and unforeseen results experienced by the artist, once determined to be revelatory and stimulating, are shared with the viewer in the form of a finished piece.

As much as white light breaks up prismatically under the experimental control to which it is subjugated by other artists and holographers, it is also recombined by them to yield chromatic harmonies that are as ethereal as the visualized dancing forms themselves. They also realize the power of holography to create images that are suspended between the imagined and the experienced, and they pursue this dance of spectral hues with a unique sensibility that unites intent, chance, and constructive or destructive decision-making.

Other artists and holographers have also created a series of installations in which they integrate holography, video, and computer animation. In these installations, the hologram is not used straightforwardly as a display device, as in Berkhout's work, or as a means to record and replay the photonic dance of colors. These holograms do not contain any kind of imagery themselves. They are what is technically known as HOEs, or Holographic Optical Elements. These are holograms that do not display a picture, but instead are used to act as a lens, mirror, or a complex optical component. Rarely are HOEs used in artwork. Their main visual arrangement comprises a video monitor, with noise or black-and-white computer animations, and a large HOE, which is placed in front of the monitor. Instead of a halogen bulb, the monitor (with its changing forms, pulsating contrasts, and moving elements) becomes the light source for the HOE. This is carefully designed to take light in and manipulate, distort, and multiply it in numerous visual echoes, further blending colors and creating an overall calm and meditative meaning. The result is a cinematic spectacle that blends the temporal linearity of computer animation with the spatial and chromatic dimensionality of holography. The artists have created many works based on this principle. HOE-TV, for example, is used only for noise on the monitor. It was created in 1990 and was shown the same year at the European Media Arts Festival in Osnabruumlck, Germany.

Some artists and holographers are also very interested in integrating their installation into indoor settings, and further investigating the relationship of new media and architecture. These artists have incorporated HOEs into walls and floors, activating the space with multiple monitors and digital animations.

Holographic Film and Video

If one were to write the history of holographic cinematography, one would have to trace its scientific roots to the early 1960s. Back then, many experimental setups were developed in laboratories around the world in which a special kind of laser (known as *pulsed laser*, due to the very short pulses it produces) was used to record moving images. Traditional cameras were adapted by removing lens and shutter, since the film had to be exposed to the scene directly. The early short films that were produced required special viewers and could only be seen through small windows by one person at a time. A laser had to be used to play back these films, normally in setups that brought back memories of Edison's kinetoscope.

In 1969, physicists in the United States produced a 70mm 30-second holographic film of tropical fish swimming in an aquarium. In the early

1980s, a team of French scientists at the Franco-German Defense Research Establishment (ISL) at St. Louis in France, intensified their previous research on holographic cinematography and showed the first film reconstructed with the technique known as *reflection* holography (invented in 1962 by Yuri Denisyuk, in St. Petersburg, to enable reflective display of holograms in white light). From 1983 on, they shot several short holographic movies in 35mm and 126mm film (12). Most of these holograms had to be seen with a laser. Working with members of ISL in France, English sculptor Alexander, who now lives in the United States, created in 1986 an 80-second 126mm fiction film called *The Beauty and the Beast*. Inspired by Jean Cocteau (*La Belle et La Becircte*, 1945) and not the Disney Studios, Alexander portrayed Beauty as a model and Beast as an artist. As the artist himself acknowledged, the film showed more the limitations of the process as an artistic form than its potential.

Before Alexander, holographic films of a nonscientific nature were produced only by a couple of French filmmakers (Claudine Eizykman and Guy Fihman) with a background in philosophy, who developed their research at the Experimental Cinema Laboratory, LEAC, Université Paris VIII. Their first films, in 35mm and 70mm, with up to 24 frames per second, were produced in 1984. In 1985, they showed a film, *Circular Flight of Sea Gulls*, at the monumental exhibition *Les Immateacuteriaux*, at the Center Georges-Pompidou. This was meant as homage to Ettiene-Jules Marey, a forerunner of the cinema who also worked with this theme. That same year, they showed publicly in Paris a new 126mm film, called *Un Nu*, a palindrome that translates literally as *A Nude*. This 5-minute film shows a mummy who slowly turns and frees itself from the bands that contain it. Little by little, the bands reveal the flesh of a female nude that emerges from within the green light that surrounds it. The viewer can push a button and play the movie backward, observing the young woman become a mummy again; hence the title in palindromic form. This movie must be seen inside an apparatus that is almost 7 feet high, 24 inches wide, and 28 inches deep, and exhibits the same technical limitations that frustrated the English sculptor Alexander.

Instead of dismissing the idea of holocinema altogether, Alexander drew from his first experiment with ISL scientists and set out to develop a new strategy for experimental holographic filmmaking (based on a simpler technique) known as an *integral hologram* (or *multiplex*). Integral holograms became very popular in the 1970s due to their ability to use film footage (shot with regular 35mm or 16mm ciné cameras) to create standalone 360-degree animated holographic scenes viewable in regular white light. Multiplex holography saw its popularity dwindle in the 1980s, and virtually disappear in the 1990s, with few exceptions.

Alexander finally decided to remove integral holograms from their cylindrical container, and stretch and splice as many as 14 of them in a roll. He developed a simple motorized mechanism that, with light source firmly in place, rolls the holograms from right to left through a 7" × 9" window, showing series of animated scenes sequentially. A lens in front of the hologram enlarges the images to approximately 12" × 16". The irretrievable linear flow of time responsible for the drama unique to traditional cinematography doesn't apply. If the viewer moves to left or right, he or she can catch glimpses of the next scene, or see again the one perceived a second ago. *The Dream* (1987), his most accomplished holographic film, has original music composed by Alexander himself and runs for 8 minutes. It shows loosely linked sequences that evoke chimeric mental states. Anamorphic ballerinas, a man walking upside down inside a big head, a fusion of a human body with a landscape, couples dancing and dissolving in space, and a child playing with falling cubes, are some of the scenes that structure the movie. The holograms were shot by Sharon McCormack, an early collaborator of Lloyd Cross.

In Japan, holographer Jun Ishikawa and filmmaker Shigeo Hiyama, from Tama Art School (Tokyo, Japan), working with engineer Kazuhito Higuchi, from Nippon Telegram and Telephone Corporation (Tokyo, Japan), have been working since 1992 on yet another approach to the same basic idea of running actual holograms before the viewer's eyes to create motion pictures. Their initial projects, however , were closer to the zoëtrope than to the kinetoscope. At first, they shot holograms of inanimate objects that were manipulated frame by frame. They sandwiched 300 10mm × 200mm holograms between two acrylic drums with a diameter of approximately 3 feet. The drum was rotated and the film illuminated with a low-power red laser at a rate of up to 24 frames per second. This 40-second animation, entitled *ORGEL—A Boy's Fantasy*, tells the story of a boy who dreams of playing a music box (ORGEL) made by a nymph. The researchers employed eight cuts and experimented with several techniques, including stop-motion, panoramic shots, enhanced depth perception, upshots, and overlaps. This piece, which also raises unavoidable comparisons with late nineteenth-century cinematic representation devices, is a good indication of the infancy of holographic cinema, both as a technology and as an art. They have developed other alternative systems, including a 35mm camera that records moving images with a pulsed laser, and what they call a *retro-directive screen*, which enables viewers to see the animations more comfortably. All of their film experiments revolve around the same theme described earlier. Their longest movie to date runs for 2 minutes and 50 seconds, at 10 frames per second.

The current scenario for the development of real holographic movies is not very encouraging. Victor Komar, a Russian scientist based in Moscow, developed as early as 1976 the first actual holographic cinematographic system using lenses to record moving images, a special projector that employs a mercury-cadmium lamp, and a holographic screen for the projection of the movies. Four people could see his first 47-second film in its monochromatic yellow hue at the same time. The subject: a young woman holding a bouquet of flowers was perceived in full four-dimensions (the three of space and that of time) with naked eyes. This was just like the films we see regularly in movie theaters, except that the viewer could move around in the seat and perceive spatial details of the scene. The viewer could, for example, move to the side and see the whole face behind the bouquet. Komar's first holographic color film, with a duration of 5 minutes, was produced in 1984. However, in 1990, Komar's research was interrupted due to budget cuts. More recently, Komar has retired and no one is carrying on his research. On the other hand, scientific research has been successfully carried out since 1989 at MIT to develop a *holographic video*. While several laboratory prototypes have been produced, no artistic or fictional work has been created. Most of the animated imagery developed so far represents arbitrary subjects or objects of interest to those entities sponsoring the research, such as automobiles.

Finally, new technologies undoubtedly open up unprecedented opportunities for artists. However, they are not in themselves an indication of the directions experimentation in visual arts will take. Photography had to wait more than a century to be accepted in art circles, and even longer to become incorporated into the artist's general repertoire of tools in equal terms with painting and sculpture. Video did not have to wait this long. Holography will benefit tremendously from the passage of time, and one day will be understood, accepted, and seen as just another tool. Discussions concerning the validity of the medium as an art form will be long forgotten, and the work of individual artists-holographers (which is what really matters) will be discussed without conceptual hindrances. What will become of holographic cinema is hard to say. Predictions about the future of holographic cinema have failed miserably, so it could be a futile rhetorical exercise to try to anticipate the state of maturity this technology will reach as an art.

In Summary

- The art of organizing a stream of audiovisual events in time is the uniqueness of the digital moving image as distinct from other forms of cinema.
- From the early nineteenth century until now, we need not have to examine in detail the technical development of automatic imaging systems to understand the creation of historical, aesthetic, and material conditions for the current digital synthesis of holographic images.
- To create complex holographic artworks of distinct beauty, many artists and holographers have used computers, video, film, and photographic techniques.
- Some artists and holographers use more than 100 different pictures in a piece in which they combine photographic prints and stereograms made from old movies around the central images in order to draw attention to the hologram's capability for storing information nonlocally rather than to produce three-dimensional images.
- The modulation of luminous structures whereby light itself is controlled as a new visual medium is one of the most intriguing and fascinating styles of holographic work.
- Not very encouraging is the current scenario for the development of real holographic movies.

The next chapter discusses the holographic art market, such as holographic gifts and novelties, worldwide wholesale distribution service, limited manufacturing, mail order, and custom production of holograms.

End Notes

1. Rudie Berkout, "Event Horizon," MIT, The MIT Museum, 77 Massachusetts Avenue, Cambridge, MA 02139-4307, USA, 2001; and, Eduardo Kac, "Beyond the Spatial Paradigm: Time and Cinematic Form in Holographic Art," Friedman & Associates, Chicago, IL, 2001.

CHAPTER 17

THE MARKET FOR HOLOGRAPHIC ART: HOW WELL ARE HOLOGRAMS REPRESENTED?

In This Chapter

- Holographic gifts and novelties
- Worldwide wholesale distribution service
- Limited manufacturing
- Mail order
- Custom production of holograms
- Pseudoscopy and retinal rivalry
- Time-domain holography

Every medium has a code, a set of rules or conventions according to which determined elements are organized into a signifying system. The English language is a code, as is perspective in painting and photography. In the first case, the elements are phonemes organized into words and sentences according to a social convention: the syntax of English. In the second case, the elements are dots and lines organized into pictures according to a geometric method.

An artist or movement can break the conventions of the art marketing medium. If this is done, the level of predictability (or conventionality) is lowered and unpredictability is increased—becoming more difficult for the immediate audience to understand it. However, once these new rules are learned and the ideas behind them widely understood, the level of unpredictability is lowered and they become new conventions that can be accepted by the audience.

Holographic artists exploring the marketing medium for holographic art (as opposed to advertisers using holography, who favor a high level of predictability) are breaking several visual and cultural conventions (see the sidebar, "The Holographic Marketplace"). In fact, holography is so new that many questions are left open about the nature of the marketing medium. Therefore, any attempt to clarify the issues raised by holography on a cultural level has a prospective (and not conclusive) tone, concentrating more thoroughly on general points and on the promise of its potentialities than on the records of its historical achievements so far.

■ THE HOLOGRAPHIC MARKETPLACE

Holos Gallery of San Francisco[1] first opened its doors to the public in February 1979. Modeled after a *conventional* art gallery, its mission was to display and sell holographic art works via an ongoing program of four group or individual exhibitions per year.

There was no retail space in the entire Bay Area where the public could go to see (or buy) a hologram. A market niche was therefore perceived to start a business to fill the void. It was Holos Gallery's goal to bring to the public this fabulous new medium that they found so fascinating, exciting, and glamorously *high tech*.

In retrospect, *naivete* could best describe Holos Gallery's initial business plan and expectations. After their first year, it was painfully obvious that they (the founders of Holos Gallery) could not sell enough holographic art to survive. To insure viability, Holos Gallery was transformed into a *full-service* operation, offering not only holographic art, but also holographic gifts and novelties, a worldwide wholesale distribution service, some limited manufacturing, retail mail order via their own catalog, and custom production of holograms in all types and formats. This approach allowed the Holos Gallery to survive, but they always regretted seeing sales of holographic art comprise a smaller and smaller percentage of their total annual sales.

By 1990, the founders of Holos Gallery had built a fairly large organization with worldwide sales of approximately $1,000,000 annually. However, the business had completely lost touch with its original goals (to sell holographic art), and they felt they might just as well be selling computers or socks as the often tacky holographic novelties and gifts were selling in huge quantities. In September 1990, as a result of growing disenchantment and frustration, Holos Gallery and all its assets and operations were sold to A.H. Prismatic of Brighton, England.[2]

A number of factors led to Holos' failure to sell significant quantities of holographic art. One of the biggest problems in the 1970s and early 1980s was the public's lack of familiarity with holograms. During those early days, almost all visitors to the gallery were seeing holograms for the first time. While many were astonished and enthused, just as many walked away scratching their heads in disbelief. To them, holography seemed at best a *trick* or fad. There was no way for them to relate to what they had seen based on any prior experiences.

Further compounding this problem was the popular notion that art should be bought as an investment. There was no historical basis, no auction house price records, and no market data with which to convince potential buyers that they were making a good investment. It was not possible to predict future increases in value, and quite impossible to guarantee any kind of liquidity with regard to reselling their purchase at a later date. Even in the last few years, most purchasers were buying the more expensive holographic art works because they liked them and could afford them, not primarily as investments.

Another major issue was the simple lack of good quality holographic art works and artists. The technical constraints were (and still are) most formidable, and the equipment extremely rare and expensive. It was a time when a *good* hologram was simply a hologram that was bright and of decent quality. Too often, a hologram's only raison d'être was the novelty of the three-dimensional image with no regard to aesthetics.

As the years passed, there was an increase in both the numbers of art works sold and the prices they commanded. However, Holos' art sales never attained a volume adequate to sustain the business. Even today, there are no hologram art galleries in the world that survive without augmenting their income, either through sales of related products and services or by the receipt of government and/or private foundation funding.

A number of factors contributed to the growth of holographic art sales. The appearance of holograms on the covers of *National Geographic* and on major credit cards introduced millions of people to holography and, at least subconsciously, began to make people more comfortable with the medium. With the ability to recite a growing list of accomplishments in the commercial realm, it became possible to *validate* holography more easily in the minds of potential art buyers.

Furthermore, the laudable increase in the availability of and improvements in holographic education, equipment, facilities, and techniques allowed more and more artists to make holograms. The result was holograms that were (and continue to become) brighter, larger, more colorful, more varied, and of greater aesthetic value. The seller of holographic art gained access to a finer and more varied inventory, resulting in a much better chance of finding a piece that could satisfy a customer's needs, tastes, and budget.

Though it may seem a trivial concern, the availability of better and less expensive illumination sources for holography also made a big difference. A good piece of holographic art must have a good light source to be properly seen and appreciated. The advent of small, inexpensive quartz halogen light bulbs and fixtures made it much easier to overcome customers' worries about properly displaying holographic art in their homes.

The most important thing, however, is that over the past 20 years, the holography community has come to accept that a high-quality, very bright holographic image does not equate to *good art*. Only by shedding reliance on the medium's novelty can the art form itself evolve. Holography must be able to compete on the same generally accepted guidelines used to judge other fine art: form, content, aesthetics, execution, communication, and inspiration. As holography matures, the blending of these traditional elements with holography's exceptional ability to project and focus light will assure its long-range viability and success.

Growth in the market for holographic art has been characterized as *slow but steady*. Numerous reasons can be rationalized for not participating: time filling out forms, fear of disclosing information, and costs of subscribing are some of them. However, if we continue to let minor obstacles interfere and stubbornly insist on carrying on our activities in a vacuum, *slow* and *slower* may well become the best adjectives available to describe future growth in the holographic art market.

In closing, increasingly fantastic and beautiful works of holographic art are being produced by more and more exceptionally talented and dedicated artists. Such perseverance is vital to this medium's long-term acceptance as a fine art form. Holography will take its place, alongside painting, sculpture, photography, and musical composition, as a valid and valuable means of artistic expression. How long this process will take remains unanswered.

The Etymology of a Hologram

The word *photograph* was first suggested by French-born pioneer Hercules Florence in 1833, as a consequence of his attempt to use light to print labels and diplomas on paper with a silver coating. While photography as we know it today was invented by Nicephore Niépce as a culmination of centuries of research in that direction, holography, as many other inventions in the twentieth century, was the byproduct of a search for something else: a method for improving the quality of images recorded on an electronic microscope.

As opposed to photography, holography did not come as a consequence of centuries of perfectibility. Dennis Gábor, its inventor, needed in 1947 what was to be later called a laser to make three-dimensional holograms, but he invented holography almost 15 years ahead of the appearance of the first laser. Even in the early 1960s, when the first three-dimensional holograms were made, the technique was labeled *a solution in search of a problem* by the press. More than 50 years have passed since the invention of holography, but the character traits of the medium as such and its cultural meaning remain incognito.

As observed before, Gábor coined the word *hologram*, which is widely used to the detriment of the word *holograph*. While the former corresponds to the nomenclature established by its inventor, the latter is used only by those who try to imply that the hologram is an extension of the photograph, or by those who are not concerned with their differences. The point to be made here is that the word *hologram* is more precise in naming its referential object.

Holos in Greek stands for *total, complete*; *gramma* means letter and writing. It has the same roots of *graphein* (*to write*). In ancient Greece, however, the letter was also used as a number (as a system) for the measurement of distinguishable unities (therefore, the current use of the suffix in *kilogram*, which does not mean writing with weight, but the unity formed by 1000 grams). Therefore, if *gram* designates the unity and *holos* (the total), the word *hologram* means the unity of the whole as well as the wholeness of the unity—which the word *holograph* could never express. Actually, *holograph*, in any dictionary, designates *a document wholly in the handwriting of its author*. So, since in a hologram each part is similar to the whole (a spatial characteristic that will be addressed later), only the word *hologram* should be used and the misleading *holograph* avoided.

A Hologram Is Not a Picture

By avoiding the word *holograph*, one is also avoiding the equivocated idea that holography is a kind of photography. After 170 years of development, photography is a medium the conventions of which are now well established and accepted by the public. Therefore, a comparison between both media might take advantage of the general acceptance of photography, not to imply kinship but to serve the purpose of examining a few questions by contrast.

Although Nicephore and Florence were also interested in photography as a printing technique, the use of the camera by Daguerre and Talbot gave shape to the medium's capability of taking accurate pictures of real things. From then on, photography evolved in two basic directions: one, the historical and journalistic approach, has foundations in the recognition of the photograph as a reliable method of documentation; the other, the creative approach, is based on the invention of nondocumental images by means of cropping, solarization, montage, negative manipulation, and so on. In both cases, though, one wouldn't be wrong in stating the truism that whether taken with a camera or not, the photograph is ultimately a two-dimensional image and photography is a picture-making technique. It is the art of fixing an image on a plane in such a way (with varying degrees of shade) that the image becomes the plane.

Here, a distinction should be made. As opposed to the photograph, the hologram is not a picture, and holography is not primarily a picture-making technique. If this is true, all attempts to analyze holography from a straight pictorial viewpoint, or taking simplistic novelty holograms as if they represented the whole world of holography, will prove to be misleading. An examination of an ordinary fact, such as shopping in a supermarket, might make us aware of something that otherwise would be unnoticeable. Every time one goes shopping in a supermarket, one realizes that the checking out was speeded by the use of a laser scanning process that *reads* the Universal Product Code (UPC) and provides price, an itemized receipt, and inventory data to the store computer. The device located inside the scanner that bends the laser light in the direction of the product and back to the microprocessor is a hologram (or a set of holograms) in a glass wheel. This hologram (or set of holograms) doesn't bear an image at all. What it actually does is just perform the function of a lens (it only diffracts light in a particular way). Holograms that do not display images but instead perform as optical elements are gaining ever more industrial and scientific applications. So, if we cannot only think of but actually use holograms as optical elements and not pictures, and if they can perform rather than bear an image, they are not extensions of photographs, but a new way of recording, storing, and retrieving optical information (information carried by light waves).

The way a hologram optically stores an image can be compared to a certain extent to the way a computer disk digitally stores an image. The digital image has to be transformed into 1s and 0s to be recorded on the disk and to be read by the software on the hard drive. The holographic image has to be codified into an interference pattern to be recorded on the film or plate. This pattern diffracts an incident beam of laser or white light so that the microscopic pattern can be translated into a visual image.

To say that the hologram is distinct from the holographic image means that the first is just the storing medium, while the second is what is stored. All the fluctuations, changes, inconstancies, leaps, turbulences, and rhythms perceived in complex holographic art pieces are the result of careful work on the level of the mutable structures by which the visual information is stored, and not on the level of the images themselves. To say that the hologram is not the holographic image means that one cannot *retouch* the image, because one can only *touch* the storage medium. The sensorial response to holographic images (and very clearly not to holograms) is ambiguous in the sense that the person who tries to grab the image knows that he or she is not looking at an apple, but nevertheless tries to clutch the luminous image only to have the unusual experience of contrast between vision and touch. The beholder is not looking for deception, he or she is not wanting to be fooled, but wants to have an acute contrasting experience based on the identification of the appearance of a familiar object and its noncorrespondent tactile contours. The distinction between a hologram and a holographic image also means that damaging the hologram does not damage the image, because the later is recorded all over the emulsion. In a photograph, as seen from the viewfinder, for every geometrical point on the surface of the object there is one and only one correspondent geometrical point on the surface of the image. In a hologram, as seen from the film holder, for every geometrical point on the surface of the film there is a complete view of all the available information of the object. If one point is missing, the other points will reconstruct the complete image of the object without any problem.

This distinction between the hologram and the holographic image (raised by the example of the supermarket scanner) might ultimately mean that, when looking at holograms that do display images (chiefly at art pieces with complex space-time relations), one should concentrate on appreciating the rhythm orchestrated by his or her own dynamic perception of the informational structure of the piece rather than try to reduce to a monoscopic vision a sensorial experience that demands extensive binocular probing. In other words, the idea of a hologram as *perfected photography*, and capable of reproducing *better* the appearance of an object (the advertising approach), gives place to the understanding of the hologram as a medium for encoding complex spatiotemporal information (the artistic approach). The complexity of the way in which information is stored in the hologram might even in certain cases challenge the binocular (or *stereoscopic*) perception of images and space by sending absolutely different images to each eye, therefore deterring the perception of three-dimensional *objects* to favor a dynamic amalgam of images.

The holographic image (as optical information) is ultimately a spatiotemporal one (and not a volumetric or in-relief one), as distinct from the stereoscopic or anaglyphic drawing or photograph. Anaglyphs and stereographs produce an immutable image of a localized relief without surrounding space (from which the temporal dimension is extracted), while holograms produce images surrounded by actual space—the parallax of which has to be perceived in duration.

To Be or Not to Be Representations in Holography: That Is the Question?

Because holograms are a way to record, store, and retrieve information, and because they can display an image or replace certain objects in their function (or even perform certain functions that objects cannot) without displaying any image, they pose a complex question regarding the way they represent (or not). It is very clear from the outset that holograms don't represent anything, in the same way that a computer disk doesn't represent anything. Holograms present things (images) or are things (diffraction gratings in a supermarket scanner). Holographic images do *represent* in a way that is peculiar: a result of the characteristic traits of the holographic marketing medium as such, but is also a consequence of our esthetic code (the conventions according to which we associate an image to something exterior to itself, as in the case of a holographic portrait). The problem comes from the fact that the syntax and the elements of the vocabulary of holography are not yet widely known, making it very difficult, even for the scholar, to *read* holographic images. The ones that are very easy to *read* are the ones that try to resemble photographs or objects. They try to take advantage of the viewer's knowledge of another signifying system to convey a straightforward message (usually for advertising or decorative purposes) for commercial rather than for inquiry aims. Those images are of no interest to this chapter, because they mislead the viewer by making him or her think that holograms are *optical illusions* rather than a means for recording, storing, and retrieving optical information. The emphasis shall be placed in the uniqueness of the holographic image, and not in its reduction to other systems that training and habit have already mastered.

Holographic content is not emptied of interpretability, but full of predictability. Those images are designed not to cause strangeness. They are designed in such a way that they can be promptly associated with well-known images so that they can convey transparently their content as pictures that resemble other pictures. Pictures in perspective, like any others, have to be read, and the ability to read has to be acquired. A simple image of an anonymous woman's face, even if it is in three dimensions, does not demand the discovery of rules of interpretation. It relies on the experience of looking at actual faces as well as the experience of looking at pictures of faces.

Not all holograms, though, try to minimize their unique characteristics so that they can be grasped as quickly as they would if they were photographs. We are yet to know what kind of objects and forms will be hologenic, because the hologram's destiny is not one of reproducing three-dimensional movies as a movie's destiny was not one of reproducing theater, and photography's destiny was not one of appropriating the concepts of painting. However, to think in terms of *objects and forms* is still to think in terms of volumes and reliefs, and not in terms of empty space or in terms of a four-dimensional space-time continuum.

So, why will the simulacrum in three dimensions be closer to the real than the one in two dimensions? It intends to be, but its paradoxical effect is, inversely, that of making us sensitive to the fourth dimension as concealed truth (secret dimension of all things) that all of a sudden assumes the power of an evidence. In short, there is no real: the third dimension is the imaginary of a two-dimensional world; the fourth, of a three-dimensional universe. Scaling in the production of a real, which becomes even more real by addition of successive dimensions: It only is truthful, it only is truly seducing that which plays with one missing dimension.

The *real*, like the object of observation in quantum mechanics, depends upon observation—it changes when observed, it is only real when it can be observed. Every observation corresponds to the cultural standard of a certain society at a given time. The images produced within this framework change together with the model employed to observe, and describe or represent any particular notion of the *real*.

The meaning, the truth, the real: they can only appear locally in a restricted horizon, because they are partial objects, partial effects of equivalence and mirroring. Every reduplication, every generalization, every passage to the limit, every holographic extension (whim of explaining exhaustively the universe) makes them emerge in their own derision.

Some holographic images said to be *realistic* will only be so if we state that our notion of the real is changing, because it will be only then that we will be able to establish a symbolic relationship between those images (that challenge our senses and our very notion of what images are and how they work) and the world. In the Renaissance, man placed himself in a vantage point to organize the sensorial (irrational) data of the surrounding world with mathematical (logical) rules. Man also interrupted symbolically its chaotic flux with a schematic picture that expressed the supremacy of his/her viewpoint. In other words, the man of the twenty-first century is experiencing the satellitization of the gaze. He or she is looking at the world as a whole from its orbit, not from inside as the Renaissance man did with perspective, but from the outside. Basically, man is looking at the world like he or she does with holography—to have a totaling view and to be able to scrutinize and control it more firmly as we try to do with the weather (forecast satellite) or with our neighbor country (military satellite). We now measure the smallest temporal unit to be ever measured: the *femtosecond*, or a billionth of a millionth of a second. We also play with atoms at our will in scanning tunneling microscopes linked to immersive virtual reality systems. Even with gravity, we seem to start to neutralize with superconductive ceramics. Whether on a microscopic or a macroscopic scale, we are redefining our models. The luminous holographic image, for example, is a clear statement about matter and energy as a continuum rather than distinct entities. Why would an image made of a less dense state of this continuum look more or less realistic than an image made with a denser state? The notion of what is or what looks realistic will follow the ongoing paradigmatic change. *Realism is a matter not of any constant or absolute relationship between a picture and its object, but of a relationship between the system of representation employed in the picture and the standard system.*

Is a holographic image more *realistic* than a photograph if it is a scientifically more accurate *optical illusion*? Is a hologram an *illusion* in any instance? In order to answer these questions, one has to clearly define what one means by *illusion*. Thus, the term *illusionism* does not connote that illusion or even that deception is the main aim of art.

The term *illusionism* was introduced by Franz Wickhoff in 1895 in his famous publications of the *Vienna Genesis* (an early Christian manuscript) to characterize the deft style of brushwork that had survived from Hellenistic times. The idea that anyone should have confused the illustrations of the manuscript with reality obviously did not enter his mind. What he wanted to convey, quite rightly, was the difference between this style and other, less illusionistic, methods.

The term *illusionism*, therefore, never meant that the viewer would be deceived by an image and think that he or she is looking at the object rather than at a representation. But an argument might be raised that the aim of illusion is to create in the viewer the same response that the viewer would have to the object in similar conditions of observation. Descartes knew that *no images have to resemble the objects they represent in all respects (otherwise there would be no distinction between the object and its image)*; furthermore, what deceives depends upon what is observed, and what is observed changes with interests and habits.

In a complementary view, Gestalt psychology explained that apparent motion, like in movies, for example, is not the result of a mistake in the observer's thinking, but a *perceptual fact* (a fact as real as the so-called real movement). Furthermore, it would be hard to convince an audience that the movement perceived in a movie is the result of thousands of mistaken judgements made in a couple of minutes. The apparent movement is *perceptually real*. It proves that visual processes resulting from local stimulations occurring in different places under particular temporal conditions do interact with each other. Likewise, one could say that an apparent tomato is perceptually real—discarding the meaningless idea that holograms are illusions. A hologram resembles another hologram more than it resembles any object, which, of course, doesn't mean that a hologram necessarily represents another hologram. Representation is a matter of context and illusion, a matter of how one understands the relationship between objects and images that represent objects. In informal discourse, the word *illusion* implies absolute pictorial fidelity or unawareness of the distinction between a representation and an object, which ultimately does not occur in any aesthetic setting such as galleries or museums.

The holographic *real image* (the one that stands in thin air between the viewer and the hologram) changes our notion of depth, leading the layperson to eventually say that there is an illusion, that *it looks as if the thing was really there*. In other words, holographic *real space* provokes a perceptual inversion; it contradicts our expectations. *Instead of being in a vanishing field for the eye, we find ourselves in an inverted depth that transforms ourselves in a vanishing point*. However, the holographic image has a spatial presence that eliminates the sense of illusion usually associated with conventional representation.

The hologram is the inverted fascination of the end of the illusion, of the scene, of the secret, by means of materialized projection of all available information of the subject and of materialized transparency. The hologram doesn't have exactly the intelligence of the trompe-l'oeil, which is that of seduction, of always proceeding, according to the rules of appearance, through illusion and ellipse of the presence.

Going further with this argument, to a certain extent, this process represents the end of aesthetics and the triumph of the holographic marketing medium. Like in stereophony, *that sophisticated in itself puts an end to the charm and intelligence of music*. However, does stereophony put an end to the intelligence of music, or does it offer another way, with a different *intelligence*, of enjoying it? Different ways of producing and recording music will result in different ways of listening, as exemplified by contemporary experimental pieces composed specifically for magnetic tape. The fact that music played live demands a deductive auditory response (for watching the musicians play influences the listener's perception of music) suggests that recorded music is more for inductive listening (for stereo-listening brings only a part of the whole spatial-acoustical experience of a live concert). The same is valid for the hologram: while tangible objects and shapes trigger deductive perception, holographic images demand a more inductive perceptual response.

The perceptual and conceptual experience of a hologram will continue to change as new technologies make possible new aesthetic adventures. As holography is perfected, it begins to exhibit the terrifying rigor mortis of all new advances toward illusion. For example, look at a life-sized portrait of former President Bill Clinton. He stands out in full volume. As the viewer moves from the left to the right, he sees a part of Clinton's shirt that was hidden before by his jacket, and the reflections on his eyeglasses change. The viewer sees his head first from the left, then from the right. So complete is the illusion of Clinton's three-dimensional presence that his immobility makes him a frightening corpse. The strength of the spell makes one ungratefully aware of what is missing. Instead of an image of a live man, the viewer sees a real ghost faking life. It will take a while before this new advance toward realism loses the power of seeming to be reality. It happened before with the motion picture, the stereoscope, and the sound film.

Holographic images possess several characteristics by means of which they represent things (like in a portrait of a person) or produce self-referential signs (like an abstract composition). The study of them all would be beyond the scope of this chapter. The fact that holographic images demand binocular perception in motion (as opposed to stereographs and anaglyphs), and the fact that they can produce a relief desirably identifiable with the relief of recognizable objects, are two prominent aspects and therefore shall be addressed. However, let's discuss these features in their reversed perceptual manifestation: the inside-out relief and the non-stereoscopic vision (but still binocular)—because of their unique holographic qualities.

Retinal Rivalry and Pseudoscopy

In the 1960s, the only existing holograms were laser transmission holograms (holograms only visible by means of a laser). Those red images displayed in normal conditions a convex relief as such. The optical relief, because it was as convex as the relief of the object, was called *orthoscopic*, from the Greek word *orthos* (right). However, if the light source was kept in place and the hologram flipped around, the relief appeared concave. The inversion also happened if two objects were used as models, one in front of the other. The one that appeared closer in the scene appeared further way in the image. This overall reversion was named *pseudoscopic*, from the Greek word *pseudo* (false) and is caused by diffracted light rays crisscrossing in front of the film plane.

The *pseudoscopic* phenomenon was discussed in 1868 by Herman von Helmholtz. Studying the image formed in a stereoscope (where two flat photos taken from a distance correspondent to the distance of human eyes are seen simultaneously to give the impression of relief), Helmholtz observed that the perception of relief and solidity was not produced by the movement of the eyes, but by binocular perception. He described an experiment that showed that the impression of relief was still produced in a stereoscope, despite it being illuminated with an electric spark of very short duration. The spark lasted for less than a four-thousandth of a second, during which there could not have been, according to Helmholtz, any recognizable movement of the eyes. He also noted that the image perceived with the right eye is different from the one perceived with the left eye, *otherwise we should not be able to distinguish the true from the inverted or pseudoscopic relief, when two stereoscopic pictures are illuminated by the electric spark.* Helmholtz describes the pseudoscopic image on the stereoscope, saying that *what should be further off seems nearer, what should stand out seems to fall back.*

The volumetric image in holography, as opposed to the stereograph, has the quality of reversing itself in all dimensions, the three dimensions of space, and that of time, too. It makes no ultimate distinction between left and right, between up or down, between surfaces of curvature positive or negative, and between temporal increase or decrease. Instead of traveling through the looking glass, the holographic image travels the other way: it draws out the image from inside the mirror to our space. The pseudoscopic and the orthoscopic images are symmetric in relation to time; when shaving or making up in front of a mirror, a person moves to the right to see the left side of his or her own image. To see the left side of a pseudoscopic image, the beholder has to move to his or her left side. To see the underside of the pseudoscopic image, the beholder has to move upward. Perception takes place in time, even if it is a reversible time.

This unusual perceptual experience has a correspondent model in Physics, where *time reflection symmetry* holds that any physical situation should be reversible in time. "*According to this principle, if time could be reversed (run backwards), the time reflection of a particular physical situation would correspond to what one would normally see by reflecting the situation in a space mirror, except that all the particles would be replaced by their antiparticles.*" In a similar but distinct way, the pseudoscopic image replaces relief by anti-relief, left-right parity by anti-parity, and direction by anti-direction. For the holographic artist, pseudoscopy might be more than a stereoscopic *relief inversion*; it might be one possible element for the creation of unexpected rhythms and images with highly unpredictable volumes.

In his investigation of the stereoscopic image, Helmholtz also noticed that binocular vision does not simply coalesce the two distinct images projected on the retinas. He advocated that if one of the pictures in the stereoscopic pair is white and the other black, the resulting picture appears to shine. This phenomenon, called *stereoscopic lustre*, according to Helmholtz proved that a *complete combination of the impressions produced upon both retinae* does not take place, because if that would be the case, *the union of white and black would give grey.*

We know that our normal perception of objects in the world out there depends on the fact that each eye sees a slightly different view of the same object, uniting both views into a coherent whole: what is called *cyclopean vision*. The regularity of this principle reflects the constancy of the conditions of observation that we have. This helps us in the elaboration of concepts about the world based on this regularity. It's more like the relationship between what is seen as solid objects and what is felt by touch as solid objects. The action of the senses as an interconnected system also reflects this regularity that grounds our binocular vision.

Although not having had any further artistic consequence before holography, what Helmholtz described as *retinal rivalry* in the stereoscopic image is a cosa mentale that challenges the regularity of our perception of three-dimensional objects in the world. This is how Helmholtz described it:

"There are some very curious and interesting phenomena seen when two pictures are put before the two eyes at the same time which cannot be combined so as to present the appearance of a single object. If, for example, we look with one eye at a page of print, and with the other at an engraving, there follows what is called the rivalry of the two fields of vision. The two images are not then seen at the same time, one covering the other: but at some points one prevails, and at others the other. Hence the retinal rivalry is not a trial of strength between two sensations, but depends upon our fixing or failing to fix the attention. If we leave the mind at liberty without a fixed intention to observe a definite object, that alternation between the two pictures ensues which is called retinal rivalry."

This phenomenon can also be observed holographically, but in a somewhat different way. First, in the stereoscope the two flat images are placed side by side, defining very clearly which eye sees what image. In the hologram, all images are recorded simultaneously on the emulsion, and it is the diffracted light that sends toward each eye a different image, making it harder to know *from where* the image is coming. Second, while in the stereoscope each image remains flat, in holography, each image can be fully three-dimensional, which makes focusing of the eyes more complex. Third, at last, vision in the stereoscope is motionless, while in holography, as in normal vision, the eyes move relative to what they see, scanning the spatial amalgam of the rivaling images.

Some holographic art pieces will display a very complex field, the components of which can only be seen, in discontinuous fashion, when the eyes of the beholder are placed in the direction of the diffracted light. In those cases, the amount of recorded information, and the way this information is presented, neutralizes the action of binocular vision as a system that unifies two different views into one. The expectations brought by the viewer are neutralized as well, and he or she cannot perceive individual solid objects but an inconsistent field, where objects and forms shift and coalesce.

Retinal rivalry becomes, therefore, one element in the vocabulary of holography, establishing a new visual parameter. In holography, it is not an exceptional structure or just a *curious experiment*, as Helmholtz describes it in the case of the stereoscope. It is a distinctive feature that checkmates our perceptual and symbolic conventions, for our response to visual stimuli depends, as we have seen, upon habits and conditions of observation. If each of our eyes perceives an absolutely different image, vision becomes a somewhat more complex and intense process, and demands with it that the beholder questions the very nature of his previous experiences.

Weaving the Photonic Webs of Time-Domain Holography

We need not examine in detail the technical development of automatic imaging systems (from early nineteenth century until now) to understand that it created the historical, aesthetic, and material conditions for the current digital synthesis of holographic images. In a clear development of painting's aspiration to truth and veracity, photography first attempted to fix images as seen in nature. The camera obscura, used by painters for centuries, became the photographer's essential tool. In the next stage, photographers tried to capture different moments of an action. As a consequence, Edison and the Lumiere brothers showed that images representing motion could not only be recorded as stills, but set to motion themselves, allowing us to see representations of the recorded events as a temporal flux. Much later, video technology instantiated the recording, eliminating the temporal gap between the action and its playing back; and therefore, reinforcing the congruity between the representation and the reference. More recently, personal computers seem to have demolished photography's truth ambition by allowing anyone to manipulate photographic images and to easily recombine them in any desired way. If photography forced painting to redefine its direction in the beginning and middle of the nineteenth century, today computers have a similar impact on photography. How does holography fit in this context? Holograms are already routinely synthesized from secondary sources, including silver photography, video, film, sensing devices, and computer graphics.

Many holograms created today involve electronic image manipulation and digital synthesis, and draw from other artistic fields such as photography, film, and video. These holograms explore time more intensively than holograms created without the computer, and reveal a very important aspect of the medium. Many of the computer holograms created by scientists or commercial holographers are motionless, or at best have very limited motion, because their images usually aim at reproducing a virtual environment or object with the visual stability typical of laser holography. Since the object in most laser holograms is three-dimensional and stationary, many holographers use the computer to make stationary virtual objects. Holograms thus produced emphasize space instead of time, and volume instead of movement.

There are seven domains of holographic temporal manifestation. Not suggesting a hierarchy or the absolute predominance of one category over the other, a critical taxonomy is proposed here based on the presence or absence of sequential imagery and the nature of its manifestation in space. While it is obvious that many of these identified time features will coexist in certain kinds of holograms, it is also true that in many cases one feature might be predominant in determining the kind of experience the viewer has with the piece. What follows charts holography as an art of time and organizes its structure according to seven principles:

1. **Symultaneism:** Holograms that present the viewer with one stationary object or construction organize the space so that it can be immediately and simultaneously perceived by two or more observers in like manner. This allows the viewer to probe a space that remains consistent for the duration of the experience.

2. **Time Suspension**: When a hologram is made that captures light phenomena as its subject matter, the dynamic behavior of light patterns is suspended at the moment of recording. Whenever illuminated, the hologram becomes active again, and in its diffractive power reactivates the patterned choreography of the original propagation.

3. **Freezing**: A holographic frozen moment implies the use of a pulse so short that it captures a symultaneist scene out of an original dynamic context, in which free motion is not a constraint.

4. **Linearity**: A linear hologram is one in which a series of images is stored in such a sequence that its frames are meant to be experienced in a particular order. Any other directional readings become distractive in this case.

5. **Time-Reversability**: Holographic images can be conceived to be seen in a flux that is bidirectional. No beginning is implied and no conclusive end is suggested. The experience of time-reversible holograms revolves around the possibility of eliminating oppositions such as forward and backward, since these two become equivalent.

6. **Discontinuity**: Discontinuous holograms break with the homogenous three-dimensional space reconstructed by a symultaneist hologram, shattering it into discrete viewing zones. These viewing zones can only be seen from restricted points of view. The space created is multifaceted, with controlled zones of visibility, gaps, and visual leaps.

7. **Real-Time**: Real-time holography is computed on-the-fly and projected freely in space in response to the viewer's command. Holographic videos can be controlled through the use of dials and other kinds of interface. Still in the early stages of technological development, this technique has not been explored by artists yet. The most prominent example is the apparatus being experimentally developed at the Massachusetts Institute of Technology. One day we will speak of real-time transmission of holographic images.[3]

Cinema is the art of organizing a stream of audiovisual events in time. It is an event-stream, like music. There are at least four media through which we can practice cinema—film, video, holography, and structured digital code—just as there are many instruments through which we can practice music. Of course, each medium has distinct properties and contributes differently to the theory of cinema. Each medium also expands our knowledge of what cinema can be and do. This observation is of particular interest, because it emphasizes holography as a time-based marketing medium, and not as a three-dimensional imaging technique. The prospect of digital holographic movies of the future notwithstanding, the multimedia nature of the computer compels us to a redefinition or, at least, to an expanded definition of what holography is. Computer holography, as practiced by a small but increasing number of artists around the world, multiplies the expressive possibilities of the medium and asserts time. It also is expressed as changes and transformations as its main aesthetic feature. Holograms become interactive events that can be perceived in any direction, forward or backward, fast or slow, depending on the relative position and speed of the viewer. Unlike the unidirectional *event-stream* of film and music, four-dimensional holograms are *buoyant events* with no beginning or end. The viewer can start looking at any point. Time is suspended from its extended continuum and can flow forward or backward.

Finally, the computer sparks a unique form of visual thinking in which visualization of concepts becomes almost instantaneous, and any hypothesis is tested by means of immediate practical experimentation. With holographic video and desktop holographic laser printers being developed today, it is clear that in the future, holography will merge with digital imaging systems in new and unforeseen ways. Design by holographic models may still be far in the future, but the concrete activity of design even now partakes substantially of immaterial techniques, or usage of artificial representations, images, and diagrams composed by image-generating machines. This is also true in the fine arts, and one can only expect this usage of immaterial techniques to increase. Or, the future of holographic art lies no doubt in its combination with video and/or computer art.

In Summary

- One of the facts that makes it difficult to discuss holography from a nontechnical point of view is the historical infancy of the holographic marketing medium.
- By trying to address some of the key issues of holography today, we are putting ourselves in the position of the essayist that, around 1869, tried to encompass the cultural meaning of photography. It is clear that by then, photography had already been popularized to a certain extent, but no argument should be further developed concerning the transformations that the medium went through in the following 120 years.
- When Nadar photographed Paris from a balloon, he could never imagine that an artificial satellite would in the future photograph Paris from outer space.
- We have already seen holographic images produced from satellite orbits. Would these images symbolize that more than 100 years from now, holography will play a major social role, comparable to what photography means socially today? It is hard to say.
- Holography is a new medium, and as such, is already remapping our perception of the world. Far from it being the object that antedates

the viewpoint, it would seem that it is the viewpoint that creates the object; and in holography, the viewpoint is an ever-changing one.

The next chapter discusses selling holographic art. It covers the problems encountered when trying to sell holographic art; holographic art and the conventional world; business considerations; improving relationships with the contemporary art scene; and ideas for increased success and recognition.

End Notes

1. Holos Gallery, 1792 Haight Street, San Francisco, CA 94117, USA, 2001.
2. Eduardo Kac, "Photonic Webs in Time: The Art of Holography," Friedman & Associates, Chicago, Illinois, 2001.
3. Ibid.

CHAPTER 18

SELLING HOLOGRAPHIC ART

In This Chapter

- Disappointments of selling holographic art in the past
- The problems of the present
- Holographic art and the conventional art world
- Business considerations
- Improving the relationship with the contemporary art scene
- Ideas for increased success

First, we must agree on a definition of what is *holographic art*. As with most art, definitions can be arbitrary, and holography seems to be especially susceptible to the problem of defining where *giftware* stops and *art* begins. For the purposes of this discussion, however, some distinctions and definitions should be made, arbitrary though they may be. Thus, most works 8" by 10" or smaller in size (produced in unlimited editions, often by companies as opposed to individuals, and retailed at around or under $200.00) will be considered *giftware*. In contrast, works generally 8" × 10" or larger (sold in limited editions or as one-of-a-kind pieces, normally attributed to a single individual artist or perhaps a collaboration between artist and technician, and retailing at or above $200.00) should in many cases be considered *holographic art*.[1] Artistic intent may also be considered:

was the holographer attempting to make an artistic statement, or was the holographer attempting to create an image strictly for commercial appeal and salability?

Of course, some larger works are definitely giftware, such as many mass-produced 30cm by 40cm film holograms, and some smaller works could be considered *art*; for example, the 4 by 5 inch three-color reflection holograms. These types of exceptions are consistent, however, with a method of definition that relies primarily on attribution of the work to an artist (as opposed to the anonymous company name identification), and size, price, artistic intent, and rarity of the work. The judging of the aesthetic content of the piece is simply too subjective a criteria to be appropriate to this chapter. For example, is a representational image of a soup can fine art (ala Warhol and Oldenburg) or simply a technical accomplishment (difficult though it may be) of little or no artistic merit?

A Disappointing Past

With the holographic art definition in mind, let us consider the early history (early 1970s) of holographic art galleries. In terms of selling holographic art (and an estimate as to what type of art sales were to be expected), this period was a definite disappointment. It soon became obvious that sales of holographic art would in no way sustain the business, thus forcing most sellers to initiate wholesale distribution businesses as a means of survival.

There are many reasons for this early lack of success in selling holographic art, one of the most important being the old cliche, *We were ahead of our time*. A vast majority of people entering the gallery in the late 1970s and early 1980s were seeing holograms for the first time. The axiom that people fear the unfamiliar is appropriate here. Holograms were totally unfamiliar to a majority of those visitors. Often, their initial reaction was that this was some type of trick or gimmicky new fad—certainly not something that was part of their everyday life or that they knew how to relate to in any meaningful way.

Compounding this problem was the ongoing perception that art should be bought as an investment, a concept that was even more in vogue in those days of high inflation and early *Reaganomics*. Not only were customers totally unfamiliar with the art form, but holographers were unable to provide them virtually with any of the types of data that support perceived value in the traditional arts. There were no auction records as a basis to compare and justify prices. Also, there were precious few reviews and critiques to help establish artistic merit and intent, and most of the criticism that did exist was quite negative anyway. Furthermore, there was no historical basis for predicting future increases in value for those concerned with the appreciation of their *investment*; and, there was no perceptible resale market that could offer potential customers any type of liquidity. Thus, many of the subtle *tools* holographers were reading about in those days as techniques for selling art were completely unavailable to them.

Another major problem at the time was the undeniable paucity of good quality holographic artworks and artists. Technical constraints were (and still are) most formidable, and the equipment was much rarer and difficult to come by. Thus, too often a *good* hologram was simply one that was bright and of decent quality. There was too strong a reliance on the novelty of the three-dimensional image, with too little attention to aesthetic content.

With so few artists at that time having the knowledge and/or equipment to make holograms, it seems in retrospect that there was little chance that one of them would have been a really great and inspirational artist capable of igniting the art world (and my customers) with his or her talent and vision. With tens of thousands of painters and sculptors in the world, odds were good that a few could be brilliant, inspired virtuosos. With perhaps only 30 or 40[2] dedicated full-time holographic artists in the world around 1980, what were the chances that one of them could be holography's Rembrandt or Da Vinci?

THE SITUATION FINALLY IMPROVED

Fortunately, as the years passed, the situation did improve. Art sales increased, both in the number of pieces sold and the prices they could command. Unfortunately, at no point in the history of hologram art galleries did sales of art ever come close to the point of being able to sustain the business. Indeed, the same unhappy situation remains to this day. There are no hologram art galleries in the world that are able to survive on sales of art alone. Virtually all that remain open are either subsidized by the government or their owners; or they have learned to augment their income through sales of holographic novelties and giftware; implementation of manufacturing and/or wholesale distribution—offering educational and consulting services; or providing custom hologram production. Nevertheless, an ongoing increase in sales was an upward trend and thus a cause for cautious optimism. A number of factors contributed to the increased market for holographic art in the 1980s.

Certainly, the big breakthrough was the sudden appearance in the early 1980s of millions of holograms on the covers of *National Geographic* magazine and the Visa and MasterCard credit cards. Granted, these were not *art* holograms, nor were they intended to be. Yet they served to bring holography, for the first time in history, into almost everyone's life on a day-to-day basis. All of a sudden, gallery visitors would look at hologram displays and comment, *Oh, those are just like the cover of the* National Geographic *I got last month*, instead of saying *What are these things? They must be some kind of fad or trick!*

A related factor that aided in raising public consciousness was the growth in venues where the public could see and experience holograms in person. Galleries and museums dedicated to holography began to appear with a certain regularity, and holographic novelties and gifts were introduced into general gift markets outside of specialty stores. Traveling holography exhibitions came into vogue and attracted excellent attendance at a wide range of venues. *Mega-exhibitions* (large group shows featuring wide ranges of at least some very high-quality holographic art works and displays) appeared for the first time and were often catalysts for an explosive growth in the public's awareness of holography, especially in Europe. Such exhibits served the added important function of providing often life-sustaining income to both the holographic artists whose works were purchased in good quantities to grace the exhibition walls, as well as to early holography companies and distributors whose goods were sold for profit at the ubiquitous *gift shop* accompanying the exhibit.

The 1983, *Light Dimensions—The Exhibition of the Evolution of Holography* at the Octagon Gallery of the National Centre of Photography in Bath, England, for example, caused an absolutely incredible explosion of interest in holography throughout all of Great Britain.[3] The association with the National Centre of Photography, combined with the attendance of Princess Margaret at the opening, created an almost overnight growth in the respect and credibility accorded the holographic medium in that country.

Thus, the critical process of exposing the public to holography and thereby transforming the frighteningly unfamiliar into the everyday commonplace had begun. Moreover, this subtle change was of utmost importance in helping the public to accept holographic art. With these new assurances of validity provided by large, well-known, trusted institutions, people could begin to accept that holography was a viable means of expression. It became much easier to *validate* the medium in the minds of potential art-buying customers as well.

Concurrently, of course, we must give tremendous credit to the dedicated holographic artists, technicians, and researchers who were working so hard during this time period, often with little or no recompense, to better the art and science of holography. Technology provides the fundamental palette upon which the holographic artist relies. Thus, important improvements during this period in holographic technologies gave artists more and better tools with which to experiment and create. New methods evolved for making multicolor images, brighter images, new types of animated and moving images, collages and multimedia creations, and generally larger images. Mass-production techniques greatly improved and proliferated as well, unleashing a torrent of new holographic stickers and novelties into the public domain.

Furthermore, equipment did become slightly more available, and more and more schools and universities were adding holography labs and instruction to their programs. While many of theses labs were in the Physics departments, some did find their way into Art departments as well. The addition of holography programs by such prestigious institutions as London's Royal College of Art and The Art Institute of Chicago was quite noteworthy. The association of holography with such bastions of the classical arts helped tremendously to further validate the nascent art medium, while also providing a new talent pool of practitioners trained in both the classical arts and the technological side of holography.

Let us certainly not overlook, too, that we were all coming to understand that a high-quality, very bright holographic image did not necessarily equate to *good art*. Throughout the 1980s, holographers were learning to shed their reliance on the medium's novelty and to pay attention to the form and content of the artistic statement being attempted. Bright and colorful was increasingly becoming a means to an end as opposed to the end itself, and thus the technology began to assume its proper function as one more tool for artists to use to create art.

Though at first it may seem trivial, there was one other technological improvement during this period that deserves mention, and that is that holographic artists were able to obtain new and better light sources throughout the 1980s. Certainly one could not ignore lighting when educating prospective customers about holographic art. By the end of the 1980s, however, a whole new generation of small, inexpensive, quartz halogen spotlights emerged. Finally, holographic artists had good, effective lighting to offer their customers that did not cost more than the hologram they were thinking of buying.

The concrete result of all of these factors was an appreciable improvement in sales of fine art holograms. The finer and more varied inventory of art works meant a much greater chance of finding a piece that would satisfy a particular customer's tastes, needs, and budget. Finally, too, holographers could recite a list of accomplishments for holography and convince the customers that they were at the least making a reasonable investment in a new form of art. They found that they were not only selling more art works, but they were commanding higher prices as well. For the first time, holographers were beginning to sell holograms they had purchased in the late 1970s for several times their original cost. Thus, the holographers could honestly tell their customers that some works were truly appreciating, though they were always cautious not to guarantee anything!

Present Problems

In spite of all of these factors, though, we now come to the present and find the news is still not totally encouraging. Since 1990, art sales have completely leveled off and actually slowed in the higher price categories ($2000.00 and above).[4] From recent informal telephone calls and interviews conducted by holographers, it seems the leveling trend has not yet reversed. The market for holographic art remains weak, and galleries striving to market and sell the best (and usually highest price) holographic art works report dismal sales results in the top price ranges. Now, let's examine briefly what seems to be the underlying factors for these problems to best suggest how we can work to improve the situation in the future.

An important factor over which we have no control is the general economic climate. As we moved into the 1990s, the United States and much of the world moved into a recession. As people found themselves with less discretionary income, sacrifices had to be made; and the choice between food on the table or a hologram on the wall was pretty easy to make. As this is one area over which we have such little control, it bears little further discussion.

Note: Sales in all types of art have leveled off at best, and the record art auction prices of the 1980s have not yet been matched in the 2000s. Let us simply try to lay the groundwork today to best attract the art buyers of tomorrow, as the world's economy improves.

Economic factors have also contributed to a decline in the number of new hologram galleries opening and the number of holography *mega-exhibitions* being assembled. We cannot deny that these two areas were prime markets for holographic artists to sell their holograms. Any new entity preparing to enter the business of displaying holography to the public had to buy holograms to fill its walls. Most of these entrepreneurs were savvy enough to realize that the best and highest quality holographic art works available at the time would offer them the best chance of attracting public attention; good publicity, possibly reviews, and the word-of-mouth advertising being so crucial to survival.

Thus, holographic galleries were all more than pleased to offer their best hospitality (and holograms) to any visitors announcing that they were planning to open a hologram gallery or exhibition, whether they were from England, Japan, California, or anywhere in between. We must realize that with the current slowdown in the startup of new display holography enterprises, a sales void has been created, and there is no strong base of dedicated individual collectors to fill the vacuum.

Another current problem is that most art holography is still presented to the public primarily in dedicated medium-specific venues. The problem here is, as mentioned earlier, that hologram galleries cannot survive on sales of holographic art alone. Thus, nonsubsidized gallery owners eventually face the reality that they must augment their income, often by adding holographic *giftware* to their merchandise selection. Holographers are all coming to realize the damage this does to their attempts to promote holography as art. However, this arrangement has in the past and will continue in the future to limit the growth in the holographic art market.

The problem is that the art works and the novelties become too closely associated simply by virtue of physical proximity in the gallery environment. The fact that the holographers knew that the *art* was on one wall and the *schlock* was on another was not often perceived by gallery visitors. Thus, the novelties and their presence ruined the atmosphere that experienced art buyers expected to encounter. It made it obvious that the gallery could not survive by selling art, which in turn could not help but force art buyers into second-guessing their own buying decisions. The subtle but damaging implication is that this must not be good art and certainly not a good investment, since it does not sell well. Granted there is the exceptional art lover secure enough both financially and in trusting his or her own instincts to go ahead with a purchase, but this remains the exception. We must accept the fact that the majority are followers, not leaders, and that they will usually require a much different atmosphere than the conventional hologram gallery can provide before they will make the higher priced purchases.

Future Outlook

Let us now look toward the future and try to illustrate some of the possibilities for transforming the weaknesses of today into the strengths of tomorrow. The main thrust of this discussion will center around a crucial goal for all of us: to work together to bring holographic art out of its own small network of holographic galleries, museums, and exhibitions and into the infinitely larger world associated with the other fine arts. Holographic art will not prosper until we make such a concerted and well-planned assault on the conventional art world.

THE CONVENTIONAL ART WORLD AND HOLOGRAPHIC ART

Before proceeding, however, we must address the concerns of those who do not believe this is necessary or desirable. Some would say that the holographic medium is so new and revolutionary that it must define its own place in the world, outside of an art establishment based on thousands of years of pre-holographic history. Yet if we accept this view, we are denying ourselves a grand opportunity. The reality is that holographic-specific venues have rarely been successful at selling enough holographic art to support themselves or the artists they show. Artists produce art but must make money to survive, and thus there must be a way for the art to find its audience. This audience defines a *market*; in this case, the fine-art market. If we insist on isolation from the conventional art world, we are isolating ourselves from an already well-established network that successfully brings art to the art-buying *market*. Thus, we can insist on *reinventing the wheel*, or we can accept that there is an already established, well-functioning mechanism in our society for selling art and realize that it can help us in our endeavors as well.

Indeed, considering some of the developments in the past year alone, holographic artists may have little or no choice in the matter. Some of the most well known and oldest galleries that featured holographic art and artists on a regular basis have, under their new ownership, dropped their rotating exhibitions of fine art holography and have become little more than another earring and gift boutique. For example, Light Fantastic, the famous London hologram gallery, has closed its doors completely. Should this inauspicious trend continue, the holographic artist wishing to display his or her works will simply have no choice but to seek out more conventional venues for exposure.

CONSIDERING THE BUSINESS

We have now crossed the delicate boundary between *art* and *business* by bringing up the issue of selling art. Some would perceive this whole area as the antithesis of why they became artists. However, an understanding of basic business concepts is important to all holographic artists and should be studied by those who are not yet fully acquainted with such information. The reality is that there are almost no agents or gallery owners out there today who will sell your works for you. Even if you are lucky enough to find such individuals, you must make sure they are not going to take advantage of you. Therefore, fundamental knowledge of rudimentary art business and sales techniques is essential for holographic artists to succeed. There are plenty of books, classes, and seminars available on how to sell and market art. All holographic artists should take advantage of these learning opportunities. Some business acumen is simply one more asset that enhances your chance for success.

Exposure to art marketing and business techniques can only help one to better prepare for presentations to gallery and museum personnel, and gives one a much better idea of what they will expect. Simple yet important details should be known and in place in order to make the professional type of presentation that offers the best chance of success. The look and content of resumes, how reviews and letters of recommendations are composed and presented, and how slides or videos of the works are photographed and displayed are important and should not be considered unworthy of attention. Take time and care with all of these details; make your entire presentation highly professional, and you will find the people you approach more willing to listen to what you have to say and to look at what you have to show. We all wish that the art itself could stand alone, but that is most often simply not the reality. In such a highly competitive world, you must make your best effort in all areas of your presentation.

Knowledge of contemporary business practices in the arts is also essential when you do find the gallery owner or agent who is willing to consider working with you. A good dealer should work hard on your behalf and will help validate what you do to the outside world. When considering entering into this type of relationship, however, such issues as exclusive versus nonexclusive representation, what type of exposure and exhibitions you are to receive, who pays for the exhibitions and how much, how the unique display and lighting requirements of holography are to be dealt with, what other services or remuneration the owner or agent will provide you in return for his or her representation, and how much the owner or agent is to be compensated must all be negotiated (a familiar concept to the artist). Do not be surprised, for instance, if the price you are used to seeing your works sell for increases twofold when your work first appears on the gallery walls. The gallery owner must recoup his or her expenses and cover overhead; and a decent mark-up on your artwork is the only way to do that. Remember, however, this higher price may not necessarily be a detriment to sales. To some customers' way of thinking, the higher price equates to higher-quality art and actually helps legitimize your efforts.

Gallery exhibits are an important addition to any resume and may help holographic artists gain entry into the important realm of museums. The two venues do have their pluses and minuses, yet each offers excellent opportunities. Museums have large followings and are typically not concerned with selling what they exhibit. Thus, museums will sometimes be more apt to try something new, especially those museums with a track record of presenting new media. As such high-visibility locations, however, they can be inundated with information and proposals, and one often needs a *connection* to get in the door. Such connections should be cultivated whenever possible, but the lack of one should not stop anyone from trying.

Conversely, galleries may be much more receptive to your artwork if you can list some museums that have shown your work. Gallery owners know that a record of prior exhibitions can aid them in their efforts to sell and market your work to their clientele. Galleries that feature multimedia and innovative art forms of all sorts are probably the best to approach. They have already demonstrated a more open, forward-looking state of mind and have probably cultivated a more open-minded clientele as well.

Indeed, the gallery's clientele is critical to its (and the artists') success and one of its most important assets. Successful gallery owners work extremely hard to build up a clientele through the years who will trust that owner's taste and judgement. We must garner the respect and attention of such owners in order to have a chance of reaching their prime customers. Such customers will be tremendously more inclined to seriously consider buying your art if they see it on the walls of a gallery whose owner they have come to trust and respect.

THE CONTEMPORARY ART SCENE: IMPROVING THE RELATIONSHIP

In addition to basic art business practices, it is critical that we not overlook the importance of educating ourselves about the history of art and the contemporary art scene. Holography is revolutionary and offers the artist countless ways of doing things not available ever before. However, if you agree that we need to take advantage of the marketing opportunities in the conventional art world, then we must be somewhat resigned to playing by their rules and following their trends. One can best come to understand these elements by acquainting oneself with the history of art and the current gallery scene.

Gaining such knowledge accomplishes several things. For one, it allows us to converse with the art establishment in its own language and with the appropriate historical and contemporary references if needed. We cannot deny that there is a stereotyped image in the art world of the holographer as a technician; and we must work to change that stereotype. To further that goal, we may need to swallow just a little of our pride. Thus, it is not out of place here to recommend that holographic artists read the art periodicals and literature, familiarize themselves with the contemporary art scene, and incorporate some of these elements into their holograms.

In other words, it is not completely inappropriate here to be telling any of the many, many fine holographic artists in the world what to put in their holograms. Simply put in the interest of a shared common goal to change widespread misconceptions, it is hinted here that we may want to add more contemporary elements to our contemporary works. Granted, holographers may never wish to play by all the rules of the art establishment, but we still need to let them know we understand their game and are at least willing to be in the same ball park.

Art curators stress that holography should engage in the issues of today and should attempt to find its own place in the current multimedia mix of film, video, audio, computer graphics, and other new art forms. They stress that without a knowledge of and concern for the current modes of presentation and fashion in the art world, holography will find little acceptance. This does not mean that holography should abandon what is uniquely holographic. Rather, holography should use its uniqueness to interpret the contemporary fashions of art in its own distinct way. Distasteful as these ideas may be to some of us, they do convey some important points of view from the contemporary art establishment with regard to how to improve the acceptance of holography in that community.

 Note: Some might view these unwavering curators and critics as our enemies. However, one must know as much about the *enemy* as possible in order to best survive and eventually triumph.

Mention is also made here of the numerous logistical problems associated with displaying holography. We must give thought to overcoming these problems and minimizing their importance. The problem of documenting one's work, too, is extremely difficult in holography. Video and stereo photography should not be overlooked as useful tools for minimizing some of these problems. Such technical issues, however, are merely roadblocks that will be put up by those looking for excuses. Artists who can successfully incorporate some of the forms, colors, and content of the contemporary art scene into a cohesive and impressive body of work will find that such minor technical obstacles will not impede their success.

RECOGNITION AND INCREASED SUCCESS

By now, it is becoming clear that the future outlook for holographic art is rather bleak if we cannot break out of our own tight display and exhibition network. In addition to the suggestions already mentioned, let's touch on a few other ideas to aid in this transition. For example, the concept of *buying oneself* a career is an extremely crass notion, but not without merit. Sad but true, you can buy credibility if you can afford it. At some point, it is feasible that a well-to-do holographic artist and/or exhibitor might wish to initiate an advertising campaign in the established art media. Entire careers in the arts have been generated through proper publicity and promotion.

Without the expense of advertising, however, every artist can and should mount a simple but effective public relations campaign. The cost is close to nothing and is thus within the capacity of all artists. Create or borrow a list of local and national media, art critics, gallery owners, and museum personnel, and keep the list up to date at all times. Then, make up brief but polished press releases for any and all of your activities, and send them to everyone on your list on a regular basis. A few years later when you feel you are ready to knock on their doors, your name and activities will already be familiar to them, and you should receive a better reception. Also, if we all make this effort together, sheer volume will indicate to these people that there is something happening here that they can no longer afford to ignore.

Critics should be included on the list, as we must get over our fears of the dreaded art review. Let's not debate here whether a bad review is better than no review at all. If we disdain all critics, however, we will never have the opportunity to get a good review either. In any case, we must be willing to listen to what the critics tell us, and in some instances we will find we can prosper from the publicity they provide.

Some artists are having good success with commission work. This is another area that will greatly increase public awareness about and acceptance of holographic art. Thus, holographic artists should learn how the system of awarding public art projects functions, and should be sure they are plugged into the system. When requests for art in public spaces appear, holographic artists should be prepared to submit sound ideas and proposals. These corporate- and government-funded projects are a good opportunity for exposure and income.

In Summary

- As we continue to look toward the future, perhaps the most important thing is that we never lose faith in ourselves and our chosen art medium.
- Remember that holography remains singularly lacking in advocates outside our own realm, and so we must be our own strongest and loudest advocates.
- We must approach conventional galleries and museums everywhere, and we must do so in the most professional and competent manner possible.
- Each small success in this venture will most likely be accompanied by myriad rejections.
- As long as we continue to believe in our own skills and holography's unique abilities, we will be able to endure the criticisms and rejections, and ultimately survive and prosper.
- In the long run, our belief will be contagious. No one will take holographic art seriously if its practitioners cannot demonstrate, at all times, a strong and inexhaustible faith in the holographic medium.

- With continued dedication, hard work, perseverance in the face of adversity, a strong consensus of shared goals in the community, and above all, a 100-percent faith in what we are doing, success is only a matter of time. It is hoped that the information presented herein will help make that time come sooner than later.

The next chapter opens Part Four, "Computer-Generated Holography," and discusses computer-generated holograms on the Web. It covers the engine memory, interface design, and implementation issues that are required to create computer-generated holograms.

End Notes

1. Gary A. Zellerbach, "Selling Holographic Art: An Analysis of Past, Present, and Future Markets," Presented at the Fourth International Symposium on Display Holography, Center for Photonic Studies, Lake Forest College, Lake Forest, Illinois, (published in Proceedings SPIE, Vol. 1600 © 1992 SPIE), Holos Gallery, 1792 Haight Street, San Francisco, CA 94117, USA, 2001.
2. Ibid.
3. Ibid.
4. Ibid.

PART IV

COMPUTER-GENERATED HOLOGRAPHY

For those of you creating 3D computer images, this part of the book discusses the ability to transfer your work into actual three-dimensional holography displays. The process presented is quick, easy, and surprisingly affordable. Some computer-generated holography customers mentioned here are the NYU Medical Center, Mitsubishi, The American Museum of Natural History, Michigan State University, and Allied.

CHAPTER 19

COMPUTER-GENERATED HOLOGRAMS ON THE WEB

In This Chapter

- Engine
- Memory issues for the master fresnel zone plate
- Translation and superposition
- Interface
- Design issues
- Intelligent parameter generation
- Accepting user-supplied images
- Implementation

There are many software programs around that quickly create computer-generated holograms of two-dimensional images. There are two aspects to these hologram programs: the computation engine and the World Wide Web interface. A hologram program, for instance, is optimized for speedy computation on a workstation, and should have enough flexibility to be able to produce computer-generated stereograms. The Web interface should be designed for user-friendly, worldwide access, and also for high durability.

The Engine

When holography was being developed in the 1960s, interference patterns were generated physically, using laser light and high-resolution photographic film. However, with the recent development of inexpensive, powerful computers, it has become increasingly feasible to generate holograms computationally.

One stumbling-block in computer-generated holography has been information overload. Since optical holograms must have detail on the order of the wavelength of light, there is an enormous amount of information to be accounted for. Creating a diffraction pattern with a 0.5 micron pixel diameter translates into 4 trillion pixels per square centimeter. If each pixel is rendered in 256-level grayscale, that's 4 terabytes to be accounted for, per square centimeter. Furthermore, in some cases, every piece of the hologram is affected by every part of the image; thus, there is an enormous amount of calculation involved.

The computation engine for most hologram programs is designed to run on a standard workstation, which implies limited memory and computation speed. Thus, the engine makes considerable compromises in order to reduce computational demand.

One tactic is to take advantage of superposition. First, the engine computes a hologram whose image is a single point. Such holograms are known as fresnel zone plates. By superimposing many fresnel zone plates, it is possible to build up an image pixel by pixel. For plane wave illumination, the same fresnel zone plate may be used for all pixels in a given depth plane. As the fresnel zone plate is translated and superimposed along the x- and y-axes of the hologram, x and y translated pixels of a fixed depth are created in the image.

Most hologram programs only calculate a single fresnel zone plate. This *master* fresnel zone plate is translated and superimposed onto the hologram. Superposition is a very computationally inexpensive task. The trade-off is that the image must be flat (all pixels in the same depth plane). Also, illumination must be done with a plane wave. Converging or diverging illumination, often used in physical holography, is not permitted.

MASTER FRESNEL ZONE PLATE MEMORY ISSUES

The first question is how to represent the master fresnel zone plate. While it is possible to keep generating the fresnel zone plate on-the-fly, we can achieve a shorter overall computation time by calculating once and then storing the pattern in memory. For simplicity, let's assume the hologram will be illuminated orthogonally, resulting in a circular, radially symmetric fresnel zone plate.

There are three formats in which the fresnel zone plate pattern can be looked up with little or no computational overhead as shown in Figure 19-1.[1] One option is to take a square, two-dimensional array and inscribe the zone plate. Another choice is to only record the first quadrant of the zone plate and take advantage of symmetry. The final approach eschews rectangular two-dimensional arrays entirely. By stringing together many one-dimensional arrays of assorted lengths, we can approximate the shape of a (quarter) circle.

In Table 19-1, fresnel zone plate diameters (2r) for typical workstation memories (N) are listed.[2] The quarter circle methods come out as winners. In addition to saving a whole bunch of memory, this coordinate frame is much more natural to deal with, with the origin at the center of the fresnel zone plate.

While a polar coordinate system (not shown) would be the hands-down winner in terms of memory, superimposing translated polar coordinate

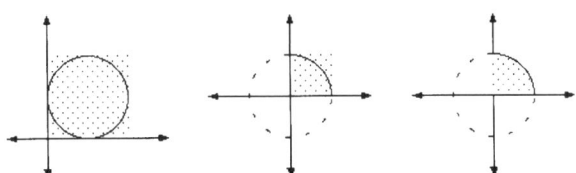

FIGURE 19-1: Fresnel zone plate pattern formats.

TABLE 19-1. Fresnel zone plate diameters for typical workstation memories

		Plate Diameter (in pixels) for Memory of				
		1M	2M	4M	6M	8M
Method 1	2r = sqrt(N)	1K	1.4K	2K	2.4K	2.8K
Method 2	r = sqrt(N) 2r − 2 * sqrt(N)	2K	2.4K	4K	4.9K	5.7K
Method 3	(1/4) * pi * r² = N r = 2 * sqrt (N/π) 2r = 4 * sqrt(N/π)	2.3K	3.3K	4.5K	5.5K	6.4K

systems is too computationally expensive. In the Cartesian coordinate system, translation and superposition are quite simple.

If the hologram is not to be illuminated by an orthogonal plane wave, some of the symmetry is lost. If the illumination beam comes in at a non-normal angle memory, requirements for methods 2 and 3 double. The coordinate systems x and y may be chosen so that the illumination beam is only offset in one axis; therefore, it is always possible to take advantage of some symmetry.

FRESNEL ZONE PLATE CALCULATION

A fresnel zone plate can be calculated by mimicking physical holography. In physical holography, a point source of light illuminates the plane of the hologram. A standing wave is created by interfering the point source's wavefront with a reference plane wave. The resulting interference pattern is then recorded on photographic film.

The phase of a light wave varies periodically with the distance it has traveled. For red helium neon laser light, a periodic 2 pi phase difference occurs every 633 nanometers (corresponding to the wavelength). If the wavefront of the point source is being interfered with the wavefront of a normal plane wave (which is constant across the plane of the hologram), we need only consider the phase of the point source's wave. Whenever the path of the point source to the hologram varies by a wavelength, there will be a periodic variation in the hologram. This explains the concentric circles in a fresnel zone plate.

Therefore, to generate the fresnel zone plate, just consider how far away each pixel of the fresnel zone plate is from the point source. From that distance, one can deduce the phase of the wavefront.[3]

$r = \text{sqrt}((x-x0)^2 + (y-y0)^2 + (z-z0)^2)$
$\phi = \phi_{ref} + \cos(2\pi/\lambda)*r$

When the illuminating plane wave is not orthogonal, the phases of both light sources must be considered. Fortunately, a non-orthogonal plane wave presents a linearly varying phase that is easily calculated.

In physical holography, the difference in phases between the two wavefronts describes whether the light will constructively interfere, destructively interfere, or partially interfere. The resulting intensity (proportional to the square of the interfering amplitudes) is recorded on film.

Nevertheless, you may instead use *bipolar intensity* by simply taking the cosine of the two phase differences. This results in negative as well as positive values, something with no physical analog since there is no such thing as negative light. The values are uniformly raised later to create a hologram. The computation engine takes advantage of this relatively fast method for generating a fresnel zone plate.

A key issue when generating the fresnel zone plate is knowing when to stop. Specifically, aliasing must be avoided at all costs. In a fresnel zone plate with orthogonal illumination, spatial frequencies increase as they get further away from the point source. Since the fresnel zone plate is essentially generating discrete values (pixels), under no circumstance may the discrete frequency exceed pi. In other words, pixel on, pixel off, pixel on, pixel off is the maximum spatial frequency representable without aliasing. Since this is a two-dimensional image, you have to back off even further since the pixels along the diagonals are spaced more sparsely than the pixels along a row or column.

In most hologram programs, the fresnel zone plate is only calculated until the Nyquist sampling limit (or an even more stringent, user defined limit) is achieved. This places a limit on the size and shape of the fresnel zone plate when other factors (distance of the point source, resolution of emulsion, angle of illumination beam, etc.) are fixed. For the orthogonal plane wave illumination, the fresnel zone plate will be circular, while non-orthogonal plane wave illumination produces an oval shaped pattern.

Note: According to the Nyquist theorem, the discrete time sequence of a sampled continuous function { $V(t_n = n * T_s)$ } contains enough information to reproduce the function $V = V(t)$ exactly, provided that the sampling rate ($f_s = 1/T_s$) is at least twice that of the highest frequency contained in the original signal $V(t)$ where: $f_s = 1/T_s$ is the sampling frequency; $V(t)$ is the value of the signal (voltage) at arbitrary time t; and, $V[n] = V(n * T_s)$ is the value of signal at time $t = n * T_s$.

SUPERPOSITION AND TRANSLATION

The next issue is the translation and superposition of the master fresnel zone plate onto the emulsion of the hologram (see the sidebar, "Aligning the Images in UV Space"). It is important to keep track of borders and boundaries during the superposition process. Missing a boundary means an incorrect hologram, or possibly software suicide. Copying information into unallocated memory at high speeds will likely result in a program crash.

■ ALIGNING THE IMAGES IN UV SPACE

You should align the images in UV space. This means transforming both the fresnel zone plate and the holographic emulsion to a common coordinate frame for the superposition. Transforming the coordinate systems allows for methodical bookkeeping for boundaries. In Figure 19-2, u is the independent variable, and v is the dependent variable.[4] For example:[5]

Transforms

$x^1 = |u|$ $\qquad x^2 = u - u0 + a/2$
$y^1 = |v|$ $\qquad y^2 = v - v0 + b/2$

Fresnel Zone Plate Limits

$u = [-r, r]$
$v = [-\text{sqrt}(r^2 - u^2), \text{sqrt}(r^2 - u^2)]$

Rectangular Emulsion Limits

$u = [u0 - a/2, u0 + a/2]$
$v = [v0 - b/2, v0 + b/2]$

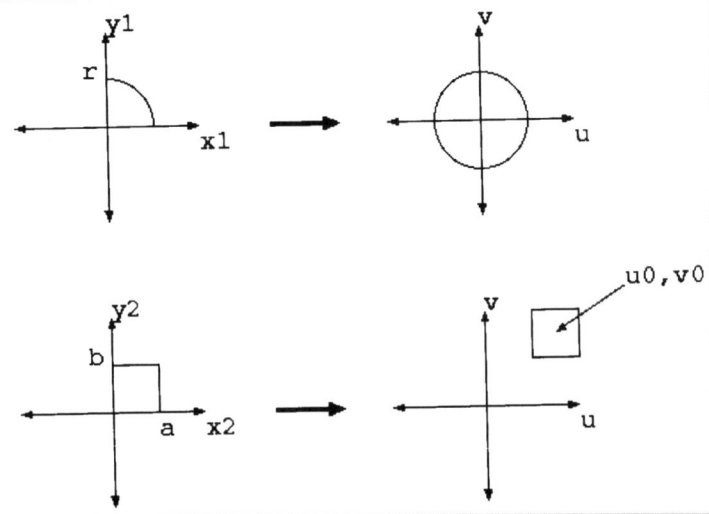

FIGURE 19-2: Independent and dependent variables.

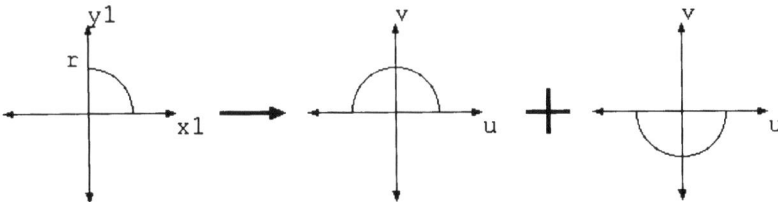

FIGURE 19-3: Independent and dependent variables.

TABLE 19-2: Boundary Calculations

Variable	u	v
Lower bounds	-r, u0 - a/2	-sqrt(r² - u²) v0 - b/2
Upper bound	r, u0 + a/2	sqrt(r² - u²), v0 + b/2

Finally, we need to optimize the inner loop for speed. By breaking the fresnel zone plate into two parts, we can simplify one of the transforms. Instead of having v = abs(y1), we consider the two cases separately—v = y1 and v = -y1. That optimization takes the absolute value operation out of the inner loop as shown in Figure 19-3.[6]

Thus, the superposition module is fairly straightforward. First, figure out the appropriate overlaps in U,V space. Then, using the transforms, find out where the actual pixels reside (in X1,Y1 and X2,Y2 space) and add them together. While boundary calculation uses computationally expensive operations, the number of boundary calculations is insignificant compared to the number of pixel superpositions as shown in Table 19-2.[7]

TRANSLATION AND SUPERPOSITION CODING

In practice, the C code shown in Listing 19-1 has been optimized so that the brunt of the superposition consists of incrementing two pointers and adding one's content to the other's.[8] It is fast enough, so that on the implementation machine, the performance bottleneck comes from the RAM propagation delay.

LISTING 19-1: Shows both pseudocode and readable (unoptimized) C source code.

```
Given: x0, y0, a, b, r
compute v limits
for u = vmin to vmax
compute u limits
for v = umin to umax
    using the transforms, find x1,x2,y1,y2
    copy the appropriate pixel

void superimpose(int u0,int v0) {
  int u,v,vmax,vmin,umax,umin;  /* for u-v space */
  int x1,y1,x2,y2;              /* for x-y space */
  int t1,t2;                    /* temporary variables */

  t1 = r;                       /* set the u-limits */
  t2 = u0 + a/2;
  umax = lesser(t1, t2);
  t1 = - r;
  t2 = u0 - a/2;
  umin = greater(t1, t2);

  for (u=umin; u<umax; u++) {

      x1 = abs(u);              /* the u transforms */
      x2 = u - u0 + a/2;

      t1 = r*sqrt(r*r - u*u);   /* set the v-limits */
      t2 = v0 + b/2;
      vmax = lesser(t1, t2);
      t1 = -r*sqrt(r*r - u*u);
      t2 = v0 - b/2;
      vmin = greater(t1, t2);

      for (v=vmin; v<vmax; v++) {

         y1 = abs(v);           /* the v transforms */
         y2 = v - v0 + b/2;

         EMULSION[x2][y2] += MASTER[x1][y1];  /* copy the pixel */
      }
   }
}
```

Web Interface

The user interface is a significant part of computer-generated holography. Besides traditional interface design issues, a user interface has the difficult task of passing many interrelated parameters to the computation engine. More importantly, the interface will determine the usability of the software, greatly affecting the overall value of computer-generated holography. Historically, computer software in general suffers from a serious flaw. Programs tend to be written and rewritten, because of either poor distribution, hardware incompatibilities, or interface obsolescence. Many programs are only used a few times by the author and then fade into oblivion. This problem, along with other related issues, is occasionally referred to as *the software crisis*.

Avoiding the software crisis was the primary motivation for a World Wide Web interface. It provides easy, global access, and the interface is expected to have a high durability. The World Wide Web interface assures that a hologram program can be run by many people for a relatively long period of time. That provides important motivation to the program author that computer-generated holography will not be a wasted programming effort.

INTERFACE DESIGN

There are many issues involved with interface design. The basic question is how to represent the computational engine in a meaningful and understandable way. The conflict of simplicity and flexibility is a major issue. Other design considerations include durability and accessibility. Finally, limitations of the Web interface need to be addressed.

The complexity of the user interface is a thorny issue. One goal is to be as general as possible, and the other is accessibility, simplicity, and ease of use. In order to meet conflicting demands, two separate input interfaces can be created—a very simple one for casual users, and a serious interface offering lots of control. The simplified interface only allows the user to pick the type of hologram being created and start the computation; thus, no knowledge of holography is required.

For the more complex and flexible interface, the first thing considered is who will be using the hologram program and why. The primary user will probably be knowledgeable in diffraction and interference phenomena, and will use the hologram program only occasionally to create a graphical image representing an interference pattern. For this purpose, most hologram programs provide a well-documented step-by-step approach. The user is guided into filling out a multisection form, sprinkled with text and drawings. While a more compact representation is possible, this method should allow a knowledgeable user to run the hologram program without any separate documentation.

After filling out the form (which comes preloaded with reasonable default values), the selection may either be sent to the engine or to an analysis program that estimates what is going to happen. All aspects of the interface use a standard, off-the-shelf Web technology known as fill-out forms to provide wide accessibility.

Working with the Web means inheriting a whole set of interface characteristics, some of which are even desirable. For example, the hologram program is set up to allow easy *bookmarking* or storage of a particular configuration. This is a convenient benefit, since entering all the parameters can be a laborious process. Unfortunately, the Web interface also has some faults. Chief amongst these shortcomings is the difficulty of returning the results of computationally intensive tasks. When a computation takes several minutes, or possibly hours, it is unreasonable to keep the user's Web browser waiting. Network connections will time out, sockets will close, and so forth. One solution is to tell the user where to pick up the completed hologram and approximately when—but this is usually not a very satisfactory result. Fortunately, a hologram program should be fast enough to provide near instant gratification when the holograms being computed are small (perhaps on the order of what is easily displayed on a computer screen).

PASSING INTELLIGENT PARAMETERS

A flexible interface allows greater control of the computation engine. However, specifying a hologram requires many parameters, and giving the user complete control might produce unoptimal or worse, inconsistent constraints. Therefore, the computer takes many parameters from the user and then calculates the rest in order to assure efficient and consistent holograms.

At this point, the user should be able to easily specify what is going on from a pixel-centric view of the world. If someone has a picture in his or her mind of what the hologram should look like, he or she should be able to produce that hologram without a lengthy trial-and-error process tuning distances, resolutions, and so forth. In addition, efficient utilization of computational resources is encouraged. Users should be able to easily avoid particularly annoying situations such as computing a uniform fresnel zone plate (such as when you just look at the center of it—highly magnified).

With these goals in mind, users are provided with select parameter choices. They can adjust the wavelength of light being simulated, as well as size and resolution of the emulsion, and the pixel size of the fresnel zone plate in the *x* direction. Units are taken from physical holography; thus, resolution can be specified in line pairs per millimeter, wavelength in nanometers, and so on.

Users may also specify a very important parameter: how close they would like to get to the Nyquist sampling limit. Since aliasing occurs if there are less than two samples per cycle, this is a very convenient parameter to control. Adjusting the sampling limit allows a great deal of control from a pixel-centric viewpoint. The user can directly limit what discrete frequencies are allowed. Most hologram programs make sure that the fresnel zone plate is calculated efficiently. Thus, the outer edges are exactly at the Nyquist cutoff specification.

Because of parameter interrelations, some control has to be sacrificed. Specifically, the user is not given direct control over the depth of the image. Since image depth is a function of film resolution, the number of pixels allocated to the fresnel zone plate, and Nyquist bravado, the hologram program computes image depth along with other dependent parameters and reports it back to the user.

For example, the user might specify that there will be no fewer than *n* samples per cycle (where *n* is presumably greater than 2). Thus, we know that at the edge of the fresnel zone plate, the spatial frequency will be:[9]

$$f = \frac{1}{n * T} \quad \text{where T is the sampling period}$$

First, the focus along the *x*-axis is determined. Since the illuminating plane wave is not allowed to be offset along the *x* direction, then:[10]

$$\sin(\theta_{out}) = m \lambda f$$

The projection depth *d* can be solved by simple trigonometry: where light starting at the edge of the fresnel zone plate will intersect the *z*-axis as shown in Figure 9-4.[11] Once the projection depth has been decided, we can

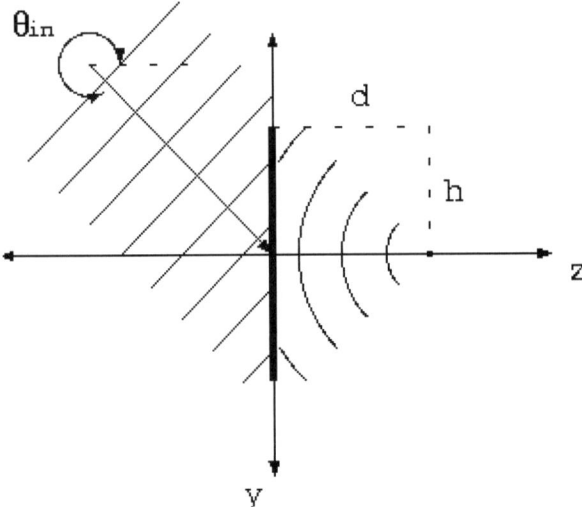

FIGURE 19-4: The projection depth.

figure out the *y* dimension of the fresnel zone plate. By considering the upper tip of the fresnel zone plate, we know that:[12]

$$\sin(\theta_{out}) = m\,\lambda\,f + \sin(\theta_{in})$$
$$\tan(\theta_{out}) = h\,/\,d$$

Since $m = -1$, we can solve for h, the length of the fresnel zone plate along the *y* direction. This can easily be converted into pixel values, since the resolution of the emulsion has been specified. Both the *y* dimension of the fresnel zone plate and the depth of the image are returned to the user.

HOLOGRAM IMAGE SELECTION

Even more important than tweaking with parameters, the user must be able to specify the subject of the hologram. There are two ways to select the image for the hologram. One way lets the user choose an example hologram from a menu. Choices include a one-point image, a two-point image, a square, a circle, and several other options.

In addition, arbitrary two-dimensional images are supported. To keep things running quickly, only binary images (where each pixel can be represented by a single superposition) are permitted. Grayscale, or varying intensity images will require a scaling operation in addition to the superposition—which is currently too computationally expensive. Besides choosing from a few canned sample images, the user is also allowed to upload his or her own picture files for immediate holographic computation. Currently, most hologram programs only accept images in the TIFF file format.

The ability for the end user to compute holograms of arbitrary 2D images greatly enhances the power of the hologram program. This added feature helps boost most hologram programs out of canned demos and the *neat toy* stage. Engineers or scientists can calculate holograms for desired output. This feature also opens up the potential of stereograms. Stereograms are composed of a series of two-dimensional images, forming a discrete approximation of a three-dimension scene. Most hologram programs can in theory compute the holograms of many different two-dimensional images that can then be combined to produce a stereogram.

INTERFACE OUTPUT

At the end of a computation, two items are returned to the user. First is the hologram itself, in the form of a graphical image. Next, and equally important, are statistics and analysis of the hologram calculated and of the computation itself.

The hologram is rendered to one of two popular graphics file formats, GIF or TIFF. While helpful, the compression offered by these formats barely makes a dent in the huge file sizes of larger holograms. In addition, a raw uncompressed image data is temporarily stored on the server machine itself, potentially offering yet another output format.

The hologram program should be designed to provide the user with a lot of computation statistics. These statistics may be either as a record of computation or an estimate of what is to come. Computation time is tracked, and broken down stage by stage. Analysis of the spatial frequencies are required and available, and the proximity to limits (such as the Nyquist limit) are provided.

The Implementation

Computer-generated holography demands computation speed and efficient use of memory. The Unix operating system and ANSI C language are normally chosen because of their relatively low memory overhead and high performance. The implementation machine is usually a 90-Megahertz Pentium-based PC, running a Linux operating system with 32 megabytes of RAM. The hologram program should be compiled using the GNU C compiler (GCC); and, using maximum optimization during compiling.[13]

 Note: The GNU C compiler (GCC) is a C compiler that offers portability to a great number of systems and fairly good optimization. There is also a C++ front end, sometimes called g++, which is often distributed with GCC. Because the GCC back end is used for multiple languages, the term *GCC* is often used to encompass those compilers as well.

By making all output available over the Web, it is not necessary to run a graphical windowing system on the host machine—thus, freeing up additional memory. Virtually all of the computer's resources should be devoted to calculating and storing holograms.

Despite the optimizations for speed, most hologram programs should be readably portable across Unix and ANSI C, especially since there is no graphical user interface component to the program. The common gateway interface to the Web server, the only other potential source of portability problems, should be implemented using a popular C library.

SOURCE CODE

The source code should consist of three sections as shown in Table 19-3.[14] The first section, dubbed *meat*, contains the brunt of the hologram computation. The second section, dubbed *fluff*, contains interfaces, disk i/o, and other administrative code. Finally, the third section, *grams*, contains

TABLE 19-3: Source code and interfaces

Source Code	Interfaces
• Makefile	
• meat.c	
• grams.c	
• fluff.c	Simplified interface
• bipolar.h	

• and additional libraries:
  • Uncgi[15]
  • Tiff[16]
	Advanced interface

code for making different types of holograms. In addition, the hologram program should make use of programming libraries for graphics handling and World Wide Web interfacing. The total hologram program size (not counting libraries) should be approximately 1,200 lines of code.

PERFORMANCE

The computational bottleneck appears to be RAM propagation delay. It requires approximately 60 nanoseconds to retrieve a pixel value from memory. For example, a hologram with a 1-million pixel emulsion might require two memory transactions for every pixel for every superposition (one for loading, one for storing). Given 1,000 superpositions, that's 2 billion memory transactions, or two minutes. Computing an image with tens or hundreds of thousands of pixels can take quite a while.

For tiny holograms (those with fewer than 1 million pixels, and with only a few hundred points in the image), the overhead associated with computing the fresnel zone plate and storing the completed hologram on disk dominates computation time. For larger holograms, the superposition time dominates. Superposition time rises at order N with the area of the hologram, and order N for the number of points on the image.

While memory fetches from RAM may not be very fast at 60 nanoseconds, other memory devices like disk-based virtual memory would be prohibitively slow. Thus, the size of the fresnel zone plate is limited by the amount of available RAM. The entire hologram could be stored in RAM as well; however, this could result in further straining the limited amount of memory available on the computer.

QUALITY OF OUTPUT

Most hologram programs can calculate small holograms. The output of a hologram program is simply a computer graphics file—transferring that pattern to a medium that diffracts light is beyond the scope of this chapter. However, most hologram programs have nowhere near the capability to calculate holograms as detailed as those produced by optical means. In summary, the laser is not out of business yet.

Fortunately, there are other applications. Holograms for larger wavelength phenomena can be modeled quite nicely (this could include acoustic waves, radio waves, water waves). The domain is not limited to light. In addition, the hologram program should be able to rapidly produce simple interference patterns that can be used in an educational setting.

Further Work

Most hologram programs can still use some improvement. Most computation engines are currently restricted in size to what can fit into memory. It should be fairly straightforward to allow unlimited-sized emulsions, which are calculated piece by piece. Also, the cleanliness of the computation engine's handling of Leith and Upatniks' holograms could be improved.

Note: In 1962, Leith and Upatnieks at the University of Michigan removed Gábor's brain child from the shelf and gave holography its rebirth. Like Gábor, they did their early experiments with a filtered mercury arc lamp. Leith and Upatnieks invented the off-axis reference beam with all its great advantages, which they did not even appreciate at the time. After the development of the continuous wave gas laser in 1960 by Ali Javan et al, Leith and Upatnieks started using the laser and discovered the three dimensionality of the images. They performed these experiments as an adjunct to their work in side-looking microwave radar. They independently discovered off-axis holography, only to find that Gábor had proposed holography 12–14 years earlier.

On the interface side, there needs to be more polishing—the user may currently specify a hologram too large for the machine to safely handle. While that's not an issue in a development environment, releasing the program to the public requires additional safeguards. Also, the interface could easily add additional services to feed into the computation engine. One example is to create a hologram of text submitted by the user.

As computers continue to increase in power and memory, hologram programs will be able to compute larger and more complex holograms. In addition, reasonably priced ultra-high-resolution printing technology may someday become available. Once these technologies are in place, hologram programs will finally start producing satisfying computer-generated holograms (or stereograms).

In Summary

- Computer-generated holography has succeeded in many of its goals.
- Holographic programs provide simple computer-generated holography to a global audience, with all calculations done on a workstation.

- There are several possible uses for a holographic program. First and foremost is a rapid visualization tool for interference. The program is capable of providing beautiful examples of interference in a short amount of time.
- Hologram programs may be used as a design tool for people creating holograms of two-dimensional images.
- Hologram programs are not limited to the domain of light—they can design holograms for any media that follows wave physics. This includes water, sound, radio waves, etc.
- Hologram programs can certainly generate enough detail to handle longer wavelength media.
- Unfortunately, hologram programs will not be challenging optical holography any time soon, for two reasons. First, even with today's inexpensive computer power, it is not feasible to calculate anything but the tiniest optical holograms. Next, even when such holograms are calculated, there is no simple and inexpensive way to render that hologram to the correct scale on an optical material.

During the last few years, the number of holography sites on the Web has increased rapidly and now includes institutions, private individuals, commercial holography companies, publishers, and enthusiasts. The next chapter reviews several holography sites and presents a case study of the development of one of those sites.

End Notes

1. Jeff Breidenbach, "Computer-Generated Hologram on the Web," Advanced Undergraduate Project, Electrical Engineering and Computer Science, MIT, 19 Hunting St., Apt #2R, Cambridge, MA 02141, USA, 2001.
2. Ibid.
3. Ibid.
4. Ibid.
5. Ibid.
6. Ibid.
7. Ibid.
8. Ibid.
9. Ibid.
10. Ibid.

11. Ibid.
12. Ibid.
13. SourceGear Corporation, 3200 Farber Drive, Champaign, IL 61822, USA, 2000.
14. Jeff Breidenbach, Advanced Undergraduate Project, Electrical Engineering and Computer Science, MIT, Cambridge, MA 02141, USA, 2001.
15. Hyperion Solutions Corporation, 1344 Crossman Avenue, Sunnyvale, CA 94089, USA, 2000.
16. SGI, 1600 Amphitheatre Parkway, Mountain View, CA 94043 USA, 2000.

CHAPTER **20**

INTERNET HOLOGRAPHY

In This Chapter

- Uses for the Internet
- Holography on the Internet
- Holography and the World Wide Web
- The digital advantage
- Design considerations
- Commercializing the Web
- The future of holography on the Web

Holography is now featured regularly on the Internet, particularly on the World Wide Web (Web), which provides text, still and animated graphics, and video and sound information on an ever-expanding catalogue of subjects to an ever-expanding audience of users. During the last year, the number of holography sites on the Web has increased rapidly and now includes institutions, private individuals, commercial holography companies, publishers, and enthusiasts. Is what they provide useful, and are there any major benefits to this system that could not be achieved with more traditional methods of communication?

This chapter provides an answer the preceding question. A review of several holography sites is provided, and a case study of the development of one of those sites is given in detail.

Internet Usage

As you know, the Internet is a computer network made up of many smaller networks all capable of communicating with each other. It started about 30 years ago as ARPAnet, an American Defense Department experiment,[1] and is now made up of government, academic, commercial, and private computer networks from around the world: a network of networks. The early uses were for the exchange of research and data electronically. It has since developed into a carrier for news, discussion, databases, sound, video, and interactive multimedia services. During 1994/5 the Internet became a popular topic for the world's media. There was hardly a newspaper, magazine, or TV program that did not refer to it and its effect on our society.

The Internet and its services have become visible and popular. Users are now attracted from outside academic circles, to include home users with personal computers who access the system via modem and telephone line. Computer enthusiasts were always there, but now it is easier for the non-specialist, with little or no computer or electronics knowledge, to get *on-line*. Internet service providers (companies who charge you a fee for providing access to the Internet and its services) have become more active and numerous. The opportunity now exists for people to use the Internet from most major cities around the world. The Internet family has grown quickly, now estimated at 370 million users worldwide.[2]

One of the main uses of the Internet is the transmission and reception of electronic mail. A message typed into a computer in Columbus, Ohio can be delivered to the recipient's computer in Melbourne, Australia automatically within seconds. Mail is stored on reception and read by the recipient when it is convenient to do so. The problems of time zones, like with paper mail, is removed, making the whole process convenient for both sender and receiver.

Large amounts of information, such as computer software, can be collected using the Internet to access a remote computer and "download" files, software, or other digital material for use on the local computer. Anything that can be digitized can be transmitted over the Internet, including sound and vision. As a communications system it is vast, accessible, flexible, global, and it works. The Internet is not owned by any one person, company, or government, and it has no political or geographical borders. Its social borders are limited to people with money who can purchase a computer and the access services they need to connect to the Internet, or to those who do not own the equipment but have access to it either via academic or community suppliers. There are also Cyber cafés where visitors can drink coffee and "surf" the Internet. Such establishments have been attractive and newsworthy for the media and helped *sell* the Internet, and the virtual world it supports, to millions.

Note: The Internet Society (ISOC) does exist with a responsibility for technical management and direction of the system. It is, however, a voluntary membership organization and does not own the Internet.

Note: Public libraries, exhibitions, and community centers are installing computer terminals, either permanently or as a promotional event, to connect to the Internet either for free or at a low cost.

The World Wide Web

There are many sections of the Internet designed to achieve particular results for specific users. One area that has grown quickly over the last five years is the World Wide Web, referred to as "the Web." This is a multimedia area that allows the transmission of text, graphics, sound, and video using an extremely simple user interface. The Web was developed at CERN,[3] the European Particle Physics Laboratories. What started life in a scientific research facility has become the most active and attractive multimedia section of the Internet. Web users receive pages of material, text, and graphics in color. Sections of these pages are interactive. Text, for example, can be highlighted (often in blue) to produce a hypertext link. If a user places his or her cursor on any of this highlighted text and then "clicks" the mouse, he or she will be provided with a new page of text or graphics. What is interesting is that this new page of information need not be held in the remote computer to which the user was originally connected. The new page can come from any other computer around the world that is currently connected.

ELECTRONIC GLOBAL MAGAZINE

It is, for example, possible to produce a Web page, made up of text from a computer in Paris, one picture stored in a computer in Tokyo, and a second picture stored in San Diego. The software program, or Web browser, used to access the system deals with all these connections, collects the individual pieces of information, and composes them for the user on their own computer. The system can be completely transparent to users, the result being a page of information that can be viewed on the screen or printed out for later reference.

 Note: One of the most popular Web browsers is Netscape, which offers the end user extensive features. It has been given away for free, using the Internet and the Web as a distribution network, and is consequently installed on many computers worldwide.

GENERAL INTERACTIVITY

Because any area of a Web page can be made to take the user to more (and different) information, the system can be considered interactive. Using only text for this example, a story can be written with several endings that can be selected by the reader. Characters in the story can have background descriptions that can be selected by the reader, should they require further information. References to a text can be hypertext linked to other texts, which are then made available to the reader in full.

If graphics are included, a particular picture can be linked to, for example, a larger version of the same image, a detail of the image, a text description, or a completely different Web site. It is also possible to produce links in the pages to audio and video files, so a picture can become animated or sound can be played back. Again, anything that can be digitized can be accessed from Web pages using these links. In effect, the Web provides simple *point-and-click* access to all of the facilities the Internet offers in one user interface.

SPECIFIC INTERACTIVITY

The Web is interesting because the flow of information is not one way, from the remote host computer to the local receiving computer. Feedback can be sent from the user to the host. Recent versions of Web browser software offer users the opportunity to fill in forms on screen, with specific details that are then transmitted back to the host computer. This is not revolutionary; it has always been possible to send electronic messages across the Internet. What is exciting is that the feedback can be automatically processed and placed back into the Web pages to be read by other users. The Web pages can be constantly updated with new information.

A simple example is the search facility. Using one of the many Web search pages, a word can be typed into the page. This is then sent back to the host computer, which searches its database for all references to it, and then returns a new page to the user with the results, often in a matter of seconds. These results will normally be hypertext linked to the full information requested. The page and its contents will be designed and presented automatically for the person who requested the search. Another user, searching for a different word, will receive a different page layout with different content.

Early Attempts at Internet Holography

There has been interest in providing an accessible, electronic, information service for holography for many years. One of the first attempts was a project aimed at constructing an online forum so that people could exchange ideas and experiences relating to holography in all its forms. Established in August 1985 at the New York Institute of Technology (NYIT), USA, the project was a conferencing system called Participate. This was a text-only system (the Web had not yet been invented) and was to cover several conference areas, including:

- Holography Bulletin Board
- Classified Ads for Holography
- Holographic Chemistry
- Society of Holographers
- Holography Questions

The preceding areas were announced at the Second International Symposium on Display Holography, Lake Forest, Illinois, and was hosted, a short time later, by NYIT on their computer system. There were several holographers signed on to the system who began using it as a discussion forum, but before it could flourish, a change of policy at NYIT resulted in a halt to the project. A second attempt by the holographers, this time in Denver, Colorado, could not proceed due to lack of funding and a loss of momentum.

Most of what holographers wanted to do now exists in different parts of the Internet using different host computers, and distribution networks. This situation has developed due to the reduction in the price of computers, the reduction in the price for connection to the Internet, the increased use of e-mail as a communications device, and the massive expansion of the Internet.

THE SOLUTION IS E-MAIL

Many people working with holography are able to use e-mail. Scientists, researchers, and educators have academic access that is free. Independent artists can use service providers for private access for a small fee. This means that questions and comments can be exchanged between colleagues and members of the holography community. Messages tend to be short, precise, and answered quickly.

Note: Institutions have to pay for maintenance of their computer networks and connection to the academic networks, but in general, the individual academic user is not charged for the use of e-mail.

SPIE'S SOLUTION

The International Society for Optical Engineering (SPIE) started an e-mail service for holography in 1993 aimed at members of the Holography Working Group and other interested parties.[4] This is a listserver that allows electronic messages to be sent to a computer server owned by SPIE and then redirected, automatically, to all *subscribers* to the system.

Response to the service has been very good and is a regular source of information for its users. SPIE provides other specialist listservers for a variety of interest groups. The advantage of this service is that it is automatic. You can send one message and it will be distributed to all members of the server. The disadvantage is that *junk* mail is also easy to send to many people—mail that has little or nothing to do with the target group. Complaints against this abuse of the system prompted SPIE to provide human intervention, meaning that messages sent to the system are read by a human before further automatic distribution. This is seen by SPIE as a way of filtering out the *junk*, rather than censoring the mail.

The Holography listserver has been a success. *The Internet and e-mail have provided a means of global communication on a massive scale. When a student in the UK posted a message asking a question from the group regarding a subject for a term paper he was working on, he was inundated with responses. That's the beauty of the listserver; people helping others within the same technological interest to grow and expand the field.*

The World Wide Web and Holography

The multimedia aspects of the Web have made it one of the fastest developing areas of the Internet. The ease of use for the end user and provider, together with facilities offered—interactivity, text, photos, video, and sound—make it attractive, impressive, and useful.

RAPID WEB HOLOGRAPHY DOCUMENTS INCREASE

Holography has found a home on the Web. There are now several holography sites, with the numbers increasing every few weeks. Some indication of the rapid increase in sites can be seen by carrying out a search request using one of the Web search services. These sites are various, ranging from dedicated holography sites, through Resumes/CVs for individuals, to research organizations, commercial marketing companies, and *holo* enthusiasts.

The increase in holographic sites is significant, but not conclusive. Search sites and search engines on the Web are not universal. They find references to information in their data banks alone, not the Web as a whole. Searching for *holography* on another site with another data bank will return different results. It is clear from these results alone that expansion in this area is rapid.

WEB HOLOGRAPHY SITE DEVELOPMENT

One of the first sites to offer a regularly updated series of holography pages was HoloCom (as shown in Figure 20-1),[5] which started its Web server at the beginning of 1995. It now receives about 75,000+ *visitors* a day, giving some indication of the interest generated in the service. Not all of these *visitors* are for the holography pages, but if only a fraction want holography, this still means that there are a great many people looking and this is only one of many Web sites.

It is possible, for example, to visit the *Internet Webseum of Holography* (as shown in Figure 20-2)[6] and look round their gallery spaces and shop. There is also the *Holos Gallery Online* (as shown in Figure 20-3)[7] with featured artists and their work. Or visit the exhibition of Russian Art Holograms that was on show in the Canadian *Royal Holographic Art Gallery*.

The MIT Museum, which now has a huge collection of holograms (purchased from the Museum of Holography in New York), also has a presence on the Web with images and information about its collections as shown in Figure 20-4.[8] These virtual museums and galleries, together with their

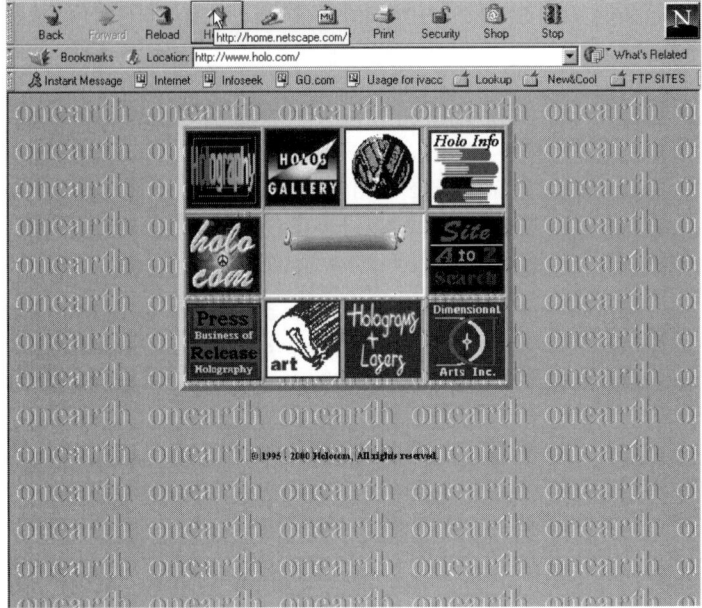

FIGURE 20-1: HoloCom site.

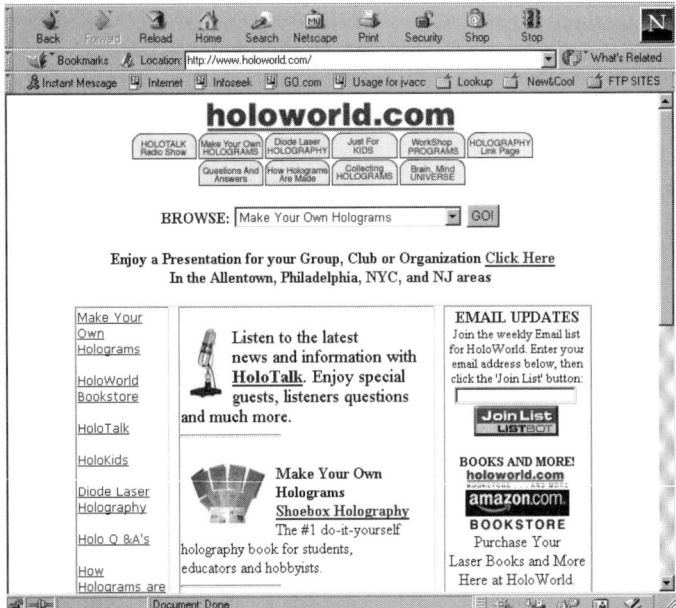

FIGURE 20-2: Internet Museum of Holography.

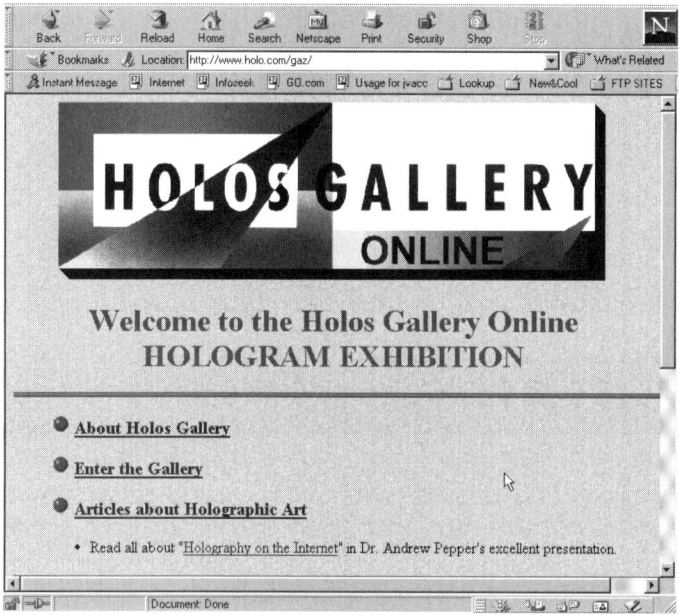

FIGURE 20-3: Holos Gallery.

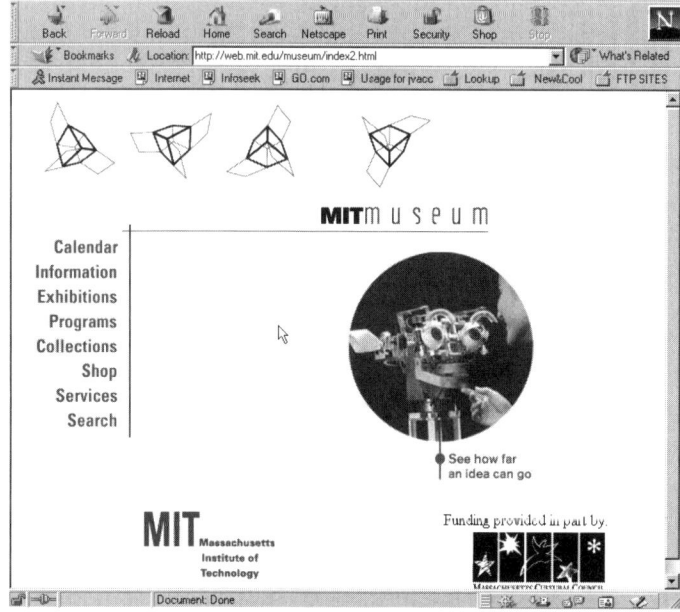

FIGURE 20-4: MIT Museum.

virtual shops, are growing. It is cheaper and quicker to put together a virtual museum than to build a physical one. But why do it? Well, one of the reasons to do it is because there is a void in holography that needs to be filled—hence the Internet Museum of Holography. As time goes on, it will function just as a museum in the *real world*, providing a source of information about the history as well as the future of the field of holography, along with exhibits and educational programs.

HOLO-GRAM: PUBLICATIONS ONLINE

There are not many publications dedicated to holography, but they are on the Internet. The *Holo-Gram*, a newsletter established in 1983, went electronic in October 1995 with its first Web issue. There you can learn about the SPIE symposium, the virtual holography classroom, the forthcoming International Symposiums, plus profiles of someone new to holography. If the computer system and peripherals (scanners and storage) are already available, then this form of publication is faster and cheaper than printing ink onto paper and mailing it to subscribers. It also means that information can be updated very quickly.

HOLOGRAPHIC NEWS

Holography News,[9] the well-known industry newsletter, uses a different strategy (see Figure 20-5). It is possible to obtain information about *Holography News* on the Web, with highlights of previously featured articles and news items; however, the complete texts are not there. This is a way of selling subscriptions to the full paper publication by providing Web users with a *taste* of the quality and quantity of information they can expect to find there. The *Holography News* site is receiving an average of 5,000 visitors each week. About 60 percent look at headlines, highlights, and specific stories, while 5 percent go to the subscription information. It still remains to be seen if this is an effective way of gaining new subscribers. The selling of new subscriptions of the paper publication via the Web appears to be about average as compared to other publications of a similar nature.

INDEX OF CREATIVE HOLOGRAPHY

The Creative Holography Index[10] was first published, on paper, at the end of 1992. It catalogs the work of artists using holography and commissions critical essays on the subject. It was launched on the Web in November 1995 as *Search The Light*, and forms the basis for the following discussion.

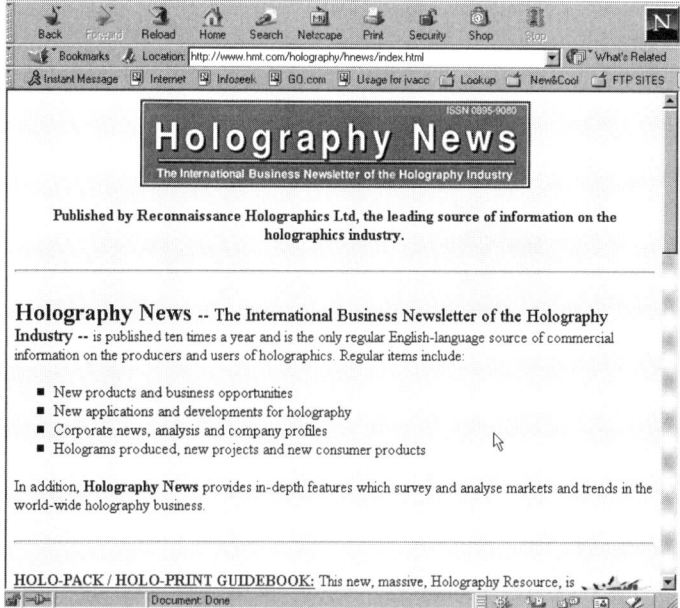

FIGURE 20-5: *Holography News.*

Search The Light

The Creative Holography Index is the only international catalog for creative holography and is collected by artists, curators, gallery and museum directors, media managers, libraries, and the general public worldwide. It currently weighs over two kilos. The cost of printing, in color, and distributing the Index is immense. It is also difficult and expensive to update. Simply updating the artist address list for the last 2000 issue took over seven months (people are slow to respond) and cost a considerable amount of money.

THE ADVANTAGE OF DIGITAL

As the Index is designed using desktop publishing systems, all the information, including its database, is digital. The transfer of this information into Web pages is quick and easy, so it was decided to experiment with an online version of the publication. This experiment was made possible with support from HoloCom,[11] who provided server space to host the site, and Monand Press, who did the page make-up and simple programming of the Web pages.

350 CHAPTER 20

FIGURE 20-6: *Search The Light.*

The advantages of an online version are considerable. The pages can be updated quickly and without great expense. When a new artist is featured in *Search The Light* (see Figure 20-6),[12] he or she is asked to send an updated biography and exhibition list to the Index electronically, where it is edited and placed directly into their electronic pages.

PRODUCTION SPECIFICS

Each page is produced on Macintosh computers in Europe. Scanned color images or graphic titles (which are produced using Adobe Photoshop) are combined with text information, which is taken from the original QuarkXPress pages that are used to produce and print the paper Index that is pasted into an HTML editor. The page is laid out and includes hypertext links for text and graphics. It is then sent to the HoloCom server in Houston, Texas, via File Transfer Protocol (FTP) over the Internet, where it is instantly accessible to Web users.

This HoloCom computer is a 240 rack-mounted purpose-built configuration with 4 gigabytes of hard disk storage, 128 megabytes of RAM running Berkeley software in a Unix system with the latest version of Pearl.

The connection from the computer to the Internet is made using a T1 link directly into the network, thus providing very fast access times for users wanting to connect to the Web sites stored in the machine.

Design Considerations

The Web is a design disaster! It is at the stage now that computer desktop publishing (DTI) found itself at the beginning. When DTI programs became available, they provided people using a computer with the opportunity to produce printed publications quickly and without professional knowledge of the printing or design industry. Very few of these computer users were designers or had any idea about design problems. Thousands of publications were produced that were design disasters. The situation has now settled down. Designers have learned to use DTI, and people with no idea about professional design have learned to employ designers who have the skills.

The Web is still currently in its early stages. Anyone with a computer and the right software can produce a *home page* and make it available on the Internet. This is both positive and negative. On the positive side, there is no one to ask for permission, and anyone can put anything he or she wishes onto the network. There is no censorship and no minimal design requirement. On the negative side, this means that there are tens of thousands of people who know nothing about layout, design, content, marketing, editing, and related subjects who are now electronic publishers. This does not mean that there are no Web sites with stunning graphics, excellent editorial ethics, and totally professional output.

Note: In academic organizations and companies, access to computer storage and the Internet may be restricted, needing authority from the system manager or owner. Anyone with a computer and modem and enough money to pay the connection to the Internet can run his or her own Web server from home.

FACTUAL DESIGN CONSIDERATIONS

The Web is peculiar in that it is so flexible. If an agency designs its pages *as ready to go out on the Web*, there is absolutely no guarantee that the layout will be seen by any of the end users in the way it was intended. The layout of the page at the user's end is dependent on which typeface his or her browser is programmed to display and the type of browser being used. This includes the size of the screen and the type of system (pages look different on a Silicon Graphics machine than on a Macintosh). Developments are so rapid now that changes to the systems can take place in a few days, making old versions of software obsolete (or at least old-fashioned). The situation will settle down and there is already software available to allow the accurate transmission and reception of Web pages being developed. One positive aspect to this rapid changing of standards is that almost all new software released is available for downloading to the user's computer via the Web and Internet. The handbooks are also online, and programming is completely accessible. It is possible to request the source programming of any Web page and see how it has been put together. Clever programming of pages and new ideas therefore spread rapidly.

SEARCH THE LIGHT DESIGN CONSIDERATIONS

Many Web sites are full of large graphic images, which take a long time to load into the user's computer once a Web page is requested. This time delay is a problem and frustrates many users. It was felt by the creators of the Web sites that although color graphics are an essential part of the Web, they should be used with caution in the Search The Light site. Graphic images for page title design within the page are kept small. The largest graphic images are the pictures of the artists' holograms. It was also felt by the creators of the Web sites that if people are visiting a site dedicated to the art of holography, they would be willing to wait for the downloading of the images of artists' holograms, and this would not be seen as a disadvantage. The location of these large images has also been considered. If a user requests a featured artist page, he or she will normally receive an introduction text first, at the top of the page, followed by details of the hologram featured. While the user is reading this, the large graphic image is being loaded in the background and much further down the page (off screen), so that the user is not aware of the time taken to load it. By the time the user has finished reading the text, he or she can scroll the computer screen to look at the color image, which is fully loaded. This gives the impression that although the site has some beautiful images, it remains fast to download; whereas, in fact, it loads at the same rate as other sites. The Search The Light site is very *clean* and simple. It has received many comments for this combination of simplicity and detailed information, not only from graphic designers, but artists and scientists, researchers, and the general public.

E-Commerce

The Web and Internet are full of places to buy things. There is a great deal of interest in using the Internet to transact business. This involves credit card companies exploring the possibilities of using the network, as well as completely electronic cash services that are being developed. People already buy things on the Internet; estimated in 1999, at 18 percent of users.

One of the subsidiary aims of the Search The Light site was to help sell the paper product on which it is based: and thus fund the cost of the electronic version. In the four years that it has been available, there have been very few sales. Either the market is too small for the product, the product is too expensive, or the selling aspect of Search The Light is so subliminal that users do not realize they can purchase something. There is, however, an *order* button on almost every page—so it is difficult to avoid.

Other holography sites have been more successful. HoloCom has sold products and services, and has reduced direct (paper) marketing activities to concentrate on electronic services and sold holograms, services, and arranged speaking engagements. Hypermedia Technologies,[13] a Web site including many holography pages (see Figure 20-7), has sold server space, and the companies who use the site have reported sales from their presence there.

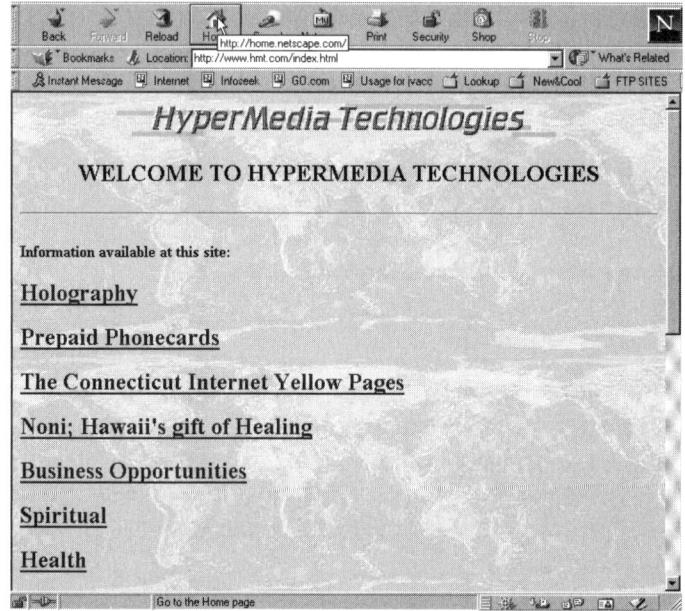

FIGURE 20-7: Hypermedia Technologies.

The Future of Holography on the Web

More and more people working with holography are being attracted to the benefits of placing information about themselves on the Web. In 1996, a new holographic service was started: HoloNet (see Figure 20-8)[14] is a project that will gather and transmit vast amounts of information about many aspects of holography with particular emphasis on the arts. In the original proposal it aims to direct its services at:

- Architects
- Curators
- Customers
- Designers
- Holographers
- Journalists
- Newcomers

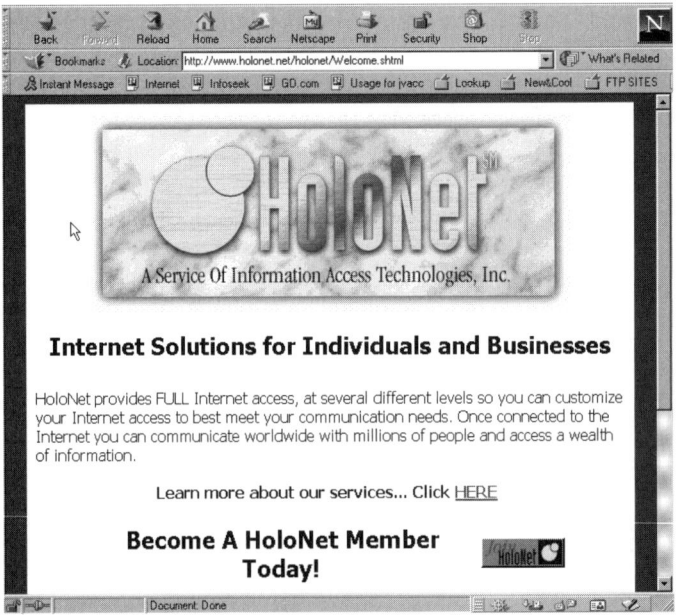

FIGURE 20-8: HoloNet.

HoloNet is a large system that allows artists to enter information about themselves, making it instantly available to other users. Research and collection of material and information for the HoloNet is well advanced. The GRAM archive project[15] has already discussed the possibility of providing data for the HoloNet database, *The Creative Holography Index*, and the *International Catalog for Holography*. Other providers are expected to allow their data to be stored there as the project develops. The main aim of HoloNet is to provide a powerful search engine to access its database of text and images. This means that users can type into the search page a word, or collection of words, and be provided with search results that will take them to specific pages or sites. Because of the nature of the Web and its hypertext system, this means that users will be directed not only to information held on HoloNet, but to other holography sites on the Internet. The HoloNet provides a global *clearing house* for holography, while allowing scientists, research facilities, commercial producers, manufacturers, artists, and enthusiasts to maintain their own presence on the Web. HoloNet is set up as a not-for-profit project and is independent from any company, academic facility, artist, or scientific grouping. It is by nature of its size and complexity, housed in an institution that helps maintain the technical requirements needed.

Finally, although the GRAM project collaborates with HoloNet, it produces its own Web site that includes information about holography (in French), together with data and images on other media within the Media Arts. GRAM has been collecting information, interviews, and images for several years, and its appearance on the Web provides an exceptional resource on the subject.

In Summary

- Holography on the Internet and particularly the Web is growing rapidly.
- There are sites that provide a high quality of information and comment. There are also those that are less successful. What is difficult at the moment is to find the right information.
- It is obviously not possible to request a Web search, receive over 3,600 references, and then look at all of them. The cost in time is

prohibitive, and the cost in telephone connections (for non-academic users) is too high. The system is, however, becoming more streamlined.

- It is now possible to search multiple databases on the Web with a single request. Search engines are becoming increasingly flexible. To narrow the search criteria to specific requests with the development of intelligent agents (software programs that search the Internet for you and come back with the results), the Internet will become more usable.
- The Internet will also become a victim of its own success.
- As more and more people use the network, it becomes slower to transfer the information, which is particularly irritating for Web users waiting for their megabytes of color graphic or video to be downloaded. In Europe, for example, it is often difficult to connect to American-based Web servers after lunch time (local time), because America is waking up and logging on to the network, thereby placing more traffic onto the system. Solutions will arrive in the form of faster connections from server to Internet and high-capacity transmission lines within the Internet. However, it appears it will get slower before it gets faster.
- Holography on the Internet is active and impressive.
- The networks and systems currently available can help the researcher locate the information he or she needs quickly, as well as often put them in direct contact with the author of papers or the maker of holograms. This can only help in the understanding of the holographic media and process and further the dissemination of information about it.
- Finally, what all of this means is that information is, perhaps, no longer power; the power will now rest in what an individual does with that information and how he or she interprets it. We can all have the information if we want it.

The next chapter discusses the use of holograms to store data in memories that are both fast and vast. It also covers the replacement of hard disks with data holograms that have 2,000 times more capacity.

End Notes

1. Andrew Pepper, "Holography on the Internet: A Useful Resource Or Expensive Distraction," presented at the 1996 SPIE Conference, San Jose, CA, SPIE Headquarters, 1000 20th St., Bellingham WA 98225-6705, USA, 2000.
2. Ibid.
3. CERN, The Laboratory straddles the Swiss-French border near Geneva, in two major sites: Swiss site: CERN, CH - 1211 Geneva 23, (Switzerland), French site: Organisation Européenne pour la Recherche Nucléaire F - 01631 CERN Cedex, (France), 2000.
4. SPIE Headquarters, 1000 20th St., Bellingham WA 98225-6705, USA, 2000.
5. Dimensional Arts, Inc., 401 Carver Rd., Las Cruces, NM 88005, USA, 2000.
6. Frank DeFreitas Holography, Allentown, PA, 18101, USA, 2000.
7. HOLOS Gallery, 1792 Haight Street, San Francisco, CA, 94117, USA, 2000.
8. Stephen A. Benton, MIT, 77 Massachusetts Avenue, Cambridge MA, 02139, USA, 2001.
9. Reconnaissance Holographics Ltd, 7105 East Powers Avenue, Greenwood Valley, CO 80110, 2001.
10. *The Creative Holography Index*, Monand Press. 46 Crosby Road, West Bridgford, Nottingham NG2, 5GH, England, 2000.
11. The Dimensional Arts Inc., The HoloCom Group, 401 Carver Rd., Las Cruces, NM, 88005, USA, 2000.
12. Ibid.
13. Martin Berson, HyperMedia Technologies Inc., P.O. Box 1282, Avon, CT 06001, USA, 2001.
14. HoloNet, Information Access Technologies, Inc., 2115 Milvia Street 4th Floor, Berkeley, CA 94704-1112, USA, 2000.
15. Georges Dyens and Philippe Boissonnet, *GRAM*, contact Université du Québec à Montréal, Département des arts plastiques, C.P. 8888, succ.A, Montréal, Québec, Canada H3C 3P8, 2000.

CHAPTER **21**

HOLOGRAPHIC STORAGE SYSTEMS: CONVERTING DATA INTO LIGHT

In This Chapter

- Dense holographic storage promises fast access
- Demonstration systems
- Mass memory applications
- Digital holographic storage system
- Holographic memories
- Data holograms of the future

Optical storage of data has been one of the bright spots in technology over the past 21 years. Compact discs, for example, dominate the market for musical recordings and are now also the standard medium for multimedia releases, which combine text, images, and sound. Video games, entire journals, encyclopedias, and maps are among the multimedia products available on CDs to users of personal computers.

Without a doubt, optical memories store huge amounts of digitized information inexpensively and conveniently. A compact disc can hold about 3.2 billion bytes—enough for six hours of high-fidelity music or more than 1.5 million pages of double-spaced, typewritten text. All indications are, however, that these large memories have stimulated demand for even more capacious and cheaper media like Digital Versatile Disc (DVD). Executives in the entertainment industry now have the capability to put one or more motion pictures on a single optical disk the same size as a CD via DVD. Moreover, so great are the data storage needs of some hospitals, law firms, government agencies, and libraries that they have turned to so-called jukeboxes that have robotic arms to access any one of hundreds of disks.

Note: Currently, DVD-Video is widely available in all formats. All major studios are now supporting the open DVD-Video format with over 6,400 titles available in North America, around 3,000 in Japan and a few thousand in Europe. Over 6 million standalone players are shipped in the United States. Even more have the ability to play the software on DVD-ROM drives. The growth rate of DVD-Video has by far exceeded those seen at the introduction of CD and VHS. Hollywood still insists on regional coding of players and software, despite that this often works against its intentions by preventing people from buying software published locally. Most players sold outside of the United States are modified to play in all regions.

Engineers have responded by trying to wring the most out of CD systems. Some are working on semiconductor lasers with shorter wavelengths (in effect, these will be finer styli that permit closer spacing of bits on a CD). Others are investigating techniques of data compression and *super-resolution* that also allow higher density (the latter at the expense of increased background noise). Another promising development has been multiple-level CDs, in which two or more data-containing tracks are stacked and read by an optical system that can focus on one level at a time. Such schemes are expected to push the capacity of CDs into the hundreds of billions of bytes within five years or so.

However, to pack a CD-size disk with much more data—a trillion bytes, say—will require a fundamentally different approach: holography. The idea dates back to 1963, when Pieter J. van Heerden of Polaroid first proposed using the method to store data in three dimensions.

Holographic memories, it is now believed, could conceivably store trillions of bytes of data, transfer them at a rate of a trillion or more bits per second, and select a randomly chosen data element in 50 microseconds or less. No other memory technology that offers all three of these advantages is as close to commercialization—a fact that has compelled such large companies as Rockwell, IBM, and GTE to launch or expand efforts to develop holographic memories.

Initially, the expense and novelty of the technology will probably confine it to a handful of specialized applications demanding extraordinary capacity and speed. Such uses are already attempting to carve out little niches—one recently offered product holographically stores the fingerprints of those entitled to enter a restricted area, permitting access when a matching finger is placed on a glass plate. If in meeting such needs the technology matures and becomes less expensive, it might supersede the optical disk as a high-capacity digital storage medium for general-purpose computing.

The main advantages of holographic storage (high density and speed) come from three-dimensional recordings and from the simultaneous readout of an entire page of data at one time. Uniquely, holographic memories store each bit as an interference pattern throughout the entire volume of the medium.

How Do Holographic Memories Work?

The pattern, also known as a *grating*, forms when two laser beams interfere with each other in a light-sensitive material whose optical properties are altered by the intersecting beams. Before the bits of data can be imprinted in this manner in the crystal (see Figure 21-1), they must be represented as a pattern of clear and opaque squares on a liquid crystal display (LCD)

FIGURE 21-1: Memory crystal could trap 100 trillion bytes of data in 3D.

screen, a miniature version of the ones in laptop computers.[1] A blue-green laser beam is shined through this crossword-puzzlelike pattern, or page, and focused by lenses to create a beam known as the *signal*. A hologram of the page of data is created when this signal beam meets another one, called the *reference*, in the photosensitive crystal as shown in Figure 21-2.[2] The reference beam, in this case, is collimated, which means that all its light waves are synchronized, with crests and troughs passing through a plane in lockstep (indeed, such waves are known as *plane waves*). The grating created when the signal and reference beams meet is captured as a pattern of varying refractivity in the crystal.

After being recorded like this, the page can be holographically reconstructed by once again shining the reference beam into the crystal from the same angle at which it had entered the material to create the hologram. As it passes through the grating in the crystal, the reference beam is diffracted in such a way that it recreates the image of the original page and the information contained on it. The reconstructed page is then projected onto an

Figure 21-2: Holographic memory stores data in a crystal of lithium niobate not much larger than a sugar cube (foreground). The hologram is created in the crystal by the meeting of a reference laser beam, shown thick and bright in this photograph, and a signal beam, fainter and thinner, which contains the data.

array of electro-optical detectors that sense the light-and-dark pattern, thereby reading all the stored information on the page at once. The data can then be electronically stored, accessed, or manipulated by any conventional computer.

The key characteristic is the accuracy with which the *playback* reference beam must match the original one that recorded the page. This precision depends on the thickness of the crystal—the thicker the crystal, the more exactly the reference beam must be repositioned. If the crystal is one centimeter thick and the illumination angle deviates by one thousandth of a degree, the reconstruction disappears completely. Far from being an inconvenience, this basic mechanism is exploited in almost all holographic memories. The first page of data is holographically recorded in the crystal. The angle of the reference beam is then increased until the reconstruction of the first hologram disappears. Then a new page of data is substituted and holographically recorded. The procedure, known as *angle multiplexing*, is repeated many times. Any of the recorded holograms can be viewed by illuminating the crystal with the reference beam set at the appropriate angle as shown in Figure 21-3.[3]

How many pages can be imprinted into a single crystal? The number is limited mainly by the dynamic range of its material: as more holograms share the same crystalline volume, the strength of each diminishes. Specifically, the percentage of light that is diffracted by each hologram (and therefore sensed by the electro-optical detectors) is inversely proportional to the square of the number of holograms superimposed.

If 50 holograms in a crystal yield a diffraction efficiency equal to 5 percent, 1,000 holograms will have a diffraction efficiency of only 0.0005 percent. This effect determines the maximum number of holograms that can be stored, because the drop in diffraction efficiency ultimately makes the reconstructions too weak to be detected reliably amid the noise in the system—fluctuations in the brightness of the lasers, scattering from the crystal, thermally generated electrons in the detector, and so on. This maximum number of holograms can be determined by measuring the optical properties of the crystal material and the various noise sources in the system. In practice, when the diffraction efficiency has dropped too low for the pages to be reliably reconstructed, the rate at which erroneous data are detected (the bit-error rate) becomes unacceptably high.

Now that we have some idea of how holographic memories work, let's look at some recently developed data storage projects.

364 CHAPTER 21

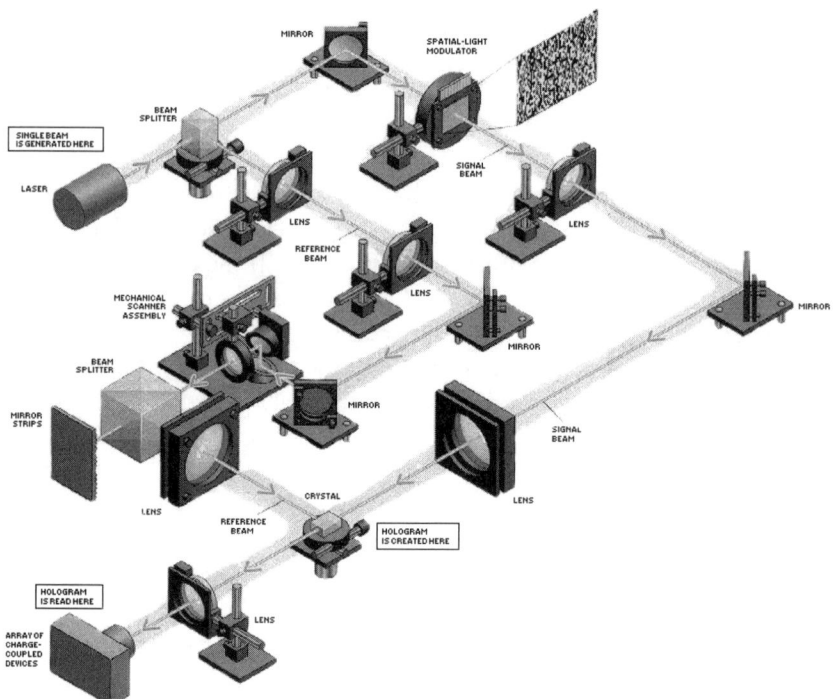

FIGURE 21-3: A system has been built that can store 800,000 holograms and read them out. It uses a combination of angle multiplexing (10,000 angles), fractal multiplexing (25 fractal directions), and spatial multiplexing (80 locations on the crystal) to store a grand total of 800,000 holograms. The optical layout shows how a crystal of lithium niobate can be imprinted with pages of data. One laser beam, known as the *signal*, takes on the data as it passes through a spatial light modulator, which displays pages as a crossword-puzzlelike pattern. This beam meets another, called the *reference*, in the crystal, which records the resulting interference pattern. A mechanical scanner changes the angle of the reference beam before another page can be recorded. Any stored page can be retrieved by illuminating the hologram with the reference beam used to record it. The reconstructed page is read by charge-coupled devices, which produce a current in response to light.

Holographic Memories: Data Storage Projects

Much of the work in developing holographic memories comes down to the application of new techniques to develop data storage projects. Better holographic data storage technologies have allowed improvements in holographic recording methods, enabling more pages to be imprinted into the crystal. Let's look at some the following projects in detail:

- **Compact Phase Conjugate Holographic Memory**: The design of a compact holographic memory for use with applications requiring high-capacity read/write data storage with fast random access.
- **Dual-Wavelength Storage Method**: A demonstration of the storage of 6,000 holograms in a memory architecture that uses different wavelengths for recording and readout in order to reduce the grating decay while retrieving data. Bragg-mismatch problems from using two wavelengths are minimized by recording in the image plane and using thin crystals.
- **Using multiple lasers for hologram reconstruction**: Multiple images can be combined in hologram reconstruction to provide a speed/capacity tradeoff for the system designer.
- **Pixel-matching experiment**: Pixels on SLM and detector array are matched in order to decrease mismatching anti-aliasing noise for very-small-pixel systems.

COMPACT PHASE CONJUGATE HOLOGRAPHIC MEMORY

Most people are familiar with holograms as a technology for displaying 3D images. The working principle is fairly simple: When two coherent beams of light intersect within a holographic medium, the resulting interference pattern causes a grating to be written within the medium in a pattern that is unique to the two writing beams. This grating, which is referred to as a hologram, has the property that if it is subsequently illuminated by either of the beams used for recording, the hologram causes light to be diffracted in the direction of the second writing beam. To an observer it appears as if the source of the second beam is still there.

Motivational Aims

While this technology has found widespread use in reconstructing 3D images, it can be used equally well to store any type of information by simply replacing the source image with a page of data and placing a detector array at the observation plane. Also, due to Bragg effects, you can multiplex many holograms within the same volume of material by slightly changing the angle of the reference beam with each new data page. Tens of thousands of holograms can be multiplexed this way in a small volume of crystal, offering the potential of very high storage densities. Furthermore, holography has the inherent advantage of massive parallelism. Unlike conventional storage media such as magnetic hard disks and CD-ROMs that access only one bit at a time, each access of a holographic memory yields an entire data page—potentially megabits at a time.

However, a common problem of many holographic systems is that various lens systems are required for imaging the signal from the Spatial Light Modulator (SLM) to the detector array or for steering the angle of the reference beam. These optics can cause holographic memories to be both bulky and expensive. The research goal here is to develop a compact, low-cost holographic memory as shown in Figure 21-4.[4]

Research Approach
The research approach is to minimize the volume of the system by making use of phase conjugate readout to eliminate the need for the imaging sys-

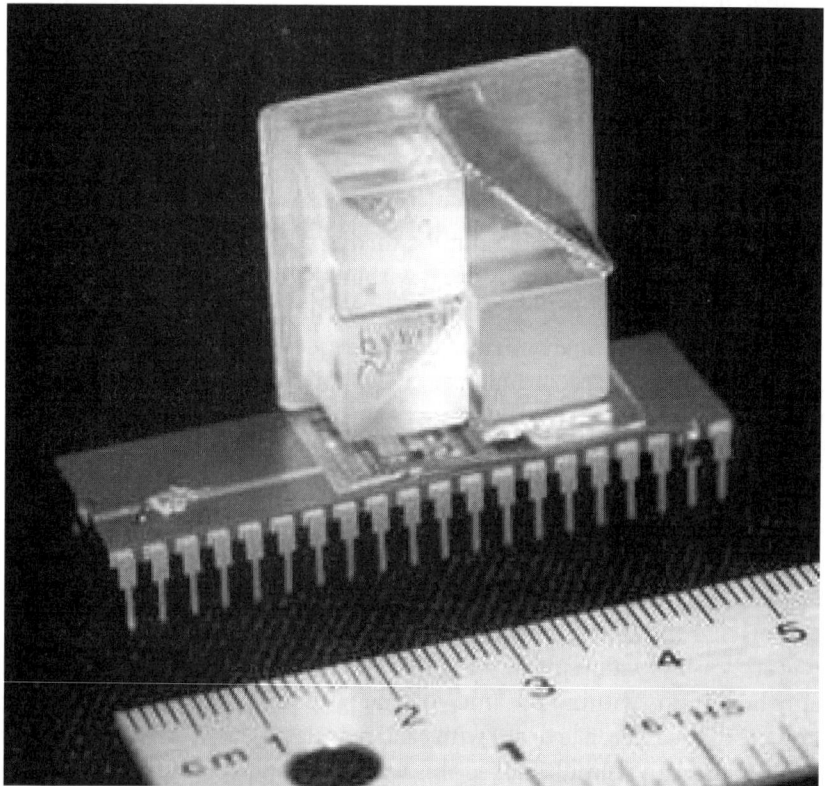

FIGURE 21-4: A compact holographic memory for use with applications requiring high-capacity read/write data storage with fast random access.

tem that is ordinarily needed between the SLM and detector as shown in Figure 21-5.[5] Rather than read out holograms with the original reference beam that was used for recording, the researchers instead used a conjugate reference beam that propagates in the opposite direction as the one used for recording. This causes the diffracted signal to propagate back along the path from which it came. In the case illustrated, the image will refocus onto the SLM. While this permits the researchers to do without the imaging optics, it does require a more sophisticated device that can serve as both SLM and detector.

Recording

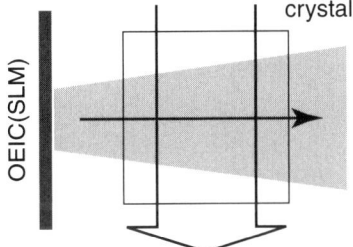

Conventional Readout

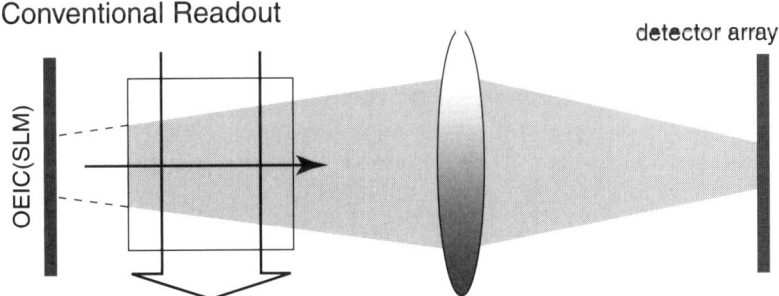

Conjugate Readout

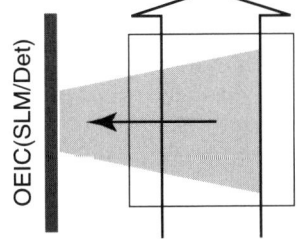

FIGURE 21-5: Conjugate readout.

A cross section of such a device that the researchers fabricated is shown in Figure 21-6.[6] The circuitry for the SLM and detector arrays are fabricated in silicon. On top of the silicon is a layer of liquid crystal covered by a glass plate. The SLM array consists of reflective metal pads in the silicon. By varying the voltage across the liquid crystal layer, the polarization of the light reflected from each pad can be controlled. Hence, when the reflected light is passed through an analyzer, each pixel in the array can be effectively turned on or off.

This device also allows the holographic memory to be periodically refreshed. Because the researchers wished to preserve the system as a dynamic read/write memory, the holograms cannot be fixed in the crystal. The researchers must therefore be able to reinforce the holograms periodically to compensate for the gradual decay of the stored gratings due to illumination. With the custom Opto-Electronic Integrated Circuit (OEIC), this operation is especially easy: When a hologram needs to be refreshed, the researchers simply read out the data page with the conjugate readout beam, latch the information into the OEIC memory, and then use this to drive the SLM array and rewrite the hologram with the forward reference beam. Because of this capability, the researchers refer to this OEIC chip as the Dynamic Hologram Refresher (DHR).

Using the DHR, the researchers can make a very compact memory module a few centimeters on a side. The angle of the reference beam can be selected by a laser array such as a Vertical Cavity Surface Emitting Laser (VCSEL) array or a laser diode array; thus, allowing very fast switching between pages of the memory. The conjugate readout beams are generated through the use of a mirror and by properly selecting which laser to acti-

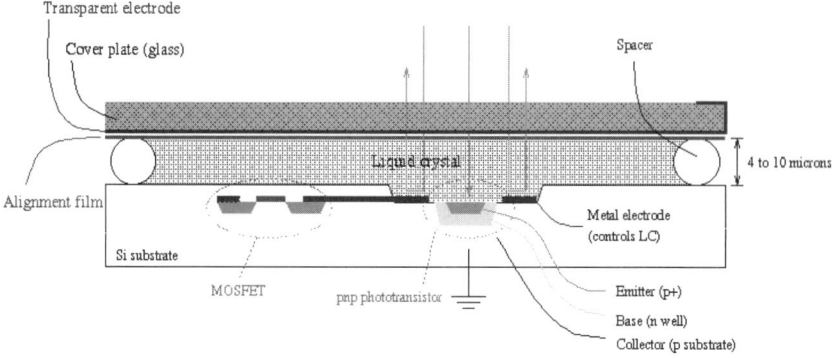

FIGURE 21-6: Cross-section view of OEIC.

HOLOGRAPHIC STORAGE SYSTEMS   369

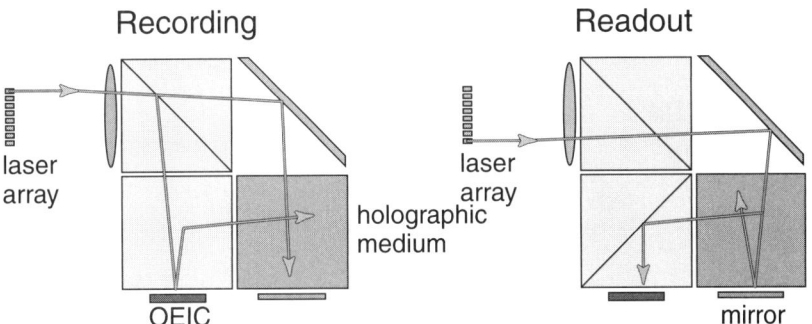

FIGURE 21-7: Memory operation.

vate. A module such as this could hold on the order of about a gigabit of data. For applications requiring large storage capacities, many of these modules can be arrayed together on a compact board as shown in Figure 21-7.[7]

Achievements

In the laboratory, the researchers assembled an experiment to test the phase conjugate readout and periodic refreshing aspects of the system. A schematic diagram and photograph of the system are shown in Figures 21-8 and 21-9, respectively.[8]

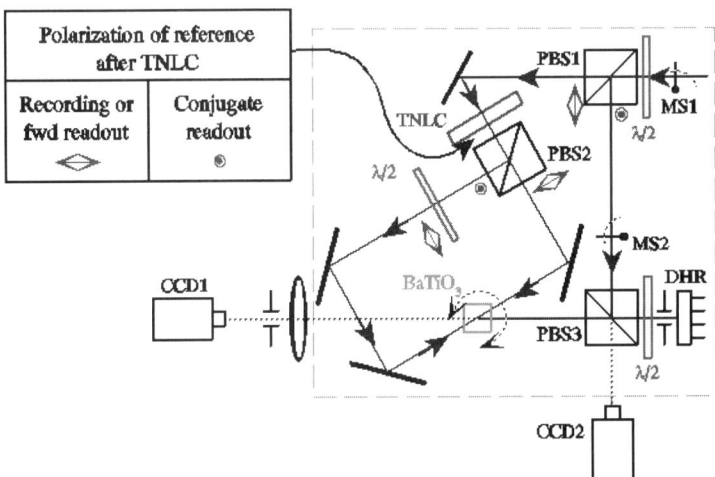

FIGURE 21-8: Experimental setup: transmission geometry.

FIGURE 21-9: Photograph of the periodic refreshing aspects of the system.

In this experiment, the researchers also demonstrated that the decay of holograms in an angle-multiplexed memory due to subsequent illumination can be compensated by a refreshing process, while maintaining the integrity of the data pages over many refresh and decay cycles. The researchers multiplexed 25 holograms of the letters *CIT* displayed on the DHR chip using a barium titanate crystal as the recording medium. Each hologram was recorded for a constant exposure time with the forward reference beam, and then gradually decayed as the other holograms were written. The holograms were then reconstructed with the conjugate reference, the data page latched by the DHR. This page was then refreshed by rewriting with the forward reference. All 25 holograms were successfully refreshed in this manner over 100 cycles with no errors. Figures 21-10 and 21-11, respectively, show the evolution of the diffraction efficiencies for all of the holograms as well as sample images and results of the numeric analysis.[9] Both visually and quantitatively, the researchers observed no degradation in the holograms as a result of the refreshing process.

HOLOGRAPHIC STORAGE SYSTEMS   371

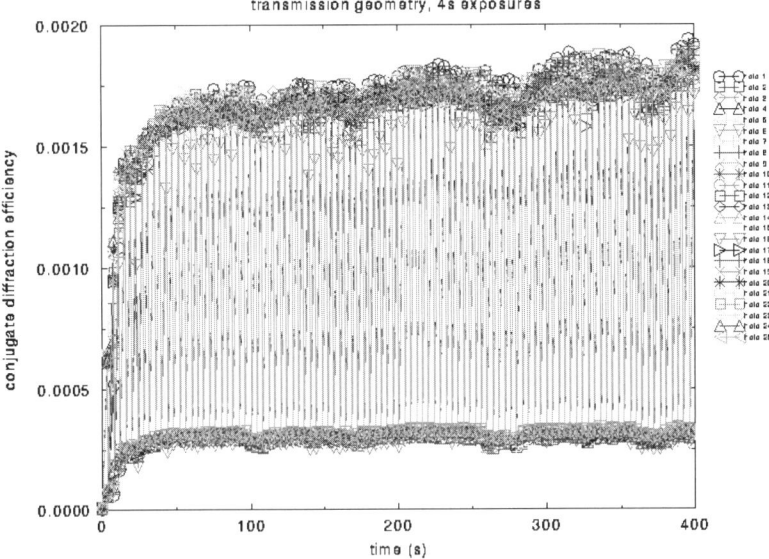

FIGURE 21-10: Refreshing of 25 holograms for 100 cycles.

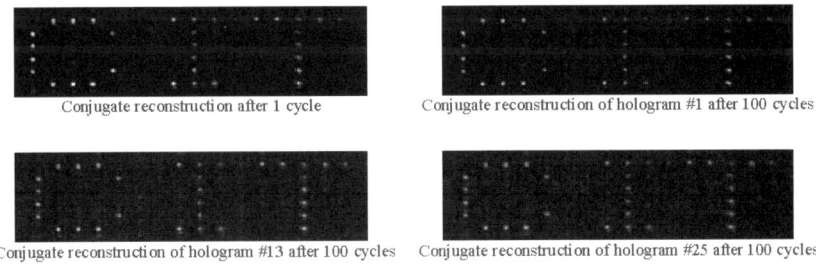

Image	SNR	P.E.
Conjugate reconstruction, 1 cycle	3.94:1	8.2×10^{-4}
Conjugate hologram #1, 100 cycles	4.28:1	1.0×10^{-4}
Conjugate hologram #13, 100 cycles	4.69:1	5.3×10^{-5}
Conjugate hologram #25, 100 cycles	5.03:1	2.9×10^{-5}

FIGURE 21-11: Twenty-five-hologram experiment: image analysis.

DUAL-WAVELENGTH METHOD

When a photorefractive crystal is used as the recording material in a holographic memory, the recorded gratings decay when illuminated by the read-out beam. Several methods have been developed to address this problem, including thermal fixing, electrical fixing, two-photon methods, and periodic copying. Another alternative is to use a dual-wavelength method. The motivation for using different wavelengths of light for recording and read-out is simple: If a crystal has an absorption spectrum with a substantial variation as a function of wavelength, then by recording at a wavelength at which the crystal is highly sensitive and reading out at a second wavelength at which the crystal is relatively insensitive, the researchers can reduce the decay of the gratings caused by the read-out illumination.

Motivation and Aim of the Researchers
Implementing the dual-wavelength method is straightforward for a single grating. Figure 21-12 shows the dual-wavelength configuration for the transmission geometry, along with the corresponding k-space diagram.[10] A grating is recorded in the usual manner, with signal and reference beams at the first wavelength. The researchers reconstructed this grating at the second wavelength, by introducing the read-out beam at a tilted angle with respect to the recording reference beam. Bragg-matched read-out occurs when the read-out beam is positioned such that the grating vector lies at the intersection of the two k-spheres.

While the researchers could easily Bragg-match a single grating, when a hologram of a complex image consisting of many plane wave components is recorded, it is generally impossible to match the entire spectrum simultaneously using a single plane wave read-out reference. However, the researchers showed that by using a sufficiently thin crystal and the peristrophic multiplexing technique, a large number of holograms can be stored and recalled with a single plane wave reference.

The Actual Research
Recording an image consisting of many plane wave components can be represented in k-space by a cone of signal vectors that interferes with the reference beam to record a cone of grating vectors, as shown in Figure 21-13.[11] When the researchers attempted to reconstruct the signal with a reference at the second wavelength, only the gratings that lie on the circle of

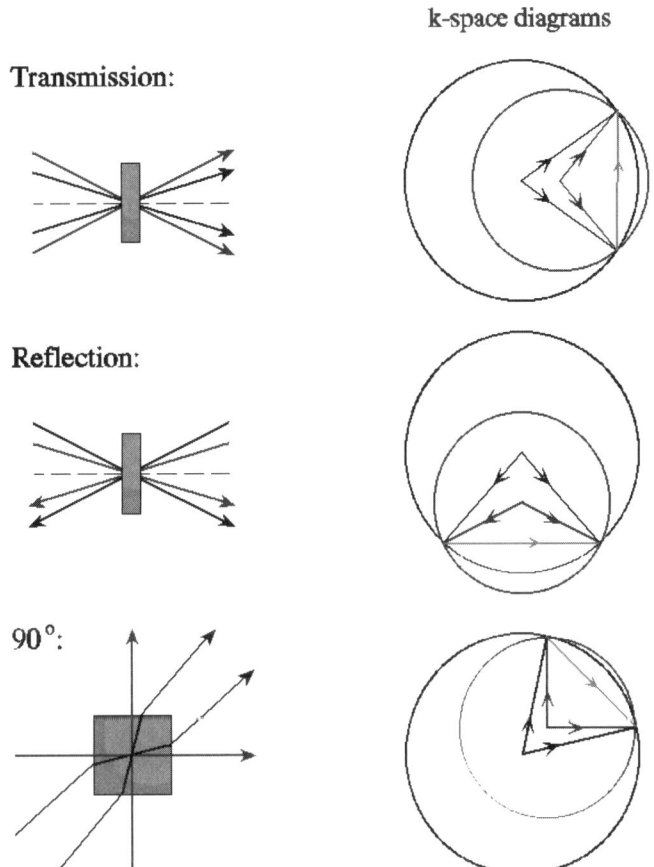

FIGURE 21-12: Bragg-matched geometries.

intersection between the two k-spheres are exactly Bragg-matched. Hence, only an arc of the signal cone will be strongly reconstructed.

The effect on the reconstructed image of limiting the bandwidth of the signal cone depends on whether the researchers recorded in the Fourier plane or image plane. When the researchers recorded in the Fourier plane, each plane wave component of the signal beam entering the crystal corresponds to a spatial location (pixel) on the input image. Hence, if the researchers reconstructed only a limited angular bandwidth of the signal cone, they expected to reconstruct a strip of the image, as shown in Figure 21-14.[12]

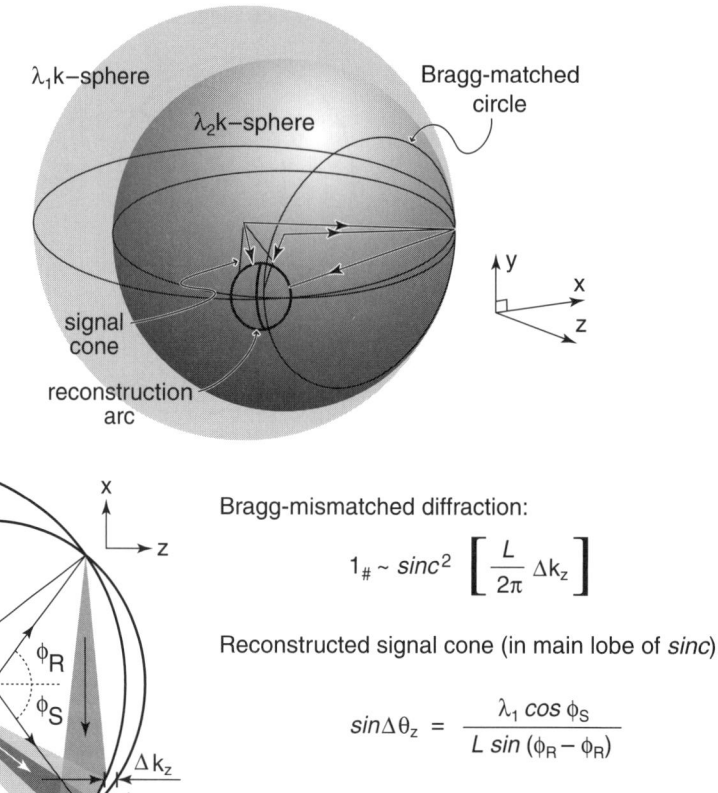

FIGURE 21-13: Bragg-match condition for two wavelengths with a complex signal.

Recording in the image plane is analogous to recording in the Fourier plane, except that in place of the input image, the researchers would have its Fourier transform. Therefore, instead of reconstructing a strip of the image, the researchers reconstructed a *strip* or band of the frequency spectrum of the image. If the researchers position the read-out reference to Bragg-match the DC component of the image, the resulting reconstruction will be a low-pass filtered version of the original in the x-dimension. In Figure 21-14, the researchers compared the reconstruction obtained by the original recording reference to that obtained by a second wavelength.

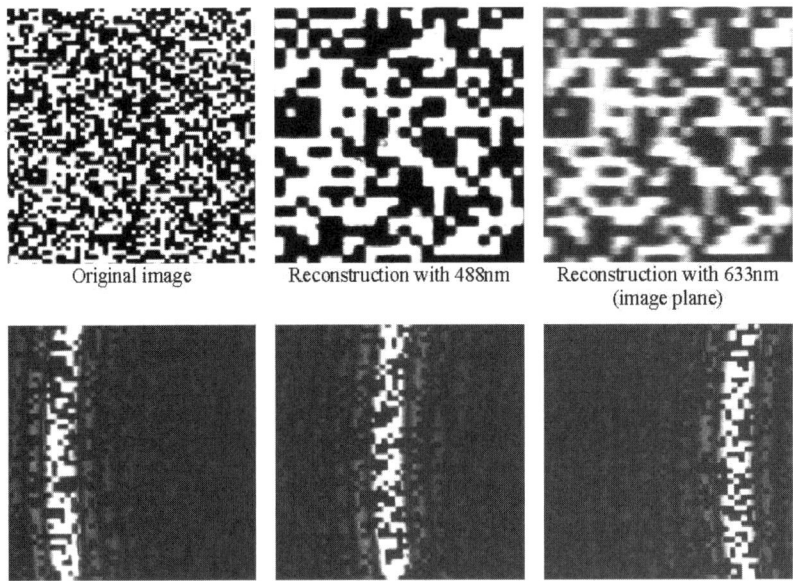

FIGURE 21-14:   Reconstructions.

NOTE: The edges are blurred in the *x*-dimension that results from the loss of the high-frequency components of the input signal.

A number of solutions have been proposed for the Bragg-mismatch problem of the dual-wavelength scheme. Most have dealt with Fourier-plane recording, such as using spherical read-out beams to Bragg-match a larger range of the signal cone or interleaving strips from adjacent holograms. The researchers can also recover all the necessary information by recording in the image plane, without the added complexity of the preceding methods, if the researchers simply adjust the system parameters according to the resolution of the images that they wish to store. The angular bandwidth of the reconstructed signal can be increased by:

- Reducing crystal thickness
- Using recording and read-out wavelengths that are as close together as possible
- Reducing the angle between the signal and reference beams[13]

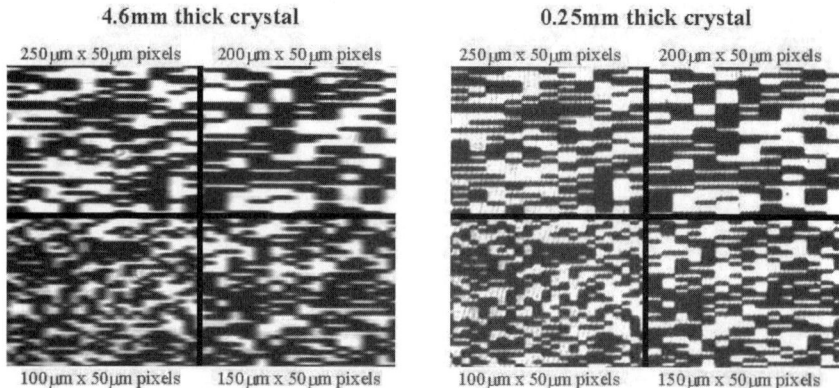

FIGURE 21-15: Holograms recorded at 488nm and reconstructed at 633nm.

Therefore, even after choosing the system wavelengths and geometry, the researchers can still set the bandwidth arbitrarily high by choosing an appropriate crystal thickness. As shown in Figure 21-15, the researchers show two image-plane reconstructions of the same pattern—one recorded in a 4.6mm-thick crystal, and another in a 0.25mm-thick crystal.[14]

NOTE: The reconstruction from the thinner crystal preserves the higher spatial frequencies, so that the edge-blurring is hardly noticeable. Rectangular pixels were used for these images to demonstrate that the pixel-size limitation is only in the *x*-dimension. Even the thicker crystal reconstructs high spatial frequencies cleanly in the *y*-dimension.

Unfortunately, as the researchers reduce the crystal thickness, the angular selectivity suffers, so that the number of holograms that they can angularly multiplex is reduced. However, the researchers can compensate for that by combining angular and peristrophic multiplexing methods. The peristrophic selectivity is independent of the crystal thickness, so it is not adversely affected by using thinner crystals.

Noted Achievements

The experimental setup used for the dual-wavelength image-plane architecture is shown in Figure 21-16.[15] The crystal was mounted on two rotation stages—one to provide the angular tilt and the other for the peristrophic tilt. Fifty angular locations and 20 peristrophic locations were used to multiplex 1,000 holograms of a random bit pattern. Figure 21-17 shows the sample reconstructions from the experiment.[16] Due to the care-

HOLOGRAPHIC STORAGE SYSTEMS 377

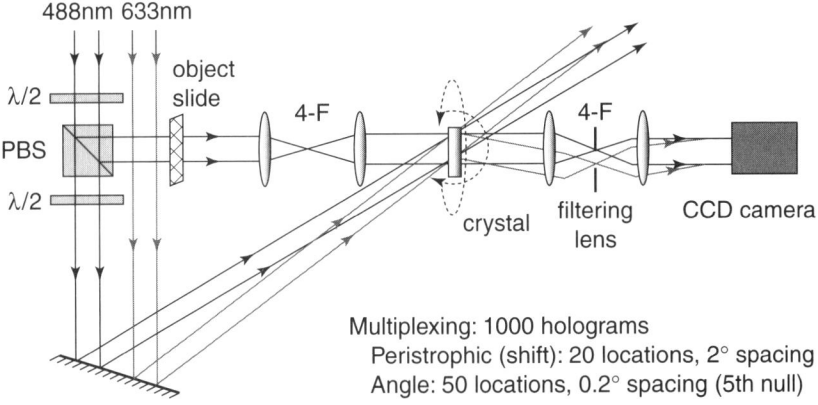

FIGURE 21-16: Experimental setup.

FIGURE 21-17: 1,000-hologram experiment: reconstructions (hologram recorded at 488nm).

ful choice of crystal thickness to suit the pixel size, the low-pass filtering effect from using two wavelengths is minimal and does not significantly degrade the image fidelity.

The researchers also conducted an experiment to examine how well the dual-wavelength method reduces the decay rate due to the read-out illumination. The researchers recorded two holograms with the same exposure at 488nm, and erased one with a non-Bragg-matched beam at 488nm and the other with an equal intensity beam at 633nm, periodically monitoring the grating strength by probing with a 633nm read-out beam. The decay rate was also measured with no erasure beam to determine the decay contribution from dark conductivity as well as from the monitoring beam. The results showed that using the dual-wavelength method with this particular combination of crystal and wavelengths the researchers reduced the readout decay rate by over an order of magnitude. However, the dark conductivity of the material remains a limiting factor of this approach. While the dual-wavelength method may minimize erasure due to the readout illumination, it does not compensate for the inherent dark erasure of the material.

MULTILASER HOLOGRAM RECONSTRUCTION

This project explores using multiple laser diodes to read out holograms in a memory. In one approach, the end-user data pages are coded (using majority rule) into holograms, which are then read out by multiple lasers turning on at once. The reconstructions from these lasers add incoherently on the detector, reconstructing the original data page (after thresholding). This approach allows a speed-up over conventional recording, since it uses multiple laser diodes at once during reconstruction. It is a source of additional errors, though, since the coding scheme relies on the statistics of large numbers of pages being read out at once to drive down the bit error rate.

PIXEL-MATCHING EXPERIMENT

For a holographic memory system, the two main advantages are the large memory capacity and the parallelism in data access. The memory storage density (bit per unit area) equals the ratio of M (number of multiplexed holograms) to the area of each pixel. The capacity per unit area data page is also inversely proportional to the pixel area. It is a challenge to increase the storage capacity and the access speed by decreasing the pixel size. One limit to the pixel size is the available pixel size of the detector array as shown in Figure 21-18.[17]

HOLOGRAPHIC STORAGE SYSTEMS

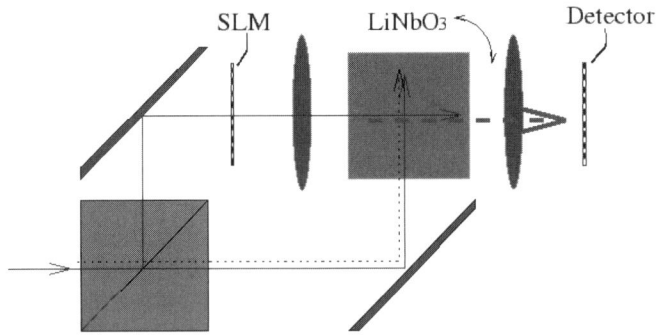

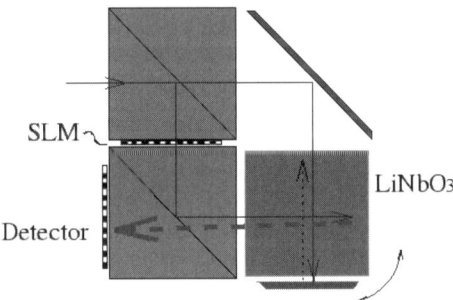

FIGURE 21-18: Pixel-matched architecture.

Research

As the pixel size of SLM (Spatial Light Modulator) is comparable with that of detect array, it needs to match each individual pixel to the corresponding detector sensor by a high-quality imaging system. For the researchers' compact phase conjugate system without lenses, it is an obligation to have the same physical pixel sizes for the detector array and the SLM.

In this experiment, the researchers used a Kopin liquid crystal SLM with pixels 24 × 24mm and a VISION CMOS detector array of sensors 12 × 12mm. The active area of each SLM pixel is actually 12 × 16mm. By using one sensor out of every 2 × 2 sensor array, the detector sensors and the SLM pixels have the same pixel period. For the conventional architecture, the researchers used a pair of customized high-quality lenses to form a one-one imaging system. For the phase-conjugate architecture, the reconstructed image is relayed to the detector simply by a beam splitter. The pixel-matched holographic image recording and reconstruction are demonstrated for both architectures as shown in Figure 21-19.[18] The phase conjugate system is proved to be robust and easy to operate.

The researchers compared the SNR (signal noise ratio) of the reconstructed binary data pages between the phase-conjugate and the conventional architectures. The histograms of the SNR of different images demonstrate a better performance for the phase-conjugate architecture.

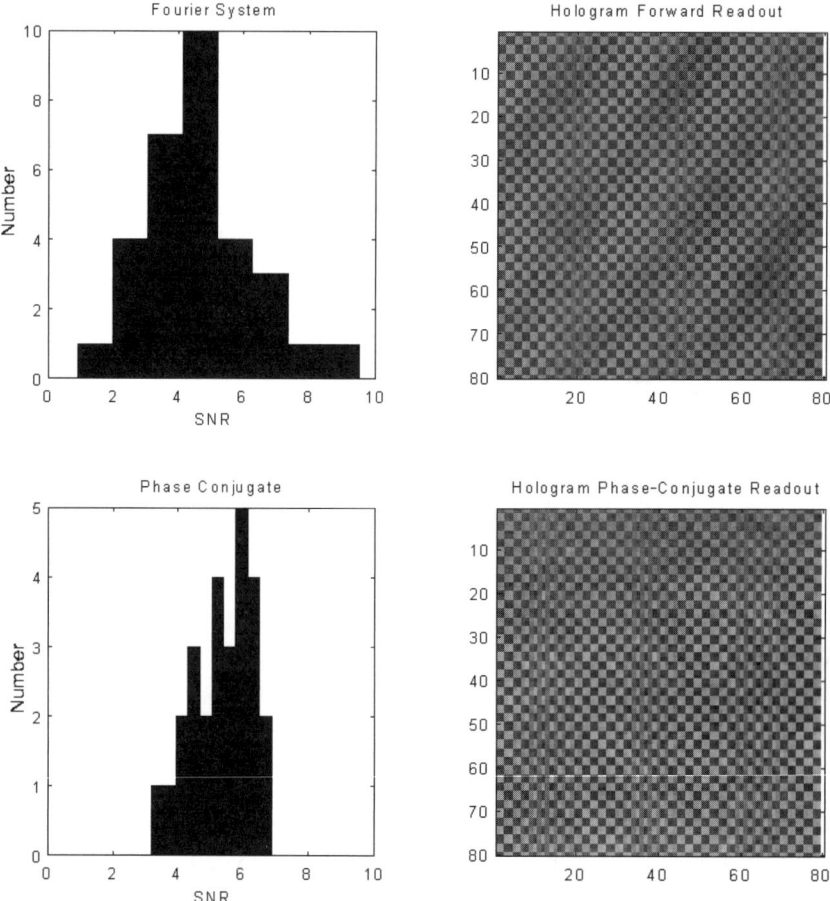

FIGURE 21-19: SNR comparison between forward and phase-conjugate readout.

Achievement
The researchers demonstrated the pixel-matched holographic image reconstruction for both the conventional and the phase-conjugate architecture. It proved the fidelity and facility of the compact phase-conjugate design.

Strengthening Signals

Much of the work in developing holographic memories comes down to the application of new techniques to strengthen, against the background noise, the optical signals representing pages of data. Better technologies have allowed fainter and fainter signals to be reliably detected, and improvements in holographic recording methods have strengthened the recorded signals, enabling more pages to be imprinted into the crystal.

The first attempts to store many holograms date back to the early 1970s. Juan J. Amodei, William Phillips, and David L. Staebler of RCA Laboratories recorded 500 holograms of plane waves in an iron-doped lithium niobate crystal. Robert A. Bartolini and others, also at RCA, stored 550 holograms of high-resolution images in a light-sensitive polymer material, and Jean Pierre Huignard's group at Thomson-CSF in Orsay, France engineered a memory with 256 locations, each capable of storing 10 holograms. Besides storing relatively many holograms, Huignard's system was exceptionally well engineered.[19]

Impressive though as some of these early efforts were, none of them led to a practical system. Semiconductor and magnetic memories were progressing quite rapidly at the time, making more exotic technologies seem unworthy of pursuit. Gradually, holographic memories fell out of the limelight.

A renaissance began in 1991 with funding from the U.S. Air Force and the Department of Defense's Advanced Research Projects Agency. A demonstration by researchers was conducted showing the storage and high-fidelity retrieval of 500 high-resolution holographic images of tanks, jeeps, and other military vehicles in a crystal of lithium niobate with trace amounts of iron.

Several new theories and experiments followed. In 1992, researchers stored 1,000 pages of digital data in a one-cubic-centimeter, iron-doped lithium niobate crystal. Each stored page contained 160 by 110 bits obtained from the ordinary electronic memory of a digital personal computer.[20] The researchers then copied segments of the stored data back to the memory of the digital computer—and detected no errors. This experiment demonstrated for the first time that holographic storage can have sufficient accuracy for digital computers.

A similar setup was used to store 10,000 pages. Each of these pages measured 320 bits by 220 bits, so all told the system could store a little less than 100 million bytes (100 megabytes). The researchers performed this experiment in 1993 at the California Institute of Technology in collaboration with Geoffrey Burr.[21]

The majority of the 10,000 stored holograms were random binary patterns, similar to the data that can be stored by a conventional computer. The raw (uncorrected) error rate was one bad bit out of every 100,000 evaluated. Such a rate suffices to store image data, particularly if they have not been compressed or manipulated to reduce the number of bits needed to represent the image. Several photographs of faces and of the Caltech logo were also included among the pages to demonstrate that images and data can be easily combined in a holographic memory. The information contained in the 10,000 holograms would fill only one-eighth of the capacity available in a conventional compact disc. However, holographic memories that have a much higher capacity can be made by storing holograms at multiple locations in the crystal. For instance, the researchers demonstrated a system in which 10,000 data pages are stored in each of 16 locations, for a total of 160,000 holograms.[22]

In 1994, John F. Heanue, Matthew C. Bashaw, and Lambertus Hesselink, all at Stanford University, stored digitized, compressed images and video data in a holographic memory and recalled the information with no noticeable loss of picture quality. They stored 308 pages, each containing 1,592 bits of raw data, in four separate locations in the same crystal. The Stanford group combined several techniques, some electronic, others optical, to keep the bit-error rates under control. For instance, they appended a few bits to each string of eight bits to correct a single erroneous bit anywhere in the group. This error-correcting code reduced the error rate from about one bit in every 10,000 or less to about one bit per million.

Another important potential advantage of holographic storage is rapid random access by nonmechanical means. For example, high-frequency sound waves in solids can be used to deflect a reference light beam in order to select and read out any page of data in tens of microseconds—as opposed to the tens of milliseconds typical of the mechanical-head movements of today's optical and magnetic disks. At Rockwell's research center in Thousand Oaks, California, John H. Hong and Ian McMichael have designed and implemented a compact system capable of storing 6,000 holograms in each of 100 locations. An arbitrary page can be accessed in less than 30 microseconds, and its data are retrieved without errors.

Polymers

As with the original experiments in the 1970s, these recent preceding demonstrations used a crystal of lithium niobate with trace amounts of iron. When illuminated with an optical pattern (such as a hologram created by the intersection of two laser beams), charged particles migrate within the crystal to produce an internal electric field whose modulation closely matches that of the optical pattern. The way the crystal then diffracts light depends on this electric field. When the crystal is illuminated again at the correct angle, light is diffracted in such a way that the original hologram is reconstructed. The phenomenon is known as the photorefractive effect.

A different type of holographic material became commercially available for the first time in 1994. This material, known as a photopolymer, was developed at Du Pont and undergoes chemical rather than photorefractive changes when exposed to light. Electrical charges are not excited, and the photochemical changes are permanent—information cannot be erased and rewritten. The medium is therefore suitable only for write-once or read-only memories. The material does, however, have a diffraction efficiency 2,500 times greater than a lithium niobate crystal of the same thickness. An experiment was conducted between the researchers at Caltech and at AT&T Bell Laboratories in which 1,000 pages of bit patterns were stored in a polymer film 100 microns thick. The researchers retrieved the data without any detected errors.

In recent years, researchers at IBM and the University of Arizona have begun experimenting with polymer films that, like lithium niobate crystals, exhibit the photorefractive effect. Promising though the developments in polymeric holographic materials are, it is too soon to count out lithium niobate, which has lately also shown greater versatility. For instance, crystals of lithium niobate doped with trace amounts of both cerium and iron (which are sensitive to red light rather than green) recently became available. They point the way to crystals that can be imprinted with inexpensive and tiny semiconductor lasers, instead of the much more costly green or blue-green ones (see Figures 21-20 and 21-21).[23]

384 CHAPTER 21

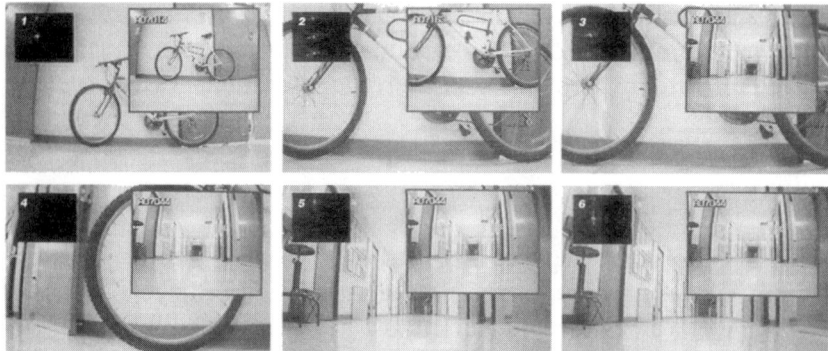

FIGURE 21-20: Vehicle steered by holograms navigated itself around the researchers' laboratory at the California Institute of Technology. Each compound photograph in this sequence shows what a video camera on the vehicle saw, along with another image (insert) that was transmitted to the little machine from a holographic memory.

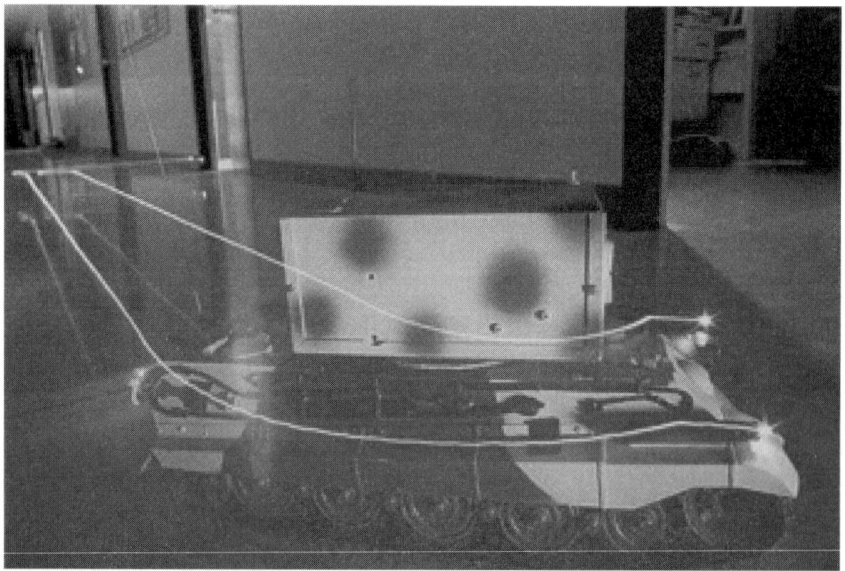

FIGURE 21-21: To navigate, the vehicle (shown at right) oriented itself until its camera image matched the one from its memory. Lights in the other, smaller inset indicate the extent to which two image sequences are in synchrony. In this series, the vehicle initially recognized and approached a bicycle. It was then prompted by the image it would see after a left turn, which it found after a bit of searching.

Holographic Data Storage Changes

The iron-doped lithium niobate crystals used in the recent preceding demonstrations are not the only surviving aspect of the early experiments more than two decades ago. The argon lasers typically used today are also the same; and angle multiplexing was relied on in the past, as now. What changes, then, have revived holographic data storage?

The most significant advance has been the emergence of a mature optoelectronics industry, which has produced the inexpensive, compact, and power-efficient devices needed to build large-scale holographic memories and to interface them with digital computers. For instance, tiny semiconductor lasers that emit red light, originally developed for fiber-opt/c communications, can be used as light sources either with a cerium- and iron-doped lithium niobate crystal, or with Du Pont's photopolymer. Large detector arrays made for television cameras, which take an optical image and convert it to an electronic signal, read the output of the memory. Liquid-crystal display screens originally designed for video projectors serve as the input devices, creating the bright-and-dark patterns that represent pages of data.

Such technological advances made possible the recent preceding memory demonstrations that, in turn, prompted new investigations into the underlying physics. For example, a long-standing problem in holographic memories is cross-talk noise—the partial, spontaneous, and unwanted readout of stored data. In practice, cross talk causes faint, ghostlike images of all the pages to be called up when only one is being accessed. Cross-talk noise and its sources are now completely understood, allowing the researchers to calculate and counteract the effect in any recording setup from such parameters as the angle between the signal and reference beams, the angle between the reference beams in a multiplexed recording, and the geometric properties of the page of data.

Another byproduct of the theoretical work has been the development of new multiplexing methods and the refinement of existing ones. These can replace or supplement angle multiplexing, giving the system designer more options. In one alternative, pursued separately at Pennsylvania State University and at Caltech, successive pages are recorded with reference beams of different wavelengths. Reference beams that are coded with a different pattern for each page have also been demonstrated at the University of California in San Diego and, independently, at the Optical Institute in Orsay, France.

Pump Up the Volume

Better multiplexing techniques are certainly welcome, but a fundamental means of increasing capacity will be needed if holographic memories are to make inroads against compact discs. Holographic memories have been shown to be significantly faster at present than are compact-disc systems, but speed alone is rarely enough for a new technology to supplant an entrenched one. What is generally needed is another basic advantage, such as greater storage capacity.

One way to increase storage in a holographic memory would be to tile a two-dimensional surface with sugar-cubelike memory crystals, a technique called *spatial multiplexing* (see the sidebar, "Crystal Cube Storage"). As expected, the capacity of such a system is proportional to the number of cubes. Data are stored in each of the cubes in the usual way, as angle-multiplexed holograms.

■ **CRYSTAL CUBE STORAGE**

Computer gurus love to muse about ubiquitous computing—seamlessly linking the physical and digital worlds so that everything and everybody is woven into one big web. The inhabitants of this future world would spend their lives accumulating digital information. One wild idea even has folks outfitted with wearable cameras equipped with computers that would record whatever they encounter. You won't forget anything—certainly nothing you've ever heard, seen, or read.

But where will all this data be stored? At the NEC Research Institute Inc., (part of NEC Corp. of Princeton, New Jersey), researchers are working with an argon laser. Its green light bounces off mirrors and splits into two beams that meet perpendicularly in what just might be the answer: a crystal cube smaller than a die. With the right tweaking, the cube exhibits properties that might herald a revolutionary storage medium—one that preserves the *on* and *off* signals of digital bits as aspects of light. In other words, holograms. Theoretically, one die-size cube could store 40 terabytes—a library's worth of data. Researchers see it as a replacement for hard disk drives, but with maybe 2,000 times more capacity. People are going to want that kind of storage in every PC.

The key to this crystal cube storage is an unusual trait associated with certain dopant atoms that are used to enhance semiconductor electrical properties. When manipulated with a laser beam, a tiny area around each atom switches from a highly refractive state to a lower refractive index.

These two optical states sound like a storage medium. After all, disk drives rely on two magnetic states to represent the ones or zeros of binary data. Under this scheme, laser light is split into two paths: the signal beam ferrying in data and the reference beam that *writes* information into the crystal. Before entering the cube, the signal passes through a glass mask similar to the ones used for photolithographic etching of integrated circuits. The mask is laced with metal, which blocks the beam in some places but allows it to pass unencumbered through areas of glass.

The resulting combination of light and dark represents the zeros and ones of the data to be stored. The data then rides the signal beam into the crystal. From there, the process is classic holography.

Inside the semiconductor, the interference between the beams forms a pattern of bright and dark regions throughout the crystal, not just on its surface (hence the term *hologram*). Since the refractive index is lowered only in regions struck by light, and stays low after the signal is turned off, what remains is a recording of the mask data. Moreover, if the reference beam reenters the crystal at the same wavelength (angle and polarization), it recreates the signal beam, allowing stored data to be *read*.

In theory, a fresh page can be written by rotating either the crystal or beam just a smidge—a mere 40 arc-seconds. So far, the researchers have stored 22 million bits (about 2.8MB) in a single hologram. By 2001, they hope to implant thousands of these *pages* onto one crystal.

Formidable obstacles remain. One is that refractive changes only last about 10 seconds at room temperature. Another is that there's no way to selectively erase data: Everything must go in order to eliminate even one bit. Undeterred, the researchers at NEC have patented a *dynamic refresh* technology designed to read out and rewrite data every few seconds. It won't be reliable enough for archival storage, but it might do for security camera footage, for instance. Moreover, a refresh scheme solves the data-erasing problem, because users can choose to rewrite only the portions they want. After all, there are always a few things we'd like to forget.[24]

The challenging part of this kind of system is the optical assembly, which must be capable of addressing any one of the cubes individually. One such assembly is the three-dimensional disk, which has many similarities to a conventional CD. The disk-shaped recording material is placed on a rotating stage. A laser-based reading and writing device, or head, is mounted above it. The rotation of the disk and radial scanning of the head make it possible to illuminate any spot on the disk. Researchers have recently built such a system at Caltech.

As in any holographic medium, data are stored throughout the volume of the recording layer of the 3D disk. The head has a detector array for reading out an entire page of data and a beam deflector for angle multiplexing. A spatial-light modulator, which imprints the page of data onto the signal beam (such as the LCD screen used in current demonstrations), could also be incorporated into the head.

Even though a 3D disk stores information in three dimensions, the number of bits that could theoretically be stored per square micron of disk surface can be computed for the purpose of comparing this areal density to that of a conventional CD. Such a comparison is reasonable because a 3D disk can be as thin as a CD. It turns out that for thicknesses less than two millimeters, the areal density of the holographic disk is approximately proportional to the thickness of the recording medium. In demonstrations at Caltech, researchers achieved a surface density of 10 bits per square micron in a disk made with a polymer film 100 microns thick (the maximum available for this particular material). This density is about 10 times that of a conventional CD.

Researchers can increase the surface density, moreover, by simply increasing the thickness of the holographic layer. Density of 100 bits per square micron would be possible with a material one millimeter thick. Such a 3D disk would be nearly identical in size and weight to a conventional CD, but it would store 100 times more information.

Among the companies pursuing this basic technology is Holoplex,[25] a small startup in Pasadena, California. The company has built a high-speed memory system capable of storing up to 6,000 fingerprints (see Figure 21-22), for use as a kind of selective lock to restrict access to buildings or rooms.[26] Its entire contents can be read out in less than one second. Holoplex is now working on another product that would be capable of storing up to a 100 trillion bits, or almost 20,000 times what can be put on a CD.

DIGITAL HOLOGRAPHIC STORAGE SYSTEM

A digital holographic storage system, one actually integrated with a computer hard drive, has also been developed by scientists at Stanford. In such a system, data is converted into light patterns. The light waves enter a photorefractive medium, where they bring about microscopic rearrangements of electric charge, which in turn affect the local index of refraction. To read out the data, a reference laser beam is sent into the medium. The refracted beam bearing the decoded data is then detected with a charge-coupled de-

FIGURE 21-22: Holographic lock stores up to 6,000 fingerprints. To gain entry to a room, a user places a finger on a glass plate. The fingerprint must match one of those in the memory, which are stored as holograms. The fast memory minimizes the delay while the system searches for a match. This type of device is being developed by the Japanese company Hamamatsu. It uses the holographic memory shown here from Holoplex, a Pasadena, California startup company.

vice. Data can be stacked up in the hologram by recording at several angles. By home-computer standards, the Stanford results so far are modest: total storage capacity of 326kB and a data transfer rate of 7.4MB per second. The researchers believe future hologram performance should be much better: terabytes of storage and transfers above 2 gigabits per second.

Holographic Memories by Association

Before a *super CD* becomes a commercial reality, holographic memories may be used in specialized, high-speed systems. Some might exploit the associative nature of holographic storage, a feature first expounded on in 1969 by Dennis Gábor, who was awarded the 1971 Nobel Prize for Physics for the invention of holography.

Given a hologram, either one of the two beams that interfered to create it can be used to reconstruct the other. What this means, *in* a holographic memory, is that it is possible not only to orient a reference beam into the crystal at a certain angle to select an individual holographic page, but also to accomplish the reverse. Illuminating a crystal with one of the stored images gives rise to an approximation of the associated reference beam—reproduced as a plane wave emanating from the crystal at the appropriate angle.

A lens can focus this wave to a small spot whose lateral position is determined by the angle, and therefore reveals the identity of the input image. If the crystal is illuminated with a hologram that is not among the stored patterns, multiple reference beams (and therefore multiple focused spots), are the result. The brightness of each spot is proportional to the degree of similarity between the input image and each of the stored patterns. In other words, the array of spots is an encoding of the input image, in terms of its similarity with the stored database of images.

In 1995 at Caltech, researchers used a holographic memory in this mode to drive a small car through the corridors and laboratories of the Electrical Engineering building. The researchers stored selected images of the hallways and rooms in a holographic memory connected to a digital computer on a laboratory bench, and communicated them to the car via a radio link. A television camera mounted on the car provided the visual input. As the car maneuvered, the computer compared images from the camera with those in the holographic memory (see Figure 21-21). Once it spied a familiar scene, it guided the vehicle along one of several prescribed paths—each defined as a sequence of images recalled from the memory. Some 1,000 images were stored in the memory, but only 53 were needed, it was found, to navigate through several rooms in the building.

Researchers at Caltech are now designing a different vehicle, which they hope to equip with a large enough memory to travel autonomously anywhere on the campus. Even with so much capacity, the parallelism of the holographic memory would permit stored information to be called up rapidly enough to let the vehicle follow roads and avoid obstacles. Indeed, navigation may be one of the specialized applications that generates the impetus needed to bring the technology into widespread use.

Finally, such acceptance may be years away. However, as the need to store vast amounts of data increases, so too will the expediency of storing the information in three dimensions rather than in two.

In Summary

- With three-dimensional recording and parallel data readout, holographic memories can outperform existing optical storage techniques.
- In its basic form, a hologram is the photographic record of the spatial interference pattern created by the mixing of two coherent laser beams. One of the beams usually carries spatial information and is labeled the *object* beam. The other is distinguished by its particular direction of travel and is labeled the *reference* beam. Illuminating the recorded hologram with the reference beam will yield or reconstruct the object beam, and vice versa.
- As the holographic material becomes thicker, the reconstruction becomes very sensitive to the particular angle of incidence of the reference beam, which allows multiple objects to be recorded in the same volume and accessed independently by using an appropriate set of associated reference beams. Such holograms would be recorded sequentially, each object beam illuminating the holographic material simultaneously with its unique reference beam.
- The angularly selective property of holograms recorded in thick materials (typically 1–10mm thick) enables a unique form of high-capacity data storage distinguished by its parallel data access capability.
- A holographic data storage system is fundamentally page oriented, with each block of data defined by the number of data bits that can be spatially impressed onto the object beam. The total storage capacity of the system is then equal to the product of the page size (in bits) and the number of pages that can be recorded.
- A holographic data storage system can be constructed to exploit this principle by using a spatial light modulator to properly shape the object beam; an optical beam scanner to point the reference beam; a detector array to convert the reconstructed output object data into an electronic bit stream; electronics to control the entire process and condition the input/output electronic information; and a sufficiently powerful laser to overcome the optical losses of the system.
- The holographic data storage system developed by Rockwell for avionics applications is based on an acousto-optic addressing scheme and contains no moving parts.

- In practice, the number of holograms that can be stored and reliably retrieved from a common volume of material is limited to less than 60,000 so that spatial multiplexing techniques must be used. Although solid-state designs are possible, it is easiest to envision a storage material formed as a volume disk in which holograms in a particular cell are stored and retrieved by angular multiplexing, and where random access to arbitrary cells is enabled by rotation of the disk.
- The holographic 3D disk being developed at the California Institute of Technology (Caltech) is a practical example of such a system. The 3D disk has a 100-tim-thick photopolymer laminated onto a glass disk substrate. By superimposing 32 holograms—each hologram consisting of 590,000 bits—in an area 1.77 mm2, an areal density greater than 10 bit/tim2 has been demonstrated. No errors were detected in the reconstructions. Thicker recording media can yield densities in excess of 100 bit/tim2.
- Another example is a system demonstrated at Rockwell that has no moving parts and is based on an acousto-optic addressing scheme. Rockwell is developing the technology for avionics applications in which resistance to vibration and shock must be provided while maintaining the beneficial features of holographic storage.
- In addition to such technology demonstrations, Holoplex (Pasadena, CA) has recently developed and delivered a commercial holographic memory product that stores up to 6,000 images, each consisting of 640 × 480 pixels, and is capable of reading out its entire contents in less than one second.
- Holographic data-storage systems use devices that are currently being refined by display and electronic imaging applications, so cost reductions and performance improvements can be expected in some proportion to the large volumes expected in those markets.
- Mass memory systems serve computer needs by providing archival data storage; emerging applications also involve network data and multimedia services. In general, such systems require high capacity and low cost.
- New compact visible wavelength laser sources such as high-power semiconductors and frequency-doubled solid-state lasers are also available for use in holographic systems. Further optical and system

- engineering, including error correction and other issues, however, is needed to integrate such devices into demonstration systems.
- A variety of hardware approaches is currently available and can be classified with respect to two important performance measures: storage capacity and effective data transfer time.
- The effective data transfer time is a measure of the time required to fully retrieve an arbitrarily located data block from the system, and is a combination of the data transfer rate (the rate at which data are transferred from the mass memory to the user or CPU) and the data latency (the time lag between the address setup and the appearance of valid data at the output).
- The effective data transfer time is given by the sum of the data latency and the data block size in bytes divided by the data transfer rate in bytes per second. For specific comparisons, the block size must be chosen with care depending on the application of interest. No matter how fast the data transfer may be, a fundamental limitation exists for all approaches involving mechanical motion of either the read/write head or the storage medium.
- Magnetic and optical tape storage systems are cost effective for archival storage, when data access time is less critical. At the other extreme, flash memory, which is a solid-state semiconductor approach, offers extremely fast data access time at relatively low packing density but at a high cost.
- Although incremental improvements to existing disk-based systems may be sufficient to address certain new applications, such as digital video CD-ROM, they will fall short of addressing applications in which both high capacity and short effective data transfer time are featured simultaneously, as is the case with network servers and image databases.
- A new storage technology must displace the incumbent in all other portions of the storage spectrum. To do this, the demands of a mature market must be met, as well as technical challenges.
- New technologies that attempt to increase only the achievable storage capacity must compete with magnetic-and optical-disk-based systems for which tremendous resources are constantly brought to bear. In particular, the areal density of magnetic disk systems can be increased by use of optical tracking and better read/write heads;

- while for an optical system, the areal density can be increased using multiple-layer CDs and variable pit depth recording.
- New data storage techniques must provide a hardware solution while conforming to cost/price constraints that have been imposed by existing technology.
- Holographic mass memory systems will have distinctive appeal for applications that require both high capacity and short data access time. This double technical goal is relevant in light of the numerous applications that exist in both commercial and military sectors for maintenance and usage of large image databases and digital video information.
- The low likelihood of incremental improvements to current storage technology fully meeting requirements for both high capacity and short access time presents an important window of opportunity for holographic mass memory systems and possibly other contending technologies.
- In contrast to the currently available storage strategies, holographic mass memory simultaneously offers high data capacity and short data access time. Consider, for example, a system in which each page of data to be recorded and retrieved contains 1 Mbit of data. The storage of 500,000 such pages using a combination of spatial and common volume multiplexing, will yield a total equivalent to magnetic systems; of which 50 such units must be spatially multiplexed. Moreover, because each page of data can potentially contain upwards of ~ Mbit of information, the parallel retrieval of a single page in a time interval as long as 100 us (the detector array—response time) yields a total data transfer rate of –0 Gbit/s 1.25 Gbyte/s.
- The electronic data can be scanned in parallel using a multiple-tap CCD array. Magnetic disks, by comparison, typically feature 10 Mbyte/s data rates, RAID systems feature greater than 100 Mbyte/s, and CD-ROMs achieve about 1 Mbyte/s.
- Researchers at Caltech are designing a compact holographic memory for use with applications requiring high-capacity read/write data storage with fast random access.
- Finally, researchers at Caltech have demonstrated the storage of 6,000 holograms in a memory architecture that uses different wavelengths for recording and readout in order to reduce the grating

decay while retrieving data. Bragg-mismatch problems from using two wavelengths are minimized by recording in the image plane and using thin crystals.

The next chapter opens Part Five, "Electro and Electron Holography," and discusses a new visual medium (electro-holography) capable of producing realistic 3D holographic images in real time. It also covers subsequent research in holovideo, which led to computation at interactive rates; full-color images; synthetic images and real-world input; and, most recently, a scale-up to a 36MB display system capable of producing images as large as about 100mm in width, height, and depth.

End Notes

1. Demetri Psaltis and Fai Mok, "Holographic Memories," *Scientific American*, November, 1995, Vol. 273, No. 5, SciDex, 415 Madison Avenue, New York, NY 10017-1111, USA, pp. 70–76.
2. Ibid.
3. Optical Information Processing Group and Psaltis Group, The Moore Building of Caltech, Room 262B, MS 136-93, Pasadena, CA 91125, USA, 2000.
4. Ernest Chuang, Jean-Jacques Drolet, George Barbastathis, Wenhai Liu, Faculty: Demetri Psaltis, Optical Information Processing Group and Psaltis Group, The Moore Building of Caltech, Room 262B, MS 136-93, Pasadena, CA 91125, USA, 2000.
5. Ibid.
6. Ibid.
7. Ibid.
8. Ibid.
9. Ibid.
10. Ernest Chuang and Demetri Psaltis, Optical Information Processing Group and Psaltis Group, The Moore Building of Caltech, Room 262B, MS 136-93, Pasadena, CA 91125, USA, 2000.
11. Ibid.
12. Ibid.
13. Ibid.
14. Ibid.
15. Ibid.
16. Ibid.
17. Wenhai Liu, Ernest Chuang, and Demetri Psaltis, California Institute of Technology, Optical Information Processing Group and Psaltis Group, The Moore

Building of Caltech, Room 262B, MS 136-93, Pasadena, CA 91125, USA, 2000.
18. Ibid.
19. Demetri Psaltis and Fai Mok, "Holographic Memories," *Scientific American*, November, 1995, Vol. 273, No. 5, SciDex, 415 Madison Avenue, New York, NY 10017-1111, USA, pp. 70–76.
20. Ibid.
21. Ibid.
22. Ibid.
23. Ibid.
24. Robert Buderi, "Data Holograms," *Upside*, Upside Media, Inc., 731 Market Street, 2nd Floor, San Francisco, CA 94103-2005, USA, (June, 1999), p. 146.
25. Holoplex Inc., 600 S. Lake Ave. #106, Pasadena, CA 91106, USA, 2000.
26. Demetri Psaltis and Fai Mok, "Holographic Memories," *Scientific American*, November, 1995, Vol. 273, No. 5, SciDex, 415 Madison Avenue, New York, NY 10017-1111, USA, pp. 70–76.

PART V

ELECTRO- AND ELECTRON HOLOGRAPHY

This part of the book discusses a new visual medium—electroholography—capable of producing realistic 3D holographic images in real time. It also examines the use of electron holography to record and reconstruct off-axis object wave (exit surface waves) electron holograms in conjunction for use with the electron microscope.

CHAPTER 22

ELECTRO-HOLOGRAPHY

In This Chapter

- Holovideo
- Interactive three-dimensional holographic displays
- Computational holographic bandwidth compression
- Rendering interactive holographic images
- Diffraction-specific fringe computation for electro-holography
- Interactive computation of holograms using a look-up table

Beginning in 1989, members of the MIT Spatial Imaging Group[1] created a new visual medium (electro-holography) capable of producing realistic 3D holographic images in real time. Subsequent research led to computation at interactive rates; full-color images; synthetic images and real-world input; and most recently, a scale-up to a 36MB display system capable of producing images as large as about 100mm in width, height, and depth. Computing generally involves transforming a 3D numerical description of the object scene into a holographic fringe pattern as shown in Figure 22-1.[2]

In 1990, newer, faster computational algorithms allowed for the first-ever interactive holography. Figure 22-2 shows the interactive holovideo display system as it appeared in 1991.[3]

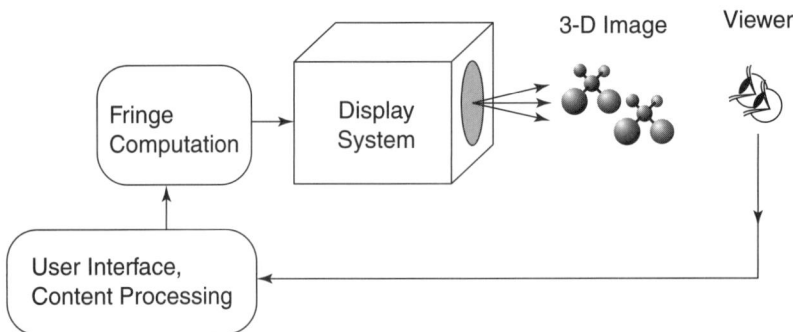

FIGURE 22-1: Overview of a holovideo system.

FIGURE 22-2: Interaction with image. MIT Media Lab Spatial Imaging Group. This composite picture shows a sequence of images (at top) that illustrates a typical manipulation of a 3D holographic image as the user changes both the form and position in a fraction of a second. The lower portion of the picture shows part of the holovideo display system and a viewer interacting with a small 3D image. The holographic fringe pattern used to create the interactive images contained 2MB of information each, and was computed on a massively parallel supercomputer (a Connection Machine Model 2 with 16 Kprocessors). A time-multiplexed scanned acousto-optic modulator transfers the fringe pattern onto a beam of light at a rate of over 100MB/s.

The holographic fringe pattern used to create the interactive image in Figure 22-2 contained 2MB of information, and was computed on a massively parallel supercomputer (a Connection Machine Model 2 with 16 Kprocessors). A time-multiplexed scanned acousto-optic modulator transfers the fringe pattern onto a beam of light at a rate of over 100MB/s.

In the most recent work, a 36MB system produces an image that is approximately 140mm wide, 80mm tall, and 150mm deep. Figure 22-3 shows schematically the layout of the current 36MB holovideo display system.[4] Figure 22-4 shows what the holovideo display system setup looks like.[5]

A new approach to fringe computation (called Diffraction-Specific fringe computation) has yielded two types of holographic bandwidth compression. Diffraction-Specific fringe computation (which is discussed in greater detail later in the chapter) is based on the spatial and spectral discretization of the hologram. The hologram is treated as a regular array of holographic elements called *hogels*. Each hogel is a small piece of the hologram and possesses a homogeneous spectrum (distribution of spatial frequencies). The first method of holographic bandwidth compression, called *hogel-vector encoding*, is a two-step process. First, an array of hogel-vectors (each one representing the discretized spectrum of one hogel) is computed from the 3D object scene description. In the second step, each hogel-vector is decoded into a hogel (the usable fringe) through the linear superposition

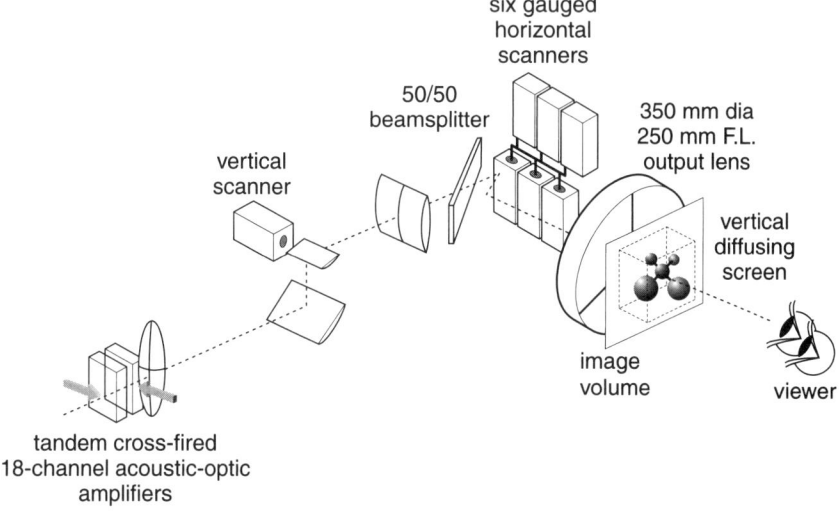

FIGURE 22-3: Layout of the current 36MB holovideo display system.

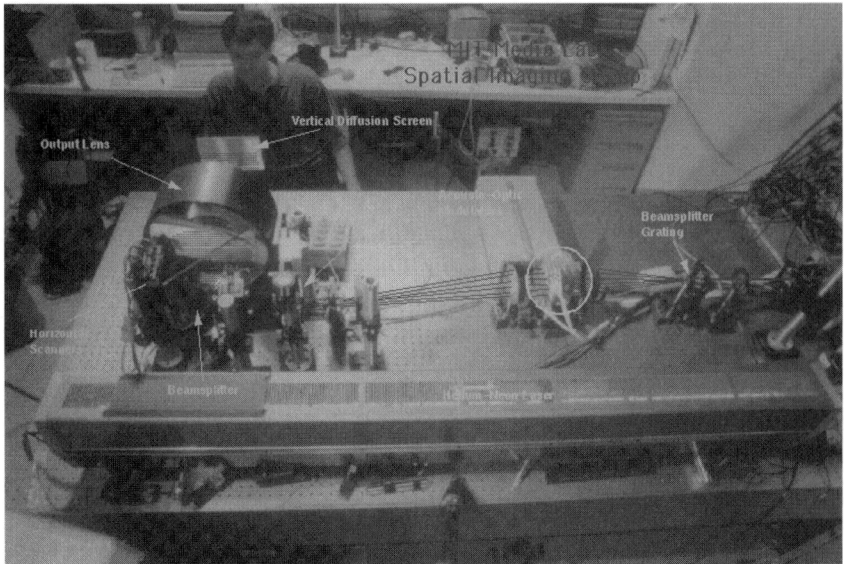

FIGURE 22-4: MIT holovideo: scaled-up 36MB display. This photo is a view (from above) of the MIT second-generation holovideo display system. The overlay graphics point out key components in the display, as well as the general beam path. The image is approximately 140mm wide, 80mm tall, and 150mm deep, with a viewing zone size of 30-degrees.

of a set of precomputed basis fringes. Hogel-vector decoding has been implemented using the Cheops Image Processing system[6] and the Splotch Engine—a superposition stream-processing daughter card (see Figures 22-5 and 22-6), as well as on many diverse computational platforms.[7] The second technique of holographic bandwidth compression is called *fringelet encoding*. A fringelet is computed for each hogel, and each fringelet is rapidly decoded to produce fringes. This method is designed to reduce the total number of calculations required per fringe sample, increasing computation speeds by over 100 times. Fringelet encoding promises to greatly simplify the design and construction of holovideo displays. Besides providing speed increases and reduction of bandwidth, the strength of these holographic encoding techniques is their simplicity, which enables implementation in simple specialized hardware, leading to further efficiencies. Hogel-vector encoding also allows holographic imaging to be combined with other forms of (2D) digital multimedia.

ELECTRO-HOLOGRAPHY **403**

FIGURE 22-5: The Cheops image processing system.

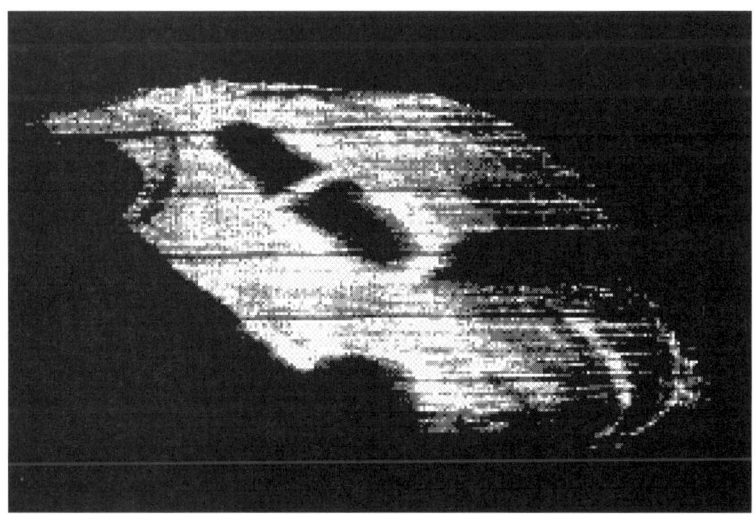

FIGURE 22-6: The Splotch Engine.

NOTE: The Cheops imaging system is a compact, modular platform for acquisition, processing, and display of digital video sequences and model-based representations of moving scenes, and is intended as both a laboratory tool and a prototype architecture for future programmable video decoders. Rather than using a large number of general-purpose processors and dividing up image processing tasks spatially, Cheops abstracts out a set of basic, computationally intensive stream operations that may be performed in parallel and embodies them in specialized hardware.

NOTE: Holographic fringe decoding has been implemented with the Cheops image processing system (see Figure 22-5) and a new daughter card called the Splotch Engine—a superposition stream-processing card as shown in Figure 22-6. The two-step hogel-vector encoding method used is a type of diffraction-specific fringe computation that allows for holographic bandwidth compression ratios of up to 20:1. An array of *hogel-vectors* (discretized spectral descriptions) is computed from a 3D object scene on an SGI workstation, and then rapidly downloaded to the Cheops P2 processor module. There, the Splotch Engine performs hogel-vector decoding, converting a stream of hogel-vectors into hologram fringes that are displayed by the holovideo system as 3D images. With two Splotch Engines operating in tandem, typical speeds are 3.0 seconds for the decoding of a 0.6MB hogel-vector array into a complete 36MB display fringe pattern. That throughput represents over 200 million multiply-accumulate operations per second!

Now, let's look at interactive three-dimensional holographic displays. Can we see the future in depth?

Seeing the Future in Depth: Interactive Three-Dimensional Holographic Displays

Computer graphics is confined chiefly to flat images. Images may look three dimensional (3D), and sometimes create the illusion of 3D when displayed, for example, on a stereoscopic display. Nevertheless, when viewing an image on most display systems, the human visual system (HVS) sees a flat plane of pixels. Volumetric displays can create a 3D computer graphics image, but fail to provide many visual depth cues (shading, texture gradients), and cannot provide the powerful depth cue of overlap (occlusion). Discrete parallax displays (such as lenticular displays) promise to create 3D images with all of the depth cues, but are limited by achievable resolution. Only a real-time electronic holographic (holovideo) display can create a truly 3D computer graphics image with all of the depth cues (motion parallax, ocular accommodation, occlusion, etc.) and resolution sufficient to provide extreme realism. Holovideo displays promise to enhance numerous applications in the creation and manipulation of information, including telepresence, education, medical imaging, interactive design, and scientific visualization.

The technology of electronic interactive three-dimensional holographic displays is in its second decade. Though fancied in popular science fiction, only recently have researchers created the first real holovideo systems by confronting the two basic requirements of electronic holography: (1) computational speed, and (2) high-bandwidth modulation of visible light. This part of the chapter describes the approaches used to address these problems, as well as emerging technologies and techniques that provide firm footing for the development of practical holovideo.

THE BASICS OF ELECTRO-HOLOGRAPHY

Optical holography (used to create 3D images) begins by using coherent light to record an interference pattern. Illumination light is modulated by the recorded holographic fringe pattern (called a *fringe*), subsequently diffracting to form a 3D image. As illustrated in Figure 22-7, a fringe region that contains a low spatial frequency component diffracts light by a small angle.[8] A region that contains a high spatial frequency component diffracts light by a large angle. In general, a region of a fringe contains a variety of spatial frequency components, and therefore diffracts light in a variety of directions.

An electro-holographic display generates a 3D holographic image from a 3D description of a scene. This process involves many steps, grouped into two main processes: (1) computational, in which the 3D description is converted into a holographic fringe, and (2) optical, in which light is modulated by the fringe. Figure 22-8 shows a map of the many techniques used in these two processes.[9]

The difficulties in both fringe computation and optical modulation result from the enormous amount of information (or *bandwidth*) required by holography. Instead of treating an image as a pixel array with a sample spacing of approximately 100 microns as is common in a two-dimensional (2D) display, a holographic display must compute a holographic fringe with a sample spacing of approximately 0.5 microns to cause modulated light to diffract and form a 3D image.

406 CHAPTER 22

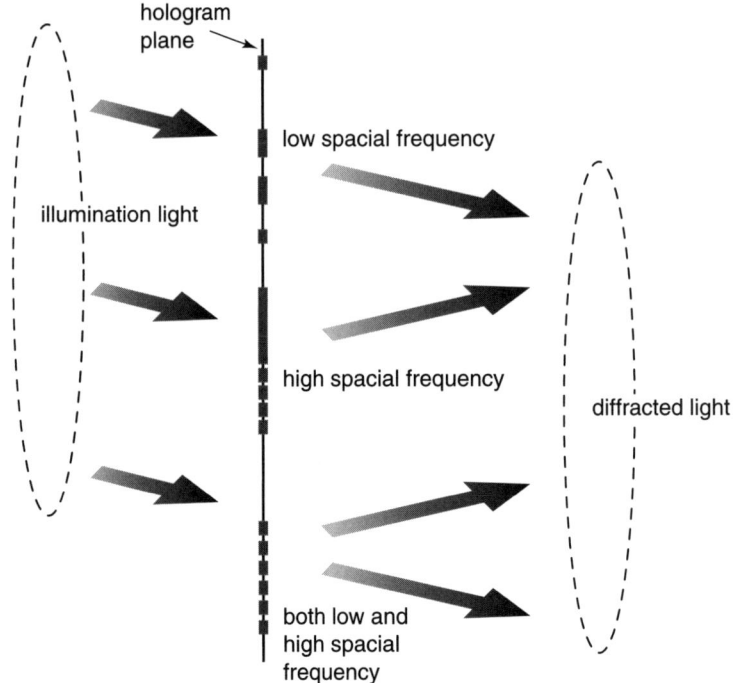

FIGURE 22-7: Diffraction of illumination light by holographic fringe patterns. Fringes with higher spatial frequencies cause light to diffract at larger angles. Fringes containing many spatial frequencies diffract light in many directions.

A typical palm-sized full-parallax (light diffracts vertically as well as horizontally) hologram has a sample count (*space-bandwidth product* or simply *bandwidth*) of over 100 gigasamples. Horizontal-parallax-only (HPO) imaging eliminates vertical parallax, resulting in a bandwidth savings of over 100 times without greatly compromising display performance. Holovideo is more difficult than 2D displays by a factor of about 40,000, or about 400 for an HPO system. The first holovideo display created small (50ml) images that required minutes of computation for each update. New approaches, such as holographic bandwidth compression and faster digital hardware, enable computation at interactive rates and promise to continue to increase the speed and complexity of displayed holovideo images. At present, the largest holovideo system creates an image that is as large as a human hand (about one liter). Figures 22-9 and 22-10 show typical images displayed on the MIT holovideo system.[10]

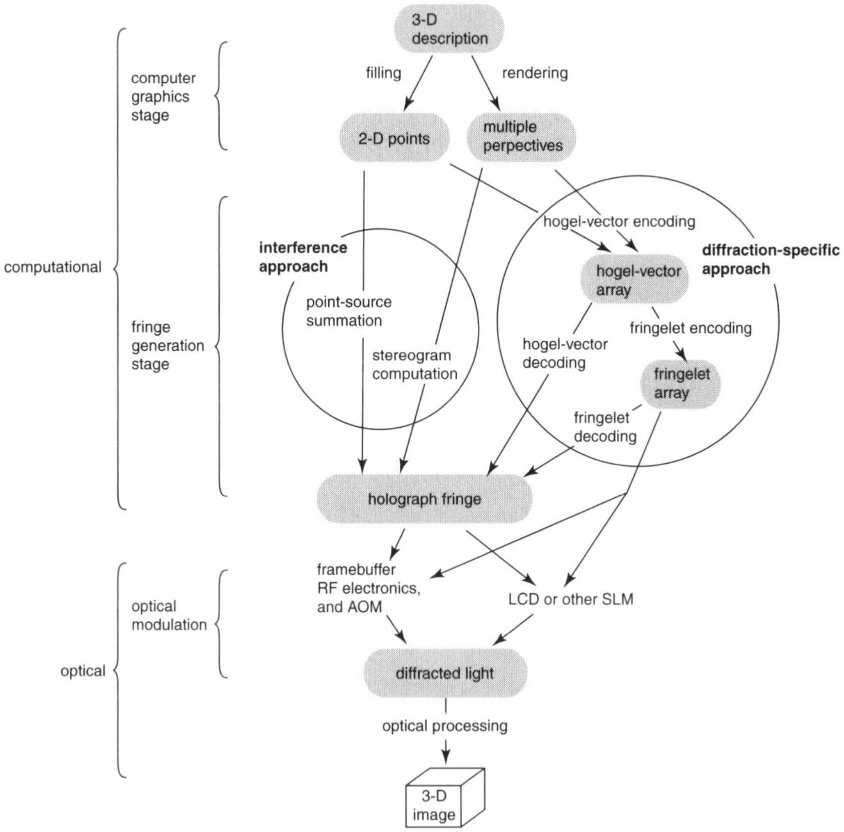

ELECTRO-HOLOGRAPHY   407

FIGURE 22-8: Information flow in interactive 3D holographic imaging. Each path traces the steps required for a particular method. Computation is generally faster for the methods that are more to the right-hand side.

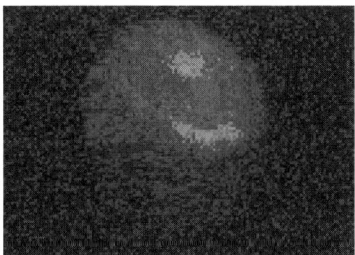

Figure 22-9: 6MB holovideo images on the MIT full-color display. Reddish apple with multicolor specular highlights, computed using hogel-vector bandwidth compression.

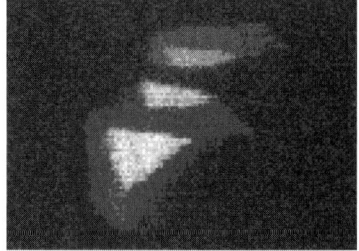

FIGURE 22-10: Red, blue, and green cut cubes, computed using stereogram approach.

HOLOGRAPHIC FRINGE COMPUTATION

The computational process in electro-holography converts a 3D description of an object or scene into a fringe pattern. Holovideo computation comprises two stages: (1) a computer graphics rendering-like stage, and (2) a holographic fringe generation stage in which 3D image information is encoded in terms of the physics of optical diffraction (see Figure 22-8).

The computer graphics stage often involves spatially transforming polygons (or other primitives), lighting, occlusion processing, shading, and (in some cases) rendering to 2D images. In some applications, this stage may be trivial. For example, MRI data may already exist as 3D voxels, each with a color or other characteristic.

The fringe generation stage uses the results of the computer graphics stage to compute a huge 2D holographic fringe. This stage is generally more computationally intensive, and often dictates the functions performed in the computer graphics stage. Furthermore, linking these two computing stages has prompted a variety of techniques. Holovideo computation can be classed into two basic approaches: interference based and diffraction specific.

The Interference-Based Approach

The conventional approach to computing fringes is to simulate optical interference—the physical process used to record optical holograms. Typically, the computer graphics stage is a 3D filling operation that generates a list of 3D points (or other primitives), including information about color, lighting, shading, and occlusion.

Following basic laws of optical propagation, complex wavefronts from object elements are summed with a reference wavefront to calculate the interference fringe. This summation is required at the many millions of fringe samples and for each image point, resulting in billions of computational steps for small, simple holographic images. Furthermore, these are complex arithmetic operations involving trigonometric functions and square roots, necessitating expensive floating point calculations. Researchers using the interference approach generally employ supercomputers and use simple images to achieve interactive display. This approach produces an image with resolution that is finer than can be utilized by the human visual system.

STEREOGRAMS

A stereogram is a type of hologram that is composed of a series of discrete 2D perspective views of the object scene. An HPO stereogram produces a view-dependent image that presents in each horizontally displaced direction the corresponding perspective view of the object scene, much like a lenticular display or a parallax barrier display. The computer graphics stage first generates a sequence of view images by moving the camera laterally in steps. These images are combined to generate a fringe for display.

The stereogram approach allows for computation at nearly interactive rates when implemented on specialized hardware. One disadvantage of the stereogram approach is the need for a large number of perspective views to create a high-quality image free from sampling artifacts, limiting the computation speed. New techniques may improve image quality and computational ease of stereograms.

The Diffraction-Specific Approach

The diffraction-specific approach breaks from the traditional simulation of optical holographic interference by working backward from the 3D image. The fringe is treated as being subsampled spatially (into functional holographic elements, or *hogels*) and spectrally (into an array of *hogel-vectors*). One way to generate a hogel-vector array begins by rendering a series of orthographic projections, each corresponding to a spectral sample of the hogels. The orthographic projections provide a discrete sampling of space (pixels) and spectrum (projection direction). They are easily converted into a hogel-vector array. A usable fringe is recovered from the hogel-vector representation during a decoding step employing a set of precomputed *basis fringes*.

The multiple-projection technique employs standard 3D computer graphics rendering (similar to the stereogram approach). The diffraction-specific approach increases overall computation speed and achieves bandwidth compression. A reduction in bandwidth is accompanied by a loss in image sharpness—an added blur that can be matched to the acuity of the HVS simply by choosing an appropriate compression ratio and sampling parameters. For a compression ratio (CR, the ratio between the size of the fringe and the hogel-vector array) of 8:1 or lower, the added blur is invisible to the HVS. For a CR of 16:1 or 32:1, good images are still achieved, with acceptable image degradation.

SPECIALIZED HARDWARE
Diffraction-specific fringe computation is fast enough for interactive holographic displays. Decoding is the slower step, requiring many multiplication-accumulation calculations (MACs). Specialized hardware can be utilized

for these simple and regular calculations, resulting in tremendous speed improvements. Researchers using a small digital signal processing (DSP) card achieved a computation time of about one second for a 6MB fringe with CR=32:1. In another demonstration, the decoding MACs are performed on the same Silicon Graphics RealityEngine2 (RE2) used to render the series of orthographic projections. The orthographic projections rendered on the RE2 are converted into a hogel-vector array using filtering. The array is then decoded on the RE2, as shown in Figure 22-11.[11] The texture-mapping function rapidly multiplies a component from each hogel-vector by a replicated array of a single basis fringe. This operation is repeated several times (once for each hogel-vector component), accumulating the result in the accumulation buffer. A computation time of 0.9 seconds was achieved for fringes of 6MB with CR=32:1.

FRINGELETS

Fringelet bandwidth compression (see Figure 22-8) further subsamples in the spatial domain. Each hogel is encoded as a spatially smaller *fringelet*. Using a simple sample-replication decoding scheme, fringelets provide the fastest method (to date) of fringe computation. Complex images have been generated in less than one second for 6MB fringes. Furthermore, a *fringelet display* can optically decode fringelets to produce a CR-times greater image volume without increased electronic bandwidth.

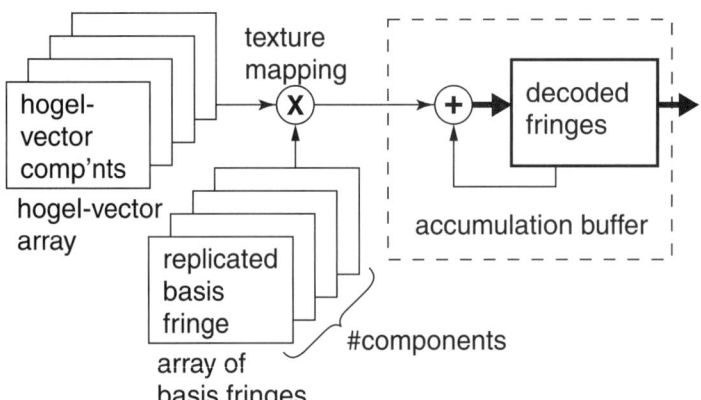

FIGURE 22-11: Hogel-vector decoding on the graphics subsystem. The inner product between an array of hogel-vectors and the precomputed basis fringes is performed rapidly by exploiting the texture-mapping function and the accumulation buffer.

PROCESSING AND OPTICAL MODULATION

The second process of a holographic display is optical modulation and processing. Information about the desired 3D scene passes from electronic bits to photons by modulating light with a computed holographic fringe using spatial light modulators (SLMs). The challenge in a holographic display arises from the many millions of samples in a fringe. Successful approaches to holographic optical modulation exploit parallelism and/or the time-response of the HVS.

Related SLMs and Liquid-Crystal

A liquid-crystal display (LCD) is a common electro-optic SLM used to modulate light for projection of 2D images. A typical LCD contains about one million elements (*pixels*). A 1-million-sample fringe can produce only a small flat image. A magneto-optic SLM, which uses the magneto-optic effect to electronically modulate light, often contains less than 1 million elements. Early researchers using LCD SLMs or magneto-optic SLMs created small planar images. The low pixel count of typical LCDs is overcome by tiling together several such modulators.

For any modulation technique, several issues must be addressed. Modulation elements are too big—typically 50 microns wide (in an LCD)—compared to the fringe sampling pitch of about 0.5 micron. Demagnification is employed to reduce the effective sample size, with the necessary but unattractive effect of proportionally reducing the lateral dimensions of the image (see Figure 22-12).[12] Holographic imaging may employ either amplitude or phase modulation. LCDs are basically phase modulators when used without polarizing optics. Phase modulation can be more optically efficient, and so is most often used. Finally, it is desirable to employ modulators possessing many levels of modulation (grayscale). Common LCDs have nominally 256 grayscale levels, sufficient for producing reasonably complex images.

Deformable micro-mirror devices (DMDs) are micromechanical SLMs fabricated on a semiconductor chip as an electronically addressed array of tiny mirror elements. Electrostatically depressing or tilting each element modulates the phase or amplitude of a reflected beam of light. A phase-modulating device was used to create a small flat holographic image, and a binary amplitude-modulating DMD was used to create a small interactive 3D holographic image.

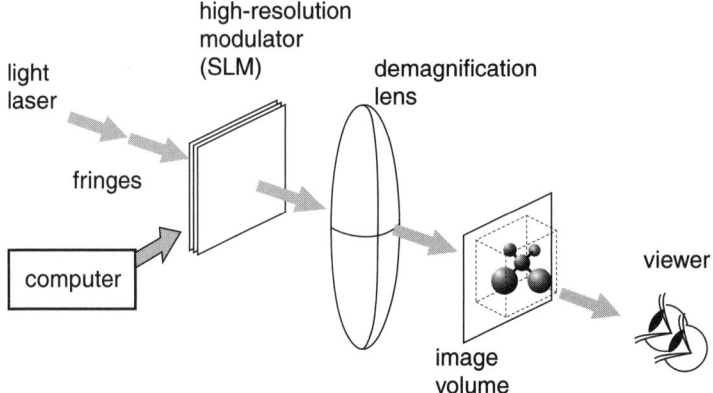

FIGURE 22-12: Holographic optical modulation using a typical high-resolution modulator (SLM). A minimum of 2 million modulation elements is required to produce even a small image the size of a thumb.

Scanned Acousto-Optic Modulator (AOM) and Other Techniques

The time-multiplexing of a very fast AOM SLM has been used in holovideo. A wide-aperture AOM phase modulates only about 1,000 samples at any one instant in time, using a rapidly propagating acoustic wave within a crystal. By scanning the image of modulated light with a rapidly moving mirror, a much larger apparent fringe can be modulated. The latency of the HVS is typically 20ms, and the eye time-integrates to see the entire fringe displayed during this time interval. This technique was invented and exploited by researchers at the MIT Media Laboratory to produce the world's first real-time 3D holographic display in 1989. A generalized schematic of this approach is shown in Figure 22-13.[13] After RF processing, computed fringes traverse the aperture of an AOM (as acoustic waves), which phase-modulate a beam of laser light. Two lenses image and demagnify the diffracted light at a plane in front of the viewer. The horizontal scanning system angularly multiplexes the image of the modulated light. A vertical scanning mirror reflects diffracted light to the correct vertical position in the hologram plane.

One advantage of the scanned-AOM system is that it can be scaled up to produce larger images. The first images produced in this way were 50ml, generated from 2MB fringes. More recently, by building a scanned-AOM system with 18 parallel modulation channels, images created from a 36MB fringe occupy a volume greater than one liter.

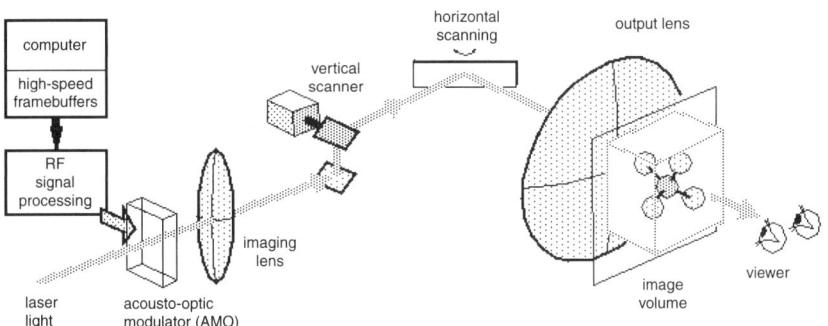

FIGURE 22-13: Schematic of the scanned-AOM architecture used in the MIT holovideo displays.

One disadvantage of the scanned-AOM approach is the need to convert digitally computed fringes into high-frequency analog signals. The 18-channel synchronized high-speed framebuffer system used at MIT was made for this application, and was a major practical obstacle in this approach. LCDs, DMDs, and other SLMs are more readily interfaced to digital electronics. Indeed, LCD SLMs are commonly constructed to plug directly into a digital computer, or are built on an integrated circuit chip. Another disadvantage of the scanned-AOM approach is the need for optical processing. Typical LCD-based holographic displays require only demagnification and the optical concatenation of multiple devices. The time-multiplexing of the scanned-AOM system requires state-of-the-art scanning mirrors that must be synchronized to the fringe data stream. Despite these obstacles, the scanned-AOM approach has produced the largest holovideo images.

COLOR
Full-color holovideo images are produced by computing three separate fringes. Each represents one of the additive primary colors (red, green, and blue), taking into account the three different wavelengths used in a color holovideo display. The three fringes are used to modulate three separate beams of light (one for each primary color).

SAW AOM
Recently, researchers have used an AOM device with multiple ultrasonic transducers. These multiple electrodes are fed a complex pattern and launch surface acoustic waves (SAWs) across the device aperture. Diffracted light forms a holographic image. Preliminary results show that this

approach may eliminate the need for time-multiplexing and consequently scanning mirrors. However, the large number of electrodes may be prohibitively expensive. Moreover, the array of SAW electrodes necessitates an additional numerical inversion transformation, making rapid computation difficult.

APPLICATIONS—PRESENT AND FUTURE

Real-time 3D holographic displays are expensive, new, and rare. Although they alone among 3D display technologies provide extremely realistic imagery, their cost must be justified. Each specific computer graphics application dictates whether holovideo is a necessity or extravagant expense.

Applications

Interactive computer graphics applications can be divided into two extreme modes of interaction: *the arm's reach* mode, and the *far-away* mode. An arm's-reach application involves interacting with scenes in a space directly in front of the user, where the user constantly interacts, moving around it to gain understanding. In this mode, all of the visual depth cues are employed, particularly motion parallax, binocular disparity, convergence, and ocular accommodation. These applications warrant the expense of holovideo and the extreme realism and three-dimensionality of its images: computer-aided design, multidimensional data visualization, virtual surgery, teleoperation, training, and education (holographic virtual textbooks on anatomy, molecules, or engines).

At the other extreme, a far-away application involves scenes that are beyond arm's reach and are generally larger. The imagery of such applications (flight simulation, virtual walk-throughs) makes adequate use of the kinetic depth cue, pictorial depth cues, and other depth cues associated with flat display systems. A high-resolution 2D display may be a more cost-effective solution for far-away applications.

Now, let's look at a novel technique to compute holographic fringe patterns for real-time display. Hogel-vector holographic bandwidth compression, a diffraction-specific approach, treats a fringe as discretized in space and spatial frequency. By undersampling fringe spectra, hogel-vector encoding achieves a compression ratio of 16:1 with an acceptably small loss in image resolution. Hogel-vector bandwidth compression achieves interactive rates of holographic computation for real-time three-dimensional electro-holographic (holovideo) displays. Total computation time for typical three-dimensional images is reduced by a factor of over 70 to 4.0 seconds per 36MB holographic fringe, and under 1.0 seconds for a 6MB full-color image. Analysis focuses on the trade-offs among compression ratio, image fidelity, and image depth. Hogel-vector bandwidth compression matches information content to the human visual system, achieving "*visual-bandwidth holography.*" Holovideo may now be applied to visualization, entertainment, and information.

Holographic Bandwidth Compression: A Computational Approach

Electro-holography, as previously mentioned (also called holovideo), is a new visual medium that electronically produces three-dimensional (3D) holographic images in real time. Holovideo is the first visual medium to produce dynamic images that exhibit all of the visual depth cues and realism found in physical scenes. It has numerous potential applications in visualization, entertainment, and information, including education, telepresence, medical imaging, interactive design, and scientific visualization. Electro-holography combines holography and digital computational techniques. As you know, holography is used to create 3D images using a two-step coherent optical process. An interference pattern (fringe pattern or simply fringe) is recorded in a high-resolution light-sensitive medium. Once developed, this recorded fringe diffracts an illuminating light beam to form a 3D image.

As early as 1964, researchers computed holographic fringes to create images that were synthetic and potentially dynamic. Both the computation and display of holographic images are difficult due to huge fringe bandwidths. A computed (discrete) fringe must contain roughly 10 million samples per square millimeter to effectively diffract visible light. Interactive-rate computation (about one frame per second or faster) was impossible. As previously stated, in 1989, researchers at the MIT Media Laboratory Spatial Imaging Group created the first display system that produced real-time 3D holographic images. Computation of the 2MB fringe required several minutes for small, simple images using conventional computation methods.

A new diffraction-specific computation technique named hogel-vector bandwidth compression achieves interactive-rate holographic computation (see Figure 22-14)[14]. The main features of this technique reported here include:

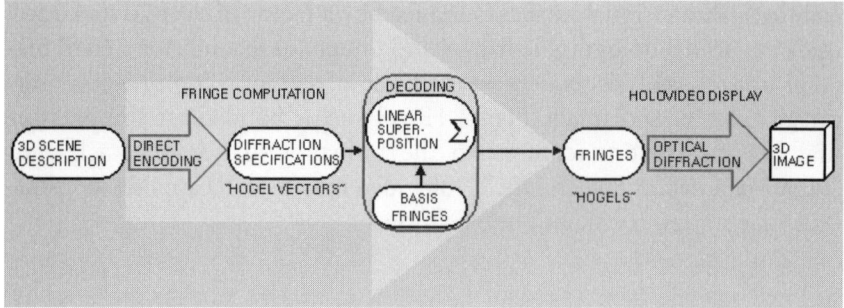

Figure 22-14: Hogel-vector bandwidth compression: direct encoding and decoding using superposition of precomputed basis fringes.

- Its architecture, based on the discretization of space and spatial frequency.
- The use of hogel-vector bandwidth compression to reduce bandwidth by 16:1 and higher, allowing for easier display, transmission, and storage.
- Fringe computation that is over 70 times faster than conventional computation.
- The trade-offs among the system parameters of image resolution, image depth, and bandwidth.

COMPUTATIONAL HOLOGRAPHY AND HOLOGRAPHIC DISPLAYS

This section of the chapter includes background information on computational holography, holographic displays, and past work in holographic information reduction. In other words, let's take a detailed look at hogel-vector bandwidth compression and its implementations, experimental results, and analysis.

Computational Holography
Computational holography begins with a 3D numerical description of the object or scene to be imaged. Traditional, conventional holographic computation imitated the interference of optical holographic recording. Speed was limited by two fundamental properties of fringes: (1) the myriad samples required to represent microscopic features (>1000 line-pairs per millimeter [lp/mm]); and (2) the computational complexity associated with the physical simulation of light propagation and interference.

In a computer-generated hologram, I define the number of samples per unit length (in one dimension) as the pitch, p. To satisfy fringe sampling requirements, a minimum of two samples per cycle of the highest spatial frequency are needed.

A typical full-parallax $100mm_{100mm}$ hologram has a sample count (also called *space-bandwidth product*, or simply *bandwidth*) of over 100 gigasamples. The elimination of vertical parallax provides savings in display complexity and computational requirements without greatly compromising display performance. This part of the chapter deals with horizontally off-axis transmission horizontal-parallax-only (HPO) holograms. Such an HPO fringe is commonly treated as a vertically stacked array of one-dimensional holographic lines.

A straightforward approach to the computation of holographic fringes resembled 3D computer graphics ray-tracing. The complex wavefront from each object element was summed, with a reference wavefront, to calculate the interference fringe. Interference-based computation requires many complex arithmetic operations (including trigonometric functions and square roots), making rapid computation impossible even on modern supercomputers. Furthermore, interference-based computation does not provide a flexible framework for the development of holographic bandwidth compression techniques.

Holographic Displays
Holographic displays modulate light with electronically generated holographic fringes. Early researchers employed a magneto-optic spatial light modulator (SLM) or a liquid-crystal display (LCD) to produce tiny planar images. The time-multiplexing of a very fast SLM provides a suitable substitute for an ideal holographic SLM. The research presented in this chapter revolves around the 6MB color holovideo display and the 36MB holovideo display developed by the Spatial Imaging Group at the MIT Media Laboratory. These displays used the combination of an acousto-optic modulator (AOM) and a series of lenses and scanning mirrors to assemble a 3D holographic image in real time (see Figure 22-15).[15] The 6MB display generated a full-color 3D image with a $35_{30\ 50}$ mm width by height by depth. The 36MB (monochromatic) display generated a $150_{75\ 160}$ mm image. By incorporating the proper physical parameters (wavelengths, sampling pitch), the fringe computation described in this chapter can be used for other holographic displays.

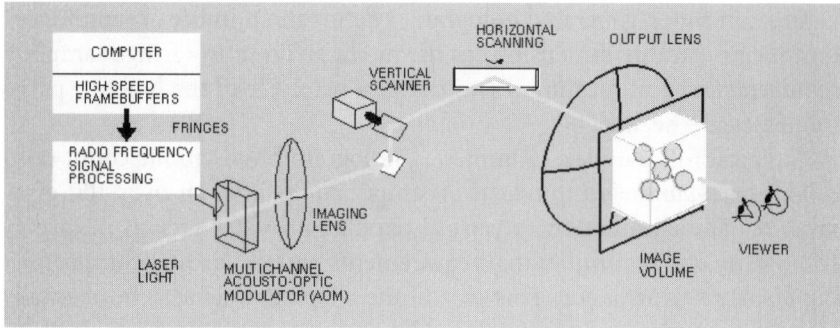

FIGURE 22-15: Schematic of MIT holovideo display, which presents a real 3D image (in front of the output lens) in the viewer.

HOGEL-VECTOR BANDWIDTH COMPRESSION

Hogel-vector bandwidth compression is a diffraction-specific fringe computation technique. Stated simply, the diffraction-specific approach is to consider only the reconstruction step in holography. In practical terms, it is the spatially and spectrally sampled treatment of a holographic fringe. Although numerical methods can compute diffraction backwards, they are far too slow for interactive-rate computation. Diffraction-specific fringe computation provides a fast means for generating useful fringes through calculations that relate the fringes to the image through diffraction in reverse. It has the following features (see Figure 22-14):

- **Spatial discretization**: The hologram plane is treated as a regular array of functional holographic elements named hogels. In HPO holograms, a horizontal line of the hologram is treated as regular line-segment hogels of width wh, each comprising roughly 100 to 2000 samples.
- **Spectral discretization**: A hogel-vector is a spectrally sampled representation of a hogel. A hogel-vector is the diffraction specification of a hogel. Three-dimensional object scene information is encoded as an array of hogel-vectors.
- **Basis fringes**: A set of precomputed basis fringes combine to decode each hogel-vector into one hogel-sized fringe.
- **Rapid linear superposition**: In the decoding step, hogel-vectors specify the linear real-valued superposition of the precomputed basis fringes to generate physically usable fringes.

By encoding the 3D scene description as diffraction specifications (an array of hogel-vectors), this technique reduces required bandwidth. Speed results from the simplicity, efficiency, and directness of basis-fringe summation in the decoding step.

Sampling and Recovery
Encoding hogel-vectors and decoding them into fringes are based on a spatially and spectrally sampled treatment of the fringe. The spatial and spectral sample spacings are selected to allow the fringe to be recovered from the hogel-vector array and used to diffract light to form the desired image. The first-order diffracted wavefront is the physical entity being represented by a fringe. Diffraction is linear, and the wavefront immediately following modulation by a fringe can be expressed as a summation of plane waves, each diffracted by a spatial frequency component.

Diffraction-specific computation treats a one-dimensional HPO fringe (at some vertical location y) as a two-dimensional (2D) localized spectrum $S(x, f)$, where x is the spatial position on the hologram and f is the spatial frequency. It possesses a continuously varying amplitude as a function of space and spatial frequency. Sij (the discrete representation of $S(x, f)$) is an array of hogel-vectors (the diffraction specifications for the one horizontal line of the fringe). When sampled and recovered correctly, $S(x, f)$ causes light to diffract and to form image points throughout one plane of the image volume (see Figure 22-16).[16]

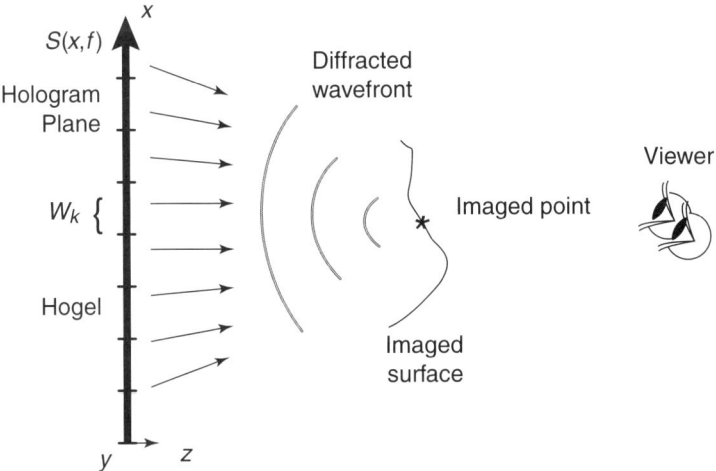

FIGURE 22-16: Computed fringe diffracts light to form an image.

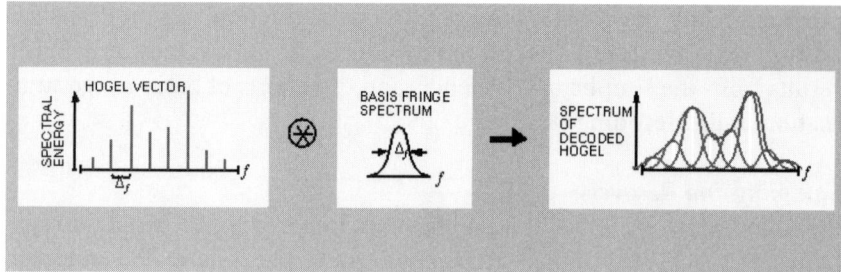

FIGURE 22-17: Spectral characteristics of hogel-vector decoding.

For the spectral dimension f, the convolution is performed in the spatial domain by the weighted summation of basis fringes, where each basis fringe represents one of the spectral regions. These convolutions are equivalent to performing a low-pass filtering. In practice, no ideal low-pass filter exists. The resulting spectral cross-talk theoretically added some noise to the image, though little additional noise was observable. A properly decoded (recovered) fringe has a smooth continuous spectrum, as shown in Figure 22-17.[17]

Bandwidth Compression: Spectral Subsampling

Conventional image and data compression starts with the desired data and then encodes the data into a *compressed* format. This compressed format is subsequently decoded into a (sometimes approximate) replica of the desired data. This approach can be applied to holographic fringes. However, total (model-to-fringe) computation speed is increased by computing the encoded format directly. This *direct-encoding* approach involves only two computation steps: direct encoding and decoding. Hogel-vector encoding is a direct-encoding technique, giving it the speed necessary for holovideo interactivity.

Basis Fringes

The spectral phase is uncorrelated among the basis fringes to make effective use of dynamic range. Spatially, a basis fringe has a uniform magnitude of 1.0 within the hogel width, and zero elsewhere. The spatial phase contains the diffractive information. The inter-hogel phase continuity is assured by constraining the endpoints of each basis fringe. The many constraints on basis fringes make their computation intractable using analytical approaches. For the present research, nonlinear optimization was used to design each basis fringe to have the desired spectral characteristics.

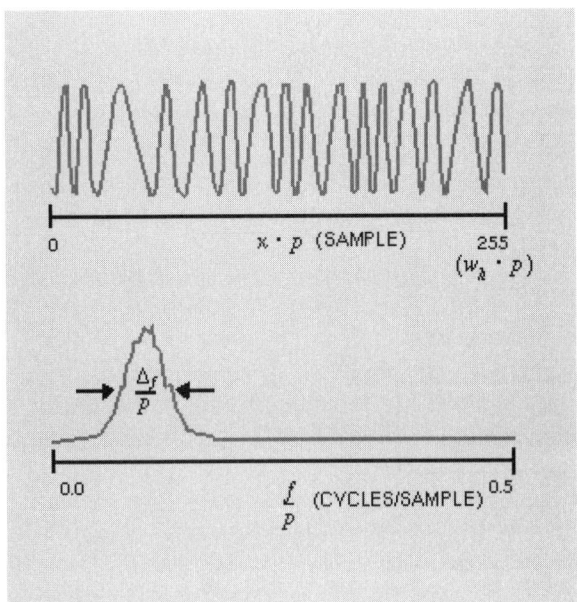

FIGURE 22-18: A basis fringe and its spectrum.

The synthetic basis fringes for given sampling spacings (w_λ) and a given display (pitch, p) were precomputed and stored for use in hogel-vector decoding. Figure 22-18 shows a typical basis fringe and its spectrum.[18]

Direct Encoding
Hogel-vector encoding converts a given 3D object scene into a hogel-vector array.

DIFFRACTION TABLES
The mapping from image element (x, z) to hogel position and vector component was precalculated and stored in a diffraction table. Indexing by the horizontal and depth locations (x, z) of each element, the diffraction table contains a spatially and spectrally sampled representation of the fringe that diffracts light to form the image element. The table lists which nonzero components of which hogel-vectors are needed to generate the correct fringe. In an HPO hologram, each line of the fringe pattern is treated independently, indexed by the vertical (y) location of the image element.

During hogel-vector encoding (shown in Figure 22-19), the diffraction table rapidly maps a given (x, z) location of a desired image element (a point) to components in the hogel-vector array.[19] The diffraction table

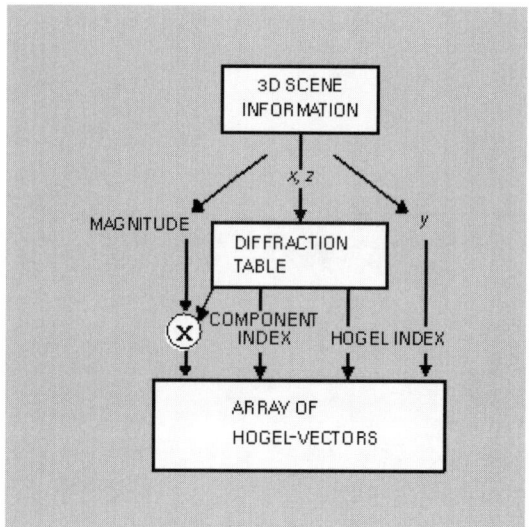

FIGURE 22-19: Generating hogel-vectors using a diffraction table.

includes an amplitude factor at each entry. This factor is multiplied by the desired magnitude (taken from the 3D scene and lighting information) to determine the amounts of contributions to each hogel-vector component. The magnitude of an image element is determined from its desired brightness. This brightness is represented as an intensity that is equal to the square of its magnitude. Therefore, in theory, the square roots of desired brightness values are used when calculating hogel-vectors. In practice, however, nonlinearities in the MIT holovideo display systems necessitated a brightness correction approximately equivalent to squaring magnitudes—canceling out the need to calculate square roots.

During computation of hogel-vectors for a particular 3D object scene, each image point is used to index the diffraction table. The contents of the table (for each indexed hogel-vector component) are summed to calculate the total hogel-vector array for this object. This direct-encoding step is fast because it involves only simple calculations.

Figure 22-20 illustrates the function of a diffraction table for image points.[20] The plane at the left represents the spatially varying fringe spectral content. The regular dot grid indicates the discretized spectrum (a hogel-vector array). Each dot is the location of one hogel-vector component. Each diagonal line represents the continuous (approximately linear) spectrum corresponding to the diffraction of light to an image point. The size

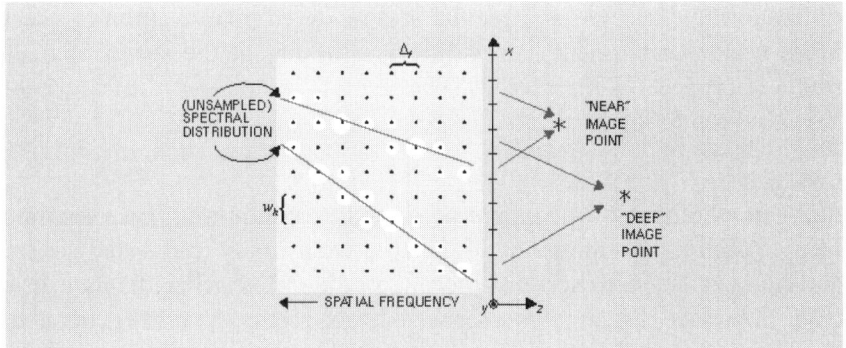

FIGURE 22-20: Precomputation of a diffraction table: typical spectral distribution for image points at different depths.

of the circular region around a dot indicates the amount of that hogel-vector component required to create the image point. The *deep* image point has a wide range of nonzero hogel-vector contributions. The *near* image point has a narrow range of contributions.

Hogel-vector encoding provides higher-level image elements. For example, if line segments of various sizes are useful for assembling the image scene, then a diffraction table is used to map location, size, and orientation of the desired segment to the proper hogel-vector contributions. Furthermore, the amplitude factors in the diffraction table allow for directionally dependent qualities (specular highlights) when a diffraction table is used to represent more complex image elements.

USE OF 3D COMPUTER GRAPHICS RENDERING

Another approach to performing direct-encoding employs 3D computer graphics rendering software. This facilitates advanced image properties, such as specular reflections, texture-mapping, advanced lighting models, and scene dynamics. A series of views are rendered, each in the direction corresponding to the center of one of the spectrally sampled regions. The view window (the plane upon which the scene is projected) must be coincident with the plane of the hologram ($z = 0$). Each rendered view is an orthographic projection of the scene from a particular view direction. The rendered views provide a discrete sampling of space and spectrum. These views are converted into a hogel-vector array using either a modified diffraction table or filtering, depending on hardware. In the modified-table method, the picture element (pixel) spacing in the 2D rendering of the scene is half the hogel spacing. This allows for subsampling. A special diffraction table uses view direction and rendered pixel location to select hogel-vector components, providing a sampling of the spatial-spatial-frequency space that matches *wh*. The second method uses the features of advanced rendering hardware.

COLOR

Full-color holovideo images are produced by computing three separate fringes, each representing one of the additive primary colors (red, green, and blue) taking into account the three different wavelengths used in a color holovideo display. Three separate hogel-vector arrays are generated, and each is decoded using the linear summation of one of three sets of pre-computed basis fringes. Basis-fringe selection via hogel-vector components proceeds using a single diffraction table, with the shorter wavelengths limited to a smaller range of diffraction directions.

Decoding Hogel-Vectors to Hogels

Hogel-vector decoding is the conversion of each hogel-vector into a useful fringe in a hogel region. To compute a given hogel, each component of its hogel-vector is used to multiply the corresponding basis fringe. The decoded hogel is the accumulation of all the weighted basis fringes (as shown in Figure 22-21).[21] Looking at the array of precomputed basis fringes as a

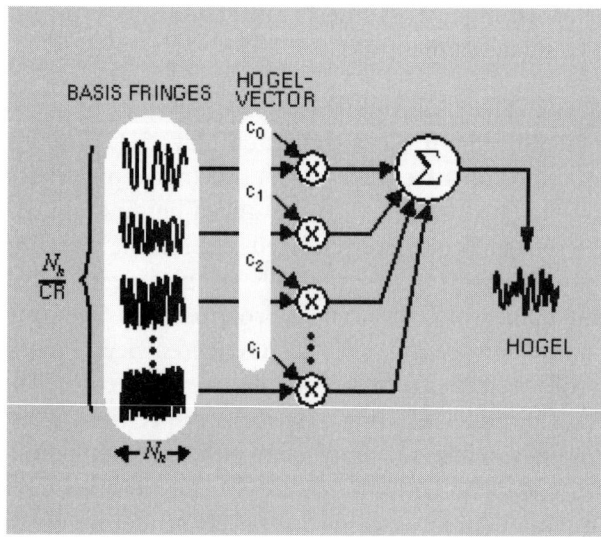

FIGURE 22-21: Decoding hogel-vectors to hogels.

two-dimensional matrix, hogel decoding is an inner product between the basis-fringe matrix and a hogel-vector.

Decoding is the more time-consuming step in hogel-vector bandwidth compression. However, the simplicity and consistency of this step means that it can be implemented on specialized hardware and performed rapidly. Various specialized hardware exists to perform multiplication-accumulation (MAC) operations at high speeds.

IMPLEMENTATION

Direct-encoding of the hogel-vectors was implemented on a Silicon Graphics, Inc., Onyx workstation, a high-end serial computer. Encoding involved a wide range of calculations, but was relatively fast. The second computation step, the decoding of hogel-vectors into hogels, was implemented on three systems: the Cheops framebuffer system used to drive the MIT 36MB holovideo display; the Onyx workstation; and a Silicon Graphics, Inc., RealityEngine2 graphics subsystem.

Implementation on Cheops

Hogel-vector encoding begins with a 3D image scene description generally consisting of about 0.5MB of information or less. After the appropriate transformations (rotations, translations) and lighting, it is direct-encoded as a hogel-vector array. For a compression ratio (CR) of 1:1 (no bandwidth compression), the hogel-vector array comprises 36MB. For larger compression ratios, this number is proportionally smaller (a CR = 16:1 gives a 2.2MB hogel-vector array).

The Cheops image processing system (see Figure 22-22) is a compact, block data-flow parallel processor designed and built for research in scalable digital television.[22] The P2 processor card communicates to the host via a SCSI (small computer standard interface) link with ~1MB/s bandwidth. Six Cheops output cards provide 18 parallel analog-converted channels of computed fringes to the MIT 36MB holovideo display. The P2 communicates data to the output cards using the fast Nile Bus. The P2 also supports a type of stream-processing superposition daughter card (the Splotch Engine) that performs weighted summations of arbitrary one-dimensional basis functions. The Splotch Engines (two were used in this research) perform the many MAC operations required for the decoding of hogel-vectors into hogels. The hogel-vector array was downloaded to the Cheops P2 card, where it was decoded using two Splotch Engines.

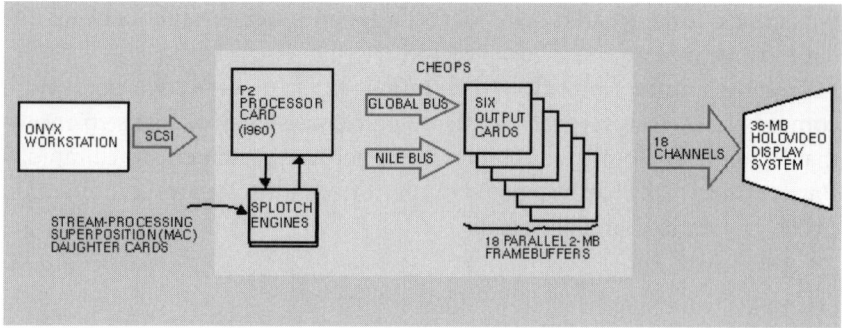

FIGURE 22-22: Hogel-vector decoding on the Cheops modular framebuffer and image processing system.

The Cheops output cards store each fringe sample as an 8-bit unsigned integer value. Computed fringes are normalized to fit within these 256 values. Normalization generally involves adding an offset and multiplying by a scaling factor. In the hogel-vector technique, normalization is built into the computational pipeline. For example, when using Cheops, the hogel-vector components are prescaled to produce useful fringes in the higher 8 bits of the 16-bit result field. Only this high byte is sent to the output cards.

The decoded 36MB fringe was transferred to the Cheops output cards and used by the MIT 36MB holovideo display to generate images. Figure 22-15, seen earlier, shows a general schematic of the MIT holovideo displays. After radio frequency (RF) processing, computed fringes (in the form of acoustic waves) traversed the aperture of an AOM (acousto-optical modulator), which phase-modulated a beam of laser light. Two lenses imaged the diffracted light at a plane in front of the viewer. The horizontal scanning system angularly multiplexed the image of the modulated light. The vertical scanning mirror positioned diffracted light to the correct vertical position in the hologram plane. Electronic control circuits synchronized the scanners to the incoming holographic signal.

Implementation on a Serial Workstation

The entire diffraction-specific computation pipeline was also implemented on the SGI Onyx serial workstation. The process for generating a 36MB fringe was the same as in the preceding, except that a simple linear loop performed the decoding step. The computed fringe was downloaded to Cheops to generate images on the 36MB holovideo display.

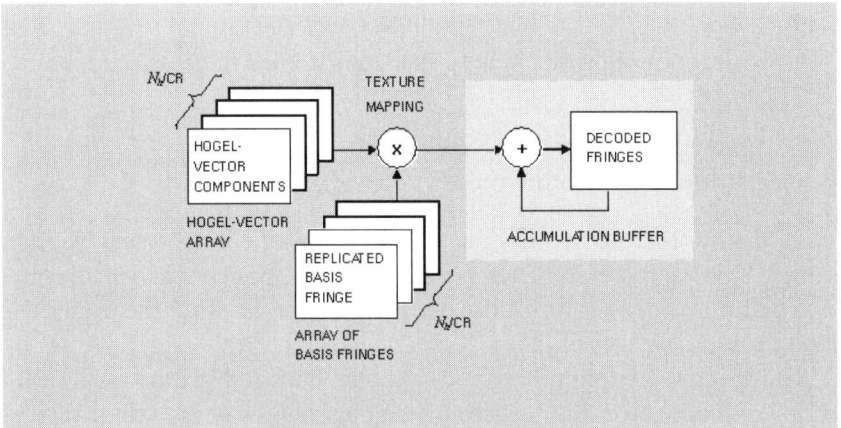

FIGURE 22-23: Hogel-vector decoding on the graphics subsystem.

Implementation on a Graphics Subsystem

The SGI RealityEngine2 (RE2) is a computer graphics subsystem generally used to render images. The rapid texture-mapping function and the accumulation buffer were used to perform rapid multiply-accumulate operations. The RE2 rendered directionally dependent 2D views of the object scene. These rendered views were converted into a hogel-vector array that was then decoded in the RE2, as shown in Figure 22-23.[23] The texture-mapping function rapidly multiplied a component from each hogel-vector by a replicated array of a single basis fringe. This operation is repeated Nh/CR times, once for each hogel-vector component, accumulating the result in the accumulation buffer. Transfer times were negligible because all computations occurred inside the graphics subsystem that included the frame-buffer to drive the display. For CR = 32:1, a 2MB fringe was decoded from a 64KB hogel-vector array for each of three colors.

MODEL OF POINT SPREAD

A fringe generated using hogel-vector bandwidth compression generally loses some of its ability to produce a sharp image. A given image point appears slightly broadened or blurred. The increased point spread results from several processes:

- Aperturing due to spatial sampling, blurspatial
- Spectral sampling blur due to sparsely sampled hogel spectra, blurspectral
- Aberrations in the display, blurdispl
- Quantization and other noise

As the number of symbols per hogel $N = Nh / CR$ decreases, spectral sample spacing increases. Each hogel-vector component carries information about a wider region of the hogel spectrum, limiting the achievable image resolution.

The spectral sampling blur and aperture blur add geometrically with other sources of blur. Blur caused by the display was measured for various z locations in the image volume.

Experimental Verification of Model

Figure 22-24 shows a point image focused at 80mm in front of the plane of the hologram (focused between the hologram plane and the viewer).[24] The

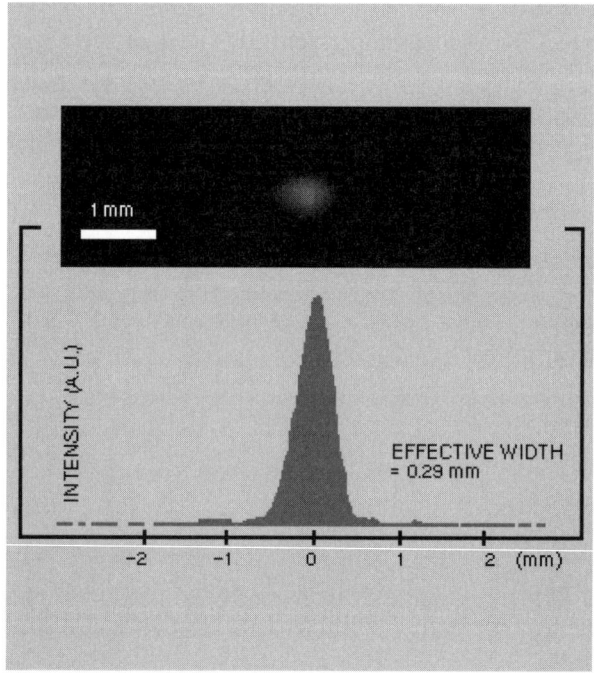

FIGURE 22-24: A point imaged at z = 80mm with CR = 1:1.

fringe used to generate this image was computed using hogel-vector bandwidth compression (CR = 1:1, hogel width $Nh = 512$ samples or $wh = 0.3$ mm). When compared to a conventionally computed point image, there is no additional noise or other artifact. The graph shows a cross section of the focused point, grabbed using a small tricolor charge coupled device (CCD) array placed at the location of an imaged point. A horizontal profile was obtained from the digitized photograph by integrating over the vertical extent of the red image component. The effective (horizontal) width of the imaged point was measured by calculating the narrowest horizontal region containing half of the total energy.

The point shown in Figure 22-24 was imaged using the MIT 36MB display. Points imaged at z = 80mm to test the worst-case resolution of the MIT 36MB display are shown in Figure 22-25.[25] These (and many other test points) were profiled and measured (using the same half-energy method) for a range of compression ratios. Blur increases with increasing CR (with f. For CR = 8:1 or less), the point is still sharp. For CR = 16:1, the

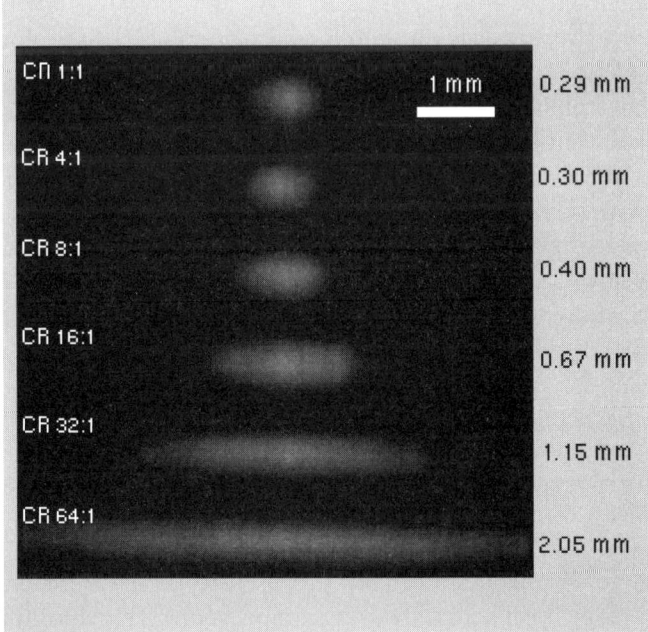

FIGURE 22-25: Hogel-vector bandwidth compression: effective width (horizontal point spread) versus CR.

effective width increases such that it is easily seen by a human viewer. Hogel width was *Nh = 512* samples, or *wh = 0.3mm*.

 Note: Fringe sampling pitch was p = 1.7 × 10$_{samples/mm}$.

Model-based Selection of System Parameters
The model for point spread provides an analytical expression that relates the various parameters of the holovideo system. This expression is used to select certain parameters such as hogel width and compression ratio. For practical imaging, blur must be below the amount perceivable to humans—about 0.18mm at a typical viewing distance of 600mm.

IMAGING RESULTS

Fringes were computed using hogel-vector bandwidth compression and used to generate a variety of 3D images on the MIT holovideo displays. Speeds were measured. Digital photographs were taken of images.

Images were generated for a variety of values of *wh* and CR. For properly selected values, point spread was not perceivable. Figure 22-26 shows a typical result.[26] The upper picture was generated from a 36MB fringe decoded (Nh = 1024) from a 36MB hogel-vector array (i.e., CR = 1:1). The 3D image (a Volkswagen Beetle car) was generated from a polygon database comprising 1,079 polygons. The image was converted into about 10,000 discrete image points using a simple lighting model. The lower picture shows the same object, computed using 2.2MB hogel-vector array (CR = 16:1). The use of only 1/16 the bandwidth causes only slight changes in the image. The discrete image points are just visibly blurred, and a slightly speckle-like appearance is added.

 Note: The picture quality suffered from artifacts present in the display. Imbalances and nonlinearities in the RF signal-processing electronics of the display system produced the unwanted horizontal streaks and bands of light and dark.

The MIT 6MB color holovideo display was used to generate images computed using hogel-vector bandwidth compression. Hogel-vector direct-encoding and decoding were performed by the RE2, using three sets of basis fringes, each precomputed for the specific wavelength used in the

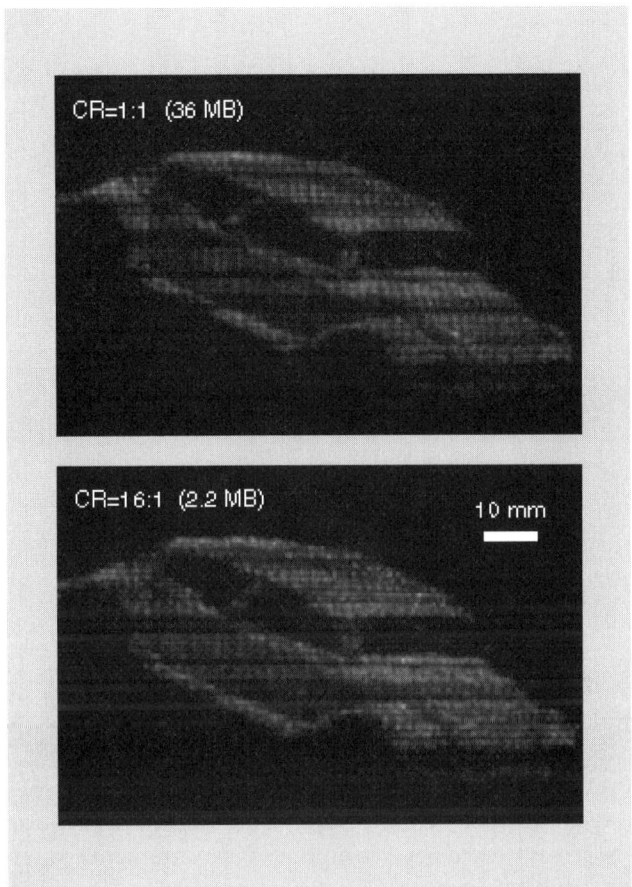

FIGURE 22-26: 36MB holovideo image: two compression ratios.

full-color display ($p = 1000mm^{-1}$). Three 2MB fringes were decoded from three 64KB hogel-vector arrays (one per color). The resulting images—computed at interactive rates—showed good quality and possessed the full range of lighting features (specular highlights and transparency). Figure 22-27 shows photographs of typical full-color images, computed with a hogel width of 256 samples (0.250mm).[27] These images were sharp despite the high CR = 32:1.

Earlier work reported the use of the RE2 to compute stereogram-style holograms. The stereogram images had noticeable blur and artifacts, especially when computed rapidly. In comparison, hogel-vector bandwidth

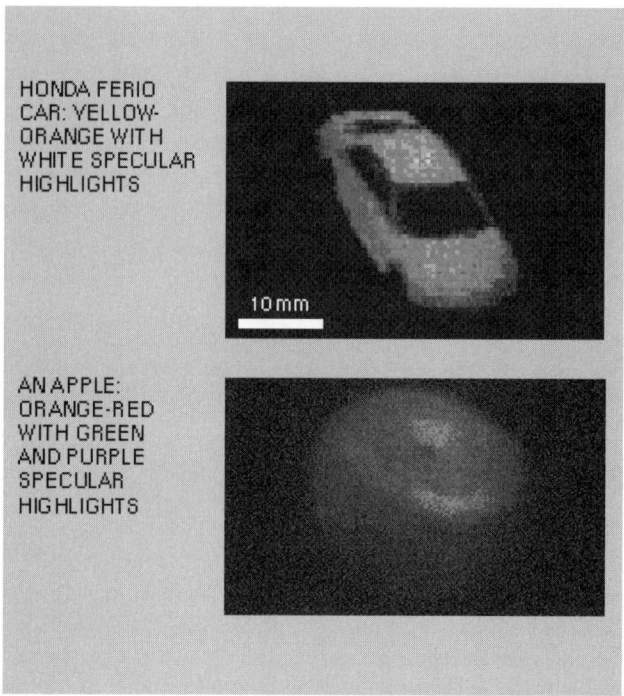

FIGURE 22-27: 6MB full-color holovideo images: car and apple.

compression maintained image fidelity, even at the high compression ratios (CR = 32:1) necessary to achieve computation at interactive rates. These images did not exhibit the artifacts of vertical dark stripes or jumps. Moreover, of course, images computed using the diffraction-specific technique produced real 3D images.

Spatial Coherence
Hogel-vector bandwidth compression added a noticeable speckle-like appearance to the image. These brightness variations at infinity likely resulted from the use of coherent light in the display. Diffraction-specific fringe computation assumes that light is quasi-monochromatic with a coherence length $Lc < wh/2 \sim 0.1mm$. This ensures that the diffracted light adds linearly with intensity. In practice, the effective coherence length of light in the holovideo displays was approximately 2.0mm. To reduce the speckle effect, a random set of phases was introduced into each hogel via the basis fringes. This reduced interhogel correlation. Implementation employed a

set of 16 different but spectrally equivalent precomputed basis fringes. To decode a given hogel-vector, each basis fringe was selected at random from the set of equivalent basis fringes.

SPEED

Hogel-vector bandwidth compression achieved an increase in speed by a factor of over 70 compared to conventional interference-based methods. Computing times were measured on three platforms: the Onyx workstation (alone); the Onyx (for hogel-vector generation) and the Cheops with two Splotch Engines (for decoding); and the Onyx/RE2. The results from three computational benchmarks are described: a conventional interference-based technique, and two diffraction-specific cases in which hogel-vector compression ratios are CR = 1:1 and CR = 32:1.

Using the hogel-vector technique, total computation time consists of the initial direct-encoding step, the time to transfer the hogel-vector array to the decoding system, and the hogel-vector decoding step. The first step (generation of the hogel-vector array on the Onyx workstation) was very fast. For most objects, typical times were 10 seconds for CR = 1:1, and 0.5 seconds for CR = 32:1. The downloading of the hogel-vector array over the SCSI link was slow. However, the use of hogel-vector bandwidth compression (CR = 32:1) reduced data transfer of the 1.1MB hogel-vector array to only 1.0 second.

Appropriate scene complexities were chosen to ensure equivalent benchmarks. For hogel-vector bandwidth compression, speed is basically independent of image scene complexity, whereas the computing time for interference-based ray-tracing computations varies roughly linearly.

Speed on a Serial Workstation
The conventional interference-based method used was to sum the complex wavefronts from all object points (plus the reference beam) to generate the fringe. Because it involved complex-valued, floating-point precision calculations, it was not implemented on either of the specialized hardware platforms (Cheops/Splotch, or the RE2). A fairly complex image of 20,000 discrete points (roughly 128 imaged points for each line of the fringe) was used. Implemented completely on the Onyx workstation, a 36MB fringe required 23,000 seconds (over 6 hours). This timing was extrapolated by computing a representative 2MB fringe.

For comparison, hogel-vector bandwidth compression was implemented on the Onyx workstation. For a 36MB fringe computed using

$CR = 1:1$ *(and Nh = 1024)*, total time was 9,600 seconds. For $CR = 32:1$, the time was reduced to only 300 seconds. Including the time for hogel-vector generation (0.5 seconds), this represents a speed increase of 74 times compared to the conventional computing method. Another advantage of hogel-vector bandwidth compression is that the simplicity of the slower decoding step allows for its implementation on very fast specialized hardware.

Speed on Cheops
When implemented on the Cheops system containing two Splotch Engines, hogel-vector decoding time was 190 seconds for $CR = 1:1$, and only 6 seconds for $CR = 32:1$. These timings are worst case (most complex image, measured using a fully nonzero hogel-vector array). In practice, typical image scenes produced many zero-valued hogel-vector components. Skipping zero-valued components resulted in faster decoding times. Typical test images ($Nh = 1024$, $CR = 32:1$) were closer to 3 seconds.

The total hogel-vector encoding and decoding time in the case of $CR = 32:1$ was 6.5 seconds, worst case. Although it is not quite fair to compare this to the conventional method implemented only on the workstation, the relative speed increase of over 3,500 times is made possible by the simplicity and efficiency of the hogel-vector decoding algorithm.

Speed on a Graphics Subsystem
When implemented on the RE2 graphics subsystem, for a 6MB (2MB per color) fringe, hogel-vector decoding time was 28 seconds for $CR = 1:1$, and only 0.9 seconds for $CR = 32:1$. The texture-mapping function of the RE2 graphics subsystem and its accumulation buffer performed rapid multiplications and additions. The 0.9 seconds is a total time, from model to image, since the MIT color holovideo display was driven directly by the RE2. Conventional methods cannot utilize this specialized hardware or achieve interactive-rate fringe computation.

Analysis of Speed
In hogel-vector bandwidth compression, the decoding step required the great majority of computing time. Decoding an Nh/CR-component hogel-vector to an Nh-sample hogel requires calculating an inner product: $Nh_{/CR}$ multiplication-accumulation operations (MACs). For example, for a 36MB fringe, a hogel width of $Nh = 1024$, and a $CR = 32:1$, the decoding step requires 1.2 GMACs (over 1 billion multiplies and adds).

NOTE: The speed increase from *CR = 1:1* to *CR = 32:1* is about 32 times (see Table 22-1).[28] This was due to the reduction by a factor of *1 / CR* in the number of time-consuming MAC calculations required. Because each fringe sample requires *Nh / CR MACs*, the speed of hogel-vector decoding increases linearly with CR. Faster speeds can be achieved by sacrificing image quality.

Further speed increase during interactivity was achieved by exploiting the scalability of a hogel-vector array (its ability to supply variable degrees of precision as required). For example, in one interactive demonstration of the 6MB display, a subset of the hogel-vector array was decoded to produce a quick-and-dirty image that was subsequently replaced by the full-fidelity image when interactivity ceased.

Next, let's very briefly look at a method for computing holographic patterns for the generation of 3D holographic images at interactive speeds. The method that is used by the researchers renders holograms on a conventional computer graphics workstation. The framebuffer system supplied signals directly to a real-time holographic (*holovideo*) display. The researchers at the MIT Media Laboratory developed an efficient algorithm for computing an image-plane stereogram, a type of hologram that allowed for several computational simplifications. The rendering algorithm generated the holographic pattern by compositing a sequence of view images that were rendered using a recentering shear-camera geometry. Computational efficiencies of the researchers' rendering method allowed the workstation to calculate a 6-megabyte holographic pattern in under 2 seconds, over 100 times faster than traditional computing methods. Data-transfer time was negligible. Holovideo displays are ideal for numerous 3D visualization applications, and promise to provide 3D images with extreme realism. Although the focus of this work was on fast computation for

TABLE 22-1: Computation times for different hardware and techniques

Platform, Fringe Size	Conventional (seconds)	Hogel-vector Bandwidth CR=1:1 (seconds)	Compression CR=32:1 (seconds)
Workstation, 36MB	23,000	9,600	300.5
Cheops'2-Splotch, 36MB		200	6.5
RE2, 6MB		28	0.9

NOTE: Transfer times are not included. Hogel-vector times are worst case and include encoding and decoding.

holovideo, the computed holograms can be displayed using other holographic media.

The Rendering of Interactive Holographic Images

The practical use of 3D displays has long been a goal in computer graphics. Three-dimensional displays are generally electronic devices that provide binocular depth cues, particularly binocular disparity and convergence. Some 3D displays provide additional depth cues such as motion parallax and ocular accommodation. Recently, some 3D displays have been used interactively. A 3D display allows the viewer to more efficiently and accurately sense both the 3D shapes of objects and their relative spatial locations, particularly when monocular depth cues are not prevalent in a scene. When viewing complex or unfamiliar object scenes, the viewer can more quickly and accurately identify the scene contents. Therefore, 3D displays are important in any application involving the visualization of 3D data, including telepresence, education, medical imaging, computer-aided design, scientific visualization, and entertainment. The merit of a 3D display depends primarily on its ability to provide depth cues and high resolutions. The inclusion of depth cues (particularly binocular disparity, motion parallax, and occlusion) increases the realism of an image.

Holography is the only imaging technique that can provide all the depth cues. All other 3D display devices lack one or more of the visual depth cues. For example, stereoscopic displays do not provide ocular accommodation, and volume displays cannot provide occlusion. Image resolution and parallax resolution are also important considerations when displaying 3D images. While most 3D displays fail to provide acceptable image and parallax resolutions, holography can produce images with virtually unlimited resolutions. In optical holography, a recorded interference pattern reconstructs an image with an extremely high degree of accuracy. A holographic pattern (called *fringes*) can be computed and used to generate a 3D image, most recently in real time. Both the computing process and the displaying process are significantly more difficult than in other 3D display systems. Nevertheless, a real-time electro-holographic (*holovideo*) display can produce dynamic 3D images with all of the depth cues and image realism found in optical holography. Therefore, holovideo has the potential to produce the highest quality 3D images. Moreover, to view holographic images, the viewer is unencumbered by equipment such as glasses or sensors.

Finally, a computer-generated hologram (CGH) must contain a huge number of samples. Furthermore, the cost of calculating each sample is high if a conventional approach is taken. Even with the power currently available in scientific workstations, researchers in the field of computational holography commonly report computation times in minutes or hours.

Now, let's very briefly examine the development, implementation, and analysis of diffraction-specific computation, an approach that considers the reconstruction process rather than the interference process in optical holography. Diffraction-specific fringe computation is a novel system for the generation of holographic fringe patterns for real-time display. The primary goal is to increase the speed of holographic computation for real-time three-dimensional electro-holographic (holovideo) displays. Diffraction-specific fringe computation is based on the discretization of space and spatial frequency in the fringe pattern. Holographic fringe encoding techniques are developed from diffraction-specific fringe computation and applied to make most efficient use of hologram channel capacity. A *hogel-vector encoding* technique is based on undersampling the fringe spectra. A *fringelet encoding* technique is designed to increase the speed and simplicity of decoding. The analysis of diffraction-specific computation focuses on the trade-offs between compression ratio, image fidelity, and image depth. The decreased image resolution (increased point spread) that is introduced into holographic images due to encoding is imperceptible to the human visual system under certain conditions. A compression ratio of 16 is achieved (using either encoding method) with an acceptably small loss in image resolution. Total computation time is reduced by a factor of over 100 to less than 7.0 seconds per 36MB holographic fringe using the fringelet encoding method. Diffraction-specific computation more efficiently matches the information content of holographic fringes to the human visual system. Diffraction-specific holographic encoding allows for *visual-bandwidth holography* (holographic imaging that requires a bandwidth commensurate with the usable visual information contained in an image). Diffraction-specific holographic encoding enables the integration of holographic information with other digital media, and is therefore vital to applications of holovideo in the areas of visualization, entertainment, and information, including education, telepresence, medical imaging, interactive design, and scientific visualization.

Diffraction-Specific Fringe Computation for Electro-Holography

Humans are visual animals. For thousands of years, humans have created visual media with which to express and communicate ideas, to record and understand our environment, and to manipulate and analyze the nature of real and imagined entities. Electro-holography (also called holovideo) is the newest visual medium. A holovideo display produces three-dimensional holographic images electronically in real time. Holovideo is the first visual medium to produce dynamic images that exhibit all of the visual sensations of depth and realism found in physical objects and scenes. Perhaps the most exciting feature of holovideo as a new visual medium is its ability to produce tangible-looking images of scenes and objects that do not exist, or cannot exist. Holovideo has numerous potential applications in visualization, entertainment, and information, including education, telepresence, medical imaging, interactive design, and scientific visualization.

The new technical field of electro-holography is essentially the marriage between holography and digital computational technologies. Holography is used to create 3D images by recording and reproducing optical wavefronts. The conception of holography began more than 50 years ago, and was applied to the recording and subsequent reconstruction of 3D images beginning only 37 years ago with the arrival of the laser. Optical holographic imaging is a two-step process.

Generally, a rigid object is illuminated with a coherent beam of light. A mutually coherent reference beam is aligned to interfere with the light scattered from the object. The resulting microscopic interference pattern (fringe pattern or simply fringe) is recorded in a high-resolution light-sensitive medium such as photographic film. Once developed, this recorded fringe diffracts an illuminating light beam to form a 3D image that can look identical to the original object. The fringe pattern diffracts lights because its feature size is generally on the order of the wavelength of visible light (about 0.5mm). Because of this microscopic resolution, the fringe pattern contains an enormous amount of information—roughly 10 million resolvable features per square millimeter.

As previously stated, as early as 1964, researchers began to consider the computation, transmission, and use of holographic fringes to create images that were synthetic and perhaps dynamic. These researchers encountered the fundamental problems inherent to computational holography: both the computation and display of holographic images are difficult due to the large amount of information contained in a fringe pattern. Approximately 10 million samples per square millimeter are required to compute a discretized (sampled) fringe that matches the resolution (diffractive power) of an optically made hologram. The possibility of computing a 10-billion-sample fringe pattern at a rate of once per second was impossible, and the possibility of modulating a beam of light with such a fringe pattern was beyond any spatial light modulation technologies available at that time. For decades, the enormous size of a holographic fringe prohibited and discouraged the pursuit of real-time electronic 3D holographic imaging.

As previously mentioned, in 1989, researchers at the MIT Media Laboratory Spatial Imaging Group created the first display system capable of producing real-time 3D holographic images. The images were small but were made possible by information reduction strategies that lowered the number of fringe samples to only 2MB, the minimum necessary to create an image the size of a golf ball. A modulation scheme based on time-multiplexing an acousto-optic modulator was used to modulate a beam of light with the 2MB of discretized computed holographic fringe pattern. Computation of the 2MB fringe pattern still required several minutes for simple images using traditional computation methods that imitated the optical creation of holographic fringes. Speed was limited by two factors: (1) the huge number of samples in the discretized fringe, and (2) the complexity of the physical simulation of light propagation used to calculate each sample value. As display size increased, the amount of information increased too, roughly proportional to the image volume (the volume occupied by the 3D image). To achieve interactive holographic computation, a new type of approach would have to be invented. That new approach (called *diffraction-specific computation*) is the subject of this part of the chapter.

DIFFRACTION-SPECIFIC COMPUTATION

This part of the chapter briefly concentrates on the generation of holographic fringes using a new method called *diffraction-specific computation*. The architecture of diffraction-specific computation is directed by two primary goals: (1) to produce fringes at a faster rate, and (2) to enable holographic encoding schemes to reduce the bandwidth required to display holographic images. Traditional computing methods achieved neither of these goals. Traditional computation imitated the interference occurring in the optical generation of fringes. In contrast, diffraction-specific computation is based on only the diffraction that occurs during the reconstruction of a holographic image. The diffraction-specific approach is a better match to holovideo, since the purpose of a real-time holographic display is to generate 3D images through the modulation and subsequent diffraction of light. The research described in this part of the chapter demonstrates that moving from interference-based methods and toward diffraction-specific methods increases fringe computation speed.

The application of diffraction-specific computation provides a means for encoding fringes to make the most efficient use of computational power and electronic and optical bandwidth. Holographic fringes contain far less usable information than is intimated by a simple measure of bandwidth. Roughly 10 million samples per square millimeter are required, whether the fringe pattern represents a simple object or a complex scene, and whether the holographic image is to be viewed through a microscope or by the less acute human visual system (HVS). However, this observation alone does not provide instruction for the reduction or compression of fringe bandwidth. The development of holographic encoding begins by taking a closer look at the nature of holographic fringes. Diffraction-specific fringes are no longer strictly physical, but are synthetically generated by specifying more deliberately their diffractive purpose. This specification leads to a more efficient use of holographic channel capacity. In other words, there is a progression from interference-based computation toward diffraction specific fringes, from physical fringes to synthetic fringes, and from inefficient use of channel capacity to a reduction in required bandwidth. Ultimately, the information-reduction strategies born of the diffraction-specific approach should allow for the design and construction of more information-efficient holographic displays.

Next, we'll look at several methods of increasing the speed and simplicity of the computation of off-axis transmission holograms, with applications to the real-time display of holographic images. The bipolar intensity approach allows for the real-valued linear summation of interference fringes, a factor of 2 speed increase, and the elimination of image noise caused by object self-interference. An order of magnitude speed increase is obtained through the use of a precomputed look-up table containing a large array of elemental interference patterns corresponding to point source contributions from each of the possible locations in image space. Results achieved using a data-parallel supercomputer to compute horizontal-parallax-only holographic patterns containing 6 megasamples indicate that an image comprised of 10,000 points with arbitrary brightness (grayscale) can be computed in under 1 second. Implemented on a common workstation, the look-up table approach increases computation speed by a factor of 43.

Interactive Computation of Holograms Using a Look-Up Table

The real-time display of holographic images has recently become a reality. The Spatial Imaging Group at the MIT Media Laboratory has reported the successful generation of small, three-dimensional computer-generated holographic images reconstructed in real time using a display system based on acousto-optic modulation of light. In any holographic display, a computer-generated hologram (CGH) must be computed as quickly as possible to provide for dynamic and interactive images. However, the numerical synthesis of a holographic interference pattern requires an enormous amount of computation, making rapid *(~1 s)* generation of holograms of even limited size impossible with conventional computers.

A holographic fringe pattern is computed by numerically simulating the physical phenomena of light diffraction and interference. Computation of holographic interference patterns often utilizes the fast Fourier transform (FFT) algorithm. Though relatively fast for images composed of discrete depth surfaces, this approach becomes slow when applied to images that extend throughout an image volume. A more general ray-tracing approach computes the contribution at the hologram plane from each object point source. This method can produce arbitrary 3D images, including image-plane holograms (images that lie in the vicinity of the hologram), a case that is more suitable for various display geometries. However, this method is slow, since it requires one calculation per image point per hologram sample. As presented in this part of the chapter, the application of several methods of reducing computation complexity leads to computation times as short as 1 s on a data-parallel-processing supercomputer. First, a bipolar intensity representation of the holographic interference pattern is applied and shown to eliminate unwanted image artifacts and simplify calculations without loss of image quality or generality. Second, a look-up table approach is described and shown to provide a further speed increase and to reduce computation to a minimum. Finally, exemplary computation times are presented.

HOLOGRAPHIC IMAGING

This part of the chapter focuses on the computation of off-axis transmission holograms possessing horizontal parallax only (HPO). It is possible to represent an HPO hologram with a vertically stacked array of one-dimensional (1D) holographic lines. The goal of this part of the chapter is to compute these 1D holo-lines as quickly as possible. Consider an HPO hologram made optically using a reference beam with a horizontal angle of incidence. Spatial frequencies are large in the horizontal direction (~1000 lp/mm) and increase with the reference beam angle. Since an HPO CGH contains only a single vertical perspective (the viewing zone is vertically limited to a single location), spatial frequencies are low (~10 lp/mm) in the vertical dimension. The number of holo-lines is, therefore, matched to the vertical image resolution. In the horizontal dimension, the sampling rate (or pitch) must be high to accurately represent the holographic information. During reconstruction of this hologram, diffraction occurs predominantly in the horizontal direction. A holo-line diffracts light to a single horizontal plane (scan plane) to form image points describing a horizontal slice of the image. Therefore, one holo-line should contain contributions only from points that lie on a single horizontal slice of the object. Essentially, the 2D holographic pattern representing an HPO 3D image can be thought of as a stack of 1D holograms or holo-lines as shown in Figure 22-28.[29]

The information content of a CGH must be reduced to a size and format that can be manipulated by existing computers. Eliminating vertical parallax reduces CGH information content by at least a factor of 100 by reducing the vertical spatial frequency content from roughly 1,000 to roughly 10 lp/mm. Sacrificing the size of the viewing zone (reducing the required horizontal pitch) and the size of the hologram also reduces the amount of data in the CGH. In the first-generation MIT system, the data content (space-bandwidth product) of the CGH was as low as 2Mbytes.

In optical holography, the object light and the reference light are incident at the plane of the hologram, represented with spatially varying complex time-harmonic electric field vectors, Eo and Er. It is assumed that both are mutually coherent sources of monochromatic light. For this analysis, the units of an electric field amplitude are normalized so that the square of magnitude corresponds to optical intensity. The polarizations are assumed to be identical, and for simplicity are not specified. The total time-harmonic electric field incident on the hologram is the interference of the

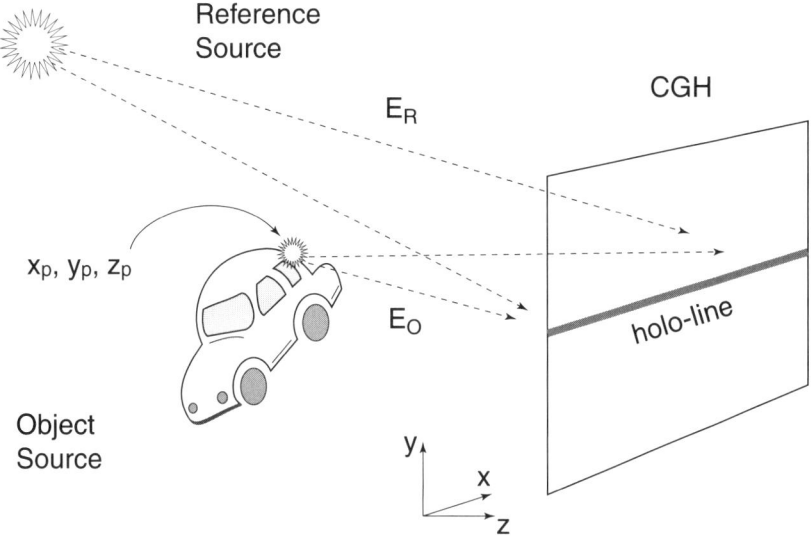

FIGURE 22-28: General geometry for HPO CGH.

light from the entire object and the reference light, $Eo + Er$ as shown in Equation 22-1.[30] The total intensity pattern

$$I_{\text{total}} = \left|Eo\right|^2 + \left|Er\right|^{2+2} \text{Re}\left\{Eo\ Er^*\right\} \tag{22-1}$$

is a real physical light distribution comprised of three components (see Equation 22-1). The first term is the object self-interference, a spatially varying pattern that is generated when interference occurs between light scattered from two different object points. During image reconstruction, this component of the holographic pattern is unnecessary and often produces unwanted image artifacts. In optical holography, a common solution is to spatially separate the object self-interference artifacts from the reconstructed image by increasing the reference beam angle to at least three times the angle subtended by the object. However, in computational holography, a large reference beam angle is a luxury that one does not have. Instead, the solution is simply to exclude the object self-interference during computation. The second term in I_{total} is the reference beam intensity and represents a spatially nearly invariant (dc) bias that increases the value of the intensity throughout the hologram. In computational holography, the reference bias can be left out, since the CGH will eventually be normalized.

 NOTE: *Normalization* is the numerical process that limits the range of the total fringe pattern by introducing an offset and a scaling factor to tailor the fringe pattern to fit the requirements of a display system.

Finally, the third term is the interference between the object wave front and the reference beam. This fringe pattern (called *Ib*) contains the holographic information that is necessary and sufficient to image reconstruction.

COMPUTATION USING POINT SOURCE SUMMATION

In this part of the chapter, the images to be generated are approximated as a collection of self-luminous points of light. The horizontal, vertical, and depth locations are specified as x_p, y_p, and z_p (see Figure 22-28). Each point has an associated real-valued magnitude (a_p) and phase (phi_p). The square of the magnitude (a_p^2) is proportional to the desired brightness of an image point, and phi_p is the phase relative to that of the reference beam. The CGH is positioned at $z = 0$. Only the x dependence of *Eo* and *Er* and other physical quantities need to be considered in computing a given holo-line.

 NOTE: Hereafter, functional dependence on x, y, or z is explicit *(Eo(x)* and *Er(x))*.

A full CGH is computed simply by generating an array of holo-lines for each value of y. The fan-shaped beams of light from each of the object points contained within a scan plane contribute to a limited width of the particular holo-line that is intersected by the scan plane. Within the region of contribution, the phase of the object wave front and *Phi_p(x)*, is approximated as an angularly truncated spherical wave front centered at the point source location:

$$Phi_p(x) = kr_p(x) + phi_p$$

where

$$r_p(x) = \left[(x - x_p)^2 + z_{p^2}\right]^{\{1/2\}} \qquad (22\text{-}2)$$

is the oblique distance to a location on the holo-line and is a function of x (see Equation 22-2).[31] The wave number is k = 2pi/lambda, where lambda is the free-space wavelength of the light. The time-harmonic representation of the total object field for a single holo-line is

$$Eo(x) = \sum_{p=1}^{Ny} e(x) \frac{a_p}{r_p}(x) \exp\{i\ Phi_p(x)\} \qquad (22\text{-}3)$$

where Ny is the number of object points contributing to this particular y-valued holo-line. Finally, to avoid singularities, it is assumed that the magnitude of z_p is never less than some small amount (10 lambda).

In Equation 22-3, the expression $e(x)$ is an envelope function used to specify the region of contribution for a given object source.[32] In its simplest form, $e(x)$ is equal to unity within the region of contribution of point p, and equal to zero for all other values of x. The envelope function $e(x)$ provides a means with which to limit the spatial frequencies of the holographic fringe pattern to avoid aliasing in the CGH at the extreme angles of object light incidence. Object *light* used to compute the CGH must have a minimum angle of incidence that is greater than that of the reference beam to produce non-overlapping real and virtual reconstructed images. At the other extreme, the total angle subtended by the incident reference and object beams cannot be so large as to give rise to spatial frequencies that cannot be adequately represented given the sampling pitch. For a horizontal sampling pitch of $1/d$, the maximum spatial frequency (f_{max}) that can be represented is $1/2d$, according to the Nyquist Sampling Theorem. Higher spatial frequencies cause aliasing and destroy image quality. In addition to providing anti-aliasing, $e(x)$ is used for the purposes of image occlusion and advanced image lighting models, resulting in a more realistic-looking image.

The reference beam Er is a point source at some specific location (xr, zr) with a horizontal angle of incidence $theta_r = \arctan(xr/zr)$ and curvature in the x dimension only; it is collimated in the y dimension. The time-harmonic representation of the reference beam field at any holo-line is

$$Er(x) = \frac{ar}{rr} \exp\{i\ Phi_R(x)\}$$

where

$$Phi_R(x) = k\left[(x - xr)^2 + zr^2\right]^{\{1/2\}} \qquad (22\text{-}4)$$

and *ar/rr* is the magnitude (approximated as constant versus *x*) of the reference wave at the center ($x = 0$) of the hologram (see Equation 22-4).[33]

BIPOLAR INTENSITY

The third term of Equation 22-5, called *Ib(x)*, contains all of the information needed to reconstruct the image in a given scan plane.[34]

 NOTE *Ib* is real valued. It represents the combined intensity variations (fringes) resulting from each object point interfering with the reference beam. Since it contains negative values as well as nonnegative values, it is a bipolar intensity. It exists physically only when superimposed on the first two bias terms in Equation 22-5.

Computationally, however, *Ib* can range both positive and negative since it is represented numerically, and is later offset during normalization. The bipolar interference pattern *Ib(x)* is numerically simpler to compute and has the advantage of containing neither object self-interference nor bias components. To apply *Ib(x)*, the expressions for the object and reference beam wave fronts (Equations 22-3 and 22-4) are substituted:

$$Ib(x) = 2\,\mathrm{Re}\left\{\sum_{p=1}^{Ny} e(x)\,a_p/r_p\,(x)\,\exp\{iPhi_p(x)\}\right\}\left[ar/rr\,\exp\{iPhi_R(x)\}\right]^*$$

$$= 2\,ar/rr\left\{\sum_{p=1}^{Ny}\mathrm{Re}\Big\{e(x)\,a_p/r_p\,(x)\,\exp\{iPhi_p(x) - iPhi_R(x)\}\Big\}\right\}$$

and finally supplied to:

$$Ib(x) = 2\,ar/rr\sum_{p=1}^{Ny} e(x)\,a_p/r_p\,(x)\,\cos\big[Phi_p(x) - Phi_R(x)\big] \tag{22-5}$$

The right-hand side of Equation 22-5 is simply a scaled sum of the real-valued *cosinusoidal* fringe pattern resulting from the interference of point source *p* with the reference beam. The bipolar intensity approach is to compute only *Ib* (instead of I_{total}), using only these real-valued *cosinusoidal* fringes.

The computational advantage of the bipolar intensity approach is readily seen by comparison to computation of the full interference pattern, I_{total}. The latter method requires the use of complex-valued arithmetic (manipulating both the real and imaginary parts of the object and reference light). In comparison, the bipolar intensity approach uses only real-valued cosinusoidal fringes. To generate the CGH, these real values are simply summed. A speed-up factor of 2 is expected.

After an intensity pattern has been computed, it must be normalized to satisfy the output device requirements of the CGH display system. Because normalization generally scales the entire pattern, the leading factor of 2 ar/rr on the right side of Equation 22-5 is hereafter excluded. The reference beam intensity (the square of ar/rr) is no longer meaningful. This makes physical sense when considering that in optical holography, the purpose of choosing an optimal reference beam intensity ratio (relative to the object light) is to provide a sufficient offset and scaling of the interference fringes to keep them within the range of sensitivity of the recording medium. Computationally, offset and scaling are provided automatically during normalization. Therefore, the factor of 2 ar/rr is set arbitrarily to unity. Substituting the definition of $Phi_p(x)$ (Equation 22-2), Equation 22-5 becomes

$$Ib(x) = \sum_{p=1}^{Ny} e(x) a_p / r_p(x) \cos\left[kr_p(x) + phi_p - Phi_R(x)\right] \quad (22\text{-}6)$$

which is hereafter called the bipolar fringe method of CGH computation (see Equation 22-6).[35] No reference beam ratio needs to be specified during computation, and bias buildup is not an issue. Compare this bipolar intensity method to the physical process occurring in some photorefractive crystals (lithium niobate), in which uniform (*dc*) intensity is not recorded due to the material's negligible response to intensity patterns with low spatial frequencies. Researchers exploit this absence of bias build-up in order to sequentially expose multiple holographic intensity patterns.

PRECOMPUTED ELEMENTAL FRINGES: THE LOOK-UP
TABLE APPROACH

Continuing with the bipolar intensity summation approach, further improvements in computation speed are gained through the use of a precomputed look-up table containing all possible elemental fringes. A table can contain the precomputed contributions to $Ib(x)$ from an image point of unity magnitude for each possible value of (x_p, z_p) in the image volume. However, the image volume is not generally discretized. Therefore, the indices X_i

and Z_i are introduced. They are defined as a set of discretized horizontal and depth locations spanning the entire range of the image volume.

Note: The image volume is already discretized vertically to match the vertical resolution of the HPO CGH.

Before the table is generated, the resolutions of X_i and Z_i must be chosen to discretize the image volume. Since the acuity of the human visual system is limited, to choose resolutions that do not visibly degrade the image is possible. The discretization steps must be sufficiently small such that images points appear to be continuous when viewed from the specified viewing distance. For example, the human visual system generally sees as continuous two points that are separated by 3 milliradians of arc. In the MIT display, since the image is viewed at a distance of approximately 600mm, and 600mm × 0.003 = 180 microns, a horizontal discretization step of 150 microns is chosen.

Equation 22-6 must be altered so that $Ib(x)$ is expressed in terms of discretized image point locations instead of x_p and z_p. The expressions $[x_p]$ and $[z_p]$ are used to represent x_p and z_p rounded off to the nearest values of X_i and Z_i. Therefore, the only change made to Equation 22-6 is to substitute $[x_p]$ and $[z_p]$ for x_p and z_p in the expression for $r_p(x)$ (in Equation 22-2).

Note: The variable x is already discretized due to the sampled nature of the CGH. The horizontal pitch is dictated by the intended display device.

An efficient look-up table should include all of the spatial dependence included in the right-hand side of Equation 22-6. Therefore, the elemental fringe look-up table $T(Xi,Zi,x)$ is defined as (see Equation 22-7):[36]

$$T(Xi, Zi, x) \underset{=}{\text{def}} e(x)/r_i(x) \; \cos\left[kr_i(x) - Phi_R(x)\right]$$

where

$$r_i(x) = \left[(x - X_i)^2 + \{Z_i\}^2\right]^{\{1/2\}}. \qquad (22\text{-}7)$$

The table looks like an array of cosinusoid-like fringes that have an approximately linear chirp in spatial frequency with respect to x. The rate of chirp is a function of the point source depth Z_i, and the horizontal position of the fringe is a function of X_i. To incorporate anti-aliasing, $e(x)$ is set to unity within the region of contribution, and zero elsewhere.

Note: This binary representation of e(x) is for simplicity. Any further variation in e(x) can be dealt with during computation, or with additional look-up tables. When utilizing the look-up table approach, phi_p is set to zero (or randomized) for simplicity.

Once precomputed, the table T is used during the computation of a CGH to represent a specific image. Rather than having to compute a cosinusoidal fringe each time one is needed (as indicated in Equation 22-6), the look-up table maps each $([x_p],[z_p])$ to the appropriate elemental fringe pattern. Thus, $Ib(x)$, expressed in terms of the precomputed table (see Equation 22-8)[37], is

$$Ib(x) = \sum_{p=1}^{Ny} a_p \, T\!\left([x_p],[z_p],x\right) \qquad (22\text{-}8)$$

Since each holo-line is computed using the same $ER(x)$, one precomputed table is used in the computation of every holo-line. Computation for a given holo-line at vertical position y is as follows (see Equation 22-9):[38]

— For every point on scan plan y:

- Round off x_p and z_p to $[x_p]$ and

  $[z_p]$ to index the desired elemental fringe.

- For each x sample in $Ib(x)$:

  Scale $T([x_p],[z_p],x)$ by a_p.

  Accumulate $a_p \, T\!\left([x_p],[z_p],x\right)$ in $Ib(x)$. (22-9)

After each holo-line is computed (at each value of y), then normalization and output are performed depending on the specific display system.

The issue of object point phase phi_p must be examined to justify setting the phase to zero for all points. Phase is invisible to the viewer. Object phase is important only in the interactions among image points. If the discretization steps of X_i and Z_i (typically 150 microns and 1000 microns) are much larger than the spot size of the display system, no overlap occurs between image points. In this case, as was the case in the MIT holographic display, restricting phi_p does not limit the size, depth, or quality of an image. Therefore, the relative phases phi_p are unimportant and can be arbitrarily set to zero.

Note: If nonzero phase values are required, two look-up tables in quadrature are sufficient for this more general case.

In some applications, phi_p can be randomized to make more efficient use of the dynamic range. A random phase factor for each point can easily be incorporated into the look-up table.

Computational Complexity

Computational complexity is dramatically reduced through the use of the precomputed look-up table. Consider the full complex computation of I_{total} that requires a minimum of five additions, five multiplications, one division, one square root, and two cosine (or sine) function calls. In addition, after the real and imaginary field components are summed, they must be squared and added together, and then the square root must be taken at all samples of I_{total}. In comparison, by using the look-up table approach, only one multiplication and one addition are required at each hologram sample for a given object point. Therefore, an order of magnitude of speed-up is expected through use of the precomputed table.

Note: The bipolar intensity simplification reduces complexity by allowing the table to contain only real values.

Look-Up Table Quantization

The greatest drawback to the precomputed table method is the enormous size of the table. For example, given the rather minimal dimensions of the first-generation MIT real-time display system (40mm wide holo-lines with a pitch of *1 micron$^{\{-1\}}$*), X_i typically has 250 discrete values, and Z_i has 50 discrete values. The table requires over 1.6 gigabytes of memory! When using the Connection Machine Model 2 supercomputer, the solution was to quantize the look-up table into only four levels. In this way, only 2 bits (rather than the standard 32-bit floating-point representation) are necessary for each sample point, and the table occupies only 8×10^8 bits of memory, or the equivalent of a manageable 100Mbytes.

NOTE: The $r_p^{\{-1\}}$ term cannot be included in the 2-bit look-up table since it requires a more continuous representation. However, for limited viewing zone widths, this term can be approximated by $z_p^{\{-1\}}$ and used to scale a_p, resulting in a negligible decrease in speed. The added quantization noise is acceptable, and by tailoring the exact waveform, the theoretical signal-to-noise ratio is 243:1.

Look-up table quantization increases image noise due to the loss of accuracy. In practice, however, table quantization has little effect on the final image quality, since $Ib(x)$ must be quantized when stored in the output framebuffer of the computing system. For example, in the first-generation MIT system, the framebuffer used was capable of storing 6M samples, each represented by 8 bits, giving the output data 256 possible quantized levels. Typically, the object light from at least 64 objects points overlaps at a single sample of $Ib(x)$. If each object point has an average amplitude of 1.0, then a typical elemental fringe pattern from the look-up table will occupy only about 256/64 = 4 different levels in the framebuffer. Therefore, a four-level look-up table is well suited to this display system.

Reduction of Table Rank from Three to Two

The precomputed table is a data array of rank three, indexed by X_i and Z_i in the image scan plane and by x (on the hologram). One way to reduce the size of the table is to precompute values for each possible $Dx = (x-X_i)$, rather than for each x and each possible X_i. In this way, the table is reduced to rank two.

During the table look-up computation, x_p and z_p are rounded off to $[x_p]$ and $[z_p]$ (the nearest values of X_i and Z_i)—making $Dx = x-[x_p]$ for a given image point p. In terms of these variables, the expression for r_p (see Equation 22-2) becomes a function of $[z_p]$ and Dx. Looking at Equation 22-6, the envelope function $e(x)$ is actually a function of Dx/z_p, since e limits the angular direction of light propagation (equal to arctan $[Dx/z_p]$). Thus, Equation 6 (see Equation 22-10)[39] becomes

$$Ib(x) = \sum_{p=1}^{Ny} e(Dx) a_p / r_p([z_p], Dx) \ \cos[k \ r_p([z_p], Dx) - Phi_R(x)]$$

where

$$r_p([z_p], Dx) = (\{Dx\}^2 + \{z_p\}^2)^{\{1/2\}}. \qquad (22\text{-}10)$$

Now only $Phi_{R(x)}$ is an explicit function of x (rather than of Dx), and must therefore be manipulated to become a function of Dx. First, by restricting the reference beam to be a plane-wave (rr goes to infinity), the reference phase is simply $Phi_{R(x)} = kr^*x$, where $kr = k \ sin(theta_r)$. The second restriction is that X_i be discretized by $2 \ pi \ / \ kr$, making every possible value of $kr \ Dx$ differ by exactly $2 \ pi \ m$ for a given value of x, where m is some unimportant integer value. Consider that when computing the cosine of total phase, any integer multiple of *2 pi* is ignored. Therefore, $Phi_{R(x)}$ can be expressed as $Phi_{R(x)} = kr\ x + 2\ pi\ m = kr\ Dx$ for all values of X_i. Finally, Equation 22-6 can be expressed as a function of only two variables $[z_p]$ and Dx (see Equation 22-11):[40]

$$Ib(x) = \sum_{p=1}^{Ny} e(Dx)\, a_p/r_p\, ([z_p], Dx)\ \cos\!\left[kr_p([z_p], Dx) - krDx\right]. \quad (22\text{-}11)$$

A corresponding look-up table of rank two can be used, with the indices $[z_p]$ and Dx.

Essentially, given the restriction of a plane-wave reference beam, the elemental fringe patterns are shift-invariant in *x* for any $[x_p]$ equal to an integer multiple of *2 pi / kr*. The first advantage is a reduction in size equal to at least the number of discrete values of $[x_p]$ used in the rank-three table (~100 or more). The second advantage is that X_i is discretized in very small steps *(2 pi/kr ~10 microns)* without increasing the size of the rank-two table. The horizontal image discretization is much finer than can be seen by the human visual system. The only drawback to the rank-two table approach is the requirement that the reference beam be a plane wave. However, this is a common case, and is applicable to many display architectures. This contraction of rank is especially useful when implemented on a standard serial-processing computer.

Application to the Computation of Stereograms

For some applications, a stereogram approach may be desirable when presenting 3D images. In general, a stereogram consists of a series of 2D object views differing in horizontal point-of-view. The views can be rapidly generated using a computer graphics rendering process, or they can be captured from a real scene by using an array of cameras. The views are presented to the viewer in the correct horizontal location, resulting in the depth cues of binocular disparity (or stereopsis) and (horizontal) motion parallax. The 2D perspective views are generally imaged at a particular depth position and are multiplexed by horizontal angle of view. A given holo-line in this case must contain a holographic pattern that diffracts light to each of the horizontal locations on the image plane. The intensity from a particular horizontal viewing angle should be the image intensity for the correct perspective view. This is accomplished simply by making the amplitude of the fringe contribution a stepwise *x*-function of the intensity of each image point from each of the views. Calculation involves reading each of the views into the computer, using the view and its corresponding segments of the look-up table to compute the fringe pattern contribution for this view, and accumulating these contributions into the final CGH.

To facilitate rapid computation of stereogram-type CGHs, the precomputed table is indexed by image x-position and view-angle (rather than by X_i and Z_i). Furthermore, the table can be reduced to rank two as described earlier, making the table much smaller. Stereogram CGHs have been computed and displayed on the MIT real-time holographic display system. A major advantage to the stereogram approach is the integration of sophisticated lighting and shading models in the computer graphics rendering process to produce extremely realistic images possessing important 3D image characteristics, such as occlusion (overlap) and specular reflections.

RESULTS

Hologram computation has been implemented for use in the MIT real-time display system using the methods of bipolar intensity summation and precomputed elemental fringe patterns. The MIT display required a CGH that contained 192 holo-lines, each of 32K samples, for a total of 6M samples. Each sample represented a horizontal physical spacing of 1 micron, or a pitch of 1000 lp/mm. A Connection Machine Model 2 (CM2) employed a data-parallel approach to perform real-time CGH computation. This means that each x location on the hologram was assigned to one of 32K virtual processors.

Note: The 16K physical processors were internally programmed to imitate 32K virtual processors. A Sun 4 workstation was used as a front end for the CM2, and the parallel data programming language C Paris was used to implement holographic computation.

Table 22-2 contains the computation times (time per image point) using different approaches: I_{total} (complex).[41] The common traditional case was I_{total} (complex), where Equations 22-1, 22-2, 22-3, and 22-4 are combined to compute I_{total}. The *Bipolar intensity* approach was simply the computation of Ib instead of I_{total}. The look-up table method employed the precomputed elemental fringe table, represented in the memory of the CM2 as 2 bits per sample. As each object point was read from an input file, the position (x_p, z_p) was used to index the table in each processor. At each x, conditional logic was performed on each of the 2 bits, and the appropriate

TABLE 22-2: The computation times (time per image point) using different approaches

Computation Method	CM2	Sun4
I_{total} (complex)	2.180 ms	943.4 ms
Bipolar intensity	1.135 ms	486.2 ms
Look-up table	0.084 ms	22.1 ms

fractions of the object point amplitude were accumulated into the register representing Ib. Since this was performed in parallel for all 32K samples of $Ib(x)$, rapid computation of images was possible. The time per point listed in Table 22-2 is the amount of computation time required (on average) to accumulate the fringe pattern contribution of a single object point source. These numbers were obtained by computing holograms of several different test images of varying complexity. For comparison, the different computational approaches were also implemented on a serial computer, a Sun 4 workstation, and the results are included in Table 22-2.

In all cases, the computation time per point was simply the total execution time divided by the number of points processed. Computation time scaled linearly with image complexity; despite the different image complexities (from 100 to 50,000 points), the time per point quotient remained within a 2-percent range. Table 22-2 shows that, as expected, the bipolar approach was roughly twice as fast as the traditional complex method of computation. This was evident on both the parallel-data and serial machines. By using the look-up table, the CM2 speed increased by a factor of 26; the serial machine was faster by a larger factor of about 43.

Note: The CM2 actually has an array of floating-point math accelerator chips that are no longer utilized in look-up table calculations. The simpler serial machine gained more from the use of a look-up table.

For practical purposes, additional procedures that are independent of object complexity must be performed, including normalizing the computed holographic pattern and moving it into the framebuffer. Therefore, a generally fixed overhead time must be added when expressing the total time to compute and output the holographic pattern. For a 6-megabyte holographic pattern, this time is approximately 400 milliseconds (or less) on the CM2. For example, the actual time to compute and display a 10,000-point image using the look-up table approach is 10000×0.084 ms + 400 ms = 1.24 s, or less.

In the current MIT display system, simple images generated from 3D computer graphics databases contain only a few thousand points. Using the fastest look-up table computation method, images are computed at a rate of over 1 frame per second.

Note: In many cases, computation time was actually less than the overhead time spent loading the computed CGH into the framebuffer!

To demonstrate interactivity, the viewer can turn any of several dials (interfaced to the computer) in order to translate the image in horizontal, vertical, and depth locations; to change its size; and to spin it along different axes of rotation. In addition, a simple drawing program was written in which the user can move a 3D cursor to draw a 3D image that can also be manipulated.

Note: The holographic patterns computed by these three approaches were equivalent, with the following exceptions: Use of the bipolar intensity eliminated object self-interference and *dc* terms, making the CGH brighter and less noisy than when using the complex Itotal method. The look-up table approach resulted in a pattern that was identical to the bipolar intensity approach, with the addition of some quantization noise if a 2-bit table was used. However, for objects of sufficient complexity, this quantization noise was comparable to that of the more straightforward approaches, due to the quantized nature of the output framebuffer device used in the system.

In Summary

- The technology of electronic interactive three-dimensional holographic displays is in its second decade. Though fancied in popular science fiction, only recently have researchers created the first real holovideo systems by confronting the two basic requirements of electronic holography: computational speed and high-bandwidth modulation of visible light.
- Currently there are no off-the-shelf holographic displays. Holographic display technology is in a research stage, analogous to the state of 2D display technology in the 1920s. What, then, does the future hold? The future promises exactly what holovideo needs: more computing power, higher-bandwidth optical modulation, and improvements in holographic information processing.

- Computing power continues to increase. A doubling of computing power at a constant cost (a trend that continues at a rate of every 16 months) effectively doubles the interactive image volume of a holographic display. Inexpensive computation (around $200 per gigaMAC) is the most crucial enabling technology for practical holovideo, and should be available in 2004.
- Although optical modulation has borrowed from existing technologies (transmissive LCDs, AOMs), new technologies will fuel the development of larger, more practical holovideo displays. Because bandwidth is most important, a good number of bits can be modulated in the latency time of the HVS (typically 20 ms). An AOM can modulate about 16MB in this time interval, at a cost of about $3,000 (or $188 per MB), including the associated electronics. The deformable micro-mirror device (DMD), a new technology for high-end 2D video projection technology, delivers approximately 100MB in 20 ms, for a cost of about $4000, or $40 per MB. Future mass-production could reduce the cost further. Reflective LCDs are another possible technology. Several researchers create small reflective LCDs directly on a semiconductor chip using VLSI technology.
- The bandwidths of computation and modulation are likely to increase steadily. Improvements in holographic information processing will likely provide occasional dramatic improvements in both of these areas. Already, holographic bandwidth compression increases fringe computation speed by 3,000 times for same-hardware implementation. Standard MPEG algorithms can be used to encode and decode computed fringes. Nonuniformly sampled fringes provide lossless bandwidth compression and promise further advances.
- User demand may be the one additional key to the development of holovideo2E. As other types of 3D display technologies (autostereoscopic displays) acquaint users with the advantages of spatial imaging, these users will grow hungry for holovideo, a display technology that can produce truly 3D images that look as good as (or better than) actual 3D objects and scenes.
- Hogel-vector bandwidth compression, a diffraction-specific fringe computation technique, has been implemented and used to generate complex 3D holographic images for interactive real-time display. The application of well-known sampling concepts to the localized

fringe spectrum has streamlined, generalized, and greatly accelerated computation. Hogel-vector bandwidth compression makes efficient use of computing resources. The slower decoding step is essentially independent of image content and complexity, and simple enough to be implemented in specialized hardware.

- Hogel-vector bandwidth compression is the first reported technique for computational holographic bandwidth compression. It achieves a bandwidth compression ratio of 16:1 without conspicuous degradation of the image, thereby eliminating transmission bottlenecks. Fringe generation was 74 times faster than conventional computation. This technique provides a superior means for generating holographic fringes—even if bandwidth compression is not needed. Hogel-vector encoding provides a foundation for *fringelet encoding*, which (as reported elsewhere) achieves further speed increases. It can also be applied to full-parallax holographic imaging.

- Hogel-vector bandwidth compression can be applied to other tasks. For example, holographic movies can be hogel-vector-encoded for digital recording and transmission over networks or television cable. As another example, diffraction-specific fringes have been recorded onto film to produce static holographic images. Hogel-vector bandwidth compression provides the speed and portability required to generate large fringes for a holographic *"fringe printer."*

- The analysis of hogel-vector bandwidth compression has revealed a simple expression relating the fundamental system parameters of bandwidth, image resolution, and maximum image depth. Hogel-vector encoding attains *visual-bandwidth holography*—it frees holographic fringes from the enormous bandwidths required by the physics of optical diffraction. Bandwidth is instead matched to the abilities of the human visual system.

- Experimental results demonstrated that a horizontal-parallax-only off-axis transmission hologram can be computed in under 1 s. Computation was fast enough to provide dynamic holographic images at interactive rates. The interactive display of holographic images added the additional sensations of depth and tangibility that accompany moving 3D images.

- The overall speedup demonstrated here is remarkable. CGH computation that traditionally would require several minutes or hours on a mainframe computer is reduced to 1 s. The look-up table approach, by eliminating the need for all mathematical functions other than simple addition and multiplication, is especially advantageous when only minimal computing power is available for CGH computation. Essentially, the table stores a large array of holographic basis functions, the weighted sum of which is sufficient to describe the CGH corresponding to any arbitrary image. Given adequate memory space to hold the precomputed elemental fringes, it is possible to design a dedicated CGH computer that requires no floating-point mathematics and uses only integer addition (and perhaps bit shifting for normalization purposes). Such a simple machine can be implemented in parallel (one computer per hololine) to achieve real-time CGH computation without the need for an expensive supercomputer. It is interesting to note that the 2-bit table approach implemented on the CM2 computes a CGH using only two conditional additions per image point per sample, emphasizing the computational simplicity made possible by the use of a look-up table.

- The look-up table is the simplest possible algorithm for computing CGH from the discrete-point object model. To achieve faster speeds, future work involves an approach in which a higher level of object description is processed during CGH computation. In addition, to increase speeds on workstation computers, it may be useful to utilize certain specialized hardware such as computer graphics rendering engines or digital signal processing boards used for image coding and decoding.

- Analytical simplification of the physical model of light interference made possible this increase in speed. These concepts can be applied to other types of holograms. The bipolar intensity method is applicable to all types, including full-parallax CGHs. The use of the bipolar intensity summation method, whether directly or through a look-up table, eliminates object self-interference noise, eliminates the need to adjust a reference beam ratio, and produces an optimally bright image by eliminating unnecessary dc intensity bias. The look-up table approach can be applied to full-parallax holograms, although by requiring both x and y indices, the precomputed table (data arrays of rank five, reduced to rank three using the method shown) would require enormous amounts of memory space. In the future, as computational power increases, the simplification of computation will continue to provide speed-up for any CGH application.

Chapter 23 examines off-axis electron holograms through a virtual electron microscope; artifacts in electron holography; holographic neural networks; reconstruction of electron holograms using simplex algorithms; high-resolution electron microscopy; and interference experiments with energy filtered electrons.

End Notes

1. The Media Laboratory, Building E15, Massachusetts Institute of Technology, 77 Massachusetts Avenue, Cambridge, MA 02139-4307, USA, 2000.
2. Ibid.
3. Ibid.
4. Ibid.
5. Ibid.
6. V. Michael Bove, Jr. and John A. Watlington. "Cheops: A Reconfigurable Data-Flow System for Video Processing," Media Laboratory, Massachusetts Institute of Technology, Room E15-351, 20 Ames Street, Cambridge MA 02139, USA, 2000.
7. Stephen Benton, Mark Lucente, Carlton Sparrell. "High-Speed Holographic Image Generation Using Cheops Daughter Cards," Media Laboratory, Massachusetts Institute of Technology, Room E15-351, 20 Ames Street, Cambridge MA 02139, USA, 2000.
8. IBM Research Division, Thomas J. Watson Research Center, P.O. Box 218, Yorktown Heights, New York 10598, USA, 2000.
9. Ibid.
10. Ibid.
11. Ibid.
12. Ibid.
13. Ibid.
14. M. Lucente, "Computational Holographic Bandwidth Compression," IBM Systems Journal, IBM Technical Journals, P.O. Box 218, Yorktown Heights, NY 10598-0218, USA, Vol. 35, No. 3 & 4 (1996).
15. Ibid.
16. Ibid.
17. Ibid.
18. Ibid.
19. Ibid.

20. Ibid.
21. Ibid.
22. Ibid.
23. Ibid.
24. Ibid.
25. Ibid.
26. Ibid.
27. Ibid.
28. Ibid.
29. Mark Lucente, MIT Media Laboratory, Spatial Imaging Group, 20 Ames St., E15-416, Cambridge, MA 02139, USA.
30. Ibid.
31. Ibid.
32. Ibid.
33. Ibid.
34. Ibid.
35. Ibid.
36. Ibid.
37. Ibid.
38. Ibid.
39. Ibid.
40. Ibid.
41. Ibid.

CHAPTER 23

ELECTRON HOLOGRAPHY

In This Chapter

- Electron holography at atomic dimension
- Artefacts in electron holography
- Reconstruction of electron holograms in real space with a neural net
- Optimized reconstruction of electron holograms using simplex algorithms
- Determination of electron-microscope parameters using neural net
- Interference experiments with energy filtered electrons

High-resolution electron holography microscopy has a wide field of applications in material science and physics. However, dedicated microscopes and special recording techniques must be used for obtaining the highest resolution. In principle, there are two ways to improve the interpretable resolution of electron microscopes toward 0.1 nm: The acceleration voltage of the instrument is increased in order to get a wider broadcasting band in Scherzer focus, or a corrector for the spherical aberration of the objective lens is installed, or a medium-voltage electron microscope equipped with a field emission gun is used in combination with a wave optical recording technique. Once the complex image wave has been stored, a subsequent reconstruction process extracts the information of the image wave, and a correction procedure makes use even of information up to the information limit of the microscope.

Electron Holography at Atomic Dimension: Instrumentation

Since the invention of the electrostatic biprism by Möllenstedt and Düker in 1954, the resolution of an off-axis electron hologram has been successfully pushed toward the information limit. Nowadays, researchers are using a Philips[1] CM30 FEG electron microscope with a high magnification projective lens system and a Möllenstedt biprism in front of the first intermediate image plane. A *2k x 2k* slow-scan CCD camera (Tietz Company, Garching, Germany) has been recently installed to provide enough pixels for correction of isoplanatic aberrations of the objective.

RECORDING AND RECONSTRUCTION

In principle, recording and reconstruction of off-axis electron holograms is a straightforward procedure. An electron object wave (exit surface wave) is propagated through the electron microscope, resulting in an image wave aberrated by isoplanatic wave aberrations of the objective lens. By means of a hologram, the aberrated image wave is fed to a computer and numerically backpropagated through an accordingly modeled *virtual electron microscope*. The reality, however, deviates from this ideal case as discussed later in the chapter:

- The hologram is sampled at discrete sampling points with a finite field of view.
- The biprism filament located between the back focal plane of the objective lens and the first intermediate image plane causes Fresnel diffraction at its edge and also causes vignetting of a small part of spatial frequencies.
- Inhomogeneities of the biprism filament and distortions of the projective lens system give rise to artificial phase shifts of the image wave.
- The parameters of the isoplanatic wave aberrations of the objective lens are not known exactly.

Therefore, a special strategy for reconstruction of the image wave has been developed in order to overcome most of these artefacts and problems, respectively. In general, two holograms are recorded: a regular hologram with the object inside the beam, and a hologram without the object (reference hologram) that should be recorded immediately after the first one.

The reconstruction of the image wave is done in two steps: First, the phase distribution of the reference hologram marked with an exponential filter in reciprocal space is reconstructed without centering of the sideband. This reference phase contains the exact position of the sideband, as well as artificial phase shifts due to inhomogeneities of the biprism filament and distortions of the projective lens system. Second, this reconstructed reference phase is multiplied to the hologram. After Fourier transformation, one finds the sideband centered in reciprocal space within a fraction of a pixel despite discrete sampling. Additionally, streaking is minimized. Finally, the sideband is isolated by means of an exponential filter, and the aberrated image wave is obtained after reverse Fourier transformation.

However, the distortion by Fresnel diffraction, which can be still present in the reconstructed image wave, can, in general, not be corrected because, due to experimental difficulties, the Fresnel diffraction distributions of both holograms do not exactly coincide. Additionally, only the Fresnel diffraction of the zero-beam could be corrected by means of a single reference hologram.

CORRECTION OF ABERRATIONS

Since the complex image wave has been restored, the researchers can perform arbitrary numerical wave optical calculations for a comprehensive evaluation of the object structure that are not possible using a standard CTEM image. In order to make use of information beyond Scherzer resolution, the isoplanatic aberrations of the objective lens such as spherical aberration, defocus, two- and three-fold astigmatism, and axial coma must be corrected. This is done by means of a numerically generated phaseplate that is multiplied to the complex Fourier spectrum of the image wave. The aberration-free object wave (exit surface wave of the specimen) is obtained after reverse Fourier transformation if the true aberration parameters are known sufficiently well. However, one of the most complicated problems is the finding of the exact aberration parameters. Several strategies for different objects have been used:

- Minimizing of amplitude contrast of the object wave of a weak phase object (e.g., an amorphous edge of a crystal) by means of automatic search routines using Genetic and Simplex algorithms.
- Numerical defocusing of sharp reflections of a crystal.
- Correction in comparison with simulated object wave images by

means of a dedicated computer program that calculates a new object wave within one second.

Now, let's look at recent applications of electron holography that deal with special aspects of high Tc superconductors. First, we'll look at off-axis electron holography artefacts.

Off-Axis Electron Holography Artefacts

Off-axis electron holography has been shown to improve the point resolution of the CM30FEG-Special Tübingen electron microscope from 0.2nm to 0.13nm. However, in off-axis electron holography, artefacts may arise specific to the method. In particular, at high resolution they must be avoided or taken into account to obtain reliable results after correction of aberrations. In the following, let's look at some artefacts that so far have been found by researchers to be essential.

VIGNETTING EFFECTS

Vignetting shows up in a hologram in that the biprism inserted between back focal plane and image plane acts as a spatial frequency filter with transmission varying with the position in the image plane: image stripes close to the biprism shadow only contain single-sideband information as shown in Figure 23-1.[2] This is not tolerable in the case of an object made up by both amplitude and phase components, because double sideband information is needed to reconstruct both amplitude and phase of the object wave.

NOTE: Less than two-thirds of the hologram at the far side of the biprism filament is free from vignetting.

FRESNEL DIFFRACTION AT THE BIPRISM FILAMENT

The biprism is located at some distance above the image plane. Therefore, in the image plane, one finds both amplitude and phase of the image wave and the reference wave modulated by Fresnel diffraction at the biprism filament as shown in Figure 23-2.[3] At first glance, the modulation seems to be negligibly small; however, it must be compared to the modulation by the

ELECTRON HOLOGRAPHY 465

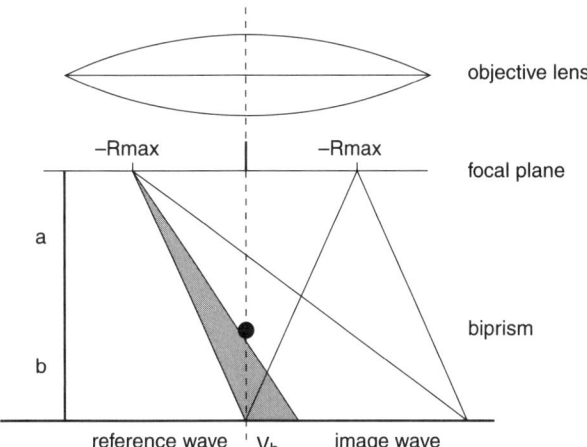

FIGURE 23-1: Vignetting by biprism filament. Applying a positive voltage to the biprism, the spatial frequency −Rmax does not contribute to the image wave in a stripe of width vb parallel to the filament. Single sideband imaging results.

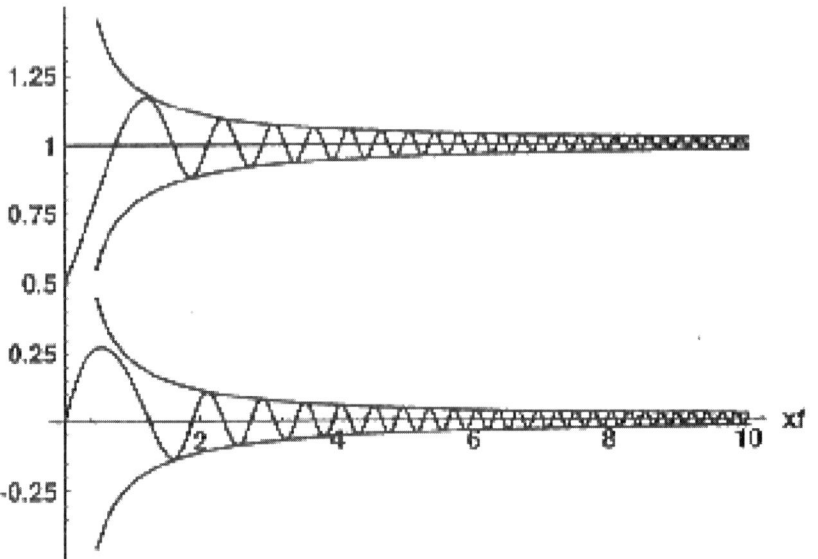

FIGURE 23-2: Fresnel modulation at the left side of the reconstructed field of view. Amplitude (top) and phase as a function of the coordinate perpendicular to the biprism filament.

object, which (in the case of single atoms) may be much smaller than Fresnel modulation. In general, Fresnel modulation cannot be corrected by means of a reference hologram, since Fresnel modulation in the object wave depends on the object structure.

WINDOWING

The hologram represents only a small window of the image wave transmitted to the final image plane of the electron microscope. Therefore, the Fourier spectrum available in the computer under reconstruction is convoluted with the Fourier transform of the window. Consequently, the numerical phase plate for correction of aberrations acts on spread-out reflections, whereas the wave aberration in the electron microscope acted on peak-like reflections. Therefore, due to the gradient of the wave aberration over the spread-out reflections, the corresponding elementary waves superimposing in the image plane are mutually shifted aside, such that the correct object wave results only in a center area of the field of view. The results show that the useful field of view is smaller by the point spread function of the electron microscope than the width of the hologram.

GEOMETRIC DISTORTIONS

In an electron microscope, geometric distortions occur as a consequence of the aberrations of the magnifying lenses, or due to local parasitic charges along the electron path. They produce deformations of the geometric arrangement of image details like atom positions, which may be well annoying but usually do not affect resolution. In electron holography, however, they do affect resolution. The reason is that distortion of the hologram fringes results in an artificial phase distribution of the reconstructed wave, which in turn produces a displacement of the reflections in Fourier space depending on real space position. Therefore, isoplanacy is reduced, and correction of aberration (assuming isoplanacy over the whole field of view) is only effective in the resulting isoplanatic patch. To reach 0.1nm resolution with the microscope, distortions must be less than 0.1 percent. This can only be reached by careful distortion correction with the help of an empty reference hologram prior to correction of coherent aberrations.

Now, let's look at the next electron holography application: the reconstruction of electron holograms in real space with a neural net.

Reconstruction Scheme for Off-Axis Electron Holograms

Off-axis electron holography uses a Möllenstedt biprism to interfere the image wave with the reference wave. The fringes of the interference pattern at the position contain the information about amplitude and phase of the complex electron image wave, with contrast and space frequency of the fringes (see Figure 23-3).[4]

The classic reconstruction scheme for off-axis electron holograms uses the linear reconstruction technique by means of Fourier transformation of the hologram, cutting out the sideband representing the image wave, and an inverse Fourier transform. This is in analogy to an optical bench where the sideband is cut out in the focal plane. Today, this process, implemented as computer software, is still in use because it yields optimum results as long as there is no noise, no difference in the detection efficiency of the CCD pixels, and the number of CCD pixels is unlimited. Real electron holograms, however, suffer from all these restrictions. One fast way to eliminate these effects is the use of an artificial neural network, because neural nets can learn nonlinear functions of many input variables and can thus use both nonlinear and linear.

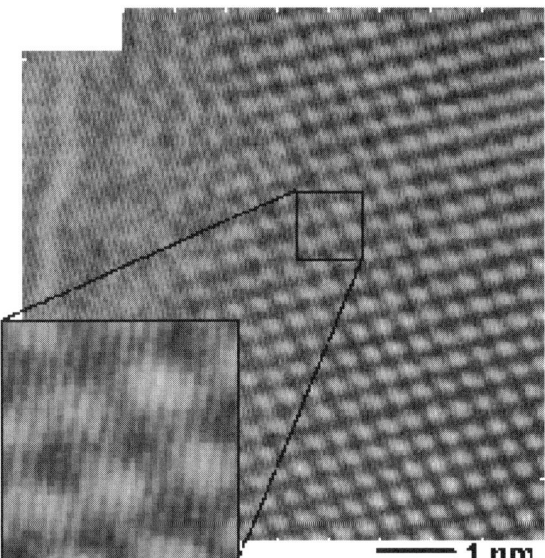

FIGURE 23-3: Electron hologram of Si in (110) orientation. The image was taken on the CM30 in Tübingen, using a 1k CCD camera. In the inset, the holographic fringes are clearly visible.

Artificial neural networks attempt to simulate the flexibility of biological networks on a computer. Researchers have used a feed-forward network, which converts the information from training patterns to matrix elements in the weight matrices of the network. A complete neural net is a combination of two or more neuron layers. The learning process starts with completely arbitrary matrices, and continues until the user is satisfied with the result. One critical point in the application of neural networks is the appropriate generation of training patterns. These patterns should contain a variety of good, representative experimental examples, although they cannot cover all the imaginable cases.

In the researchers' experiment, the complex image is reconstructed in real space. Since the correlation between the values of two different pixels in the image wave vanishes quickly with their mutual distance, it is sufficient to use a small area of hologram pixels to calculate the image wave in the center of this area. Researchers usually use a 7 × 7 area and thus have a vector of intensities as input information for the neural net. The neural net of the feed-forward type (see Figure 23-4) was simulated with the Stuttgart Neural Net Simulator (SNNS) on a Power PC.[5] During training, the researchers used input intensities resulting from well-known simulated holograms, including all the disturbances mentioned in the preceding. After training the net, the weight-table was translated to fast C code. Using this program, the researchers were able to reconstruct simulated and real holo-

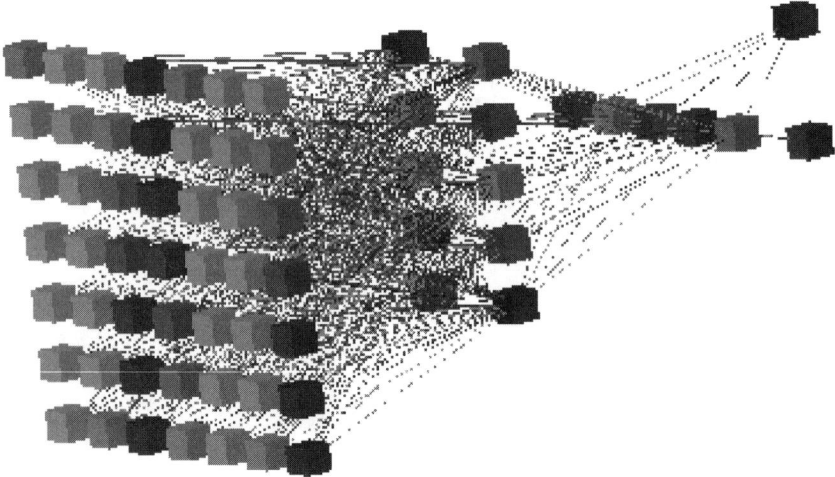

FIGURE 23-4: Neural net for the reconstruction process. The input layer shows the applied fringe pattern.

grams with speeds comparable to the conventional reconstruction process but with a better performance in noise reduction and artefact elimination.

A first result is shown in Figures 23-5 and 23-6.[6] The electron hologram of a specimen (see Figure 23-3) was used as input to the neural net shown in Figure 23-4. The net is built up by 49 input neurons, 10 neurons in the first hidden layer, 5 neurons in the second hidden layer, and 2 output neurons, which represent real and imaginary parts of the image wave. The layers are completely connected (the matrices have no zero elements).

Now, let's look at how to use Simplex Algorithmus for optimized reconstruction of electron holograms.

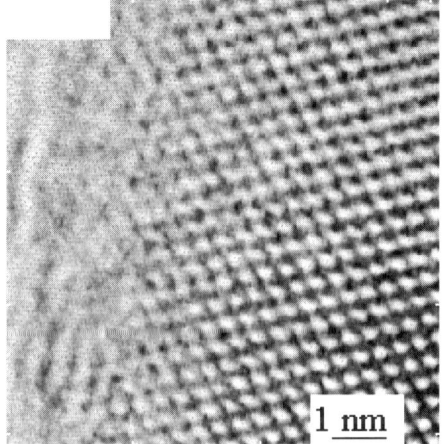

FIGURE 23-5: Amplitude of the electron wave, reconstructed by the neural net shown in Figure 23-4 using the hologram (see Figure 23-3) as input.

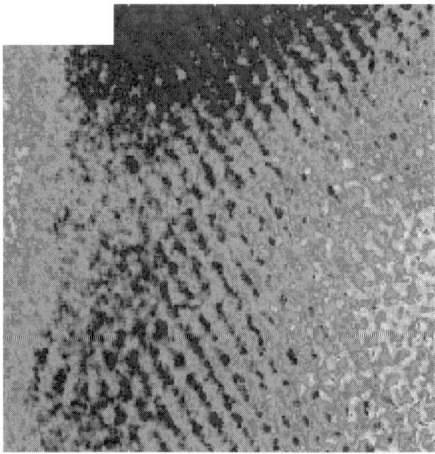

FIGURE 23-6: Phase of the same area.

Using Simplex Algorithmus for Optimized Reconstruction of Electron Holograms

Off-axis electron holography uses a Möllenstedt biprism in the electron beam to interfere the object wave with the reference wave. Quantum mechanics describes the electron wave at the position and with the de Broglie wave vector. The image wave is modulated in amplitude and phase by the complex object transparency. In the image plane, the intensity of the electron beam results from the interference of the image wave and reference wave, depending of course on the coherence of the source and mechanical and electrical imperfections. The researchers do not consider the image transfer function, and thus it does not affect further discussion.

In the case of finite numbers of detected electrons, noise has to be added by yielding the measured intensity. The researchers' goal is to identify the amplitude and the phase of the image wave. Therefore, a system of intensity equations can be solved. In the experimental situation, researchers measure the intensities at the CCD pixel positions as shown in Figure 23-7.[7] If the image wave is constant over the input area of n pixels, the researchers can collect the different intensities to an intensity vector (see Figure 23-8).[8] This vector describes a position in the dimensional input data space, which should point to a two-dimensional surface, with the

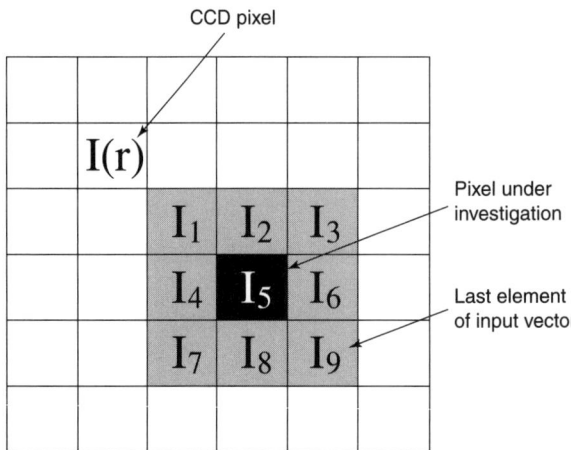

FIGURE 23-7: Detail of the CCD array. The pixel under investigation measures the intensity. The simplex uses the intensity information to find amplitude and phase of pixel 5 in the hologram.

ELECTRON HOLOGRAPHY 471

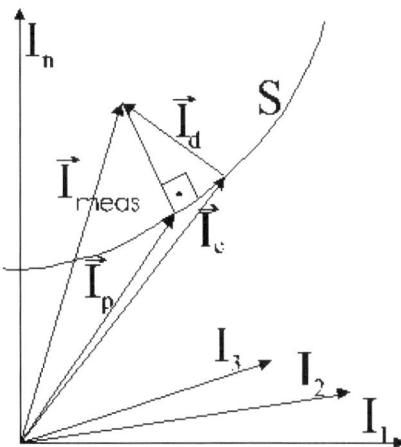

FIGURE 23-8: The possible intensities without noise lie on the surface S. The point is the surface point closest to the measured value, which is displaced by the noise vector.

parameters amplitude and phase. Due to the noise, however, the intensity vector misses the surface. In contrast to classical reconstruction, the reconstructed image wave is a nonlinear function of the intensities. This yields more information because the whole hologram (not only the sideband) is used.

There are different approaches to find the point on the surface. One approach uses neural nets; the researchers use in this case the simplex algorithm. The downhill simplex method starts with a simplex in the parameter space and then approximates numerically the point on the surface that satisfies the condition.

The search ends when the decrease of the distance to the surface between consecutive iterations becomes smaller than a given. For reconstruction, the input area is moved over the entire hologram, and the values are calculated and written to an array.

In simulated holograms subject to poisson and thermal noise, this process of reconstruction shows the best results. The disadvantage is the immense amount of numerical processing. For the determination of one pair, the simplex algorithm takes about 100 iterations; and for each iteration, the intensity equation has to be calculated.

This gets even worse if the assumption that the image wave is constant over the input area no longer holds. In this case, additional parameters have to be introduced (the linear and quadratic taylor expansion coefficients of the image wave). With increasing number of parameters, the efficiency and reliability of the simplex algorithm rapidly decreases. However, one can take advantage by easy parallel implementation of this reconstruction process from further development in computer architecture.

Next, let's look at the determination of electron-microscope parameters using a neural net.

Detecting the Parameters of the Transfer Function

Off-axis electron holography has proven to be a powerful tool in high-resolution electron microscopy, because it supplies the information for after-the-fact aberration correction by computer. The process of image generation in electron optics can be described by means of the complex wave transfer function in Fourier space. The object wave in Fourier space is described with space frequency; and the Fourier transformation is distorted in the image plane. In electron holography, the complex image wave is available; therefore, the object wave can be reconstructed if the wave transfer function is known.

There are different ways to find the transfer function. In crystalline specimens, the method of analyzing crystal reflections looks promising in finding the coefficients of the transfer function. In the case of an amorphous and thin specimen, the analysis of the diffractogram is well established.

There are some well-known techniques to detect the parameters of the transfer function by analyzing the intensity. The pitfall is that none of this process uses all the information available in the diffractogram; therefore, the optimum precision is not reached.

The researchers investigated the application of neural networks to determine the wave transfer function. To concentrate the information, the intensity of the diffractogram is projected into the space, with the basis vectors (see Figure 23-9) and the scalar product.[9]

The resulting scalar products contain most of the information in the diffractogram and are used as input in a fed-forward neural net. During training of the network, simulated diffractograms resulting from well-known transfer functions are projected and fed to the net. The a-priori known result is fed back to the used backpropagation neural network to train the interconnections of the neurons. After some thousand training cycles, the net responds to the input signal in the correct way.

The researchers trained the net with a defocus in the range, and a spherical aberration in the range. After training, the researchers tested the output of the net with new simulated diffractograms, not part of the training

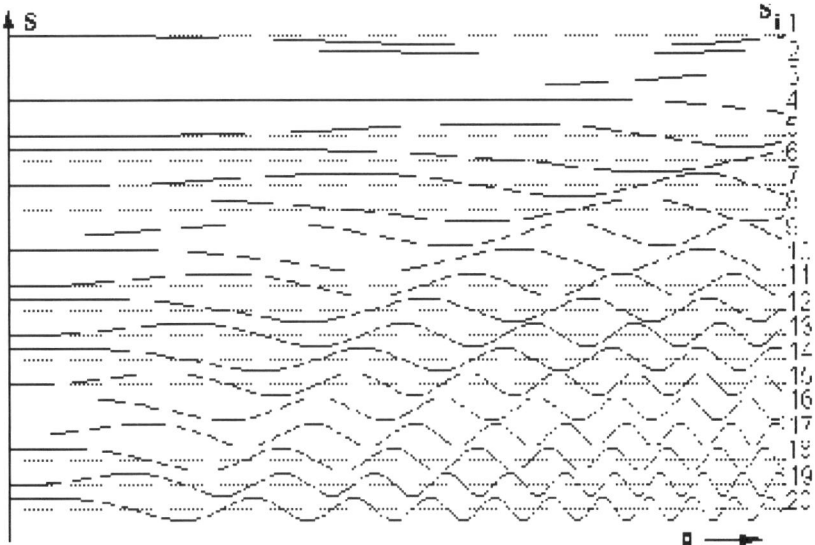

FIGURE 23-9: Basis system for analysis of diffractograms. The set of 20 basis vectors was used as a mathematical coordinate system to reduce the input space of the neural network.

set. The difference between output of the net and simulation parameter is plotted over the absolute value of the input. The resulting error is shown in Figure 23-10.[10] At the border of the training area, the error is larger due to the typical fact in neural net behavior that they are good for interpolation but worse in extrapolation.

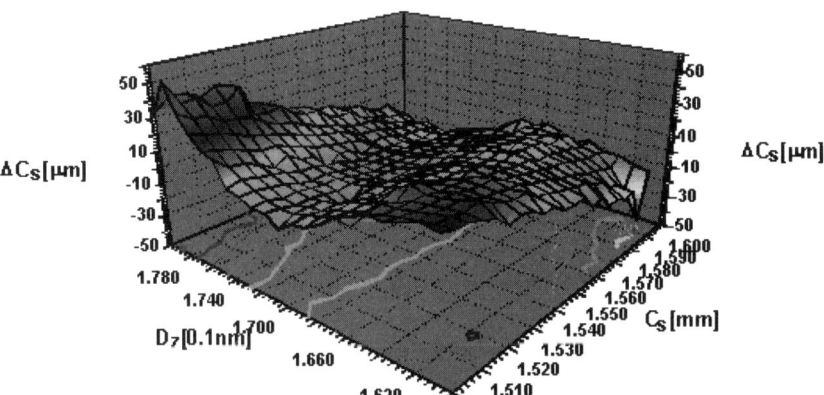

FIGURE 23-10: The error of the neural net determining spherical aberration. In the center of training, the error surface is nearly flat and with an error below 0.01mm.

Finally, let's look at interference experiments with energy filtered electrons.

Interference Experiments

Any conventional off-axis electron hologram always includes the contributions of unscattered, elastically scattered, and inelastically scattered electrons. The reconstructed wave only contains information about the elastically scattered electrons. Inelastic scattering leads to a decrease of fringe visibility, which reduces the signal-to-noise ratio in the reconstructed wave. Energy filtering, however, provides a method to separate inelastically scattered from elastically and unscattered electrons, so their respective contribution to the hologram can be examined in more detail.

Therefore, interference experiments were performed on Zeiss EM 912 OMEGA by the aid of a Möllenstedt Biprism inserted in the selected area aperture plane. All the Bragg scattered electrons were removed with the objective aperture diaphragm:

- In the unfiltered beam without any specimen present, a sufficient number of interference fringes with good visibility was present in the image. This demonstrates that the microscope meets the stability and coherence requirements.
- Zero-loss energy filtering leads to an improvement of interference fringe visibility, if either one or both waves are transmitted through the specimen (see Figure 23-11).[11]
- If only plasmon loss electrons pass the energy filter, Fresnel fringes as well as faint cosinusoidal interference fringes can still be observed (see Figure 23-12).[12]

These experiments show that inelastically scattered electrons are coherent to a certain degree. Coherence can only be preserved at all if the interfering waves have excited the same inelastic scattering event. The extension of an area in the specimen plane, where the interfering partial waves have excited the same inelastic event, is related to the width of the interference pattern in the image plane. Thus, the extension of Fresnel and hologram fringes will be a measure for the lateral extension of plasmons in the material.

ELECTRON HOLOGRAPHY 475

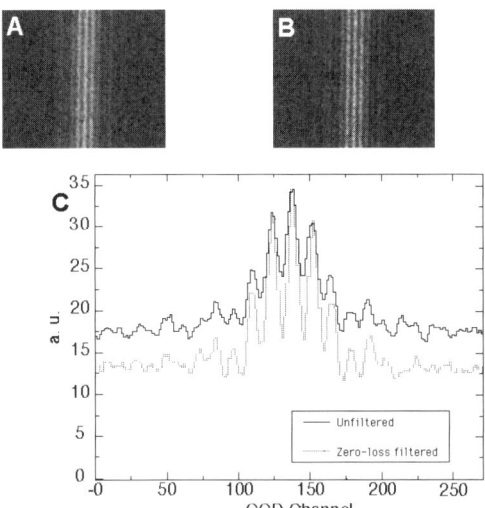

FIGURE 23-11: Interference fringes of unfiltered (A) and zero-loss energy filtered (B) electrons after passing thin aluminum foil. C shows profiles of A and B. For noise reduction, the profiles were averaged over 100 lines. Zero-loss filtering leads to an improvement of fringe visibility.

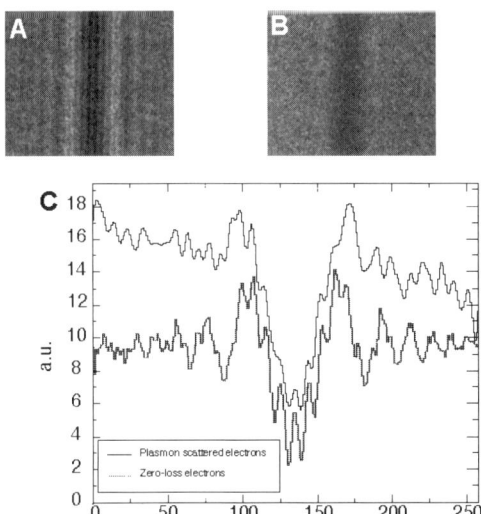

FIGURE 23-12: Interference pattern of zero-loss energy filtered (A) and plasmon scattered electrons (B). Profiles of both data (averaged over 100 lines) are shown in C. Clear Fresnel and faint hologram fringes are present in the case of plasmon scattered electrons.

In Summary

- Recording and reconstruction of off-axis electron holograms is a straightforward procedure.
- In off-axis electron holography, artefacts may arise specific to the method.
- The classic reconstruction scheme for off-axis electron holograms uses the linear reconstruction technique by cutting out the sideband representing the image wave, an inverse Fourier transform, and means of Fourier transformation of the hologram.
- To interfere the object wave with the reference wave, off-axis electron holography uses a Möllenstedt biprism in the electron beam.
- Because off-axis electron holography supplies the information for after-the-fact aberration correction by computer, it has proven to be a powerful tool in high-resolution electron microscopy.
- So that the electron respective contribution to the hologram can be examined in more detail, energy filtering provides a method to separate inelastically scattered from elastically and unscattered electrons.

Chapter 23 examined off-axis electron holograms through a virtual electron microscope; artefacts in electron holography; holographic neural networks; reconstruction of electron holograms using simplex algorithms; high-resolution electron microscopy; and interference experiments with energy filtered electrons. The next chapter begins Part Six, "Custom Holography," and shows you how to make your own hologram right in the confines of your home.

End Notes

1. FEI Company, 7451 N.W. Evergreen Parkway, Hillsboro, Oregon, 97124-5830, USA, 2000.
2. H. Lichte, D. Geiger, A. Harscher, E. Heindl, M. Lehmann, D. Malamidis, A. Orchowski, and W.-D. Rau. "Artefacts in Electron Holography," Institute for Applied Physics, University Tübingen, D72076 Tübingen, Germany; Institute for Applied Physics, Dresden University of Technology, D01062 Dresden; Institute for Semiconductor Physics, D 15230 Frankfurt/Oder, Germany, 11th

European Congress on Microscopy, EUREM 96, Dublin, Ireland, 26.–30. Aug. 1996.
3. Ibid.
4. R. Meyer, E. Heindl, "Reconstruction of Electron Holograms in Real Space with a Neural Net," Institute of Applied Physics, Auf der Morgenstelle 10, D-72076 Tübingen, Germany, 2000.
5. Ibid.
6. Ibid.
7. E. Heindl, R. Meyer, "Optimized Reconstruction of Electron Holograms using the Simplex Algorithmus," Institute of Applied Physics, University Tübingen, Auf der Morgenstelle 10, D-72076 Tübingen, Germany, 2000.
8. Ibid.
9. E.Heindl, "Determination of Electron-Microscope Parameters Using a Neural Net," Institute of Applied Physics, Auf der Morgenstelle 10, D-72076 Tübingen, Germany, 2000.
10. Ibid.
11. A. Harscher, H. Lichte, and J. Mayer. "Interference experiments with energy filtered electrons," Institut für Angewandte Physik, Auf der Morgenstelle 12, 72076 Tübingen, Germany; Institut für Angewandte Physik und Didaktik, Zellescher Weg 16, 01062 Dresden, Germany; MPI für Metallforschung, Inst. für Werkstoffwiss., Seestr. 92, 70174 Stuttgart, Germany, 11th European Congress on Microscopy, EUREM 96, Dublin, Ireland, 26.–30. Aug. 1996.
12. Ibid.

PART VI

CUSTOM HOLOGRAPHY, SECURITY, RESULTS, AND FUTURE DIRECTIONS

This part of the book discusses *practical holography*, or how to make your own holograms. It also examines silver halide hologram emulsions, a medium particularly suited for custom pieces. Transmission and reflection holography are also discussed. Transmission holograms are rear lit and can be produced as either laser viewable or rainbow-colored white-light-viewable images. Reflection holograms are lit from the same side as they are viewed, and are viewable in white light as a single-color image.

We also examine the world of holography security—counterfeiting, anti-piracy technology, and countermeasures against holographic counterfeiting.

Finally, we look at holographic results and future directions—holographic neural networks; aeronautical engineering; the dynamic structure of holographic space; autostereoscopic holographic displays and computer graphics; a digital holography system; and interactive virtual reality: holodecks.

CHAPTER **24**

HOW TO MAKE YOUR OWN HOLOGRAMS

In This Chapter

- The laser
- Movement
- Different types of holograms
- Coherence length
- Construction of a laboratory setup
- Making holograms
- Experiments
- Bleaching procedure

This chapter is for those who want to get right down to creating their own holograms. Nevertheless, unless you are savvy in the ways of hologram creation, it is highly recommended that you read the first 23 chapters.

Before we get into the actual task of how to make your holograms, we must first cover some basic concepts for those of you who are not so savvy. With that in mind, some of the questions about holography that come to mind immediately might serve as a good starting point for our discussion. They are: *What is a hologram, and how does holography work?*

NOTE: The preceding two questions were asked and answered in the previous chapters. Nevertheless, the process is referred to as *holography* while the plate or film itself is referred to as a *hologram*. The terms *holograms* and *holography* were coined by Dennis Gábor (the father of holography) in 1947. The word *hologram* is derived from the Greek words *holos* meaning whole or complete, and *gram* meaning message. Older English dictionaries define a hologram as a document (such as a last will and testament) handwritten by the person whose signature is attached.

The theory of holography was developed by Dennis Gábor, a Hungarian physicist, in the year 1947. His theory was originally intended to increase the resolving power of electron microscopes. Gábor proved his theory not with an electron beam, but with a light beam. The result was the first hologram ever made. The early holograms were legible, but plagued with many imperfections because Gábor did not have the correct light source to make crisp, clear holograms as we can today, nor did he use the off-axis reference beam (which will be described later). What was the light source he needed? The LASER, which was first made to operate in 1960.

Laser light differs drastically from all other light sources, man-made or natural, in one basic way that leads to several startling characteristics. Laser light can be coherent light. Ideally, this means that the light being emitted by the laser is of the same wavelength, and is in phase. These might be new terms for some of you, so let us form an analogy that might clarify the term *coherence*.

Let's say that you are flying over a freeway at rush hour, and directly below you is a long tunnel through which all the cars must pass. Nothing is strange about the fact that all different styles and makes of motor vehicles emerge from the tunnel at differing velocities. A Cadillac at 86 mph, a Volkswagen at 56 mph, a motorcycle at 71 mph, a truck at 51. The distances between vehicles also vary. Thus, you have different types of vehicles at varying speeds, and at constantly changing distances between each other.

Then something very strange takes place; you see that more and more 1974 Cadillac Coupe de Villes are emerging. No, wait, look! All the cars coming out of the tunnel are 1974 Cadillac Coupe de Villes, gold with tinted windows, exactly alike (a situation not totally uncommon in some carefully chosen Southern California suburbs). Not only are they the same year, make, and color, but they are all traveling at exactly the same speed and all bumper to bumper, never changing. So, if you just happened to have a stopwatch, you would find that the cars are exiting at a rate of one car per second. If you were to leave, or more likely, pass out from the fumes, you would observe upon reawakening that the exit rate of the cars is still exactly one car per second. The cars are in phase.

The way in which coherent light is emitted from a laser is analogous. Keep in mind that although absolute 100-percent coherence is rarely, if ever, attained, there are certain types of lasers readily available that have sufficient coherence to make excellent off-axis holograms.

The light emitted from a laser is all exactly the same type, or make, depending upon the characteristics of the substance that is lasing. Lasing will be explained later in the chapter: what the term *laser* means, and how the laser works to give coherent light. Right now, it is important to remember that the frequency of laser light is unvarying, and that in the same medium, all light (light of different wavelengths of frequency) travels at the same speed).

It's true that all electromagnetic radiation, including the very small portion called visible light, travels in a vacuum at the approximate finite speed of 186,000 miles per second. Light waves can oscillate at different frequencies and with correspondingly different wavelengths, so that for any given amount of time—say, one second—a greater number of shorter wavelengths of (blue) light would be emitted from a laser than longer wavelengths of (red) light. This does not mean that different wavelengths travel at different speeds. Back to the freeway analogy; given the same speed and same distance between cars, more Volkswagens (short wavelength) than Cadillacs (long wavelengths) would pass by a point in the same amount of time.

 NOTE: The velocity of light in a vacuum is one of nature's constants and is referred to by the letter c.

Now is a good time to define some terms used previously, but that you will see throughout this explanation. *Wavelength*, usually symbolized by the Greek letter λ for lambda, and frequency, symbolized by the Greek letter $v\lambda$ pronounced nu, have a reciprocal relationship $v\lambda = C$. The shorter the wavelength, the higher the frequency, and vice versa. The *amplitude* is the height or intensity of the wave. For example, a laser rated at 5mW (milliwatts, or thousandth of a watt) would give off light at the same frequency and wavelength as another laser of the same type rated at 1mw. However, the wave heights of the 5mw laser light would be five times higher than that of the 1mw laser.

The *wavelength* is the distance from one crest to the next; this is also one cycle. It seems logical that you would need some constant measure of time in order to count the cycles. This constant unit of time is usually one second; thus the term *cycles per second*, or cps, which is often referred to as Hertz or Hz (in honor of the German Physicist Heinrich Rudolph Hertz, who discovered radio waves). The oscillation frequency of electromagnetic radiation in the visible region is approximately 10 Hz. Wavelengths of visible light are between 400 and 700 nanometers or billionths of a meter in length.

Light has been described here as energy that travels through space in a waveform. For the purpose of this chapter in talking about holography, this is the case. However, the theory of light has unfolded miraculously, involving such great minds as Isaac Newton, Thomas Young, Christian Huygens, Max Planck, Niels Bohr, and, of course, Albert Einstein.

Still, the dual characteristic of light remains one of the many puzzles of nature. The particle wave problem that is referred to was clarified somewhat in the year 1900 when Max Planck proposed that all electromagnetic energy is radiated in discrete packages that he called quanta, or singular quantum. Einstein later confirmed Plank's theory via the photoelectric effect, and used the word *photon* to refer to these energy packages. Scientists today refer to light sometimes as particles (photons or quanta) and other times as continuous waves, depending on the situation or experiment. The problem is not with nature, but with our models or concepts of nature. It is always very important to remember not to let your idea or model of the way anything should be, usurp the place in your mind of the way it is or might be. That place should always be open for new information, whether it agrees with theory or not.

Light travels in a waveform—more precisely, a transverse waveform. The crests and troughs of the waves (which in the case of light are electromagnetic fields) are rising and falling in a direction at right angles to the direction of travel as shown in Figure 24-1.[1] A swell or wave in the ocean is a good example of transverse wave motion. You'll notice that the rising and falling action of the wave is at right angles to the direction of travel.

A simple proof of the wave theory was first demonstrated by an English physician named Thomas Young in 1802. He forced the light from a single light source to pass through a narrow slit, and then forced that same light to pass through two more narrow slits placed within a fraction of an inch of each other. The light from the two slits fell on a screen. Surprisingly, he saw not just the simple accumulation of the light from both slits on the screen, but a pattern of light and dark lines as shown in Figure 24-2.[2]

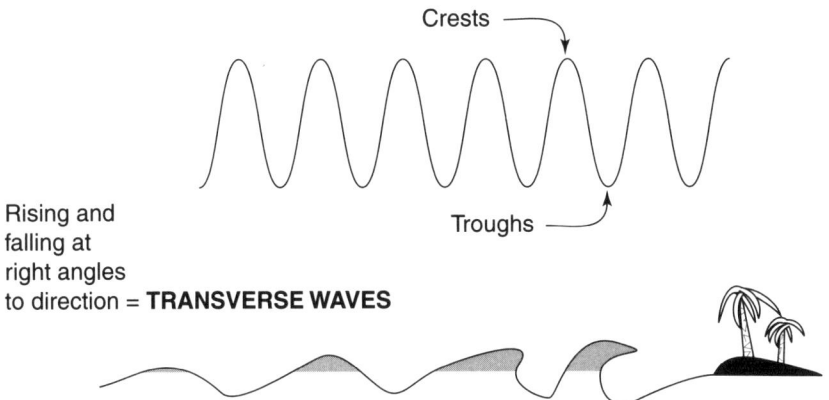

FIGURE 24-1: The crests and troughs of the waves are rising and falling in a direction at right angles to the direction of travel.

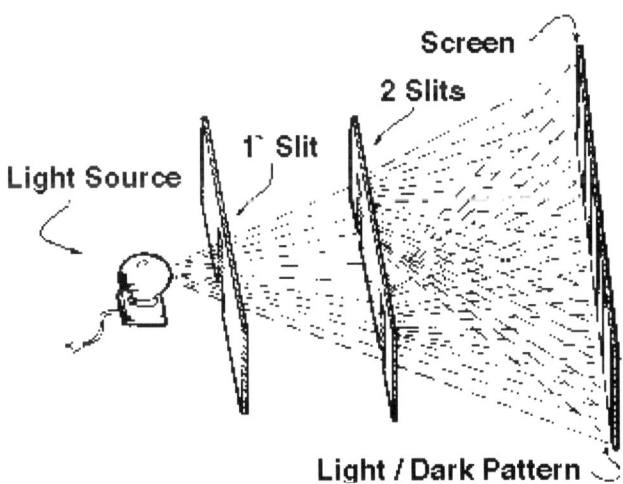

FIGURE 24-2: The pattern is the result of the mixing of the waves of light emanating from the respective slits.

He believed the pattern was the result of the mixing of the waves of light emanating from the respective slits. At the time, it was very difficult for the many justifiably avid fans of Issac Newton to incorporate this new discovery into Newton's particle theory of light. Newton tried to explain optical phenomena such as refraction and reflection in terms of gravitational-like effects. As it turned out later, in a way, Newton's theory was given partial confirmation by the Quantum Theory.

The lines or *fringes* that Young saw are called the *interference pattern* of the two light waves. When a crest interferes with a crest, it is positive, or constructive interference, resulting in a bright spot. On the other hand, when a crest meets a trough, you have a dark area or destructive interference.

As mentioned earlier, light waves oscillate at approximately 10 Hz, or a million billion times per second. There is no machine known to man sensitive enough to record the individual fluctuation's additive effect of the light waves of which at each second 10 to the power of 15 wavelengths are interacting on the screen. This number, like so many numbers you may encounter in such fields as physics, astronomy, and electronics, is incomprehensible. Yet precise measurement is part and parcel of the advancement of science. Suffice it to say that 1 billion seconds, for example, equals roughly 30 years, and 10 to the power of 15 seconds is 1 million times that.

If the number is so fantastic and if even today you can't measure the waves individually, how did you discover that light was electromagnetic radiation? It's all thanks to the amazingly successful mathematical theory of J. Clark Maxwell, developed in 1864. He predicted not only the electromagnetic nature of light, but also the speed at which it travels. Einstein used these same equations as a basis for his theory of special relativity. However, in order to understand exactly how light waves are formed, one had to wait until the year 1900 and the aforementioned Quantum Theory.

The Laser

Now that you know a little something about light in general, let's consider the light source needed to perform holography: the laser, which stands for *light amplification by stimulated emission of radiation*. The understanding of the stimulated emission of light, or how a laser works, will greatly aid in conceptualizing the holographic process.

Without the laser, the unique three-dimensional imaging characteristics and light phase recreation properties of holography would not exist as you know them today. Two years after the advent of the continuous wave laser, Leith and Upatnieks (at the University of Michigan in 1959 to 1960) reproduced Gábor's 1947 experiments with the laser, and launched modern holography.

A laser is a light amplifier with very special characteristics. The laser was designed and made to work after two very useful theories had come on the scene. One is Niels Bohr's atomic theory, and the other is the Quantum Theory. Niels Bohr, a Danish physicist, in the year 1913, proposed a model of the relationship between the electron and nucleus of the hydrogen atom. Bohr utilized the newly developed Quantum Theory in proposing that an electron circling the nucleus can assume certain discrete quantized levels of energy. In the lowest level, called the *ground state*, the electron is circling closest to the nucleus. However, if the atom is exposed to an outside source of energy, the electron can be raised to a higher energy level, or an excited state, which is characterized by the electron carving a circle of greater circumference around the nucleus.

Note: The electron can't go just anywhere when it is excited, but has to assume certain levels.

Also, not just any energy would suffice in raising the electron's orbit. The energy must be equal to the energy difference between the ground state and the excited state the electron assumes. The frequency is the energy difference divided by h, or Planck's constant. There are actually a number of different energy levels that the electron may assume, but that is not essential to this explanation of how a laser works.

Energy is radiated in discrete packages, and these packages interact only on a very selective basis. There are two important reasons why lasers work. The laser depends on the very special emission characteristics of certain atoms whose electrons have been raised to the excited state. When the electron falls back to its lower energy level (as all electrons eventually do), it in turn emits a package of electromagnetic or radiant energy that precisely equals the energy difference between the two levels, ground state and excited state. In a sense, what goes in comes out. This fact alone doesn't suffice in making a substance lase, for if too many electrons are in the ground state, the energy input would merely be absorbed by the electrons in the ground state, which then might spontaneously emit a quantum of the correct size sometime in the future, and that would be the end of that. You don't want to have an atom emitting its photon at just any old time, so you need to stimulate the atom to emit its energy package when you want it to. A package would not be absorbed by another atom in ground state. However, it would stimulate an atom already in an excited state to emit its own photon. In order to maintain the stimulated emission of photons that produce laser light, you must initiate and maintain a population inversion.

In lasers, electronic principles are applied to the visible portion of the spectrum. In electronics, oscillation is achieved with feedback around an amplifier. The feedback circuit determines the frequency of oscillation. In a laser, the tube of excited atoms is the amplifier. The mirror or resonator is the feedback circuit. Oscillation occurs at those wavelengths where the product of gain equals the loss, for a round-trip, say starting from one mirror and coming back again. The gain of a laser is determined by population inversion, or having many more excited electrons than electrons in the ground state (electrons at their lowest energy level).

The helium-neon laser, which is probably the most common laser in use today (due to its relatively low cost) is the laser you will probably use most. The laser tube itself contains approximately 10 percent helium and 90 percent neon. Of these two inert gases, neon is the active agent in the lasing process. You could term helium the *catalyst* insofar as it facilitates the energy input to the neon. Before more energy is purposefully forced in the system, there is some action among the atoms and molecules comprising the gases. Although some vary, very few of the electrons are already in the excited state, or upper energy levels; and, when they fall down, as they all tend to do, they emit a photon, only to be quickly absorbed.

The gain or loss of a photon or quantum of energy, which is defined by a change in electron orbit, takes place on the order of 10 to the power of 15 seconds, or 100 millionths of a second. You might ask how even some of the atoms might have electrons in the excited state if there's no energy input (before the laser is switched on). The answer is purely statistical. For example, if you have a church filled to capacity for a Sunday morning mass, say 360 people, someone has got to cough or sneeze during the sermon. If you take the number of times some two or three people cough and compare that with the amount of times everyone in the church inhaled and exhaled without occasion, it would give you some idea of the situation in the laser tube before excitation. A few atoms are excited and then fall back to emit energy. This energy in turn goes off spontaneously to another atom whose electron almost certainly is in the ground state. The photon is absorbed. This is the key to the laser. If you have enough atoms with electrons in the excited state, the photon not only would not be absorbed, but when it did reach another excited atom it would induce it to cough up its own photon. You go from one, to two, to four, to eight, to sixteen photons very rapidly. You have achieved population inversion (many more electrons are in the excited state than in the ground state).

Remember, you should be considering only the helium-neon laser at this time. It is the most economical laser and probably the one you would be using. There are other lasers such as the argon-ion laser that is able to lase in both blue and green; and better yet, a mixed gas argon-krypton ion laser that is able to lase in blue, green, and red. The problem is that the prices of these lasers begin at around $11,000. If you have access to these lasers, you probably would not be reading this chapter anyway.

There is also the pulse ruby laser that allows you to make holograms of animate objects. In the ruby laser, chromium ions locked in a sapphire host are the sources of stimulated emission. The chromium atoms are excited by a light flash from a special flashlamp.

Let's backtrack slightly and talk briefly about the job helium performs before going on to the more mechanical aspects of laser operation. It so happens that helium has a metastable (or long-lived) energy level that coincides quite well with one of the energy levels of neon that we need to obtain for lasing action to commence. Scientists discovered that it is much easier to raise helium to the excited state and let it transfer the correct energy packets to the neon when they meet inside the tube (which is at the correct pressure to assure their close acquaintance). So, the helium is used as sort of a messenger, or filter, if you will, to store the correct high energy input origination from the laser power supply for the neon. Although the neon is the active ingredient in the laser, the helium greatly facilitates the process.

Virtually all we have so far, then, is a glorified light tube such as you might find lighting the streets of any late night hot spot worth its salt. The difference from this light tube stage of development to the functioning laser is essentially more of a mechanical characteristic (the precise geometrical relationship of its optical components.)

The photons are emitted from the atoms inside the tube in all different directions. However, a very small percentage, around 2 percent, begin traveling in a horizontal direction within the tube as shown in Figure 24-3.[3] They naturally stimulate already excited atoms along the way to emit their photons in the same direction. This would actually mean nothing if you did not then place mirrors at both ends of the light tube in order to induce the light to start moving back and forth along the horizontal line of the tube. Eventually, this induces a large number of photons to travel in

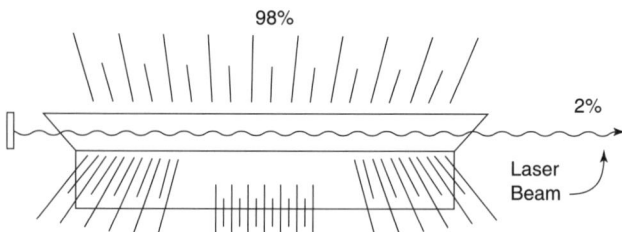

FIGURE 24-3: Photons traveling in a horizontal direction within the tube.

the same direction; and one of the mirrors is only partially reflective, which lets the light leak out.

Some of the characteristics of laser light were introduced earlier. Now, let's discuss the properties of the laser with this further explanation: The source of the light is the energy given off when an atom's electron falls back down toward the ground state. There is only one type of atom taking part in the actual coherent, laser light-giving process; therefore, according to the law of Quantum Mechanics, the energy given off by identical energy shifts in each atom must be exactly the same. In other words, each photon has precisely the same amount of energy. It will also have the same frequency and wavelength, and will be coherent light. It is the mirror setup, sometimes called the *resonant cavity*, that induces this fully saturated, monochromatic light to exit the tube in a straight, narrow beam; for then, not only do they contain the same amount of energy in stimulated emission, but the photons travel in the same direction.

Actually, the precise wavelength emitted by a laser is determined by the mirror separation: the lasing transition gives a band of wavelengths over which the laser can emit.

The diameter of the exit beam varies with the bore of the tube, but most helium neon beams are around 1.5mm diameter at exit and do not spread nearly as quickly as incoherent light would. Thus, laser light is coherent because it is radiated by a homogenous collection of atoms under precisely the same conditions. The mirrors at both ends make the small percentage of photons that hit the mirrors return in a straight line. This develops a cascade of light along the horizontal line of the tube. If you were to remove the laser casing, you would see the same monochromatic, saturated light; however, the straight beam, so distinctive of laser light, would only be emitted from the end with the partially coated mirror. Now, let's go on to see why you need these properties of coherent light, and how you should use them in holography.

The Basic Hologram

The hologram—that is, the medium that contains all the information—is nothing more than a high-contrast, very fine grain, black-and-white photographic film. There are other photosensitive materials such as photochromic thermoplastics and ferro-electric crystals, but its very unlikely you will be working with these materials in the beginning Therefore, you will be dealing with silver halide emulsion much like the black-and-white film you can buy in your neighborhood drug store. Yet there is one very special difference: the film designed especially for holography is capable of very high resolution. One way of judging resolution of film or lenses is to see how many distinguishable lines can be resolved within a certain width; in this case, it's a millimeter. Relatively slow film such as Kodak Pan X can resolve 90 lines per millimeter (depending on processing), while a good film designed for holography, such as AGFA Gaevert 8E75, is able to resolve up to 3,000 lines/mm. Holographic film is also especially prepared to be sensitive to a certain wave length of light. Each type of film is given a code number AGFA 8E75 is sensitive in the red region, and thus is used with ruby or HeNe lasers. Kodak 649F is also, however, about 10 times slower. Kodak 120 plate or SO173 film is very similar to AGFA 8E75, but not quite as sensitive.

Why the need for such special resolving power? The answer is that the hologram is not a recording of a focused image as in photography, but the recording of the interference of laser light waves that are bouncing off the object with another coherent laser beam (a reference beam, which will be described later). The wavelengths of light from a HeNe laser are approximately 24 micro-inches or 24 millionths of an inch long; thus the need for such fine grain or high resolving power.

You will be able to understand the very important difference between holography and photography, with a discussion of what happens when you take a photograph and what happens when you make a hologram. A photograph is basically the recording of the differing intensities of the light reflected by the object and imaged by a lens. The light is incoherent; therefore, there are many different wavelengths of light reflecting from the object, and even the light of the same wavelength is out of phase. Your emulsion will react to the light image focused by the lens, and the chemical change of the silver halide molecules will result from the photon bombardment. There is a point-to-point correspondence between the object and the emulsion. What is meant by this, is that a point or collection of points that reflects light on your object—for example, a white hat—would be a source of more light to be focused by your lens, than a black hat, which could absorb rather than reflect light. The white object would expose more silver halide after the development procedure, especially after you print your negative, which will naturally be lighter in the positive.

Any object to be recorded can be thought of as the sum of billions of points on the object, which are then reflecting more or less light. The lens of the camera focuses each object point to a corresponding point on the film, and there it exposes a proportional amount of silver halide. Thus, your record is of the intensity differences on the object, which form a pattern that one may ultimately recognize as the object photographed. In holography, one is working with light waves, and with most likely, a silver halide film; yet, beyond that, it is very difficult to compare the two. As you well know by now, the light sources are vastly different.

The sun or common light bulbs give off light of all different wavelengths. The laser emits a single wavelength coherent light. If you were to simply illuminate the object with laser light and take a photograph, you would still only be recording the different light intensities of the object. You would not have captured any information about the phase of the light waves after bouncing off the object.

How can you capture this vital information? You need a standard or reference. In the same way that a surveyor needs a reference point in order to make his or her measurements, you need a standard or a reference source in order to record the phase difference of the light waves, and thus capture the information that supplies the vital dimensions and depth to the holographic presentation. This standard is called a *reference beam*, and the laser light itself supplies it.

The reference light is emitted in what is called a plane wave as shown in Figure24-4.[4] By enlisting the aid of a beamsplitter, you should be able to form two beams. The reference beam is allowed to hit the film directly. It might be spread with a lens and aimed at the film by a mirror; but for all practical purposes, this does not affect the light waves.

The other beam is referred to as the *object* or *scene beam* as shown in Figure 24-5.[5] It is also usually spread by a lens and guided by a mirror, but it is directed at the object being holographed.

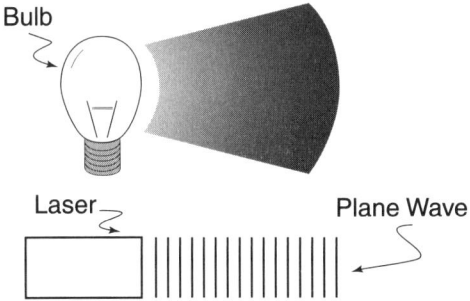

FIGURE 24-4: Plane wave.

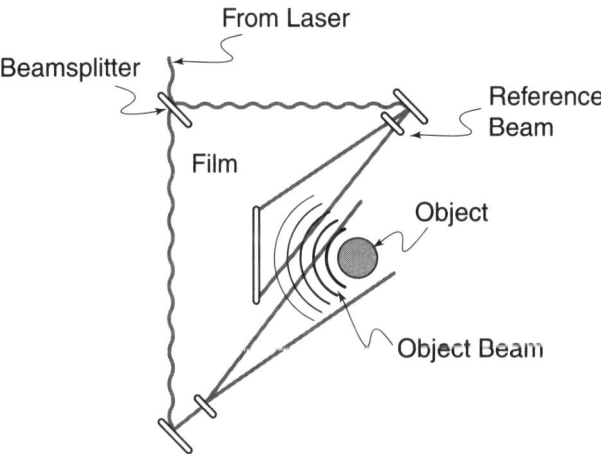

FIGURE 24-5: Object beam.

Up until the instant in time that the object beam strikes the object, it too is a plane wave. As soon as it hits the object, it is changed, or modulated according to the physical characteristics and dimensions of the object. So, that the light that ultimately reaches the film plane after being reflected by the object now deviates in intensity and phase from the virtually unhampered reference beam. That difference is a function of the object. What once began as a plane wave is now bouncing off the object in a complex wavefront, which consists of the summation of the multitude of infinitesimal object points reflecting light (see Figure 24-6).[6]

Using a laser in order to have this added information about the object would do us no good if the reference and object beams were not allowed to interfere at the film plane. The simplest interference that could take place

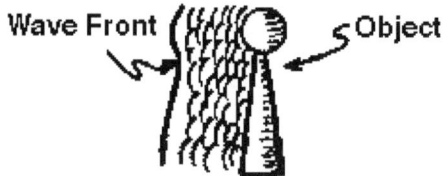

FIGURE 24-6: Wavefront.

on the film would be between the reference beam and the object beam, but with no object at all—you actually have simply two plane waves coming from different directions and interfering on the film. Obviously in this case, it does not matter which you call the reference or object beam, for neither carries any information about an object. And yet, something very definitely is recorded. If you get a good understanding of this, in the simplest case, it will be easier to understand as you move on to more complex situations.

The two beams are interfering with each other as they pass through one another. The crest of one plane wave (see Figure 24-7) meets the crest of another, or perhaps the crest meets a trough.[7] This is reminiscent of the Thomas Young experiment described earlier (see Figure 24-2), but with much more coherent light.

When a crest meets a crest, it gives constructive interference; and when a crest meets a trough, it gives destructive interference as shown in Figure 24-8.[8] Naturally, where crest and crest meet, there is more energy present,

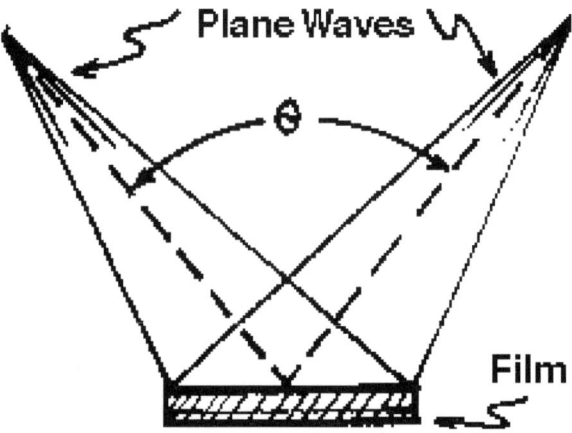

FIGURE 24-7: Plane waves.

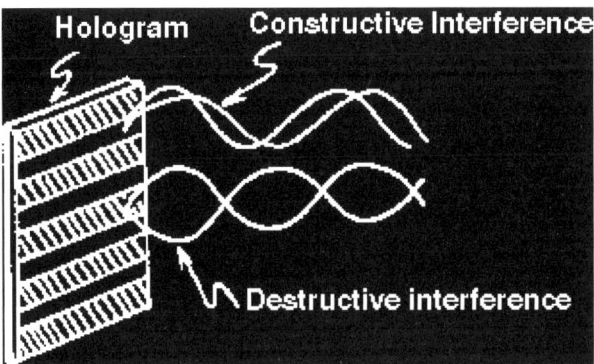

FIGURE 24-8: Constructive interference and destructive interference.

and more of the atoms in the silver halide are affected or *exposed* than at a point on the film where a trough and crest meet. The accumulation of these points sets up a very fine stationary pattern or grating throughout space. The scheme of the pattern is a function of the wavelength of light, but more importantly, the angle difference between the two plane waves. We will get back to this point when we talk about the different types of holograms. It is important to remember that the direction of the light, phase of the light, and so forth, are preserved and coded in the emulsion by the very process of reference and object beam interference. Therefore, if you were to shine the number-one beam back through the plate at the same angle you had in construction of the hologram, you would reconstruct the image of the number two-beam, and vice versa.

The very fine pattern that the emulsion assumes is a recording of the wavefronts as they interfere in the emulsion. It is definitely not a direct point-to-point recording of the image of the object, but rather a recording of the interference between the coherent light that hit the object and that which did not.

You know that light, traveling in a waveform, can be bent or diffracted along its path of travel. One way to bend light is by the use of a lens. You may consider a hologram a very complex lens. It is a bending and forming part of light of the reference beam, which is used for reconstructing the image into the wave fronts of the original object, so that you may perceive the object as if it were really there. All the infinitesimal little points that reflected light (which interfered with the reference beam on the film) are neatly focused to their respective positions in three-dimensional space.

In most cases, the object will reconstruct its original size, regardless of the size of the plate, and the same distance from the film that it was when the hologram was made. The reference reconstruction beam will be focused by the complex hologram lens, so that the front of the object appears closer, the back further away, and all the points in between are filled in accordingly. This might sound like a point-to-point correspondence very much like photography. However, there is a very special difference that makes holograms so wonderful.

Most of us have heard that if a hologram is broken or cut up, each small portion contains information about the whole object. This is because the light bouncing from each point on the object is not focused to a point on the film; rather, it is allowed to spread out through space between the object and the film, thus covering a large portion of the film and interfering with the reference beam throughout that whole portion of the film as if each point were a spray of light—each with a certain angle of divergence, so that every point is coded into a large area of the hologram. It might be easier to understand with this simple example. Let's say you have a very fine 11" × 14" hologram of a George Washington bust, complete with hat and plume (see Figure 24-9).[9] Two museums want this hologram (there is no other, and the bust was destroyed in an earthquake). After much ado, they decide to cut the hologram horizontally and exactly through the middle. Each museum then has a representation of the whole bust, unchanged in size, but from different angles. It will be easier to understand this if you think of the hologram as a window into a room containing the bust. If the window is made smaller, the object does not shrink. You merely have a narrower angle of view of the object. You would be able to see, for example, the plume, even from the bottom portion of the hologram; however, you may not be able to see the very tiptop of the plume from the reconstructed angle of view of the lower part. This is because the light from that point was not able to spread enough to reach and interfere with the reference beam in the lower extremities of the plate. The holder of the upper portion would not get an especially good look under Washington's chin. One simple remedy would be to move the object back from the plate, and thus give the light more space in which to spread. However, as the object is placed farther back from the film, it recedes from your personal three-dimensional world.

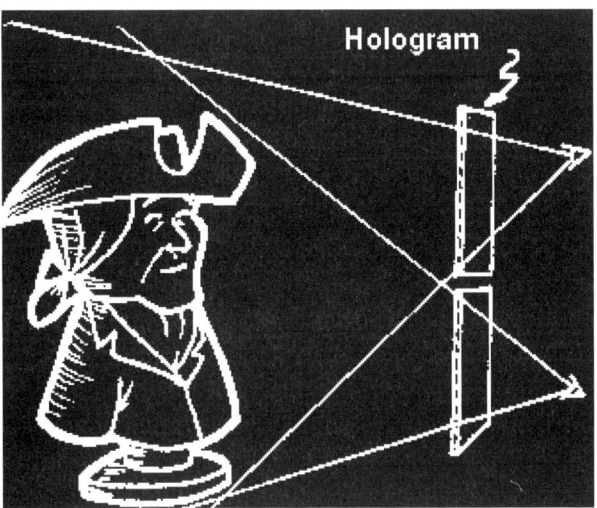

FIGURE 24-9: George Washington bust, complete with hat and plume.

Movement

One of the very prevalent practical problems in the making of a hologram is object movement. Unless you are lucky enough to own or have access to a pulsed ruby laser, you will have to look into the various methods of achieving object isolation. Later in the chapter, we'll deal very specifically with the design and construction of isolation tables and optical mounts. We need to spend a little time in this part of the chapter trying to explain why you have to isolate the object from even the slightest movement in order to make a hologram.

If you are taking a photograph with relatively slow shutter speed, say 1/30th of a second, and your subject is moving rapidly across your line of sight, the photograph will be blurred. Sometimes this is done intentionally and with excellent results; however, for most photographers, it means a ruined piece of film. In holography, the slightest movement of the object does not blur the image, but completely obliterates it. Remember, you are not recording a focused image of the object, but the interference of two wavefronts of light, the reference and object beams. The time needed to expose a hologram correctly is dependent on many things: the power of your laser, the sensitivity of your emulsion, and the reflectance of your object, among others. An average exposure for a common hologram is very roughly anywhere from a second to a minute. Let's say for the sake of argument that it is 10 seconds. During those 10 seconds, the plane waves from the laser are being reflected and diffracted by the object. The resulting complex waves are then interfering with the reference beam in the emulsion. In essence, you are recording interference fringes or patterns whose lines may only be separated by several wavelengths of light. This is an ongoing process taking place during the exposure. Remember that when the laser beams are coherent, the interference pattern is stationary in space and thus can be recorded on film. If anything moves that is involved in this train of waves by more than a fraction of a wavelength, the interference pattern will also move and the pattern is obliterated.

Any movement of the object, the film, or the optics caused by acoustic vibration has the same fatal results. Obviously, one way of reducing the chance of movement is by making the exposure very quick, say a billionth of a second, thus completely alleviating the need for isolation. Without the costly pulse ruby laser, you will need to construct an isolation table of relatively dense material, and try to isolate all components from all source of movement. For holograms that reconstruct an object clearly and brightly, all elements of your holographic setup should be stationary to less than one-tenth of a wavelength of light.

Different Types of Holograms

There are numerous types of holograms. It is important to learn the basic differences between the various types and what terms are used in referring to them so that you will understand immediately what someone means if they say, for example, they have just made a reflection hologram or transmission hologram or in-line hologram. Holograms can differ in the way in which they are produced, and they can incorporate and store the information for playback. The latter difference is the simplest to explain, so let's begin with that.

Under normal conditions, you would be using a silver halide-type film, so let's talk about that specific case. The holographic information is coded in the emulsion according to the localized microscopic differences in the absorption of light, or by the amount of silver halide converted to silver atoms during exposure and development. This is referred to as an *absorption hologram*. The absorption pattern on the film corresponds with the amount of light incident on the plate during exposure. If that same hologram is put through a bleaching process, it will then be termed a *phase hologram*. Bleaching is discussed in detail later in the chapter. The absorption index is created by changing the different residues of silver to corresponding thickness of transparent substance. The hologram is then played back by the refraction of the reference beam dictated by changes of refraction in the emulsion. In a phase hologram, the reference beam is phase modulated in order to reconstruct the wavefronts of the original objects. In absorption holograms, the reference beam is diffracted by the small patterns of exposed emulsion in the form of silver residue.

Many holographers bleach all of their holograms, because phase holograms absorb less valuable reconstructing laser light than the absorption type, and thus create a brighter image. However, some holographers do not bleach regularly, especially if they have made a perfect exposure in their original hologram. This is due to the fact that there is a slight loss of resolution along with the gain in brightness. Also, a poor bleaching technique increases the amount of noise and can greatly reduce the resolution. The source of the controversy, if any, is merely personal taste.

It is important to remember that the terms *absorption* or *phase hologram* have nothing to do with way the hologram was exposed. However, in the case of silver halide, *emulsion* refers only to a bleaching process that follows exposure and development (although you may alter your development process if you know in advance you are going to bleach). The following different types of holograms have special terms because they are actually constructed using different beam arrangements. Remember that all of the different types we are about to describe can ultimately be either absorption or phase type holograms.

The inherent difference between holograms has caused scientists and holographers to develop special terms or adjectives for them. In the construction stage, the difference is usually nothing more than the beam and object beam, as they interfere on the film. This angle difference can produce very pronounced differences between holograms in the playback stage. For example, a plane transmission hologram has to be reconstructed with laser light or a specially filtered light that approaches coherency in order for the reconstructed image to be crisp. A white-light reflection hologram can in comparison be viewed quite clearly with sunlight or under ordinary incandescent light sources.

Very simply, as the angle difference between reference and object beam increases, the tiny patterns in the emulsion exposed by the crest-crest

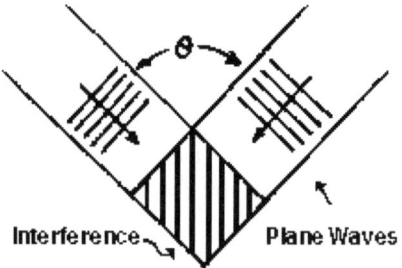

FIGURE 24-10: Tiny patterns in the emulsion exposed by the crest-crest interference of light waves are set up closer and closer together.

interference of light waves are set up closer and closer together (see Figure 24-10).[10] The resulting properties of the varying distances between fringes or the dark exposed areas in the hologram emulsion will be discussed later in the chapter, but first, let's get some terms straight.

The first hologram ever made by Dennis Gábor, in 1947 was an in-line, plane, transmission type. Remember at this time, the laser was still yet to be developed, so Gábor had to make due with the quasi-coherent light gained by squeezing light from a mercury vapor lamp through a pinhole and then color filtering it (he used the 0.546 micron mercury green line). *In-line* means that the reference beam and object beam are coming from the same direction or are the same beam. Gábor had to do this in order to maintain the little coherency he had gained.

All in-line holograms are also single-beam setups. The same beam acts both as reference and object beam. This was made possible by using a transparency as the object. The light that went through the transparency before reaching the plate was modulated by the transparency; the light that went through it and was not affected by the transparency was the reference beam. The diffracted light and reference light interfered on the emulsion of the hologram, and thus fulfilled one basic requirement for the construction of a hologram.

When the reference beam was later shown back through the hologram at the same angle of relationship it had with the plate in the reconstruction stage, an image appeared. A poor image due to the lack of coherent light, but worse still, the reference beam shone directly into the viewer's eye, thus greatly compromising the viewing of the reconstructed object. Although it was a poor image, it was there in all its dimensionality. A new medium had been born—alas, a little prematurely—and in 1948 was placed on the shelf until the advent of the laser.

 NOTE: Through his experiments, Gábor proved that an interference pattern carries all the information about the original object, and that from the interference pattern, one can reconstruct the object. For the discovery of these now well-accepted concepts, Dennis Gábor received the 1971 Nobel Prize in Physics.

TRANSMISSION HOLOGRAMS

As previously mentioned, in order to play back a hologram, the reference beam must be shone back through the hologram at the same angle of relationship as it had in construction. This is where the term *transmission hologram* arises. Transmission merely means that the reference beam must be transmitted through the hologram in order for the image to be reconstructed.

In 1962, Leith and Upatnieks at the University of Michigan removed Gábor's brain child from the shelf, and gave holography its rebirth. Like Gábor, they did their early experiments with a filtered mercury arc lamp. Leith and Upatnieks invented the off-axis reference beam with all its great advantages, which they did not even appreciate at the time. After the development of the continuous wave gas laser in 1960 by Ali Javan, Leith and Upatnieks started using the laser and discovered the three dimensionality of the images. They performed these experiments as an adjunct to their work in side-looking microwave radar. They independently discovered off-axis holography only to find that Gábor had proposed holography 12 to 14 years earlier.

The term *off-axis* means that the reference beam and object beam are not coming from the same direction. Naturally, in order to perform this feat, you must have two different beams; thus the term *twin beam*. Because the laser gives a homogenous beam of coherent light, you can extract a beam from the original beam as mentioned earlier. This is done with the aid of a beamsplitter, which could be nothing more than a piece of optical glass. A part of the original beam goes through the glass, and a part is reflected at the same angle as its incident. This allows one to bring in the reference beam from an infinite number of angles in relation to the object directed beam, thus avoiding the inconvenience in playback of having to look directly in the reference beam as with the in-line transmission hologram.

PLANE AND VOLUME HOLOGRAMS

This is a good time to point out the differences between a plane hologram and a volume hologram. As the angle difference between the object beam (or the wavefronts bouncing off the object) and the reference beam changes, so does the spacing of the patterns in the emulsion. As long as the angle difference remains less than 90 degrees, the hologram is called a *plane hologram*. *Plane* meaning that the holographic information is primarily contained in the two-dimensional plane of the emulsion.

Although the emulsion does have a thickness, usually around seven microns or 7 millionths of a meter, the spacing between fringes is large enough, when the angle is under 90 degrees, for us to imagine that the depth of the emulsion isn't really being utilized in the recording of the hologram. At 90 degrees, which is really a convenient but arbitrary point, the angle is great enough and fringe spacing has become small enough for us to say that the recording process is taking place throughout the volume of thickness of the emulsion. A point to remember is that although there are different thicknesses of emulsion put on celluloid or glass plates, seven microns is an average. One can use the same emulsion—say, seven microns thick—and make both plane and volume holograms depending on the angle difference between reference and object beam.

Thus, if you imagine your film in a fixed plane and your object in a stationary position, as you rotate the incidence angle of the reference beam (see Figure 24-11), you can determine whether you are making plane or volume holograms.[11] If your angle is under 90 degrees, it's a plane hologram; from 90 degrees to 180 degrees, it's a volume hologram. Naturally, past 180 degrees, you merely begin coming back the other way through the volume to the plane, and when you finally reach 360 degrees, you are back at the in-line plane and transmission hologram where you can collect your $200 (a bad Monopoly game joke).

A very important point for differentiation occurs as the reference beam swings around its arc of possible positions. In a plane transmission hologram, the reference beam is hitting the film from the same side as the object beam. In a volume reflection hologram, the reference beam has made an arc clear around so that it hits the film from the opposite side as the modulated object beam. When 180 degrees difference is reached, you are constructing an in-line volume, reflection hologram.

A transmission-type hologram means that the reference beam must be transmitted through the hologram in order to decode the interference pat-

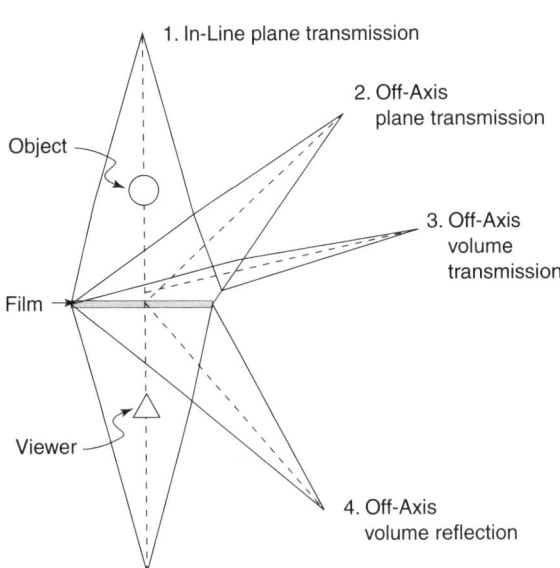

FIGURE 24-11: Reference beam angles.

terns and render the reconstructed image. The light that is used for playback must be coherent or semicoherent, or the image will not be sharp. If a noncoherent source, such as the light from a common, unfiltered slide projector, is used, then the hologram will diffract all the different wavelengths. The interference pattern or grating etched in the emulsion is not particular as to which wavelengths it bends or focuses; therefore, you end up with an unclear overlapping spectrum of colors that somewhat resemble your object.

A hologram will play back just as well with laser light of a different color or wavelength than the light with which it was made. However, the object will appear to be of a different size and/or distance from the plate. For example, a hologram of an object made with neon or red light will play back that object smaller or seemingly farther away if a blue color laser is used. This is because the grating will bend the blue or shorter light less severely than the red with which it was made and with which it is meant to be decoded.

REFLECTION HOLOGRAM

Unlike a plane hologram (sometimes called a *thin hologram*), which requires a coherent or highly filtered playback source, a reflection—or thick—hologram can be viewed very satisfactorily in white light or light that contains many different wavelengths. The one requisite is that the light be from a point source and be a somewhat straight line, such as a slide projector light or penlight, or the sun on a clear day. The reflection hologram can do this because in a way it acts as its own filter.

In a reflection hologram, the fringes are packed so closely together that they constitute layers throughout the thickness of the emulsion. The spacing between fringes remains constant. If d, or the distance between fringe one and two, is two microns, for example, then the distance between the remaining layers of fringes will also be two microns. This distance is a function of the wavelength of light used in constructing the hologram, and also the angle difference between reference and object beam. This layered affair allows the reflection hologram to absorb, or not reflect, any of the colors or wavelengths of light that are not the correct length. The wavelength that matches the fringe spacing will be reflected: the crests of the wavelengths that are too short or too long will eventually miss one of the planes and be absorbed into the darkness of the emulsion.

In a reflection-type hologram, the playback light or reconstruction beam comes from the same side of the hologram as the viewer. Some parts of the incident light are reflected, some are not, depending on the interference pattern. If the hologram was made correctly, the result should be a visible three-dimensional image. As was previously mentioned, in the transmission type, the reconstruction beam must pass through the hologram and come toward the viewer from the opposite side of the hologram, while in the reflection type, the playback source comes from the same side of the hologram as the viewer.

Incidentally, just as very few transmission holograms are made in-line or 0 degrees, so are very few reflection holograms made in-line, or else you would have to hold your point light source in your teeth or perhaps invest in a miner's cap. Most reflection holograms are made at a less severe angle, perhaps 160 degrees, so that the light can come in at an angle without being blocked by the person who is trying to see the hologram.

The image produced by the hologram can either appear to be in front of the holographic plate or film, or behind the film. In the former case, it is called a *real image* (projection), and the latter, a *virtual image* (see Figure

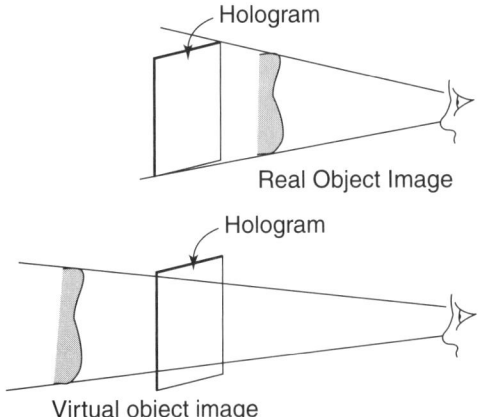

FIGURE 24-12: Real image (projection) and virtual image.

24-12). If you imagine your position as viewer to be constant, then you can easily determine whether an image is real or virtual. If the image appears between you and the hologram, it is a real image; if the hologram is between you and the apparent object, then it is a virtual image.

In general, it is easier to view a virtual image because you can see through the hologram as if it were a window. As with other windows, if you change the size of the windows, the object or objects you are viewing do not change their size. For example, let's say you are lucky enough to have a window in your house that looks out on a beautiful tree. If for some terrible reason you have to make your window smaller, your tree luckily does not shrink—you merely have a more confined view or less possible angles of view of the tree.

To view a virtual image, you must look through the hologram to perceive the object floating in the space behind it. In order to see a real image, you look at the hologram and see the object in free space in front of the hologram. It is a little more difficult to view a real image because you have to find the image or focus your eyes in front of the hologram; in this case, the hologram is less capable to act as a guide for your eyes. You may move a screen or sheet of paper back and forth in front of the hologram in order to find where the object is focused, and then, keeping your eye on that place in space, remove the sheet and look straight into the hologram.

The real image is very exciting, but there are a number of drawbacks. The object holographed should be quite a bit smaller than the size of the film you are using; if not, you will not be able to see the complete real

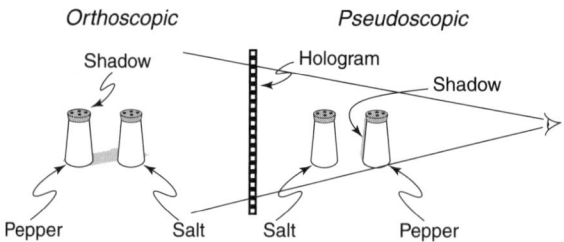

FIGURE 24-13: Pseudoscopic image.

image of the object all at once. It will necessitate craning your neck and stretching in all which ways to see parts of the whole object or objects. Moreover, unless you take special precautions in the construction of the hologram, the real image will be pseudoscopic (see Figure 24-13).[13] This means simply that everything that was closer to the film when the hologram was made will now be further away, and vice versa. This includes both individual objects in a shot or the different planes of space of an individual object. The pseudoscopic image is made by reversing the direction of the reference beam, or by turning the completed hologram around until seeing the image in front of the plate. For example, if in making your hologram, you placed a salt shaker closer to the film than a pepper shaker (let's imagine the salt shaker is even casting a shadow from the object beam onto the pepper shaker), then in a pseudoscopic playback, a real image pepper shaker will appear to be closer to you than the salt shaker, which is no longer there.

Naturally, if you play back the virtual image of the same hologram, the shakers would resume their original positions. Later in the chapter, a discussion will take place to further reveal the real image hologram, show possible ways of making holograms specifically for real image playback, and also touch lightly on why there is a real image.

MULTIPLEX HOLOGRAM

In addition to the previously mentioned types of holograms commonly made today, there are multiplex holograms and image holograms. These types of holograms are being used more commonly today.

Very simply, the multiplex hologram is the holographic storage of photographic information. In the first stage, a series of photographs or a certain amount of motion picture footage of the subject is exposed. The number of stills or frames taken depends on how much of an angle of view you want of the subject in your finished hologram. For example, if you want a 360-degree view of the subject, you might expose three frames per degree of movement around the subject (usually the camera remains stationary and the subject rotates), which will result in the exposure of 1,080 frames.

When your film is developed, you proceed to the holographic lab and (using a laser) make a series of *slit* holograms using each frame of film as a subject for each slit of holographic film. The slits are usually about 1 millimeter wide and are packed so closely that there is no *dead space* in between. In addition, the hologram is bleached so that the strips disappear. Usually, a multiplex hologram yields horizontal, not vertical, parallax. This is because the camera usually moves around (or the subject moves around in front of the camera) and doesn't usually pass over the subject. Also, psychologically, horizontal parallax is much more desirable, and the lack of horizontal parallax to humans is much more noticeable than the lack of vertical parallax. The multiplex hologram is usually, though not always, made on flexible film coated with the same holographic emulsion as the plates.

The procedure can be totally mechanized so that a machine can expose a slit hologram per each frame of footage at a very rapid pace. The advantage of this type of hologram is that you can now have a hologram of almost anything you can capture on ordinary film without the need of the expensive, clumsy, pulse ruby laser. The disadvantage is that it is not truly a hologram, but photographic information holographically stored. It seems that it will have a very solid place in the growing field of display and advertising holography.

The image hologram that was mentioned earlier also has an advantage that will make it one of the types widely used in display holography. The image hologram can be played back with ordinary *white light* from an uncoated incandescent bulb. An image hologram can be either a reflection type or transmission type. However, it is more impressive as a transmission type, because unlike an ordinary transmission hologram, the image transmission hologram can play back well with an unfiltered white light source. The image hologram can be formed by placing the correct lens between the subject or scene and the holographic film plane. Thus the subject is focused directly onto the film plane, and a hologram is made of that focused image.

This type of hologram is very pleasing because the object seems to come out at you like a real image, but it is not pseudoscopic.

The real advantage is that the image transmission hologram has much less color dispersion or spectral smear than an ordinary transmission hologram. Moreover, when you play back with an ordinary uncoated light bulb, there is a rainbow effect, but the image remains very sharp. Another way to make an image hologram is by copying the focused real pseudoscopic image of an original or master hologram. The result is a second-generation image transmission hologram whose virtual image is orthoscopic.

Coherence Length

Every laser has what is termed a *coherence length*. It is related to the length of the laser tube and the purity of the phase of light emitted and the wavelength itself. The more pure the light, the greater the coherence length. That is, not just any kind of laser can be used in making a hologram. On the specification sheets of most quality lasers manufactured today, you will see the term *TEM* (infinity). This means that the laser is operating in the lowest transverse mode, which is the most uniform across the beam and is preferred for holography. A laser intended for making holograms must ideally be lasing in just one longitudinal mode. Both of these qualifications (spatial coherence and longitudinal coherence) define the purity of the light.

In twin-beam holography, it is extremely important to measure the paths of the reference beam and object beams(s)—for even if you are using the prescribed laser for holography, its coherence length is not infinite (see Figure 24-14).[14] The coherence length also places an upper limit on the size, especially depth, of the object that can be holographed by setting definite bounds to the path difference of the reference and object beams.

These concepts are subtle and can be quite difficult to understand, so let me explain a little further. First, a laser ideally is emitting all of its light absolutely only on wavelength, with all of those wavelengths completely in phase from the point of exit to infinity. This would be wonderful, but unless you have a 50mW or more laser with a special attachment called an etalon, your coherence length is probably around six to eight inches. This is the approximate length of an average 5mW, HeNe laser. This means that once you separate your original beam and secondary beam, the path dif-

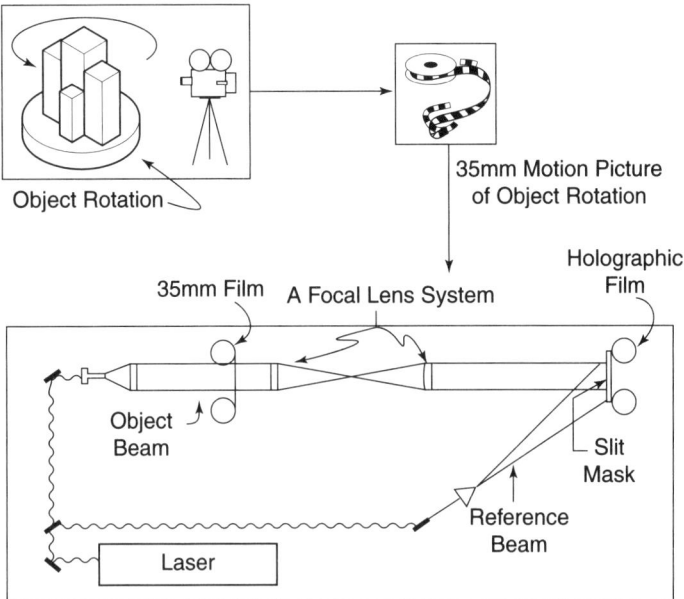

FIGURE 24-14: Twin-beam holography.

ference that they travel cannot exceed six inches. If the distance of the reference beam from the beamsplitter to the hologram is 48 inches, then the distance of the object beam from the beamsplitter to the object to the hologram must also be 48 inches. Then your available path difference can be utilized totally by your object.

In a way, this can be related to depth of field in photography, but in holography, outside of that depth of six inches, the object drops off into nothingness. In a way, you *focus* on your object by making sure the lengths of your object's beam(s) and reference beam are measured correctly.

NOTE: The coherence length is vitally affected by the type of laser used. Depending on the type of laser you are working with, you may have to adjust the length (the difference can be sizable). When in doubt, check with the manufacturer.

The light being emitted by lasers has what you might call a *coherence curve*. It is a bell-shaped curve that shows, in distance from the exit point of the laser, where the wavelengths are most in phase. This is usually a

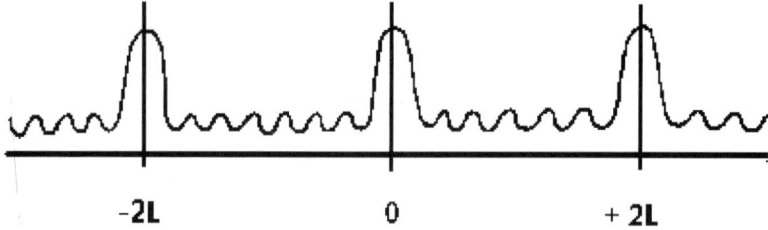

FIGURE 24-15: Coherence curve.

constant integer and depends on the wavelength or substance that is lasing, the size of the laser, as well as how purely it is emitted. This number, let's say eight inches, remains constant. At the peak of the curve, or every eight inches, the light is most in phase. You would make the path lengths of your beams multiples of twice the cavity length of your laser. In addition, the coherence function repeats itself. It is at maximum again, at a distance of twice the mirror separation in the laser. It repeats itself every 2L distance as shown in Figure 24-15.[15]

Before buying a laser for holography, it is always wise to inquire about all the pertinent characteristics of its functions. More time will be spent on this topic later in the chapter on coherence length; reference and object beam intensity ratios; and all of the practical information one needs in order to perform holography. At this point, however, you have been offered enough of the basic theory of holography, and now we'll begin applying all of this to the construction of a lab and construction of holograms. Naturally, as your practical experience grows, you will be able to absorb more theory, but now let's start with the isolation table, or in more colloquial terms, from the ground up.

Construction of a Laboratory Setup

Construction of an isolation table is relatively simple and will be much simpler if you have the 3/4" plywood and the 4 × 10" lumber cut to size by the lumber yard as shown in Figure 24-16.[16] The roofing tar is used to fill in any air spaces that may occur between joints, thus making a completely solid unit. The tar can be purchased by the gallon at a hardware store.

Obtain the inner tube from a tire store and the sheet metal from a sheet metal fabrication company (look in the Yellow Pages). You will also need the following materials:

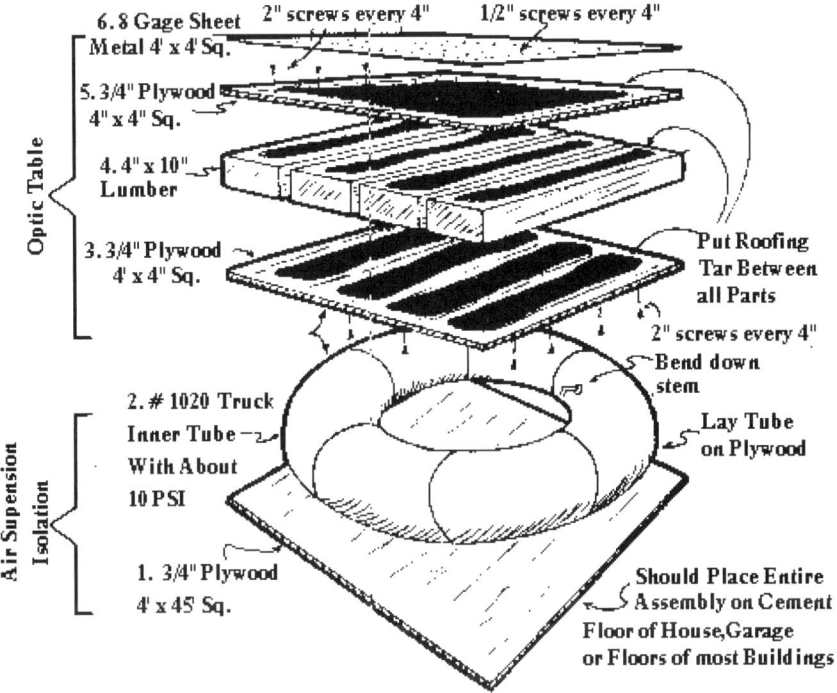

FIGURE 24-16: Isolation table.

- Five dozen 2" #8 flat head wood screws
- Five dozen 1/2" #8 flat head wood screws
- One gallon roofing tar
- One # 1020 inner tube
- One 4' × 4' 8-gage sheet metal
- Three 3/4" × 4' × 4' interior plywood
- Four 4" × 10" × 4' lumber
- Two cans flat black spray paint.[17]

This table, suspended on the inner tube, is an air flotation system that will settle out vibrations rapidly, and can be used successfully in isolating an object and optical components from vibration up to half the wavelength of 6328Å light.

Note: It is important that your isolation table be placed on a solid floor, such as cement slab in a garage. Wooden floors are not recommended. If your table is not quite stable enough after testing it with the interferometer setup (described next), try putting a 3" level of newspaper under the whole table as a primary cushion. Or try putting it on a table with legs in buckets of sand.

INTERFEROMETER

To test your isolation table for stability, an interferometer can be set up. By positioning the beamsplitter, the two mirrors, and the lens, you will produce visible fringes that show a fraction of a wavelength variation in the spacing of the mirrors. Please use the following procedure and refer to Figure 24-17:[18]

1. Mirror #1 should be positioned to redirect the beam back down the laser tube.
2. Insert the beamsplitter and touch up mirror #1's position.
3. Position mirror #2, and tilt both it and the beamsplitter so that all the spots on the cardboard screen are on the same horizontal line (there will be two sets with two bright beams in each set).

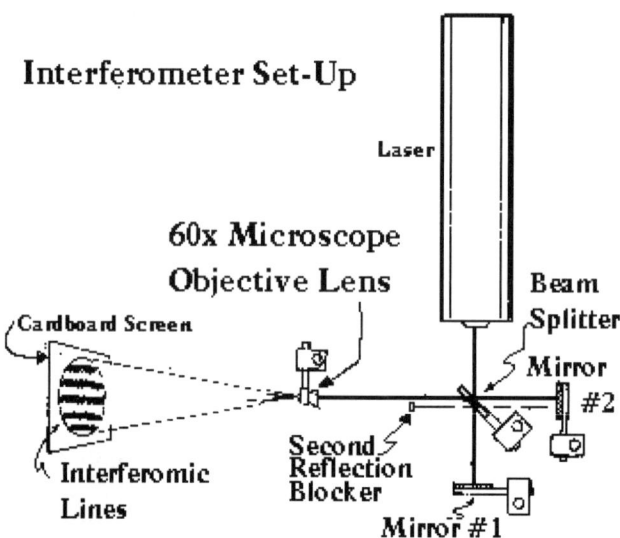

FIGURE 24-17: Interferometer setup.

4. Superimpose the two sets as close as you can by rotating mirror #2.
5. Insert the 60x microscope objective lens[19].

The black and red lines that appear on the screen represent constructive and destructive interference between the two plane waves from the two mirrors. These are the types of interference that must be recorded to produce a hologram.

Walk on the floor, stamp on it, yell, sneeze, clap your hands over the table; these vibrations will probably cause the fringes to blur. If the fringes blur during exposure of the film, you will not get a hologram. If they blur part way through the exposure, you will get a hologram. However, the image will have very little contrast because only part of the exposed information is interference; the rest is just light exposure with no information.

You may notice that the center fringe pattern appears to *breathe*. This is all right and is caused by changes in air pressure over the table. You have now assured yourself of the most important consideration in holography: *stability*.

DARKROOM CONSTRUCTION

For those of you who do not have a suitable space to use as a darkroom, here's an example of a 4' × 8' darkroom. If construction is not possible, an alternate could be to sew a few strips of black plastic that can be bought by the yard from a large fabric store and hang them from the ceiling. This type of darkroom could only be used in the evening or in a very dimly lit area because of the light leakage problem.

OPTICS AND OPTICAL MOUNTS

Here are the minimum optics necessary for making transmission and reflection holograms and the interferometer test setup (see the sidebar, "Binary Optics: What Are They, and What Are They Used For?"). The list of supplies contains the names of manufacturers and distributors of optics in various sizes and types that you may want or need to make other setups. You will need the following quantity of material:

- Three 2" × 2" front surface mirrors (Edmund #40, 040 51 × 76 mms)
- One 2" × 2" flat glass for beamsplitter (Edmund #2269 52 × 113mms)
- Two 60x microscope objective lenses (Edmund #30, 049)[20]

■ BINARY OPTICS: WHAT ARE THEY, AND WHAT ARE THEY USED FOR?

Binary optics is a new technology that has already led to thinner, lighter lenses and optical devices that were once impossible to make. The method for manufacturing binary optics devices explains the technology's name: workers use the same equipment that fabricates digital circuits to etch patterns onto optical materials. Whereas conventional lenses, using the principle of refraction, require several millimeters of material, binary optics, employing diffractive effects accomplished in a space only a fraction of a wavelength in size, can facilitate devices with surface dimensions as small as half a micron and depths as shallow as a nanometer. The etched patterns, usually made up of microscopic staircases, are designed to break up the incident light into individual wavefronts, which interfere in such a way as to create a new wavefront that travels in a desired direction. Virtually any optical transformation can be achieved with binary optics: for example, workers at Hughes converted circular patterns into straight lines, a geometry that is much easier for pattern-recognition machines to decipher. The production method offers another advantage: optical components can now be etched directly onto integrated circuits. This makes it feasible to produce systems that mimic biological vision, with light-gathering devices on top of individual processing elements.

Note: Advanced Computer Generated Holograms (CGHs), often termed *binary optics*, can be used to reduce the cost and weight of refractive lens-based systems, and to increase the performance and functionality of laser-based optical systems. For example, CGHs can be used as collimators, focusing elements, beam combiners, and splitters. They can be used to direct optical signals in free space interconnect systems, or to couple light into or out of optical waveguides and fiber optic cables. They can also be used in conjunction with refractive elements to reduce aberrations in optical imaging systems.

Micro-Optics

Micro-optics (diffractive optics, binary optics) is a revolutionary new technology that allows the fabrication of optical elements using semiconductor fabrication techniques. These techniques allow the custom design of optical surfaces. Optical processing previously impossible can now be produced precisely, repeatedly, and cost effectively.

Micro-optics can improve existing optics-based products to enable higher performance and lower cost implementation. In addition, completely new optical applications are now feasible due to the economic and design benefits provided by micro-optics.

Applications

The following are just a few of the potential new and existing applications for micro-optics:
- Lenslet arrays
- Diode laser couplers
- Fan-out gratings
- Diffraction gratings
- Beam shaping and homogenizing
- Aberration
- Compensation
- Laser printers
- CD players
- Bar code scanners
- Adaptive optics
- Fill-factor enhancement
- Aperture multiplexers
- Wavelength division multiplexing
- Laser disk players
- Copy machines
- Wafer inspection
- Systems
- Laser pointers
- Optical interconnects
- Telescopes
- Industrial process
- Control
- Fiber optics
- Communications
- Optical computing
- Vision correction
- Camcorders
- Beam multiplexing

Two pieces of ground glass, spaced 10cm apart, can be used as an object beam spreader, therefore eliminating one objective lens. This eliminates lens cleaning and beam alignment problems, and is less expensive. If you use a lens, you should take the lenses apart and use only the final, very small lens to spread the beam.

Because of the high cost of optical mounts made especially for holography, some of you may have over the past 10 years fabricated all your own optical mounts. These mounts have all the stability and versatility of mounts that are produced today costing 10 times as much.

All of the mounts are made from heavy aluminum stock available at surplus stores (check the Yellow Pages under *Metals* or *Aluminum*). These places will usually cut the aluminum to size. The large holes can be drilled by a machine shop, leaving the small drilling, tapping, and end filing to you. To keep reflections down, it is best to spray paint everything flat back. Epoxy everything together with the new 10-second epoxy that is now available from hardware stores. Aircraft surplus has the nylon screws.

The quantity indicated in Figure 24-16 is for the minimum number needed. You will probably want to make or order different kinds for various other experiments. Some mounts can be made from wood; however, remember that stability is a prime ingredient of successful holography.

LIGHT METER

Part of your holographic setup requires a very sensitive light meter. The light level of a low-power, spread laser beam is not enough to activate a conventional light meter; hence, you need a simple, accurate *light meter*, a Triplet V-O-M meter with a cadmium sulfide photo cell obtained from most electronic supply stores.

To use, first attach the photocell wires to the meter leads and plug the leads into the *COM* and *V-O-M* receptors. There are no polarity requirements, so either wire can go to either lead, and either lead to either receptor. To operate, perform the following steps and refer to Figure 24-17:

1. Move red switch to X1K OHMS.
2. Put photocell very close to a bright light, and rotate the knob on the side until the needle is right on the *0* on the very top scale.
3. To measure beam intensity, it is very important to remember that the meter's scale will read in such a way that the BRIGHTER the light falling on the photocell, the LOWER the number on the scale. Example: A reading of *50* is twice as bright as a reading of *100*.[21]

Actually Making Holograms

In most holograms, the object is displayed against black space; therefore, the hologram looks best with white or very light-colored objects. Using a securely mounted piece of wood painted white makes a good white background to display dark objects. Don't forget to place it within the coherence volume, or it will not appear in the hologram.

Object size is dependent upon depth and isolation. The coherence volume of 6" has a height of several feet.

Holograms have been made out of objects 18" tall with no problems; however, the objects were never more than 6" in depth. Lighting such a large object is also difficult because of limited table space.

A usual practice with large objects is to increase the distance between the object in the film so that the whole object can be easily seen. But remember, no matter what size the film or the object, the object will always reconstruct its original size. You may have to look through the hologram at some extreme angles, but as long as the object is properly lit and in the coherence volume, it will all be in the holographic image.

Try different textures and materials. Glass, metal, wood, ivory, objects from Styrofoam or clay, ping pong balls, toothpick sculpture—anything. It is recommended that you use objects that reflect a good amount of light.

Be careful of materials that are highly reflective, because the meter reading may be deceptive when it reads those highlights that are direct light reflections back to the film plane. You might want to spray on dulling spray that can be obtained from a photographic supply store.

Try putting a magnifying glass or a mirror in with the object. Some surprising optical illusions can be created. The most important thing to remember when choosing an object is stability.

BASIC TRANSMISSION HOLOGRAM

The word *basic* is used here because there are as many different configurations as there are holographers (see Figure 24-18).[22] Following is a beginning setup (see the sidebar, "The Basic Transmission Hologram Setup Checklist"). Later in the chapter, you will be shown others that are used for different reasons.

Note: As you can see in Figure 24-17, the components are placed with specific measurements. This is because of a coherence length of 6328Å light. It is a good rule of thumb to have the final distance from the first beamsplitter through the setup to the film be a multiple of twice the length of your laser cavity.

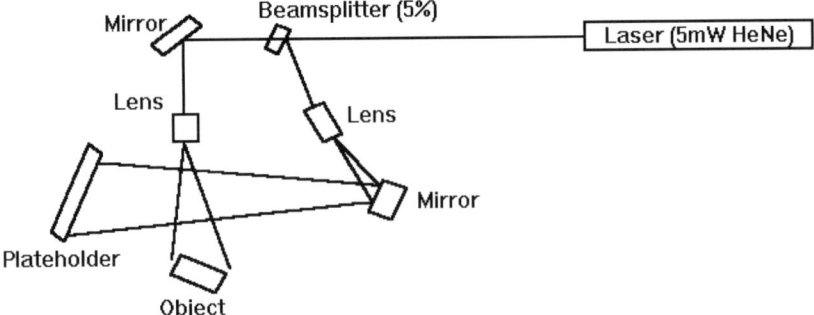

FIGURE 24-18: This diagram shows the basic setup for a transmission type hologram.

Note: The coherence volume is roughly six inches in depth, meaning that any object placed within this space will make a hologram—outside this area, the object is too far out of phase with the reference beam.

■ THE BASIC TRANSMISSION HOLOGRAM SETUP CHECKLIST

To set up, perform the following steps and refer to Figure 24-17:

1. Place the laser, direction mirror, object direction mirror, beamsplitter, reference direction mirror, and the film holder in their approximate positions. The film holder should have the thumb screws on the opposite side from the object area.
2. Measure these components for proper placement using a piece of tape for the Center of Object Area. Both the reference beam path and the object beam path must be of equal length from the beamsplitter through the setup to the film. In this setup, 48" is used.
3. Turn on the laser and adjust components for beam landing. Put a piece of white cardboard in the film holder for reference beam landing—it should hit the center of the cardboard.
4. You will notice two beams reflecting off the beamsplitter. These are first and second reflection. Use a piece of black cardboard to block the second reflection.
5. Position a 60x lens in the reference beam.

NOTE: It can be difficult to direct a 1.5mm beam through a 2mm diameter lens. First, lower the lens and adjust the mount so that the beam glides across the top of the lens holder in the direction of the lens. Then, carefully raise the lens until you see the spread beam exit. Adjust the lens so that the beam fills the cardboard in the film holder. You may notice concentric rings on the cardboard. These are either imperfections in the lens or dust on the lens. Try cleaning with lens tissue. If rings still exist, carefully rotate the lens in its holder to put the rings off the cardboard.

6. Place the object so that it is in the center of the coherence volume. Solid-white objects about as big as your fist work best for beginning holograms. Other types of objects will be discussed later.
7. Position the other 60x lens or pair of ground glass diffusers in the object beam with the same technique as in the note under step 5. The spread beam may not completely light the object. This may mean that you will have to place the lens before the object direction mirror instead of after it. Look through the film holder, rotate the object until you like the composition, and secure the object to the table with a drop of 10-second epoxy or nonhardening clay. It will take a few minutes for the epoxy and clay to solidify.
8. Measuring the beam intensity ratios is most important to ensure good contrast in the final image. Take the light meter and make the necessary preparations as described in the Light Meter part of the chapter. The optimum beam ratio is between 2:1 to 5:1—the reference beam should be 2x to 5x brighter than the light reflecting from the object toward the film plate.
9. Put a piece of black cardboard in front of the laser to use as a shutter. To measure, perform the following steps:
    a. Turn the lights off.
    b. Block the object beam with a piece of black cardboard somewhere between the beamsplitter and the object direction mirror.
    c. Put the photocell against the cardboard in the film holder tilted toward the reference beam coming from the lens.
    d. Read the meter's top scale (you can use a penlight).
    e. Block the reference beam with cardboard somewhere between the beamsplitter and the reference direction mirror.
    f. Put the photocell flat against the cardboard in the film holder facing the object.
    g. Read the meter's scale. Remember that the meter reads backwards—the brighter the beam, the lower the number.
    h. If ratios are over 5:1 or under 2:1, move the object lens closer or farther

from the object, and/or move the reference lens closer or farther from the film holder. Check to be sure that the object and film are still completely lit.
10. In the dark with only the safe light on, insert the film into the film holder with the emulsion facing the object. Gently tighten the nylon screws, and step back for a minute to let the table stop moving and allow the air to settle. You can tell which side the emulsion is on by lightly wetting your lips and touching the film between them for a few seconds. As you part your lips, one side will stick—that is the emulsion side.
11. Gently pick up the black cardboard shutter and suspend it directly in front of the laser beam for a mental count of 10.
12. Raise the cardboard all the way, allowing the beam to illuminate the film and object.

NOTE: Exposure time is by experiment based on laser output; reflective characteristics and size of object; brightness of beams; age of film and chemicals; and temperature of D-19. With a 5mW laser, a reading 2x to 5x brighter than this for the reference beam; fresh film and chemicals; D-19 at 68 degrees; and a light colored object with a 20-second exposure. It is not recommended that you turn your laser on and off with each exposure. Use your cardboard shutter to begin and end your exposure. This will help to ensure that your laser is operating at its highest output. Your can mentally count the seconds: One thousand. . . Two thousand. . . Three. . . thousand. . .

13. Lower the cardboard shutter, remove the film, and place in the D-19 developer emulsion side up and agitate, checking every 30 seconds to see if the exposed part of the film is a light to medium gray. This check can be made by holding the film over the safelight and observing the contrast between the exposed part and the clear edge that was hidden from the light by the film holder. If the film does not turn gray within five minutes development time, your exposure was much too short (or you may have left the cardboard block in the reference beam). Conversely, if your film goes too dark within the first minute of development, your exposure was too long. Optimum time in the developer is between two to three minutes.
14. When the film is dark enough, place it in the stop bath of water and agitate for 30 seconds.
15. Remove from stop bath and agitate the film in the fixer for two minutes. If you plan to bleach your hologram, you would not fix for two different techniques.

16. You can quickly check to see if there is any image at all by holding the film over a small light bulb and tilting the film around while looking through it. If everything was correct, you should see a rainbow of colors. If you do not see the rainbow, it is a sure bet that there is no image, and skip to step #18.
17. Wash film for 10 minutes under cold water faucet, agitate in Photo-Flo solution for 30 seconds, and let dry, making sure the emulsion is not in contact with any surface.
18. No image or poor image can be caused either by vibrations during exposure, insecure object, or poor contrast of beam ratios.[23]

TO RECONSTRUCT

Put the film back in the film holder with the emulsion toward the object. Remove the object. Replace the beamsplitter with the object direction mirror. This will put all the laser light into the reference beam. Adjust the components so that the reference beam through the lens illuminates the film. Look through the film and see the image of the object in its original location on the table. You may have to slightly rotate the film holder in relation to the reference beam because of emulsion shrinkage.

You can achieve very good reconstruction using a point light source, such as a slide projector or a high-intensity lamp with 5770Å narrow band pass filter appendix taped behind a piece of cardboard with a 1/4" hole punched in it. This assembly taped over the light source will result in a semicoherent playback source.

You might want to try the following experiment. After you have made a good hologram, do not remove the object from its position on the isolation table. Place the hologram back in the film holder with a reference beam incident on the hologram. The reconstructed image of the object is seen superimposed on the object. If the reference beam is blocked, the original object is seen through the hologram. Now proceed to block the reference and object beams alternately. You will find that in a completely darkened room, it is difficult to distinguish between the real object and the reconstructed holographic image. Now, adjust the hologram in the holder until the two are perfectly superimposed. Try pressing gently on the object with your fingertip. Notice the fringes at the point of contact. You have just performed stored beam holographic interferometry, a subject beyond the

522 CHAPTER 24

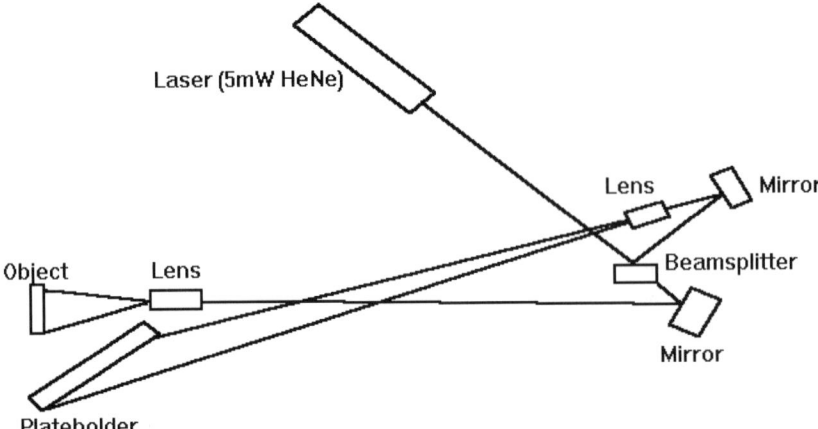

FIGURE 24-19: This diagram shows the basic setup for a reflection type hologram.

scope of this chapter, but a very important process used widely today in the testing of mechanical components.

BASIC REFLECTION HOLOGRAM

The word *basic* is also used here because there are also as many different configurations as there are holographers (see Figure 24-19).[24] The following is a beginning setup (see the sidebar, "The Setup Checklist"), and later in the chapter, you will be shown others that are used for different reasons.

■ **THE BASIC REFLECTION HOLOGRAM SETUP CHECKLIST**

To set up, perform the following steps and refer to Figures 24-17 and 24-19:

1. Roughly position the laser, direction mirror, and film holder (with the nylon screws on the opposite side of the object area).
2. Measure overall distance from laser to film as 32", making sure of the 30-degree angle of reflection off the mirror. This will make reconstruction easier, for the reconstruction beam will not be in the way of the person viewing the hologram. You can try smaller angles, but 30 degrees to 45 degrees is the extreme because of the tolerance of the emulsion.

3. Put a piece of white cardboard in the film holder and position the lens to fill the cardboard using the technique described in the Transmission setup.
4. Place your object (a solid-white object no larger than 3" deep, 3" tall, and 4" wide) as close to the film holder as possible and within the film frame.
5. Take out the white cardboard and see that the object is within the film frame, rotate the object until you like the position, and secure it with a drop of epoxy or nonhardening modeling clay.
6. Put a piece of black cardboard in front of the laser as a shutter. In the dark with only the green safe light on, place the film in the holder with the emulsion facing the object. Step back for a minute to let the table stop moving.

   NOTE: No beam ratios to figure—there is only one beam. The light passing through the plate or film toward the object is the reference beam. The light scattered back toward the emulsion by the object is the object beam. The counter flowing waves interfere with one another to produce the stationary interference pattern recorded by the hologram. Using the Reflection setup, be sure to keep your object as close to the film plane as possible, but be careful not to scratch the emulsion when you are inserting or removing the film.

7. Lift the black cardboard, suspend it in the beam for a mental count of 10, lift the cardboard all the way out exposing the film, count mentally the exposure seconds, and replace the cardboard shutter to stop exposure. Exposure is short, between 5 and 10 seconds.
8. Develop for about five minutes in D-19, checking with the safelight for a very dark, almost black exposed area on the film. Remember, you want light reflecting back to your eyes from the film, not transmitting through it.
9. Do not fix! Wash in running water for 10 minutes, rinse in Photo-Flo for 30 seconds, wipe with a sponge, and let dry. There is no stop-gap test for spectrum as in transmission holograms. You will not know whether there is an image until the film is almost completely dry.[25]

RECONSTRUCTION

Reconstruct by shining a point light source (the sun, high-intensity lamp, flash light, slide projector, etc.) at the film with the emulsion facing away from you and at the same angle as the original beam that lit the film. You will notice that the image is quite clear and has a green color to it. Turn the film around to see the real image projection. If there is no image, it is usually because of object movement.

Experiments

In all these examples, no attempt at scale has been made, and a certain amount of component fabrication is necessary on some of the setups.

DIVISION OF AMPLITUDE TRANSMISSION HOLOGRAM

This type of hologram uses only one lens and a large mirror. Part of the spread beam hits the object and goes to the film forming the object beam; and part of the beam hits a mirror that directs that portion to the film forming the reference beam. These holograms yield greater resolution. However, with this resolution, you get less depth of field or coherence volume. Also, the reference beam/object beam angle is very small, making this almost an in-line hologram that puts the reconstruction beam almost in your eyes.

MULTIPLE OBJECT BEAM TRANSMISSION HOLOGRAM

By splitting the object beam, you can light the object from many different angles, achieving lighting effects more akin to conventional photographic lighting techniques. However, remember that the distance from the first beamsplitter through each of the other beamsplitters to the film must be equal.

REAL IMAGE PROJECTION TRANSMISSION HOLOGRAM

Any hologram has the real image waves emerging from it. These waves are usually very distorted. Here is a setup that can produce a good real image. The telescope mirror collimates the beam so that instead of a cone of light for the reference beam, you have a tube of light. After setting up to view the virtual image, the film is rotated 180 degrees horizontally to view the real image.

NOTE: By using the dish mirror (approx. 50" f.l.), you create a slightly converging reference beam. You can play back with filtered light source as usual. The dish mirror can also be used to make real image reflection type holograms.

GREATER DEPTH OF FIELD (COHERENCE VOLUME)

This is a multiple lighting setup to produce a longer coherence volume. The distance from the first beamsplitter through each of the object beamsplitters is equal, forming multiple areas that, in reconstruction, seem to be a large single coherence volume.

THREE-HUNDRED-AND-SIXTY (360) DEGREE TRANSMISSION HOLOGRAMS

There are two ways to achieve 360-degree views of an object, and both use either the simpler Division of Amplitude technique in which your object and reference beam would both come down from the top, or split beam technique to independently light the object and produce a reference beam. This experiment uses four 4" × 5" pieces of film, and the second uses 5"-wide roll film that has the same emulsion as the plates and can be obtained from AGFA and Kodak. The object area is smaller with the four plates, but stabilizing of the roll film is difficult (try using a Plexiglas tube, tape the film, emulsion facing in, on the outside, and let it stabilize for five minutes or more before exposure).

NOTE: In the Division of Amplitude technique for 360 degrees, the light that reflects from the object to the film is the object beam. The light that misses the object and goes directly to the film is, by definition, the reference beam.

IMAGE TRANSMISSION OR IMAGE REFLECTION HOLOGRAMS

By placing a large lens between the object and the film, with the center of the object and the film plane at the focal point of the lens, an image is formed that, when reconstructed, lies half in and half out of the hologram plate.

TWIN-BEAM REFLECTION HOLOGRAMS

This is pretty straightforward, but remember to balance the beam ratios as in transmission holograms and measure the distances. Exposure times will be dependent upon the brightness of the beams after the ratios have been established. Here, object movement is critical. This is a good setup to try the real image.

Bleaching

Holograms that are bleached are termed *phase holograms* and are often considerably brighter than conventional amplitude holograms. The following bleaching process has been developed by AGFA for their Scientia emulsion 8E75. This reversal process utilizes the desensitized silver halide residue for making the phase holograms. This method produces bright, low-noise holograms that do not easily fade. The image contrast is almost as high as that of conventional amplitude holograms. No special developers are required. Nevertheless, you should perform the following standard bleach procedure steps:

1. Develop for five minutes with average darkening.
2. Agitate in stop bath for two minutes (one percent solution of acetic acid).
3. Wash for five minutes (check over light for spectrum).
4. Bleach for two minutes in five g of potassium bichromate ml of concentrated sulfuric acid in one liter of distilled water.
5. Wash for five minutes.
6. Clear for two minutes in 50 g of sodium (anhyd.), one g of sodium hydroxide in one liter of distilled water.
7. Wash for five minutes.
8. Desensitize in:
    a. 880 g ethyl alcohol.
    b. 100 g distilled water.
    c. 20 g glycerin.
    d. 120 mg of potassium bromide.
    e. 200 mg phenosafrinine.
9. Rinse briefly in ethyl alcohol and let dry.[26]

DON'T TRY THIS AT HOME!

You might want to try another new method of bleaching. Carefully prepare some bromine water by following this simple but very dangerous formula:

Dilute approximately two ounces of liquid bromine in one quart of water. Be very, very careful, for bromine can easily burn the skin and the vapors are extremely toxic. The bromine will require up to 72 hours to be dispersed sufficiently to produce a solution of approximately 6 percent. Extreme caution should be maintained while handling the liquid. After you have developed and fixed your hologram, immerse it in a standard Photo-Flo solution (Photo-Flo is available at any camera store) for about one minute. Then, while the hologram is still wet, place it into the bromine water, and observe the plate or film as it becomes transparent (depending on the strength of the solution, more or less time will be required). Then return to Photo-Flo and dry as usual. It is strongly suggested that you use bromine outdoors, because the vapor tends to linger in a closed area, causing a very unpleasant odor.

 NOTE: All photographic films come from the manufacturer with development inhibiting coating, not unlike the preservatives that are added to food to prevent or retard spoilage. This coating keeps the film from spoiling while it is stored. If you wish, you can remove this coating and thus hypersensitize the plate of film for a few hours by the simple procedure described here. This procedure temporarily increases the effective speed or light sensitivity of the film about three or four times without a gain in *noise*. In practical terms, this will make it possible to increase the size of your film, so as to make holograms up to 8" x 10" with a low-power laser. The procedure is as follows: Before exposing your film, immerse it for approximately two minutes in a solution of water containing three drops of 28% ammonia per liter of water containing a wetting agent such as Photo-Flo. The solution should be kept at room temperature. After the film is completely wet, remove it from the solution and let it dry without sponging. Remember, that the film must be handled in complete darkness. The film must be exposed before more than a few hours have elapsed. Proceed to expose your treated film, keeping in mind that the exposure time can be decreased for a small hologram or, conversely, that you can make a larger one in the same amount of time it usually takes to make a smaller one. If you wish to increase your speed even further (approximately 10 times), you can develop your film in Dektol that has been heated to 100 C. This latter procedure will add considerable *noise,* but it is worth experimenting with for unusual situations.

In Summary

- Laser light differs drastically from all other light sources, man-made or natural, in one basic way that leads to several startling characteristics.
- In lasers, electronic principles are applied to the visible portion of the spectrum.
- The hologram—that is, the medium that contains all the information—is nothing more than a high-contrast, very fine-grain, black-and-white photographic film.
- One of the very prevalent practical problems in the making of a hologram is object movement.

- Many holographers bleach all of their holograms, because phase holograms absorb less valuable reconstructing laser light than the absorption type and thus create a brighter image.
- In twin-beam holography, it is extremely important to measure the paths of the reference beam and object beams(s), for even if you are using the prescribed laser for holography, its coherence length is not infinite.
- It is important that your isolation table be placed on a solid floor, such as a cement slab in a garage.
- In most holograms, the object is displayed against black space; therefore, the hologram looks best with white or very light-colored objects.
- In all of the experiments presented in this chapter, no attempt at scale has been made, and a certain amount of component fabrication is necessary on some of the setups.
- Holograms that are bleached are termed *phase holograms*, and are often considerably brighter than conventional amplitude holograms.

Chapter 24 showed you how to make your own hologram right in the confines of your home. The next chapter examines the world of holography security.

End Notes

1. Christopher Outwater and Van Hamersveld, "Practical Holography," Dimensional Arts, Inc. 401 Carver Rd., Las Cruces, NM 88005, USA, 2000.
2. Ibid.
3. Ibid.
4. Ibid.
5. Ibid.
6. Ibid.
7. Ibid.
8. Ibid.
9. Ibid.
10. Ibid.
11. Ibid.
12. Ibid.

13. Ibid.
14. Ibid.
15. Ibid.
16. Ibid.
17. Ibid.
18. Ibid.
19. Ibid.
20. Ibid.
21. Ibid.
22. Matt Amberg and Tim Hecox, "Holography," University of Washington, College of Engineering, Box 352180, Seattle, Washington 98195-2180, USA, 2000.
23. Christopher Outwater and Van Hamersveld, "Practical Holography," Dimensional Arts, Inc. 401 Carver Rd., Las Cruces, NM 88005, USA, 2000.
24. Matt Amberg and Tim Hecox, "Holography," University of Washington, College of Engineering, Box 352180, Seattle, Washington 98195-2180, 2000.
25. Christopher Outwater and Van Hamersveld, "Practical Holography," Dimensional Arts, Inc. 401 Carver Rd., Las Cruces, NM 88005, USA, 2000.
26. Ibid.

CHAPTER 25

HOLOGRAPHIC SECURITY

In This Chapter

- What security holograms offer
- Level of protection
- Holographic counterfeiting
- Techniques used to counterfeit holograms
- A deterrent to contact copying holograms
- Holographic anti-piracy technology
- Enhanced CD systems
- 3-D i.d. security band holograms
- Countermeasures against hologram counterfeiting
- Holographic technology for anti-counterfeit security: present and future

Holograms belong to a class of images known as Diffractive Optical Variable Image Devices (DOVIDs), and are rapidly becoming the quintessential method of protection against counterfeiting. Highly valued for security, DOVIDs can only be produced using expensive, specialized, and technologically advanced equipment. They cannot be replicated by color copiers, computer scanning equipment, or by standard printing techniques—analog or digital—since they are governed by different physical properties (holograms diffract light, whereas print reflects it). Given that counterfeiters inherently select the path of least resistance, encountering a security hologram will frequently cause them to abandon their efforts to copy your product and move on to easier targets. Consequently, the use of hologram security devices virtually guarantees product authenticity.

Security holograms have become well established in many industries and are commonly found on a host of products and packaging, including compact discs, computer software, cosmetics, watches, and sporting goods. Other security uses include clothing hang tags, certificates (automobile registration, fine jewelry, etc.), tickets (concert, sporting event, etc.), and many kinds of identification and membership cards. You probably have a security hologram in your pocket—there has been one on every VISA® and MasterCard® produced for the past 12 years! DOVIDs also grace federal currency and passports in countries such as Austria, Bulgaria, Finland, Kuwait, Switzerland, and Russia. In the United States, you'll find high-security DOVIDs on driver's licenses in many states, as well as on postage stamps and a variety of government-issued bonds and certificates.

Security Benefits

The decision to integrate a security and authentication system into your organization's product line or printed documents is a positive marketing step, as well as a prudent economic choice. According to IHMA (International Hologram Manufacturer's Association), when properly carried out, the system should accomplish all of the following:

- Assure your customers that the product is genuine and will perform according to specifications.
- Increase or preserve sales by reducing the sale and use of counterfeit products.
- Enhance the visual appeal of the product or document and its packaging.
- Make exact product counterfeit more difficult and unlikely.
- Provide potential forensic information for prosecution.
- Establish defensive evidence against defective product and negligence claims.[1]

For a printed item, such as an identity card or document with intrinsic or exchange value:

- Ensure that only the correct individuals carry the item or document.
- Ensure that only genuine items or documents are in use.
- Make counterfeiting more difficult and unlikely.
- Provide potential forensic information for prosecution.[2]

 NOTE: See Appendix A, "International Hologram Manufacturers Association (IHMA) and List of Holographic-Related Sites," for detailed information on the IHMA.

Choosing the Right Level of Protection

The importance of adding security features to a product or document depends upon the magnitude of the potential loss. Simply adding a hologram to your product will not prevent counterfeiting, but it significantly ups the ante for the counterfeiter.

The security a hologram provides is to some extent tied to the complexity of the image; and, since not all products need the same protection, DOVIDs are produced in varying security levels. The required level can be determined by examining several factors, including actual or perceived product value; cost (in money or breach of security) if your product is bootlegged; and the likelihood that a counterfeiter will go to extreme lengths to copy both the product and its DOVID authentication. A useful analogy might be: To determine the height of a fence needed, you must consider whether its purpose is to keep school children from trampling the flower patch or to thwart professional mercenaries hired to steal your gold!

Security Levels

Depending on your security requirements, four different levels of security holographics can be obtained:

- **Fully custom holography**: Offering the highest level of security for large volume projects.

- **Custom etched holographic diffraction material**: For tightly budgeted projects.
- **Overprinted holographic diffraction foil**: For tightly budgeted projects.
- **Customized stock holography**: Offers both extremely high security and value for small to medium-sized projects.

FULLY CUSTOM HOLOGRAPHY

Fully custom holography involves the design and creation of unique, proprietary images in three dimensions. However, with setup costs ranging from $11,000 to $21,000 (depending on image complexity), fully custom holograms only become cost effective in volume orders around $21,000. Figure 25-1 shows some custom security hologram examples.[3]

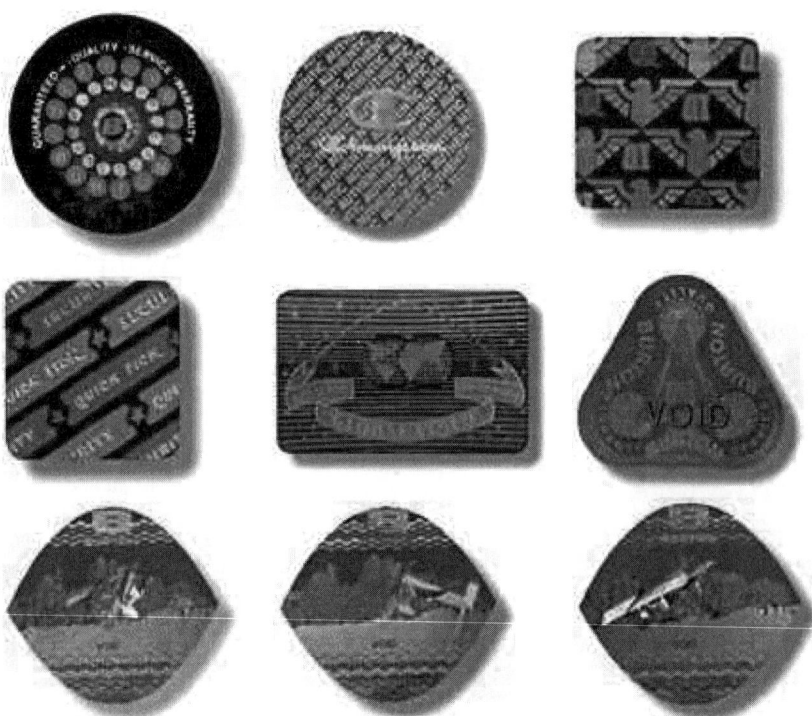

FIGURE 25-1: All International Hologram Manufacturers Association (IHMA) members subscribe to strict industry guidelines regarding the production and sale of security holograms.

CUSTOM-ETCHED AND OVERPRINTED MATERIALS

These DOVIDs are not dimensional holograms, but complex holographic diffraction patterns that emit bright rainbows of light. They can be holographically etched or overprinted with type and line art, and though a significant step down in security from customized stock images, are quite suitable for limited budget projects.

Customized Stock Holography Options

Stock security images are ready to use *as is* for most light security needs. The following part of the chapter describes a variety of customization options. The prices on the pricing page are for regular unimprinted labels. Customization price add-ons are shown separately.

CUSTOM IMPRINTING

Most holographic manufacturers offer a sophisticated and powerful way to customize the look of stock holograms, while adding significantly to their security value. For a one-time setup cost of $7,826, the manufacturer will etch your logo or ad copy into the holographic master shim, so that each hologram has your message showing as a 2D surface graphic. After setup, the prices are the same, and there usually is no setup charge on reorders. Customized images belong solely to the end user.

TAMPER-EVIDENT ADHESIVE

Tamper-evident security adhesive is available on orders of 1,000 images or more from most manufacturers. When a tamper-evident hologram is applied to a smooth, clean surface (plastic, metal, or glass), it will self-destruct if removed. The remaining portions of the original hologram will be invalid and unusable. This is an excellent treatment if a product can be outdated or canceled after issue (ID cards), and without tamper-evident adhesive, the DOVID could be removed and applied to a counterfeit product.

SEQUENTIAL NUMBERING

Sequential numbering, available on orders of 1,000 images or more, appears in black on top of the hologram in sizes of 1/16" or 1/8". Consecutive numbers can serve a variety of purposes, such as product tracking, inventory control, and multiple uses of a single stock hologram.

SPECIAL DIE CUTTING

HoloBank stock security holograms are provided on rolls as kiss-cut square or round labels ready for hand application. However, many images can be die-cut in varying shapes and sizes for an even greater level of customization. Cutting a label can be a complex issue and deserves special attention. For automatic application, the accuracy of cutting and uniformity of the repeat down the web is critically important. The prices in this catalog all assume that labels will be applied by hand. Please advise your sales representative if the project will require machine application, and they will provide a job-specific quote. All labels are supplied with acrylic pressure-sensitive adhesive.

CUSTOMIZED STOCK IMAGERY

HoloBank offers a wide range of stock security images created using state-of-the-art holographic techniques. When employed correctly, most manufacturers' stock security holograms provide the same functional level of security as fully-custom images—but, at a fraction of the cost, and for most projects, offer an unbeatable combination of protection and value.

Most manufacturers' stock security images may be purchased *as is* in various sizes, or the manufacturer can add your logo or text for a truly custom look. There are very few manufacturers in the world that offer these specific security images—which makes them extremely difficult and expensive for counterfeiters to acquire, and virtually impossible to copy exactly outside of the laboratory in which they were originally created. Furthermore, if the manufacturers customize the image with company-specific information, the customized image will never be produced or distributed without customer authorization.

Since both overt and covert security devices can be incorporated or added on to fully custom images, custom holography offers the highest possible level of security that is available, and in large volume can be extremely cost effective. You should submit a project specification (image size, quantity, use, and whether manual or machine application will be used) to your manufacturer for a quote.

All holograms employ a combination of overt and covert security components. Overt DOVIDs are visible to anyone and everyone who views them (no special experience or equipment is necessary to see the image and/or information contained in the DOVIDs). Covert elements can take many forms, from highly secure machine-readable authentication, to proprietary graphics that can be identified only by a trained eye.

Holographic Counterfeiting

Since the earliest days of market trade, counterfeit goods have existed. In order to deter the counterfeiting of currency, passports, driver's licenses, brand name products, and other instruments of value, sophisticated technologies, including holography, are employed in their manufacture and packaging.

Nevertheless, all of these technologies can be counterfeited by the 21st Century counterfeiter (see sidebar, "Counterfeiting: As Easy as Surfing the Net"). Another factor contributing to the counterfeiter's success is that there exists no readily available machine readable hologram technology to verify the authenticity of a holographic image.

■ COUNTERFEITING: AS EASY AS SURFING THE NET

They looked like typical New York City detectives, carrying guns and flashing their badges with authority. With their credentials, they gained access to secure areas at two airports and 19 federal agencies, including the CIA, the Pentagon, FBI headquarters, and the Justice and State departments. The rub: The badges and credentials were phony. And the *cops* (despite their thick New York accents) were really undercover agents from the General Accounting Office trying to determine the vulnerability of some of the nation's most sensitive sites. Their findings: It's a snap to get into even the most secure buildings with fake law enforcement badges and credentials readily available on the Internet and from other public sources.

GAO officials told a House panel that they got into 18 of the 21 allegedly supersecure buildings on the first try, and the other three sites on a second visit. At 15 sites, the agents conned their way to just outside the suites of cabinet secretaries or agency heads. To gain entry, the agents presented themselves as armed law enforcement officers; one agent always carried a suitcase. Their bogus credentials were never challenged. The agents said they could easily have carried weapons, explosives, and listening devices into the buildings. At Ronald Reagan Washington National and Orlando International airports, agents were waved around metal detectors. Attorney General Janet Reno said she was *surprised* so many federal buildings, including her own, flunked the security test. She was told that GAO operatives got into her office suite by pretending to be friends of hers from Miami.

Shaky Fakes

Besides New York detectives' badges, agents used a counterfeit federal badge and a fake drug task force badge. They also concocted phony law enforcement IDs from commercially available software packages and information downloaded from the Internet. They used a standard computer graphics program, an ink-jet printer, and photos—and then laminated them. Perhaps most troubling: The faux credentials looked nothing like the real McCoy.

The Internet offers bogus passports, birth certificates, Social Security cards, driver's licenses, college diplomas, and press credentials, along with police and FBI identification. Law enforcement officials estimate that about a third of phony identification comes from the Internet, compared with just 5 percent in 1999. The fakes can look so real that users can rip off other people's identities and fraudulently get bank loans. And, of course, Internet fake identification makes it easy for teens to get into bars illegally. During the past decade, government agencies have added security devices, such as *holograms* and bar codes, to ID documents, but Internet sites have copied those features. The Internet allows those specializing in the sale of counterfeit identification to reach a broader market of potential buyers than they ever could by standing on a street corner in a shady part of town.

It is the objective of any security device to be able to reliably differentiate the counterfeit from the authentic goods. Because of the homogenous appearance of most embossed holograms, it becomes imperative to incorporate machine-readable elements in any holographic security system.

This part of the chapter addresses the technologies used by the 21st century holographic counterfeiter, and identifies an innovative new method to defeat him or her. The inventions and technologies discussed herein provide for an economical means of defeating the counterfeit of anything of value. This technology is impossible to counterfeit!

TECHNIQUES USED TO COUNTERFEIT HOLOGRAMS

The level of sophistication in today's security hologram market is still a significant deterrent to the counterfeiter. The use of three-dimensional models, multiple exposure, and hybrid techniques, provide for holograms that are exceedingly difficult to recreate from scratch; however, all samples of security holograms that have been evaluated to date can be readily counterfeited by a competent holographer. There are three ways to optically counterfeit a hologram, and one method to counterfeit the surface relief pattern of an embossed hologram:

- Contact copy counterfeit technique
- Air copy counterfeit technique
- Re-origination counterfeit technique

Contact Copy Counterfeit Technique

Contact copying is the simplest way to optically copy a hologram. In this case, the holographic emulsion is brought into direct contact with the hologram. The appropriate wavelength of laser light is directed though the plate to expose the emulsion.

This type of counterfeit hologram is relatively easy to create with little equipment and training necessary. The primary advantage to the counterfeiter is that the contact copied hologram can be produced within a day. The disadvantage to counterfeit holograms made with the contact copy technique is that the hologram produced is not as bright as the original and contains more optical noise.

A trained eye can easily tell the original from a fake when compared side by side. An untrained eye will be fooled almost all of the time by a contact copy.

Air Copy Counterfeit Technique

The air copy counterfeit technique is more complicated to produce than the contact copied hologram, requiring a sophisticated laboratory and a competent holographer. The advantage to this type of counterfeit copy is that it appears visually indistinguishable from the original. All encrypted information and other standard security features incorporated in the original hologram are replicated.

Re-Origination Counterfeit Technique

This is another advanced counterfeit technique requiring a skilled staff and comprehensive laboratory facilities. Here the hologram is recreated from scratch. The recreation of two-dimensional graphic elements is straightforward. The recreation of three-dimensional elements may be a difficult, however, not an insurmountable task to achieve.

For some types of holograms, this would be the best method of counterfeiting. For some other types of holograms (3D), the air copy would be the counterfeit technique of choice.

MECHANICAL COUNTERFEITING

Unlike the volume hologram that most people are familiar with, embossed holograms consist of a surface relief pattern. If the relief is accessible, it can be used as a master plate and electro-formed via the Edison Process. The resulting hologram is, in fact, an original master.

This type of counterfeiting does not require any holographic technology, providing for wide market access to the necessary equipment and training. To prevent this type of counterfeiting, the surface of the hologram is coated with an adhesive layer.

CREATING HOLOGRAMS THAT CANNOT BE COUNTERFEITED

The need clearly exists for a holographic technology that is impossible to counterfeit. A similar need exists for a machine-readable technology to differentiate an authentic hologram from one that may be visibly indistinguishable from the original. These two technologies when combined together create a totally secure environment.

MACHINE-READABLE HOLOGRAM TECHNIQUES

There are two distinct categories of machine-readable holograms; those that reconstruct and compare phase information, and those that reconstruct and compare amplitude information. For security purposes, those technologies that are dependent upon phase correlation are significantly more difficult to counterfeit than systems based upon amplitude correlation.

A DETERRENT TO CONTACT COPYING HOLOGRAMS

A number of different holographic coatings have been suggested to combat the contact counterfeiter, including evaporated *oil slick* coatings and colored lines. These techniques may provide for a unique appearance to enhance marketing value; however, these measures only provide for partial security, as the machine-readable element will never be totally protected. Similarly, other display elements in the hologram can be readily counterfeited using the preceding referenced techniques.

PUTTING AN END TO CONTACT COPYING

To effectively combat the 21st-century counterfeiter, a filter that transmits light only at 632.8 NM (red) light and longer wavelengths is employed as a lacquer coating on the surface of the hologram. This filter cuts off all practical laser wavelengths useful to the counterfeiter.

The drawback with this *red only* hologram is that it is visually unattractive and becomes visually homogenous to other *red only* holograms in the marketplace. Nevertheless, while the display element may be counterfeited with the preceding referenced counterfeiting techniques, the machine-readable elements cannot be re-originated, or counterfeited using any other known optical technologies.

UNITING TECHNOLOGIES TO DEFEAT THE COUNTERFEITER

The preceding referenced technologies when combined together provide for a hologram that is impossible to counterfeit. Furthermore, the incorporation of a machine read device provides for complete accuracy in ascertaining the hologram's validity. The machine and hologram together create a totally secure holographic environment.

The cost of creating a hologram utilizing this technology is not significantly higher than other conventional holographic technologies. Because it is the only known totally secure holographic imaging technique, this technology is the one of choice for all security hologram images.

Holographic Anti-Piracy Technology

When Nimbus CD International, Inc. began producing CDs at its Wales facility in 1983, it was only the third operation of its kind in the world. Since then, the company has continued to move aggressively in the direction of new technologies, and has already made significant investments in Enhanced CDs and DVD.

Aside from its plant in Wales, where an expansion took place in 1997, Nimbus has CD manufacturing facilities in Charlottesville, Virginia, Provo, Utah, and seven lines in the newest facility located in Sunnyvale, California. The Sunnyvale facility evolved as a result of the August 1995 purchase of HLS Duplication, Inc., a provider of turnkey software duplication services. That operation, which produces 500,000 high-density floppy disks per day, functions as a subsidiary of Nimbus under the name Nimbus Software Services.

ENHANCED CD SYSTEMS

With all of the confusion about Enhanced CD (a disc that contains both audio and CD-ROM content) in the marketplace, there is no confusion at Nimbus. The company has been offering Enhanced CD mastering and replication since 1997. Although the jury is still out on Enhanced CD, Enhanced CD will gain momentum as some of the hardware issues are resolved.

The company's early entry into this market is owed to the selection of Nimbus as a beta test site by mastering lathe software developer DCA in 1997. Up until 1994, Enhanced CD could only be made using desktop CD-R equipment, because no manufacturing plant had the ability to master multisession discs.

PIRACY AN AREA OF CONCERN

Nimbus is also working diligently in the area of piracy, where its 3-D i*d hologram technology, developed in conjunction with Applied Holographics PLC, is beginning to make inroads. According to research by various international and national recording and software industry associations, the losses due to piracy and counterfeiting for the software industry as a whole, total several billion dollars annually. For example, The Software Publishers Association placed the amount of losses at $4.0 billion in the United States, while the new Business Software Alliance reports a staggering $37.4 billion lost worldwide.

The impact these losses have on consumers is staggering, due to higher prices of legal sales and cuts in funding for the development of new products and projects. When combined with steps Nimbus already takes to help support the integrity and security of their customers' work (including source identification coding of their manufacturing equipment, onsite verification of disc content to customer project, and bonded delivery services), the use of 3-D i*d provides one of the most effective piracy deterrents available in the CD business.

3-D I*D HAS MANY USES

During the mastering stage, a hologram can be put on the inner mirror band of a disc. Another option is full-face holograms applied to the disc during the replication stage.

Just like artwork, the 3-D i*d edge-to-edge application covers the entire surface of the disc without taking up capacity. Moreover, since the hologram is part of the disc, it cannot be effectively copied. Nimbus will not only license the technology, but also lease the holographic equipment under the 3-D i*d brand name. A stand-alone hologram unit for edge-to-edge applications is available for a one-time lease payment of $450,000, plus a royalty of 7 cents per disc. Fully integrated units are on a custom quotation basis. For the inner mirror band hologram application, Nimbus furnishes the masters for a $560 mastering fee, plus $60 for each required stamper and a 5 cents per disc royalty.

In April 1998, San Francisco, California-based Sirius Publishing became one of the first companies to use the 3-D i*d hologram with its *Treasure Quest* multimedia game mystery. An inner security band was to be incorporated into the disc to deter counterfeiting as well as to help verify winners. More than 250,000 *Treasure Quest* CD-ROMs containing the 3-D i*d application were manufactured in the initial run. In addition, an accompanying audio CD containing additional game clues also has the hologram.

Thus far, Nimbus has hosted four forums in different areas of the country where they have invited manufacturers, record labels, and software companies to view their hologram technology. Interest has been overwhelming, for both the piracy deterrent and the stunning graphics capabilities.

A number of ways to counterfeit holograms are described next, along with proposed countermeasures. Probable effectiveness of the countermeasures is also discussed.

Hologram Counterfeiting Countermeasures

It is fashionable to think of holograms as being virtually uncounterfeitable. The fact of the matter is that holograms in the form currently used in security are almost trivially easy to counterfeit. Although there are no known instances of counterfeited holograms in the security business per se, there are undeniable cases that commercial hologram designs have been counterfeited. Thus, the purpose of this part of the chapter is to explore the more straightforward methods by which holograms can be counterfeited, and to offer ways that hologram counterfeiting may be prevented.

TYPES OF HOLOGRAMS

There are two types of holograms currently produced in substantial quantities: embossed holograms and Denisyuk holograms. Embossed holograms are the type used in most security applications, such as credit cards and passports. Embossed holograms are easily mass-produced at a low cost. They are very thin when used in the form of hot stamping foil, but are highly durable. An embossed hologram is a microscopically detailed surface relief pattern analogous to a phonograph record or a compact disk.

Denisyuk holograms are another type of hologram that has been considered for security applications, but has not yet received wide use. Each Denisyuk hologram must be made with a laser in a high-resolution holographic recording medium, so this type of hologram is unavoidably more expensive and slower to produce. Also, Denisyuk holograms are relatively thick in their present form and are rather easily damaged.

COUNTERFEITING METHODS: EMBOSSED HOLOGRAMS

From the point of view of a counterfeiter, embossed holograms are potentially the path to easy money. There are four straightforward ways to counterfeit embossed holograms, and you can be sure that these ways will be learned by counterfeiters.

- Holographic copying in one step
- Holographic copying in two steps
- Mechanical copying
- Re-origination[4]

These methods will be examined in greater detail later. Each method is relatively simple and can be used to produce a passable copy of any embossed hologram that exists today.

Denisyuk Holograms

Denisyuk holograms are more difficult to counterfeit in some ways than embossed holograms are. First, Denisyuk holograms cannot be mechanically replicated or counterfeited, because their effect depends upon a complex three-dimensional structure that extends through the thickness of the recording material. Denisyuk holograms can only be copied by optical methods:

- Holographic copying in one step (contact printing)
- Holographic copying in two steps
- Re-origination[5]

Again, it is relatively simple to make a passable counterfeit of a Denisyuk hologram by these methods.

COUNTERMEASURES

The counterfeiting methods exploit certain properties that today's holograms have. If holograms for security purposes are made differently, counterfeiting by those methods can be effectively prevented. Briefly, these methods are:

- Use materials in hologram copies that make the laser an ineffective tool for the counterfeiter.
- Covertly store machine-readable information in the holograms to make re-origination ineffective.
- Use materials and structures in and with holograms to make mechanical copying ineffective.
- Use imagery that cannot easily be re-originated.
- Include variable information in holograms.
- Alter the processing of the hologram to introduce distortions that are difficult to copy.[6]

PRESENT MANUFACTURING AND USE OF HOLOGRAMS IN SECURITY

Embossed holograms are made by the process illustrated in Figures 25-2 and 25-3.[7] The embossing is normally done into either a plastic film such as PVC or polyester and used for adhesive labels, or into a hot stamping foil base. The holograms are then applied to a document or product, and the

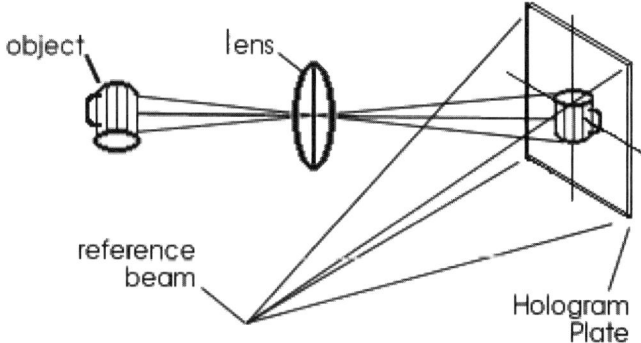

FIGURE 25-2: Forming a hologram.

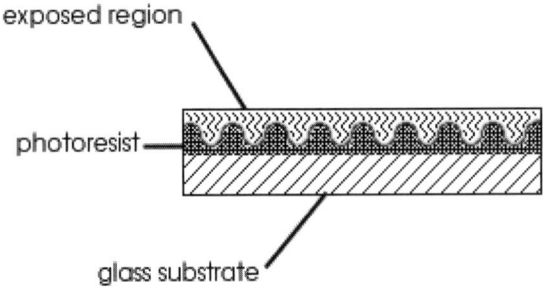

FIGURE 25-3: Exposed photoresist plate.

presence of the hologram is intended to be a reliable indication that the document or product is valid.

Denisyuk holograms are recorded into silver halide film, photopolymer film, or dichromated gelatin film. The resulting hologram is finally laminated into a card or applied to a product as a tag or label.

COUNTERFEITING METHODS AND COUNTERMEASURES

Today's holograms are rather easily counterfeited by the following processes: First, re-origination is the time-honored method of counterfeiting. It involves creating a faithful imitation of the original artwork, using a valid copy as a guide. The Visa hologram, for example, could be very straightforward to copy by re-origination. The image is there on every card for a skilled sculptor to recreate as a solid model. Once a good model is made, it might be easy for any skilled holographer to make holograms and copies that will pass even a careful inspection. Visa, seeking a second source for their holograms, gave Light Impressions the original Visa model after it had been broken. Visa repaired it and made holograms that American Banknote has described as indistinguishable from their own holograms made from the unbroken model.

Kinegrams, the type of holograms made by Landis & Gyr, are significantly more difficult to re-originate, because they are not made from a model. They are instead a type of *stereogram*, made from a sequence of two-dimensional images. A sophisticated counterfeiter, however, would be able to re-originate a kinegram. These holograms, being embossed holograms, are subject to the same counterfeiting methods as other embossed holograms.

Second, one-step copying is accomplished by simply contact copying the hologram. In one approach, a holographic film is placed against the security hologram and illuminated with a laser. For a Denisyuk hologram, the film is merely developed and used directly. For an embossed hologram, the film is developed and metalized, and the result is a passable imitation of the original hologram. In another approach, a photoresist coated film is placed against the hologram and illuminated with a laser. The photoresist is developed and electroplated. The electroform is then used to emboss counterfeit copies of the hologram.

Third, two-step copying is accomplished by illuminating the original hologram with a laser and recording the diffracted light in a second hologram. Next, the second hologram is illuminated to produce a replica of the image in the original hologram, and the image is used to make a duplicate of the original hologram. For a Denisyuk hologram, the duplicate is the counterfeit. For an embossed hologram, the final hologram is recorded in photoresist so that it can be electroplated and used to emboss counterfeit copies of the hologram. This process is only slightly more complicated than contact copying, and produces excellent copies.

Fourth, mechanical copying is accomplished by uncovering the embossed surface of the hologram and using it as a form to produce an electroplated copy. The electroplated copy can then be used to emboss counterfeit copies of the original hologram. If the counterfeiter can obtain unused original holograms, the process is trivially easy. If he or she only has a Visa card to work from, the process is a bit messier.

Countermeasures: Complex Imagery or Hidden Information

Re-origination can be prevented only by including hidden information, or by making the image so complicated that it is not worth the counterfeiter's time to duplicate it. Hidden information is of value only if the counterfeiter cannot find it or duplicate it. Hidden information is of only limited value unless it is detectable at the point of use of the security document or credit card. Therefore, effective use of hidden information requires some sort of relatively simple and inexpensive reading device or decoding device.

Complicated imagery can be produced by the stereogram process: the kinegrams of Landis & Gyr are an example. These stereogram processes all involve making a complex image as a combination of simple two-dimensional images.

Without a reader, complex imagery only works if it is simple enough that an untrained observer can detect an imperfect copy. Most manufacturers, though, are not convinced that the requirements of complexity and simplicity leave anything in between. Many systems that require a decoding device can be beaten by a counterfeiter who steals the device, because the method of encoding is usually implied by the method of decoding.

VARIABLE PROCESSING PARAMETERS

A way has been proposed to produce a Denisyuk hologram that would be extremely difficult to copy using either one-step or two-step copying. The proposal requires that you randomly vary the exposure, development time, or other processing parameters to produce variable shrinkage across the hologram. The result is a hologram whose color varies from point to point. A variation of this method has been demonstrated by Applied Holographics, resulting in a hologram whose color or brightness varies according to a predetermined pattern.

Because a laser is monochromatic, the brightness of a simple optical copy of a variable shrinkage hologram varies dramatically with the shrinkage of the film. In fact, if the film is thick enough or the shrinkage is extreme enough, the copy will have no image at all in most of its regions. None of the methods discussed can prevent direct copying, simply because even the most complex imagery, even if hidden or encoded, is faithfully copied by one-step, two-step, or mechanical methods.

It is probably possible to copy a variable-thickness Denisyuk hologram by using a tunable dye laser to record a series of copies in different wavelengths, but it would be a difficult task even for a highly skilled holographer. However, the greatest counterfeiting risk for this type of hologram would be the ease of re-originating the hologram. Thickness variations, unless controlled carefully in ways that are difficult to simulate and easy to recognize, could be produced easily.

VARIABLE INFORMATION

Holograms are riskier to counterfeit if they include variable information such as serial numbers, dates, or encoded personal information. A liquid crystal display may be used to record variable information in a Denisyuk hologram. Because each Denisyuk hologram is separately recorded, this type of variable information is completely practical.

Because embossed holograms are mechanically reproduced, it is not practical to holographically record variable information in them. However, it is possible to record variable information in the process of applying holographic hot stamping foil to a substrate. It is also possible to demetalize an embossed hologram selectively, and therefore perform a sort of *reverse printing* operation for providing variable information.

It would not be very difficult for a counterfeiter to use these simple variable information techniques, if the variable information is in the form of ordinary numbers, bar codes, or words. In order for these techniques to be effective, the variable information must be recorded in a form that is not easily reproduced. To be highly promising as a high-security feature in future holographic security systems, let's consider a variation on this approach, which will be discussed later.

SPECIAL MATERIALS

Optical methods of copying holograms depend upon the ability of a single laser beam to reconstruct and record the image from a valid hologram. The use of variable processing parameters to produce variable-thickness Denisyuk holograms as described previously is one way to greatly complicate the copying process.

Another class of methods for complicating the optical copying process is to make the hologram of materials that alter the polarization of light or selectively filter out or scatter particular wavelengths of light. Alternatively, the hologram can be overcoated or laminated under such materials.

The use of special materials can complicate the optical copying process, but cannot make optical copying impossible. The combination of tunable dye lasers and polarization rotating optics in the hands of a dedicated and highly skilled holographer is enough to permit counterfeiting.

MULTIPLY CONNECTED HOLOGRAMS

Mechanical copying is the simplest method for counterfeiting. Denisyuk holograms cannot be copied mechanically, but embossed holograms as they are made today can be mechanically copied rather easily. There may conceivably be special formulations of the plastics for hologram embossing or for making credit cards that will prevent the use of ordinary organic solvents for removing the substrate and leaving the embossed hologram surface intact.

However, the most effective approach to preventing mechanical copying of embossed holograms is to make the holograms multiply connected. Multiple connectivity means that the hologram is composed of dots or else is punched full of holes.

Multiple connectivity makes the use of solvents or other means to uncover the embossed surface of a hologram. It is apparent that an attempt to dissolve one side of the hologram away to uncover the aluminum surface would very likely result in obliterating parts of the hologram by undercutting its dots, even if the aluminum were an effective barrier to the solvent. Any unevenness or patterning in the thickness of the coating would compound the complications.

Multiply connected holograms would be virtually immune to mechanical copying, but would not be intrinsically immune to any of the other counterfeiting techniques. Optical copying by one-step or two-step techniques would work just as well on a multiply connected hologram as any other hologram, but then to make a multiply connected hologram from the copy would be a complicated task.

Combined Countermeasures

None of the countermeasures described in the preceding, taken alone, is immune to all of the counterfeiting methods. Certain countermeasures in combination, however, promise to be highly effective against all of the counterfeiting methods foreseen.

FOR DENISYUK HOLOGRAMS

Denisyuk holograms having patterned variable thickness and containing encoded and hidden variable information will be immune to mechanical copying and difficult to copy optically. If the variable information is hidden or encoded and machine readable, invalid or re-originated counterfeit holograms will be detected at the point of use by a decoding and validation system.

Denisyuk holograms tend to be relatively thick, and are not hot stampable. They must be protected from humidity by overcoating or laminating. They are relatively expensive, compared to embossed holograms. On the other hand, a Denisyuk hologram can (in principle) be originated in a few minutes at low cost by a relatively uncomplicated system.

FOR EMBOSSED HOLOGRAMS

Embossed holograms combining the features of multiple connectivity and machine-readable encoded variable information (overcoated with a polarization-varying material) will be extremely difficult to copy mechanically or optically. Re-originated counterfeits or invalid holograms will be detectable at the point of use by a decoding and validation system (see the sidebar, "The Ideal Security Hologram").

■ THE IDEAL SECURITY HOLOGRAM

Embossed

1. Will include:
   - Variable information
   - Buried multiply connected hologram
   - Encoded information
2. Will be machine readable.
3. Will not be *human readable* without a reading device.

Denisyuk

1. Will include:
   - Variable Information
   - Encoded information
   - Controlled variable thickness
2. Will be machine readable.
3. Will not be *human readable* without a reading device.[8]

A disadvantage of embossed holograms is that the production of an embossing master is a time-consuming and relatively expensive process. For small-volume applications, embossed holograms have no apparent security advantages over Denisyuk holograms.

For large-volume applications such as credit cards and telephone cards, embossed holograms have a decisive cost advantage. Embossed holograms may be applied by hot stamping, and they are thin enough to be suitable for use on thin, flexible items such as paper currency and travelers' checks. The greatest advantage of embossed holograms in some applications may be the fact that they are more durable than Denisyuk holograms.

With that in mind, as you know, holograms have served as anti-counterfeit security devices since the late 1970s. This final part of the chapter explores the factors influencing the effectiveness of holograms in security applications, considers probable technological advances in the field of holography, and projects the impact these advances may have on anti-counterfeit security in the future. This part of the chapter will not deal with specific techniques of hologram counterfeiting or anti-counterfeit techniques used today. Rather, it focuses on the kinds of new advances necessary in order to keep holography at the forefront of anti-counterfeit technologies.

Holographic Technology for Anti-Counterfeit Security: Present and Future

The subject of this part of the chapter is holographic anti-counterfeit security technology. It is helpful to begin by pointing out that anti-counterfeit security is not a product or a technology; it is a dynamic, evolving system. Anti-counterfeit security is an attempt to prevent valuable products or documents from being copied or falsified. The term *copyright owner* is used generically to mean anyone who has an interest in preventing counterfeiting, such as a government, a name brand manufacturer, or an intellectual property owner. The copyright owner, the product being counterfeited, the end users of the products, the distribution channels, the methods or technologies used to deter counterfeiting, and, finally, the counterfeiter, are all part of the anti-counterfeiting security system.

Any analysis of anti-counterfeit security (see the sidebar, "Anti-Counterfeit Security Is an Evolving System") should begin with the assumption that nothing humankind can make is uncounterfeitable. Given enough time and money, people can counterfeit anything that people can make. All of the factors and issues in anti-counterfeit security boil down to counterfeit deterrence, which in turn boils down to two economic principles: If the counterfeiter cannot make a profit by counterfeiting a product, counterfeiting will not be a problem; and if a counterfeit deterrent costs more than the losses due to counterfeiting, copyright holders will not pay for the deterrent.

> ■ **ANTI-COUNTERFEIT SECURITY IS AN EVOLVING SYSTEM**
>
> Intellectual property owners continually introduce new technologies and procedures to prevent known methods of counterfeiting.
>
> Counterfeiters continually learn to copy those technologies and circumvent those procedures.[9]

These two principles reduce to one fundamental principle that drives the competition between counterfeiters and copyright holders: Products that attract counterfeiters are expensive to create and cheap to produce, but have high value. The counterfeiter usually avoids the cost of creation and sometimes can produce a cheaper product. The ratio of value to cost must be high for the copyright holder and low for the counterfeiter, or else the copyright holder will lose to the counterfeiter.

Anti-counterfeit technologies are (or should be) designed to ensure that counterfeiting is too expensive to be profitable because of the risk of getting caught and/or the cost of overcoming the anti-counterfeiting technologies. The counterfeiter often has an advantage, though, because he or she does not have all the costs that the copyright holder has in creating and producing the product. An anti-counterfeit technology needs to reverse the counterfeiter's advantage: it must make the counterfeit product more expensive to produce than the valid product.

Time is a crucial parameter in the anti-counterfeit/counterfeit cycle (see the sidebar, "The Counterfeit/Anti-Counterfeit Cycle Is Driven by Economics"). In the long run, a counterfeiter will always be able to duplicate any security device such as a tag, label, seal, or special packaging material, and in the long run, the counterfeiter will be able to produce it just as inexpensively as the copyright holder. To the counterfeiter, time and cost are directly connected. It is normally vastly more expensive and risky to counterfeit a new security device in a few days than in a few months. Then, if the copyright holder changes the security device every few weeks, the counterfeiter is unlikely to be able to make a profit; and the copyright holder wins the race.

■ THE COUNTERFEIT/ANTI-COUNTERFEIT CYCLE IS DRIVEN BY ECONOMICS

- Products for which counterfeiting is a problem are those with a high value-to-cost ratio.
- Typically, the creator of the product has made a large investment to create the product, but the counterfeiter can copy the product by reverse engineering with a much smaller investment.
- If the counterfeiter can make a high-quality counterfeit at low cost, he or she has a big advantage over the copyright owner: his or her profit margin can be much higher.
- The end user of the product often wants counterfeit products because they are less expensive.
- The counterfeiter can afford to spend more money overcoming an anti-counterfeit security system, than the copyright owner can afford to spend for anti-counterfeit security! [10]

PAST AND PRESENT HOLOGRAPHIC ANTI-COUNTERFEIT TECHNOLOGY

Holography offers certain advantages as a security device. First, producing an original hologram is expensive and time consuming, but mass-produced replicas from the original are relatively inexpensive. Second, the equipment and skills necessary for mass-producing holograms has been fairly difficult to obtain in the past. Third, holograms look distinctively different from printed labels. Fourth, the tools that counterfeiters have traditionally used for counterfeiting (the camera and the printing press) do not work on holograms.

At the same time, holography has disadvantages. Because most holograms are embossed, mechanical copies are relatively easy to make. A skilled holographer can make optical copies of holograms that are very difficult to distinguish from the originals (see the sidebar, "What the Copyright Owner Wants").

■ WHAT THE COPYRIGHT OWNER WANTS

Examples:
- Security tags or labels that are inexpensive and mass producible.
- Security tags or labels that are impossible to counterfeit.
- A security system that does not affect his or her business procedures much.[11]

Once a good master copy has been made, high-quality replicas are very easy to make. Despite some disadvantages, holography is currently an effective counterfeit deterrent. However, if holography companies want to continue to enjoy a high share of the anti-counterfeit security market, they must look to the future. They must realize that today's hologram technology has become relatively easy and inexpensive to obtain; and hologram counterfeiting will surely become both feasible and profitable in the relatively near future. Holographic anti-counterfeiting technology needs to make new advances, comparable to the initial advance in 1980 when Light Impressions introduced embossed holograms to the world.

FUTURE HOLOGRAPHIC ANTI-COUNTERFEIT TECHNOLOGY

An effective anti-counterfeit security device will take longer to reverse engineer than is required to create it in the first place. It will be accompanied by an automated and reliable validation process that allows for updating of security tag or label designs in a time that is short compared to reverse-engineering time.

Some excellent techniques have been developed in the data encryption field, by which a message can be distributed in an encrypted form, along with a public key that allows the message to be read. Although the message is easy to read, the method for encrypting the message in the first place is dependent on a separate key that is kept secret. Therefore, it is very difficult to produce a fake message containing new information.

Current holographic anti-counterfeit security methods usually use mass-produced, identical embossed holograms. An advanced holographic security label or ID card will not be mass produced. Instead, each hologram will be unique, bearing a unique encrypted *signature* whose validity can be checked electronically. This advance will require several significant advances in holographic technology.

In order to be affordable, the new hologram labels will need to be recorded on an inexpensive, high photographic sensitivity, rapid-development medium. Photo polymers are currently too expensive, but they are almost sensitive enough and they develop very rapidly. Silver halide films are too expensive, they are environmentally unstable, and require toxic chemicals for developing. Either lower-cost, higher sensitivity photo polymers or some other advanced recording material will be developed in the near future.

The new hologram labels will need to be recorded using a high-speed optical system capable of creating three-dimensional, multicolored holographic images from computer-generated imagery in a small fraction of a second. There are *one-step* holographic stereogram recording techniques that will make this possible. Although there will be some security in the materials and in the recording systems used to make these labels, the biggest security factor will be an encrypted signature built into the holographic image. Since each hologram is created *on the fly*, each hologram will have its own unique, traceable signature.

This combination of technologies will become available within the next five years. It will cost more than embossed holograms, but will produce tags, labels, and ID cards that are relatively inexpensive and extremely difficult to counterfeit. Learning how to make the recording material or having the recording system will not be enough; the counterfeiter will still have to crack the encryption key. In addition, the copyright owner can change the key for each production run. Modern data communication systems used to verify credit cards by telephone can easily be used to track labels and ensure not only that counterfeits will be detected, but also that many types of fraud such as re-used labels can be detected (see the sidebars, "Proposed Specifications for a Future Ideal High-Security ID Card System" and "Proposed Specifications for Future Product Authentication Security System").

For lower-security applications, the security value of embossed holograms can be improved without greatly adding to the cost by improving the properties of adhesives and substrate and laminate materials. Several techniques can be employed to make mechanical or optical copying much

---

■ PROPOSED SPECIFICATIONS FOR A FUTURE IDEAL HIGH-SECURITY ID CARD SYSTEM

Each card should gave:
- A holographic 3D portrait of the individual—further containing a one-way encrypted machine-readable hologram of the variable card information.
- A magnetic stripe containing variable information in a one-way encrypted form

The card would be supported by the type of computer network used with credit cards: local magnetic stripe/hologram readers and central servers.

Advantages:
- Copying is useless, since the holographic portrait must match the cardholder's face.
- A holographic version of the card's variable information prevents alteration of the printed variable information.
- Onsite validation provides high security even if data link is out.
- One-way encryption of variable information into the hologram and the magnetic stripe detects alteration of the information on the magnetic stripe.
- Electronic communications enable cross checks to track card usage.

Feasibility:
- Requires advances in holographic imaging systems; these have been demonstrated experimentally.
- Requires advances in holographic recording materials; these have been demonstrated experimentally.
- Probable! Will be commercially available in three to four years.[12]

> **■ PROPOSED SPECIFICATIONS FOR FUTURE PRODUCT AUTHENTICATION SECURITY SYSTEM**
>
> **Each tag or label should have:**
>
> - A very high spatial frequency hologram incapable of being manufactured on today's narrow-web embossing equipment.
> - Visible features distinct from ordinary holograms.
> - Serial holographic number.
> - Covert features.
>
> **Hand-held reader should:**
>
> - Read the serial number.
> - Detect the covert features.
> - Record or transmit the serial number and covert features.
>
> **Advantages:**
>
> - Cannot be mass-replicated on common hologram embossing equipment.
> - Serial number allows distribution channel tracking as well as detection of replicas.
> - Covert features complicate the counterfeiter's task.
> - Hand-held reader enables connection to data collection/analysis system.
>
> **Feasibility:**
>
> - Requires development of improved embossing system (currently under development).[13]

more difficult. For example, extremely high frequency holograms can be made that are only embossable on special equipment. If these improvements are incorporated, embossed holograms will continue to be useful, low-cost anti-counterfeit security devices for decades into the future.

In Summary

- There are several methods for counterfeiting today's security holograms.
- Though there is no indication that those methods have been used yet by counterfeiters in the security field, the proliferation of holo-

graphic technology and the high level of activity in the field will surely lead to hologram counterfeiting.

- In the commercial sector, hologram counterfeiting and copying have been occurring for years.
- Straightforward methods exist for preventing the counterfeiting of holograms by these methods.
- In particular, holograms made with machine-readable variable information, either as variable-thickness Denisyuk holograms or as multiply connected embossed holograms, should be able to provide a level of security well above that of today's optical security devices, and should provide protection against sophisticated counterfeiters for many years.
- Holograms have been useful anti-counterfeit devices since the late 1970s.
- Hologram embossing, developed in 1980, made holograms inexpensive enough that they have now been used on everything from passports to breakfast cereal packaging, and from currency to cosmetics, to deter counterfeiters.
- The spread of hologram technology will soon reduce the value of ordinary embossed holograms as anti-counterfeit devices. On the other hand, advances foreseen in recording materials, hologram recording systems, and embossing substrate materials should greatly increase the security value of holograms within three or four years.

Chapter 25 examined the world of holography security. The final chapter of this book looks at holographic results and future directions. It looks at holographic neural networks; aeronautical engineering; the dynamic structure of holographic space; autostereoscopic holographic displays and computer graphics; digital holography systems; and interactive virtual reality: holodecks.

End Notes

1. Peter Scheir, "HoloBank Security Holograms," AD2000, Inc., 780 State Street, New Haven, CT 06511, USA, 2001.
2. Ibid.
3. Ibid.

4. Steve McGrew (New Light Industries, 9713 W. Sunset Hwy, Spokane, WA 99204, USA), "Countermeasures Against Hologram Counterfeiting," Holographic Dimensions, Inc., 16115 Southwest 117th Avenue, #21-A, Miami, FL 33177, USA, 2000.
5. Ibid.
6. Ibid.
7. Ibid.
8. Ibid.
9. Ibid.
10. Ibid.
11. Ibid.
12. Ibid.
13. Ibid.

CHAPTER **26**

SUMMARY, CONCLUSIONS, AND RECOMMENDATIONS

In This Chapter

- Holographic neural networks
- Aeronautical engineering
- The dynamic structure of holographic space
- Autostereoscopic displays and computer graphics
- Digital holography system
- Interactive virtual reality: holodecks

Among other topics presented in this final chapter, let's first look at an analysis of a prototype neural network system of multifaceted, planar interconnection holograms and opto-electronic neurons. This analysis shows that a hologram fabricated with electron-beam lithography has the capacity to connect 6,700 neuron outputs to 6,700 neuron inputs, and that the encoded synaptic weights have a precision of approximately five bits. Higher interconnection densities can be achieved by accepting a lower synaptic weight accuracy. For systems employing laser diodes at the outputs of the neurons, processing rates in the range of 45 to 720 trillion connections per second can potentially be achieved.

Holographic Neural Networks

A key element of most neural network systems is the massive number of weighted interconnections (synapses) used to tie relatively simple processing elements (neurons) together in a useful architecture. The inherent parallelism and interconnection capability of optics make it a likely candidate for the implementation of the interconnection process. While there are several optical technologies worth exploring, this chapter has concentrated on an architecture of opto-electronic processing planes interconnected with multifaceted, planar holograms. In this chapter, the operational parameters of a prototype system based on this architecture are examined. The parameters investigated include the neuron density of a processing plane, interconnection capacity of the hologram, dynamic range of the synaptic connections, connection rate, and output power requirements for each opto-electronic neuron.

SYSTEM ARCHITECTURE

The general system layout is depicted in Figure 26-1.[1] In this system, a 2D array of neuron outputs is interconnected to a 2D array of neuron inputs through a multifaceted planar hologram. An individual neuron within the 2D array emits a light beam that illuminates a single subhologram. The diffraction pattern produced by that subhologram provides the interconnections from that neuron to other neurons in the network. An array of subholograms, one subhologram for each neuron, encodes all the inter-

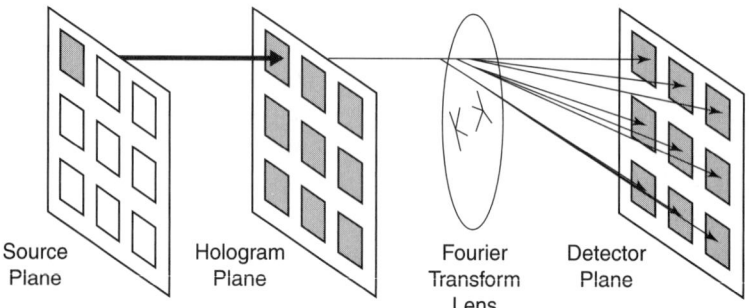

FIGURE 26-1: Illustration of the general system architecture. Each neuron output in the source plane illuminates a single subhologram, which forms weighted connections to the other neuron inputs in the detector plane.

connections in the network. In Figure 26-1, the outputs and inputs of the neurons are shown in separate planes. This could represent, for example, two layers of a feed-forward network. For a feedback system, the inputs and outputs would be combined into a single integrated device, and the outputs directed to the inputs with the appropriate optics.

Holographic Interconnections
Each subhologram in the interconnection hologram is designed as a Fourier transform hologram, so that the connection pattern is independent of subhologram position. For this architecture, the connection weight values are fixed, and learning or specification of the synaptic weights is done prior to building the network. The holograms are computer generated and designed to be either binary transmission masks or multilevel phase masks. The hologram mask has a finite number of resolvable elements (finite space-bandwidth product (SBWP)), which limits the amount of information that can be encoded in a subhologram and the number of subholograms that can placed on the mask. Since each subhologram connects a single neuron output to an array of neuron inputs, it must be able to encode a number of interconnections equal to the number of neurons in the array, and therefore, the SBWP required must be at least equal to the number of neurons. Additional SBWP may be necessary to provide sufficient accuracy in the interconnection weights; to encode a carrier; to provide light confinement in each detector cell; to encode the neuron and synaptic weight polarities; and to separate individual subholograms so as to reduce cross-talk.

The goal of the computer-generated hologram (CGH) algorithm is to produce a hologram that meets the fabrication constraints (finite SBWP, discrete number of phase levels, minimum feature size) and the diffraction plane constraints (proper diffraction geometry, low error in the interconnection strengths, high diffraction efficiency). Algorithms that iteratively produce phase holograms include those based on the Gerchberg-Saxton technique, the simulated annealing technique, and a modified technique that consists of hologram generation using the Gerchberg-Saxton technique followed by a random-search error minimization. This last technique, which we call the Gerchberg-Saxton preconditioned random-search (GSPRS) technique, produces the lowest interconnection error and highest diffraction efficiency of all the techniques investigated. The GSPRS design algorithm was used exclusively in the studies reported in this chapter.

Opto-Electronic Neurons

Each neuron is composed of an input summing port, a nonlinear transfer device, and an output port. In an opto-electronic system, a differential pair of detectors is operated as the input to the neuron; signals with positive (excitatory) weights arrive at one detector, and signals with negative (inhibitory) weights arrive at the other detector. These detectors sum up the intensity of each optical signal arriving at the neuron. The neuron's activation function is electronically applied to the detected signal to produce an output signal. The output signal drives either an optical source (or pair of sources) or a spatial light modulator. Figure 26-2 illustrates an opto-electronic neuron that uses a pair of laser diodes to encode the neuron's output signal.[2] Figure 26-3 illustrates a prototype feed-forward system.[3]

OPERATIONAL PARAMETERS: INTERCONNECTIVITY

In the architecture depicted in Figure 26-1, neurons are laid out in a two-dimensional grid, with Nx neurons in the horizontal direction and Ny neurons in the vertical direction for a total of Nx Ny neurons. Each neuron consists of one or two inputs depending on synaptic weight polarity (W), and one or two outputs depending on neuron signal polarity (P). Zero padding around synaptic weights is described by the integer factors Zx and Zy, and subhologram replication is described by the integer factors Rx and Ry. Expansion of the SBWP for carrier-frequency encoding is described by Cfx and Cfy. Any additional SBWP used to improve the accuracy in the interconnection weights is described by the expansion factors Ex and Ey.

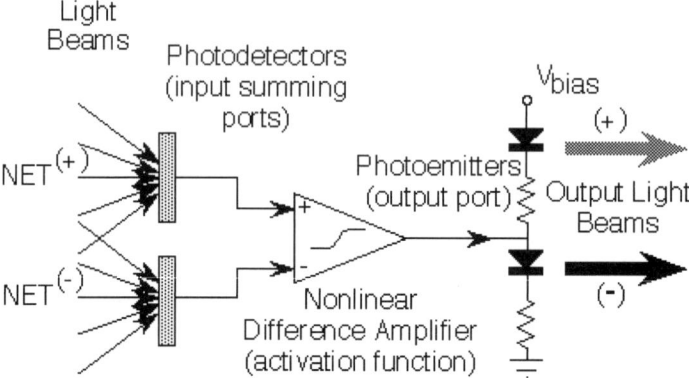

FIGURE 26-2: Prototype opto-electronic neuron constructed from detectors and emitters.

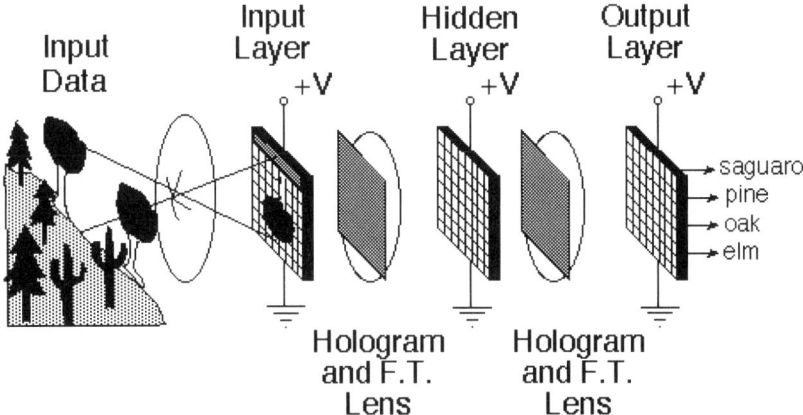

FIGURE 26-3: Prototype opto-electronic neural network system implementing the feed-forward architecture.

Combining these factors yields an equation describing the number of pixels in a single subhologram.

$$Sx = WxNxCfxZxRxEx \quad \text{by} \quad Sy = WyNyCfyZyEy \quad (1)$$

For a hologram mask of physical dimensions Hx by Hy, you need a resolution elements of size $[[\Delta]]x$ by $[[\Delta]]y$; a fractional separation factor of G between subholograms to reduce cross-talk; and the total number of neurons that can be laid out in each dimension of the neuron plane. This is given by

$$Nx^2 = Hx/(PxWxCfxZxRxEx(1+G)[[\Delta]]x))$$

and

$$Ny^2 = Hy/(PyWyCfyZyRyEy(1+G)[\Delta]y)). \quad (2)$$

The total number of neurons in a plane is the product of Nx and Ny, and the total interconnection capacity of the interconnection hologram is $(NxNy)^2$.

Analog Dynamic Range
The dynamic range of a holographic interconnect can be approximated by dividing the maximum interconnection weight value by the average root-

mean square error of the interconnection. For a hologram that generates N interconnections, the dynamic range, D.R., is calculated from the desired connection weight values, wij, and the values reconstructed from the hologram, vij, and is described by

$$D/R/ = \max(wij)(\text{sum}(vij - wij)2)^{-1/2}. \quad (3)$$

An approximate digital precision can be calculated by taking the base-2 logarithm of the dynamic range.

Connection Rate

The connection rate between two neuron layers is the product of the number of interconnections achieved by the hologram and the update rate of the opto-electronic neuron. The neuron update rate is set by the opto-electronic components. For detector/modulator neurons, this rate is limited by the modulator material to about 1 kHz. For detector/emitter neurons, update rates well in excess of 1 MHz are realizable. Here, the main limitation is set by signal detectability considerations.

Neuron Output Power

The output power of a neuron is set by the characteristics of the detector and interconnection hologram. The detected power required to achieve a particular signal-to-noise ratio (SNR) at an individual detector is given by

$$P_{det} = (A\Delta f)^{1/2} \, SNR/(D^*[\eta]d) \quad (4)$$

where D^* is the specific detectivity of the detector, A is the detector area, and $[[\Delta]]f$ is the temporal bandwidth of the detector. For a diffractive interconnection system, the minimum required neuron output power is given by

$$\begin{aligned} P_{output} &= P_{det}(\text{fanout/fanin})/\eta d = \\ &\quad (\text{fanout/fanin})(A\Delta f)^{1/2} \, SNR/(D^*\eta d) \end{aligned} \quad (5)$$

where ηd is the diffraction efficiency of the hologram. The fanout and fanin factors refer to the number of neurons in a single neuron that connects to the number of neurons that feed into a neuron, respectively.

RESULTS

To examine the potential performance of the opto-electronic system illustrated in Figure 26-1, a prototype system consisting of an array of surface-emitting laser diodes interconnected via a phase-only hologram to a silicon-based detector array was considered. A Monte Carlo technique was used to examine the interconnection hologram. In these simulations, the GSPRS algorithm was used to generate a large number of test holograms. Holograms encoding on-axis and off-axis diffraction geometries were generated with various discrete levels of phase (2, 4, 8, and 16). Systems with subhologram replication factors of 2 by 2 and 4 by 4 were also examined. By simulating the optical system, the expected dynamic range of the synaptic connections and the diffraction efficiency were determined. Table 26-1 lists the dynamic range and approximate digital precision of the synaptic weights for the different hologram configurations.[4] To date, the construction of physical systems has been limited to small-scale feedback systems of 64 neurons and to fabrication of test subholograms for large-scale systems (4096 neurons). The small-scale experimental system implemented the Hopfield auto-associated memory that was reported earlier in the chapter.

To approximate the upper limit of hologram capacity, an electron-beam lithographic mask with dimensions of 125mm by 125mm and with a pixel size of 0.5um by 0.5um was considered. To provide adequate separation between subholograms, a cross-talk separation factor, G, of 15 percent was used. Bipolar synaptic weights ($W = 2$) and bipolar neuron signals ($P = 2$) were implemented. For the on-axis holograms, an expansion of 2 by 2 ($Ex = 2$ and $Ey = 2$) in SBWP was used. For the off-axis holograms, an SBWP expansion of 4 by 4 was used to encode a carrier ($Cfx = 4$ and $Cfy = 4$). From these specifications, neuron plane sizes and interconnection densities were determined and are listed in Table 26-1 for the various hologram configurations.

A neuron update rate of 1 MHz was chosen after a radiometric analysis of the prototype system was performed. This establishes the processing rate for the prototype system. For silicon detectors operating at this rate, a detectivity of 10^{11} $cmHz^{-1/2}/W$ can be achieved. To determine the output power of a single neuron, the extreme case in which a single output of one layer is connected to all the inputs in another layer was considered with a SNR of 10:1. However, this extreme fanout/fanin case will rarely be encountered, and typically several orders of magnitude less neuron output power will be required.

TABLE 1: Operational parameters of the opto-electronic neural network for various hologram configurations

Diffraction Geometry	Hologram Replication Factors	SBWP per Connection	Neurons per Layer	Processing Rate (CPS)	Phase Levels	Analog Dynamic Range	Digital Precision (bits)	Neuron Power (uW)
on-axis	2 × 2	16	26,912	7.2E14	4	14.9	3.9	220
(Ex = 2)			(Nx = 232)		8	18.2	4.2	190
(Ey = 2)			(Ny = 116)		16	18.6	4.2	180
	4 × 4	64	13,448	1.8E14	4	20.0	4.3	110
			(Nx = 164)		8	27.8	4.8	95
			(Ny = 82)		16	30.0	4.9	90
off-axis	2 × 2	64	13,448	1.8E14	2	19.1	4.3	260
(Cfx = 4)			(Nx = 164)		4	20.7	4.4	140
(Cfy = 4)			(Ny = 82)		8	20.7	4.4	115
	4 × 4	256	6,728	4.5E13	2	30.2	4.9	130
			(Nx = 116)		4	36.4	5.2	71
			(Ny = 58)		8	37.5	5.2	57

CPS = Connections Per Second

DISCUSSION

Table 26-1 illustrates that neural network systems designed with planar holograms and detector/emitter based opto-electronic neurons can potentially achieve processing rates from 45 to 720 trillion interconnections per second, and that the encoded synaptic connections have an approximate digital precision of 4 to 5 bits. The larger capacities and processing rates are realized when less SBWP (less subhologram replication and an on-axis geometry) is used to encode the connections, although this reduces the accuracy of the stored synaptic weights. For systems constructed with modulator-based neurons, this processing rate generally drops by a factor of 10^3. While much higher processing rates can be achieved with this optical architecture than with electronic systems, this advantage will be less significant unless methods for high-speed data input and output are developed. Another observation from Table 26-1 is the modest improvement in dynamic range as the number of phase levels is increased. Since fabrication errors are likely to be greater for holograms that encode many phase levels, the modest improvement in dynamic range gained with more than four phase levels will be lost during the fabrication process.

Aeronautical Engineering

One of the fundamental problems of fracture mechanics is the thin plate with a small hole of a given shape, usually circular, situated far from the boundaries of the plate. This problem has numerous practical applications; for example, in aeronautical engineering, in container design, or in civil engineering. If the plate is subjected to in-plane loading, stress concentrations with large tangential normal stresses appear along the boundary of the hole.

This is made visible in Figure 26-4 by using the well-known Moiré holography technique.[5] With the help of computer-based image processing and quantification of the phase distribution through the fringes, the displacement field near the hole boundary can be determined quite accurately.

In ductile metallic plates, if the load is increased beyond the limit of elasticity, plastic deformations occur in the immediate vicinity of the hole. In the first theoretical phase of the present project, the stress and deformation fields in the plastic zones and the changes induced by the appearance of the plastic zones in the elastic fields around them were computed by applying asymptotic methods developed earlier at the Louisiana Tech

FIGURE 26-4: Moiré fringes in a thin aluminum plate containing a circular hole and loaded by a tensile force uniformly distributed at its left and right boundaries.

University Institute for Micromanufacturing (IfM) for this kind of elastic-plastic problem. Thus, analytic expressions for stresses and displacements as well as for the shape of the elastic-plastic boundary were obtained both for circular and elliptical holes. Using these expressions, the residual (plastic) displacement fields after unloading of the plate were theoretically predicted. If one compares holographic pictures of the plate before loading and after loading-unloading, one obtains Moiré fringes (see Figure 26-5) that can be quantified to get the residual displacement fields in the vicinity of the hole.[6]

Figure 26-6 shows both the theoretical predictions and the measured values of the residual displacements in the loading direction at two locations 1 and 2 near the hole boundary (see also Figure 26-5).[7] Since all material parameters were determined by independent experiments not connected with the holographic measurement, the excellent fit of theoretical and experimental results is a confirmation both for the accuracy of the holographic measurement and for the prediction capacity of the theoretical developments.

With that in mind, let's look at Voxel, Inc.'s Digital Holography System™. The Digital Holography System consists of four elements, the Voxpad™ digital interface, the Voxcam® laser camera, the Voxbox® holographic light box, and Voxfilm® media, which together enable a user to produce and examine Voxgram®s—the world's first true three-dimensional images. In the medical context, these volumetric holograms have been described as a

SUMMARY, CONCLUSIONS, AND RECOMMENDATIONS 571

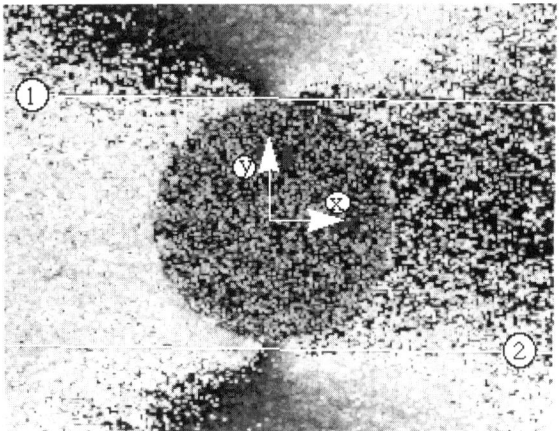

FIGURE 26-5: Moiré fringes around a hole obtained by comparing an unloaded rectangular plate with the same plate that was loaded in the x-direction by tensile stresses of 114+/-2 N/mm² and then unloaded.

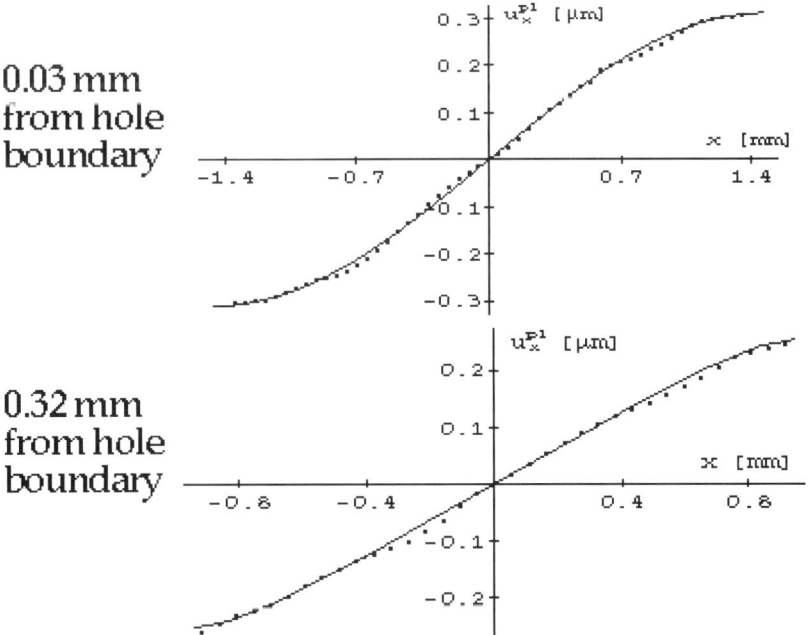

FIGURE 26-6: Residual displacements in the loading direction: measurements (dots) and theoretical predictions (solid lines) at a distance of 0.03mm and 0.32mm from the hole boundary (corresponding to line 1 and 2 in Figure 26-5).

life-size twin of the patient's anatomy because they project out in 3D space with the exact same spatial and depth resolution of the real anatomy.

Digital Holography System

The Voxpad is the user's connection between his or her remote 3D data acquisition device, usually a CT or MR system, and the Voxcam. The unique feature of the Voxpad is a patent-pending 3D display software that simulates the look and feel of the final Voxgram on a flat computer screen. Using the knowledge of how the hologram is made, the proprietary Voxplan rendering algorithm allows the user to select the optimal view of the 3D anatomy via a variety of special tools, including window, crop, and orient. Voxplan also allows the user to move and orient the volumetric image, just as if he or she were moving the Voxbox to get different orientations of data set. The remote Voxpad is connected to the Voxcam holographic printing facility over the Internet and parsed for production via a Voxqueue managed by Voxel.

The Voxcam is a special holographic printer capable of generating volumetric holograms of 3D digital data via a patented multiple exposure process called Digital Holography. It is a laser-driven optical instrument that receives electronic images collected by CT and MR scanners via the Voxpad (or any device that outputs sectional digital data), projects them using a liquid crystal display (similar to those used in certain video projection systems), and records them on film. Each image from a scan is individually exposed in sequential fashion and in its proper location. All of these exposures are recorded on a single piece of holographic film. The film is then developed in an automatic processor utilizing chemistry specially designed for this film and process.

Once the film has been processed, it is displayed on a special light box called a Voxbox. The Voxbox is an inexpensive, compact, self-contained portable unit. It is similar in appearance to the light boxes currently used throughout medicine for viewing conventional two-dimensional film; it can be installed and used in the same way. The Voxbox uses ordinary white light to reconstruct the hologram instead of requiring a laser as is usually the case. The Voxbox can be rotated or tilted, allowing the user to view a Voxgram from a variety of different perspectives.

The film used to make Voxgrams shares many characteristics with ordinary X-ray film, but its composition and the proprietary process used to encode information make it possible to record, when appropriate, more than 200 slices from a CT or MR examination on a single sheet. An exposed and developed Voxgram is a 14 by 17-inch sheet of film that reveals only a faint glow when held up to ambient light. When placed upon a Voxbox, it displays a detailed three-dimensional image stretching through a 6-inch deep volume.

HOLOGRAPHY

As previously explained in earlier chapters, holography is similar to photography. In both, light illuminates an object, and the reflected light is recorded in a thin photosensitive layer on a sheet of film. There are, however, substantial differences between photography and holography. The lens of a photographic camera forms a single image of an object, flattening it to a two-dimensional picture. When looking at a photograph, one can only see this flat picture as it appeared from the original vantage point of the camera lens. A holographic camera, on the other hand, uses a laser in place of a lens and records the entire wave front emanating from an object. This wave front contains intensity, distance, and directional information. The object is recorded and displayed three-dimensionally, and one can view the object as it would have appeared from a whole range of different vantage points by simply moving one's head.

Photography and holography are different on very fundamental levels as well. The photographic process records brightness (intensity) and color (wavelength) on photographic emulsion. The image is viewable in white light. The holographic process also is able to record the distance (as phase—the third fundamental component of a light wave) that each point of the object is from the film. This phase information, stored as an interference pattern on ultrahigh-resolution emulsion, is replayed using the appropriate laser beam or a Voxbox. In short, a hologram is capable of storing all three components (rather than just two) of the electromagnetic spectrum that is reflected off an object.

Conventional Holography
A conventional hologram is made by splitting a laser beam in half, shining one part, the *reference beam*, onto the holographic film, and directing the other, the *object beam*, to reflect off the object to be holographed. Information about the object's appearance and position are recorded when the beams intersect on film.

Multiple-Exposure Holography

Reflecting light off of an object is not feasible when the *object* is inside a patient's body. Thus, a Voxgram cannot be a conventional hologram. Instead, the *object* for a Voxgram is a projection screen that sequentially shows slices from a CT or MR scan as shown as Figure 26-7.[8] A hologram of the first slice is made; then the screen is moved back a distance corresponding to the gap between the slices; the next slice is projected onto it; and a second hologram is superimposed on the first. More slices are encoded by repeating this procedure with appropriately greater distances between the screen and the film. The process is similar to multiple-exposure photography, in which several exposures are made on a single piece of film, but the holograms encode the location as well as the intensity of each exposure. When reconstructed, each holographic exposure appears at a different distance from the film. The slices are suspended in space with the correct separation between them. A large number of exposures (up to 200 or more) can be combined with proper selection of holographic angles, intensities, and processing chemicals.

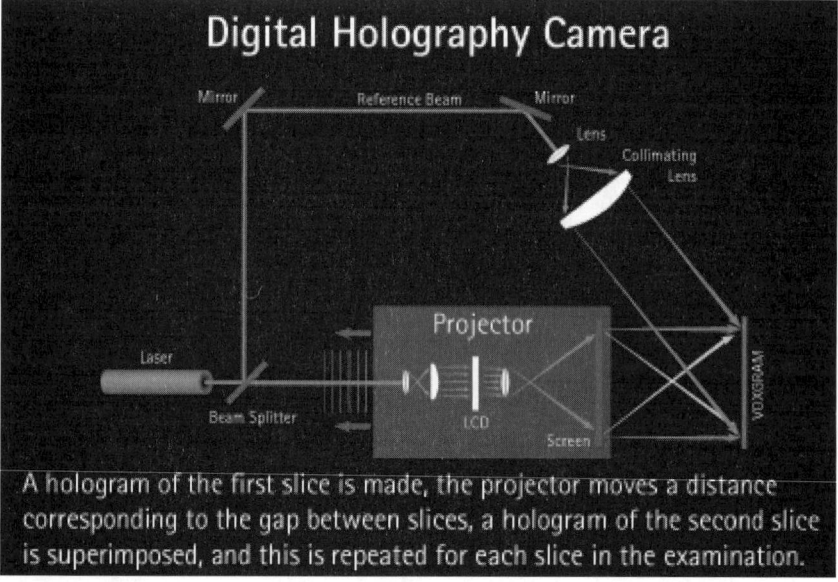

FIGURE 26-7: Schematic for making a volumetric multiple-exposure hologram of an MR or CT scan (simplified).

Holographic Display Systems

A hologram must be illuminated correctly to be seen as a bright, undistorted image without any compromise in its three-dimensionality. The classic method to reconstruct a hologram is to use laser light identical to that with which the hologram was made. However, suitable lasers are expensive and inconvenient, and generally are not found outside optics laboratories. Several alternatives use white light instead of a laser.

A hologram will split ordinary incoming light into a colorful spectrum. One solution uses the hologram as a filter to reduce the incoming light to one color where the image becomes sharper and dimmer. Medical images require a sharp, bright image. Another approach is to sacrifice the vertical information in a hologram, which causes the image to disappear when viewed from above or below. Medical images must be viewable from vertical as well as horizontal perspectives. The Voxbox pre-splits the light into an inverted spectrum, and the hologram recombines the light into a bright and sharp image, preserving both vertical and horizontal perspectives (see Figure 26-8).[9]

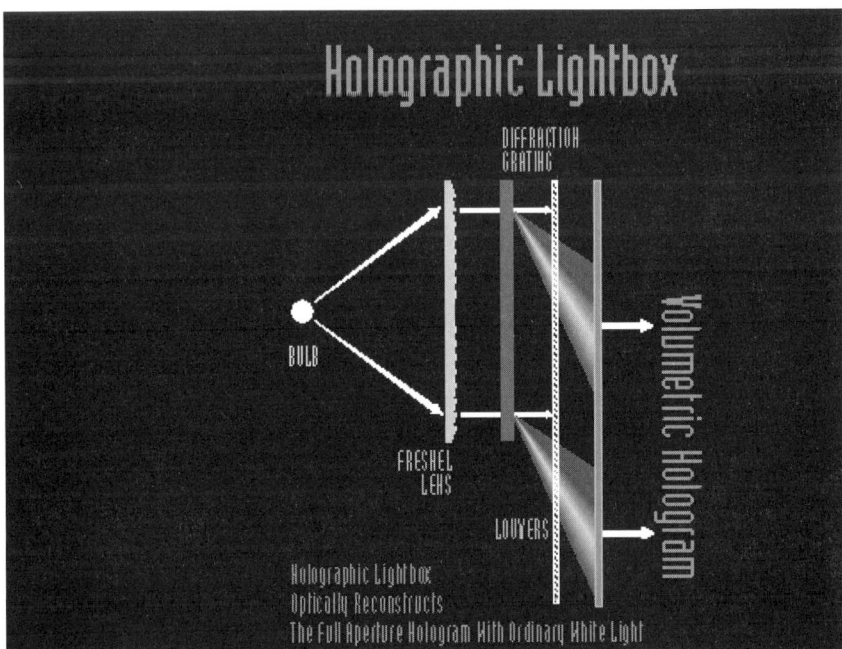

FIGURE 26-8: Schematic for the prototype Voxbox, a white-light viewer for holograms (simplified).

Production of Voxgrams

Assimilating Voxel's technology does not require a change in a hospital's equipment or procedures. CT and MR data are collected in the format required, thereby eliminating the need for the physician to modify his or her routine practice. Expensive scanner time will not be wasted, and the patient's radiation dose will not be changed. The data will be prepared for printing using a technique similar to that used to prepare two-dimensional images. After scanning the patient, the technician will adjust brightness and contrast on the Voxpad to get the most useful images, which will then be printed. The Voxel System as it is used in a typical hospital environment suite is shown in Figure 26-9.[10]

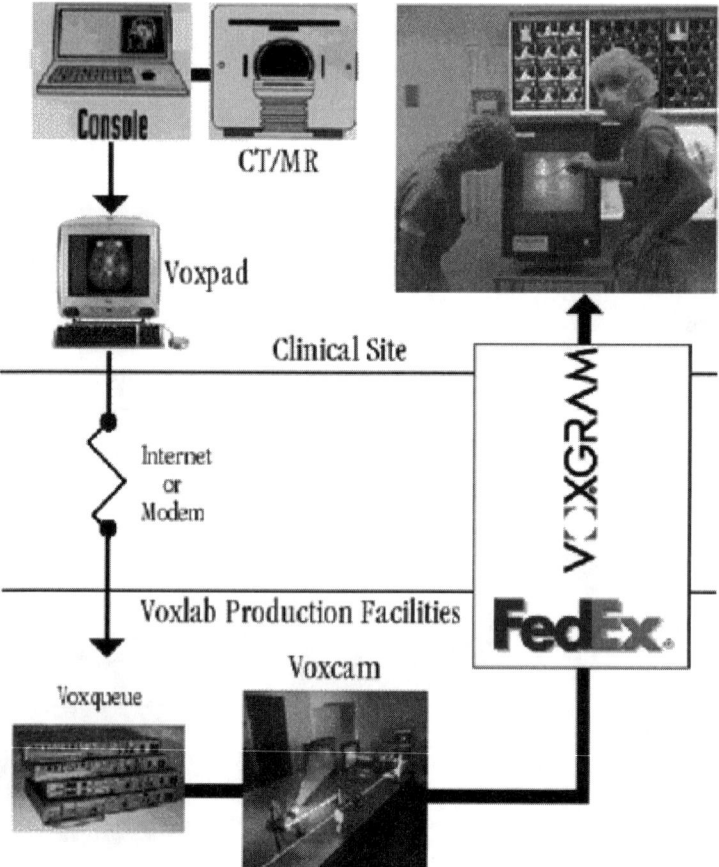

FIGURE 26-9: The Voxel System as it will be used in a typical radiology suite. The Voxcam is analogous to conventional two-dimensional cameras and light boxes.

Display Alternatives

One currently available alternative approach to volumetric display is to use computer graphics technology. A single pseudo three-dimensional picture is synthesized from the individual two-dimensional images.

Traditional artistic effects (shading, perspective) are used to convey an impression of depth. One of the limitations of computer-generated pseudo three-dimensional images is that only psychological depth cues can be used. Psychological cues depend on surfaces. Computer-generated surfaces introduce certain compromises and distortions. Moreover, substantial computer hardware and software skill experience, time, and effort are required to produce these pictures. A variety of suppliers of CT and MR scanners as well as independent companies offer software and workstations that generate pseudo three-dimensional pictures. These systems also have other uses; they enable radiologists to interact with the data that have been collected by scanners, and they may facilitate the production of *what if* images that some surgeons have found useful in surgical planning. When used for that purpose, the workstations can serve as a source of data for Voxgrams, so true volumetric displays can be made after the data have been manipulated. Voxgrams may, therefore, add value to users of workstations.

Medical Evaluation

Evaluations of Voxel's proposed system have been conducted and are continuing at medical centers throughout the United States. These programs have concentrated on the parts of the body (the head, spine, skeleton, and blood vessels) and the kinds of conditions (tumors, trauma, and vascular abnormalities) that are commonly examined with CT and MR scanners. The studies were designed to determine if the Digital Holography System would:

- Allow diagnosis of conditions that are extremely difficult or impossible to detect with existing technology.
- Provide for more accurate and comprehensive diagnosis and understanding of conditions that are difficult to characterize fully with existing technology.
- Increase the radiologist's confidence in the diagnosis made.
- Reduce the time required to arrive at a diagnosis.
- Facilitate communication of relevant information to referring physicians.
- Improve surgical planning.
- Allow for more fully informed patient consent to treatment.[11]

Studies of the system began in the Spring of 1992. Participating institutions transmit data from study cases to the company by sending relevant tapes or disks. Using a prototype Voxcam, Voxel's personnel produce the corresponding Voxgrams and return them to the institutions by overnight courier. Because each site has been provided with Voxboxes, physicians can examine the Voxgrams in the same manner as if they had been produced on site. The company does not pay for the CT or MR scan or the interpretation. No payments have been made to investigators or their institutions other than, on rare occasions, a modest research stipend. Participants have been free to publish their findings.

Voxel's research collaborators have presented over 300 oral papers and scientific exhibits at domestic and international meetings of medical and scientific organizations, and have published several articles in prestigious medical journals, all reporting on the usefulness of the company's technology. The largest forum for scientific reports on medical imaging is the annual meeting of the Radiological Society of North America (RSNA), which is attended by more than 36,000 physicians and health professionals. Many of the aforementioned papers and exhibits have been presented at RSNA meetings, and others have been given at meetings of radiologic subspecialists (American Society of Neuroradiology, European Society of Neuroradiology, British Society of Neuroradiology, and American Society of Emergency Radiology). Referring physicians, especially neurosurgeons, orthopedists, and cardiologists, have also made presentations to their peers at their own professional meetings. Presentations have been made at the Congress of Neurosurgeons, the American Academy of Neurological Surgeons, the American Academy of Orthopedic Surgeons, the American Heart Association, and the American College of Cardiology.

The reports by Voxel's research collaborators at domestic and international meetings suggest that the company's technology is useful in diverse settings and in support of the indications most commonly targeted by CT and MR procedures, including neurological, orthopedic, and cardiac conditions. Most of the data for these submissions were collected using traditional CT or MR techniques, but these submissions also included studies using leading-edge scanner applications such as:

- Noninvasive MR visualization of blood vessels
- MR visualization of the nervous system
- High-speed, high-resolution CT
- Three-dimensional ultrasound[12]

Now, let's *go where no hologram has gone before: The holodeck.* Online interactive virtual reality brings us another step closer to *Star Trek*'s holodecks. Let's take a look at the future.

Interactive Virtual Reality: Holodecks

As everyone surely knows, those holodecks work by rendering reality only in the user's close vicinity. No need to recreate all of Victorian London for Data's Holmesian fantasies, or even the Whitechapel district. That would drain too much of the Enterprise's power.

As Data or Picard stroll through London, the system generates a reality only a few feet in diameter around them—just enough to convince them it's real. After all, reality is only within the reach of our senses (see the sidebar, "Emergency Medical Holograms").

■ EMERGENCY MEDICAL HOLOGRAMS

The EMH A.K-1 was designed by Dr. Lewis Zimmerman with the assistance of Lt. Reginald Barclay at the Jupiter Station Hotoprogramming Center, as the spitting image of his designer. He was first activated on star date 48308.2. The doctor (from *Star Trek Voyager*) has access to information of more than 3,000 medical *reference* sources and the expertise of 58 medical officers, which give him unparalleled diagnostic possibilities.

His original function was that of an emergency physician if the human chief medical officer was unavailable. Since Voyager's doctor didn't survive the abduction by the Caretaker, he became the de-facto Chief Medical Officer aboard Voyager. At first, his manners were rough—a character trait he had inherited from his programmer, and his impatience with his patients did not earn him much popularity. Influenced by Kes, however, his character gradually softened.

Kes also helped him develop a sense of self-worth and identity. At first treated like an object by the crew, he was soon accepted as an important member of the crew. Initially, he could only exist in Sickbay or on the holodecks; but, ever since Voyager is in possession of a portable holoemitter, the doctor has been able to walk about freely, and to join away missions.

The doctor is a remarkable EMH. Soon after his first activation on Voyager, he expanded beyond the confines of his originate programming, displaying typical human behavior. He has the capacity of having emotions, even of experiencing the feeling of love.

> Ever since Kes suggested he should find a name for himself, the doctor has been attempting to do so. He had settled for 'Schweizer' once, but that name reminded him too much of the loss of his first love, so he abandoned it, the Vidiian Dr. Danara Pel had her own name for him: Shmollus.

HOLODECK SYSTEM

Fractal robots (shape changing robots) are ideally suited to provide a first approximation synthesis of a holodeck system. *Holodeck* is short for holographic projection deck—a widely discussed concept in holography made popular in the *Star Trek* series where you can enter a chamber and interact with holograms as through they were real. One cannot build such holodecks at present, but in the not too distant future, a mass entertainment system that offers all the delights of a real system could be built.

Note: Fractal robots start at one size to which half-size or double-size cubes can be attached; and to each of those, half size or a double size can be attached, respectively, ad-infinitum (see Figures 26-10 and 26-11);[13] and that is what makes them fractal. So, a fractal robot cube can be of any size. The smallest expected size is between 1,000 and 10,000 atoms wide.

So, what is a holodeck? It's the addition of hardware reality (HR) to a virtual reality (VR) helmet. If you could wear a virtual reality helmet, walk into a room (built of shape-changing fractal robotic cubes), command a

FIGURE 26-10: Demonstration prototypes.

SUMMARY, CONCLUSIONS, AND RECOMMENDATIONS 581

FIGURE 26-11: Fractal-sized robots.

chair to appear, walk up to it and sit down, then we have the makings of a very good approximation of the holodeck system!!

Can you do this without the virtual reality helmet? At present, this is not possible. For instance, if you look at the sky, the birds will not fly in the sky just because you command it! Neither will it turn into a sunny day. If two people were in the same vicinity of each other, then the spaces occupied would immediately start to clash because of the huge size of the spaces involved.

You will always end up needing a virtual reality helmet or something equivalent to feed your eyes with information that fools you into believing that you are somewhere you are not. You need to be sure that you can feed the eye with sufficient information, so that when you are looking at a bird flying in a clear blue sky, the eye is fooled without having to actually construct such a huge environment around yourself. The virtual reality helmet allows the physical environment size to be reduced.

Having built the environment, you must now add substance to all the elements with which you could interact. Shape-changing fractal robots and their tools linked to the virtual reality helmet is certainly good enough to create objects such as chairs, tables, floor boards, steps, lifts, windows, doors, and so forth, that would allow the sensory system to be fooled into making one believe that he or she is interacting in an environment that necessarily does not physically exist in its entirety.

Every time you get close to an object with which you could interact physically, the robots change shape to approximate that object so that you can interact with it properly. Only the objects that you are close to need to be synthesized to save energy, while the other objects can be masked out using the virtual reality headset.

Fractal robots can provide the tactile feedback you need to sustain the illusion. The cubic robots can be made to react fast enough so that you could walk up to a chair and find it exactly as you see on your visor. The resolution of the robotic cubes determine the finest features resolved to the human touch. The smaller the fractal cube, the better the approximation.

By installing sensors in the cubes, they can move according to pressure applied. The sensors within the robotic cubes give the robot reactive behavior, so that if you think you were looking at a mattress and it was built of fractal robots, then it can be made to sag to much the same degree as you would expect a normal mattress to sag! The smaller the cubes, the better the approximation. But you don't need to go that far straight away. In well defined environments, where the options to view and interact have been pre-limited, tooling such as real mattresses can be brought in at the time of use to provide the realism that may not be available straight away using finer resolution robotic cubes.

Body Suits of the Future

It should be pointed out here that if the robots are around 0.1mm to 1mm in resolution, and provided they have adequate reaction speed, then they can be worn as a suit, and all the additional robots can be dispensed with(!). Most of the time, the suit does not make contact with the body except where it is attached as a garment. It is held within a millimeter reach of the skin. You don't need to wear it as such—imagine it as a robotic goo poured all over you, but as soon as it hits the ground, it finds that as its support and keeps itself 1mm away from you in all directions. When you reach out and touch something in the virtual world, the suit deforms, allowing you to make contact with itself. This gives, for example, a hand the sensation that it is touching something without the real object having to be present. Providing the suit can react fast enough to deformation requests, you could stroke the virtual object and the suit would react against the skin as though it were in contact with the actual object. Everywhere else, it keeps its distance to maintain the illusion.

Adding Extra Realism

Regardless of which way you implement the system, a fundamental problem with these systems is the missing sensations. You need heat, wind, sound, and vibration feedback, which can all be present in the environment. Fortunately, fractal robots can tool up with actuators that generate these stimuli, which enhances the interaction with the environment to a greater extent, providing it's not a body-wide suit. Body-wide suits need to carry tools to generate these, which then makes the suit exceptionally bulky.

If you are interacting with the scenes, say in a cinematic experience in which you pick up weapons and shoot at enemies/lend a hand to your hero, then those tools must be placed into the scene and made available for the user. The feedback in these systems could be very good. For example, you could ring a bell in a bell tower; operate a machine gun and receive the vibration feedback as you would with a real weapon; see the bullet/laser trails and witness the damage done; and operate a winch and pull someone up. Weapons and tools do not have to be high tech—a simple barge pole could be used to row a virtual canal boat, an axe could be used to chop down a virtual tree. The list is endless!

Adding Extra Levels of Interaction

Even after all this, there is still something missing: the interaction of the user with the objects that are found in the virtual world. For example, you could get wet, but that is not a problem if an actuator pumped fluid at you to give that wet feeling. Free-flowing fluid becomes a problem if it spills into the environment, and you had to re-image and present it back to the user, so that he or she is aware of how much is spilt and where its boundaries are if it's lying in a puddle. To overcome these problems, cameras must be mounted on the user and around the user to determine what the interaction between the user and the fluid object result set looks like.

If the user is carrying an object and drops the object, it is necessary to re-image the object so that when he or she walks away from the object, he or she can look back and view what he or she has dropped. You have to install weight detection equipment in the shoes of the user, so that cameras mounted on the user and around the person can detect weight gain/loss and focus in on the objects that have been picked up or dropped. You should, for example, be able to pick up a piece of paper and read its contents, because the camera has become aware of weight gain and allows you to focus in on what you have picked up. Afterwards, you could crunch and drop it, and view the crumpled piece of paper as it should look through the VR helmet.

Nano Technology
Nano technologists dream about creating a holodeck that could fulfill your desires to interact with environments not possible in conventional reality. Whatever they do, something close to a VR helmet must be provided in order to complement the synthesized reality with that of your expectations such as huge expanses of space. You can't go around creating birds and deep blue skies, except in the VR helmet. If fractal robots can provide this functionality with technology that is infinitely cheaper than nano technology, then that begs the question, why do you need nano technology to accomplish the same goals? Herein probably lies another nail in the coffin that may put back the development of nanotechnology as you envision it today by a few more decades. A holodeck is a principle target for implementation, using utility fog and/or conventional nano technology. However, that can never happen while a cheaper alternative in the form of fractal robots exists.

Telepresence and Teleporting
The kind of telepresence offered by this technology is quite different from conventional robots. With this system, you can:

- Teleport objects
- Teleport yourself
- Grow and shrink in size[14]

However, you can do all this and interact with the environment that you create. For example, you can appear inside a car in a car factory; remove a defective radio and fit a new one; teleport into another vehicle to fix its exhaust; and shrink to get inside the engine through the exhaust and wrench the spark plug gaps apart on a defective spark plug—all without ever leaving the comfort of your home.

You could do some of this using conventional robots, but you are limited with such technologies in what you can do and how you interact with your environment. With this new technology, you won't face those limitations. With a spark plug example, one-millimeter robots carrying cameras can enter the engine through the exhaust, present a digital image of what is on view, and provide the necessary robotic tooling to repair the fault. You can miniaturize yourself, and once inside, touch the walls of the engine cylinder, and you will find it exactly to scale.

Carrying this analogy further, if you knew that you needed a special tool to affect repair, you can look for it on the factory floor and *teleport* it to where you are. The robots will deform and fetch the object to where you are. Once a teleport command is executed, the object disappears in flash from your visor and reappears to where you are. It will take some time, of course, while it jumps from one point to another, because the object has to be physically transported across the factory floor, but the concept of teleporting yourself and/or objects around you will become a familiar concept when these machines are finally rolled out.

Virtual Design Studio
The possibility of telepresence gives you the ability to be present anywhere you want to be without having to be actually present. The marriage of HR and VR allows the design of machines that can cater to every kind of whim you may have about manufacturing. You could build a system whereby you design a product, and fully test and assemble the product without ever having to be at the manufacturing site.

One such facility may be a machine shop. This machine shop of the future as a facility has a wire through which you log in, design your product, make it, assemble it, and package it ready for shipping. After you have paid, the product is released and is shipped to your premises exactly as you packaged it.

How much time you log on to each piece of machinery and how much material you use is billed to you. The greatest advantage is that you can get things made extremely quickly. For low-volume products, the design process and the manufacturing process is one and the same. If you did the operation once, then you merely download the project file into the manufacturing site and wait for the manufactured product to be delivered.

It may be that products from several different producers have to be manufactured, assembled, and shipped to your facility for final assembly and integration. All this could also be automated. All you would be expected to do is to spend a lot of time designing and putting together the components as design engineer. If you have a team and other people at the manufacturing facility who can log in to your program, then a great deal of *devolved* experience could be brought into play to manufacture your product more efficiently with better design decisions. Parallel manufacturing activity could take place, such as test engineers going on line to test your product for tolerance checks, interference, and so forth, as it is being manufactured.

On top of all this, it is possible that over a period of time, other parameters such as materials selection and clearance parameters could be added to the design suite to avoid making serious design mistakes. It also implies you need less participation by other engineers. As an example, if you wanted to make a plastic bearing, a materials menu option could walk you through the advantages and disadvantages of using common materials. It means you won't need an engineer to run you through the material selection process.

Cinema Systems of the Future

It's quite easy to see that the holodeck is going to be one of the mass entertainment systems of tomorrow competing with the likes of cinema and theater. People go out to the cinema/theater for new releases and for the fun of going out. A lot of the time, this kind of entertainment could be piped into the home through television, but the experience is not the same.

The holodecks of tomorrow provide, if anything, even more exhilaration, as users wearing these helmets would be required to *interact* with the environment. If you could imagine a complex about the size of a football stadium, where visitors enter at one end and are transported through a maze system to the other end in groups of 10 people, then you could begin to see what kind of entertainment is on offer here. The users would be confined to a space of about 4 x 4 meters and herded along. Right behind them would be another group in their own 4 x 4 meter *bubble*. These people could be crossing giant ravines, they could walking along cliff edges, or in a maze and interacting with the animated creatures and mechanoids as though they were real. They could get into a car and participate in a car chase; get inside a train and be transported to the other side of the world; fly in a space ship and dock into a mothership; and they could be in a fighter plane dodging missiles. There is no experience (included horror and x-rated) that could not be experienced through this system.

Of course, if you had the actuators that were previously discussed, then you would be able to blow wind through the system to indicate bad weather, for example; radiate heat to indicate you are approaching a hot engine perhaps; or drip water and rain to indicate a very gloomy setting; drop small pebbles to indicate falling debris. The content for these systems have to be built using a mixture of computer graphics and real photography that has been texture mapped onto virtual objects and scenes. Content has to be created in systems that are practically all digital. More than one camera would be needed on most occasions, and depending on the levels of interactions that have been programmed for the system, a huge array of computers would also be needed. The computer systems have to work out what the VR helmet should see and what kind of sounds should be heard. The kind of content that would go into these systems could range from action interactive movies to quiet visits in the country for the less energetic, the beach for the holiday maker, cartoon worlds for the kids (something Walt Disney—if he were alive—would probably snatch from the kids!!!!), Manga world for the techno freaks, and dangerous worlds for military training exercise and simulations.

Soundscaping

The sound systems will convey some information about where you think you are. Within a holodeck, that could be anywhere. To make the sounds real, you could place speakers, above, behind, in front of, and below the audience, and operate them according to which directions the audience should be looking when interacting with their objects. This is called *soundscaping*. For example, if they were being chased by a dinosaur, then you would put the noise behind them—and if they were about to be run over by a speeding car, you would place it ahead of them. If you are looking at birds flying in the air and singing, then you would place it above. You could place it below if you were crossing a ravine or a cliff edge.

There are ways of phasing sound such that it appears to come from those different directions, but there are technical problems. You need sensors to dynamically vary the phasing, which would be extremely difficult to achieve in real time. You need narration and music, which is probably best played through speakers fitted in the VR helmet—but, if you do that, you must leave sufficient *openings* in the system to ensure that sound from outside the helmet reached the insides of your ear. This, of course, also implies that the robotic cubes and actuators are fairly quiet—or that you have vents in the helmet that open and close to let outside sound into the ear.

The entirety of soundscaping is an art—not only do you have to place the sound in the landscape correctly, but you also have to provide content for it that is synchronized with the video systems. If you use the bubble

Holodeck Movies

Public entertainment systems where holodecks are a place to gather and enjoy the holodeck experience will likely start with five-minute movies that are very exhilarating. The first systems are likely to be extensions of PC computer games like *Doom*, which have good rendering engines for 3D. These rendering engines put onto cheap and powerful network PCs and linked to the user via a VR helmet are likely to be the first contenders that realize a commercial holodeck system. A series of plots and features of such holodecks is described in the sidebar, "Holodeck Movie Production."

■ HOLODECK MOVIE PRODUCTION

This sidebar is provided as a guide and perhaps a taste of the holodeck movies that could be produced. A PC computer game called *Doom* has been chosen to explain game play screen shots taken to explain how a holodeck version of the game could be produced that would change immensely the nature of entertainment.

One of the first points to remember about the PC game *Doom* is that it synthesizes pictures in first-person perspective in REAL TIME. The rendering of surfaces, removal of hidden lines and surfaces, lighting, collision detection, texture mapping, and above all, near/far sound integration and digital mixing of sounds, is all done in real time using an extremely powerful graphics and sound engine that runs on an ordinary IBM-compatible PC. *Doom* was the first proper fully featured engine, and since then there have been many released. The PC game *Decent 2* probably has one of the best rendering engines at the moment, because it not only allows full 3D tunnels and terrain, but also a proper 3D map viewing technique that is not available on other engines. Figures 26-12 to Figures 26-15 illustrate what is possible using *Doom*'s original engine.[15]

Figure 26-12 shows a view you might see if you went upstairs. First, the place is a lot brighter because of the sunlight. You can also see mountains in the distance through a window, and you should be able to reach up to the ledge—though not fall through the window glass. In actual fact, when wearing the VR helmet, the window would just be a brick wall in the real world (!). It's the total immersion visor that paints the scenery into your eye to make you believe that you are looking outside.

Figure 26-13 shows how objects such as the ball can float in the air, defying physics in the virtual world. At that rate, you could, for example, stand under a

SUMMARY, CONCLUSIONS, AND RECOMMENDATIONS 589

FIGURE 26-12: What you might view if you went upstairs.

FIGURE 26-13: How objects such as the ball can float in the air defying physics.

waterfall and not get wet. What you need to do is link some sort of reality with the virtual world so that these virtual and physically impossible objects have meaning. If you walk up to the ball in Figure 26-13, a sound is made and the ball disappears, indicating that you have taken the ball. Similar schemes will be employed in the holodeck to link virtual objects to real responses in the real world. You could, for example, have a windmill floating in mid air. As the blades swing past, you could generate a noise in the audio system of the helmet and use a wind actuator to blow wind in the direction of the user.

Figure 26-14 shows a picture of a wall where there is a slight contrast difference in the middle. In *Doom*, this is usually a secret panel, and Figure 26-15 shows the panel opened.

The secret panel is only opened when the user goes up to the wall and pushes on the wall. In the holodeck, this action could be fed from a glove worn with the helmet or a switch held on either hand.

As well as secret switches, *Doom* also has doors that can only be opened when a key is found. The keys and doors are color coded—and they could just as well be numbered or display patterns that must match on both key and door. The keys are picked up in a similar manner to the blue ball in Figure 26-13. In the holodeck, these keys could be real keys or they could be virtual objects as in *Doom*. In *Doom*, the players often have to find keys in a sequence, because some keys are in locked rooms for which you must find the other keys before you can retrieve these.

FIGURE 26-14: A picture of a wall where there is a slight contrast difference in the middle.

SUMMARY, CONCLUSIONS, AND RECOMMENDATIONS   591

FIGURE 26-15: The panel opened.

Doom has a lot of monsters. You open a door and you never know what to expect! This is the thrill of playing the game. In a holodeck, a similar thrill is entirely feasible.

Adding noise, vibrations, and smells, and controlling the lighting conditions can all contribute to a terrifying atmosphere. You should be careful that you are not creating too much of a terrifying atmosphere—people could walk away with that fear instilled into them! For example, if you just landed in a snake pit or a spider-filled cave, it's guaranteed that you could develop a morbid fear of these things when you leave the holodeck. These fears get into your dreams and then into your real world. Therefore, you need to be a bit careful about how many monsters and how much gore you are prepared to screen as with any movie.

Doom-Style Plots

Many other plots could be used. A list of examples are as follows:

- You could walk through an entrance, and as you do, you activate doors and a monster that comes chasing after you.
- You could fly through the air by jumping onto an aero bike, let's say. The aero bike would be moved in sync with the scene to make you think you are flying.
- You could get into elevators and go up/down levels. They could be real elevators made for robotics technology.

- You could fly a space ship and dodge asteroids. As the asteroids go past, thundering noises from all three directions could be made; and if you get too close, hammers could knock on the ground to give you the sensation that you hit debris.
- You could travel on a train, get off in the center of New York, and walk in that environment if you wanted to! The trains would be rocked to imitate the motion of going from A to B; and hammers hammer the ground to generate the vibration effects of travelling in a real train.
- You could be shrunk or enlarged, which means the objects in the scene are scaled to fit the perspective of your present stature.
- You could teleport, which simply means making a lot of bright flashes and noises while you change the scene using the robots.
- You could be in fog and navigate by sound only. Fog, fire, smoke, and other objects could hide monsters from where they lurch forward.
- Similarly, you could be in a dark tunnel and navigate by a light source only.
- You could be in a swamp or a farm, and the smell actuators make sure there is no olfactory mistake!
- You could be wading through water or taking a rest beside the seashore—and although water is not a good thing to have with robots or electronics equipment, in the future, waterproofed versions and controlled baths could be used to imitate this functionality.

Generic Plots

There are many plots you can take from movies and computer games, and add a lot of terrain details and variations. The users are guided through the terrain in circles as new terrain is synthesized in front of them while the old ones are *erased* behind them. There should be at least two levels, and that is sufficient for the users to go up buildings and climb down into pits. You can easily make a Esher type of environment, where as you go in circles, you appear to be climbing all the time; whereas in reality, every now and then you descend very gentle slopes.

It's always better to take existing plots and known formulas and re-tell them in your movie rather than create ingenious plots. When you have a unique plot, it's always better again to reserve just one per movie, as that is the only thing that will then be remembered. If you do too many new things, the user is just going to be confused while he or she tries to work out what happened!

Finally, since the movies are only a few minutes, they should be exhilarating. In the future, they could be more relaxing, but the thrills should be high so that the users need to relax before going on to the next movie—which might involve a 10 to 15 minute wait.

Holo-Hotels

Virtual hotels are living space spaces that you take with you on a *floppy disk*. Imagine your house and what it looks like—you could take that with you to a holo-hotel, put the disk in, and the room transforms to what you appreciate as home.

Would you need to wear the VR helmet? That depends on your taste and the levels to which technology will evolve over the coming decades. It is quite probable that the holo-hotel will provide you with a room layout and color scheme that you like with objects in that scene that you would appreciate in a hotel room.

There could be a water bed if you prefer it; local swimming pool just beside your bed, which can disappear when you're done with it; and all manner of other things that normally would take up a lot of space for any particular visitor. So, if you are a busy executive, your office with its own computer, fax, tables, chairs, and projector could be built for you from the time you sign in to the time you open the door. Consequently, you could hold meetings in hotels as though you were at your own business premises. Then when you are done, you kick out your guests and turn it into your sleeping room. The potential of this technology to revolutionize the holiday and hotel industries is very high indeed.

Holo-Objects: The First Steps into Holodeck Technology

The first steps to developing the holodeck is in the formation of objects that can be viewed as different objects through the VR helmet. The robots in development at the Open University could be linked to a standard VR helmet so that you can see them through a virtual world and have them perform shape-changing functions as illustrated in Figure 26-16.[16]

The Global View

Holodeck technology can be enjoyed at its best when it's miniaturized. Entertainment is also a big repeating source of revenue. Together, it can fund rapid miniaturization of fractal robots. The smaller machines can in turn be used for micro-factory applications, medical applications, civil applications, and space applications. The holo-objects can be used for design and evaluation, speeding up the lead time for products to enter the market. The large numbers of robots needed (for the larger holodecks) can be a source of industrial activity that leads to the faster development of the smaller machines.

Finally, in the future, you may be able to touch and smell what you buy online.

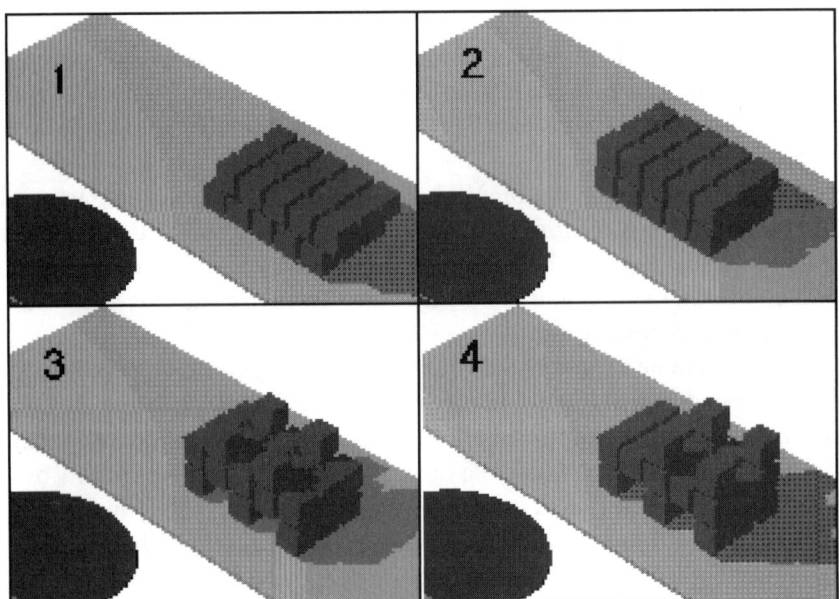

FIGURE 26-16: These basic *holo-objects* provide all the functionality of a development system needed to develop the complete holodeck systems of the future.

Building the Holographic Internet of the Future

For a glimpse of the Internet's future, step inside the holographic *cave* at Indiana University—Bloomington. There, in a tiny room with wall-to-wall computer screens, is a series of created dreamscapes. In one, the virtual traveler, wearing 3D goggles and manipulating a joystick, is confronted by terrifying, lizard-like creatures.

What makes these creatures come alive is a network that projects them from a computer at blazing speeds. The network in question is the backbone of Internet2, a joint university-business effort aimed at delivering the next generation of gee-whiz services that Web surfers have long awaited.

GRIDLOCK

Universities may have created the first Internet, but they turned it over to the public in 1994. As anyone who has ever experienced the World Wide Wait can attest, the Net quickly grew more congested than a Los Angeles freeway. Meanwhile, universities no longer had a high-speed, low-cost network of their own. Consequently, progress on new applications has been slow, and the ultimate promise of the Net—high-quality video, concert-like audio sound, remote-education programs, and telemedicine—hasn't been delivered.

Now, some 370 universities in partnership with companies such as Cisco Systems, Qwest, Nortel, and IBM are trying to remedy that by supporting Internet2 and its network, called Abilene (after the railhead in Abilene, Kansas that opened the West). Abilene operates at warp speed—some 2.4 gigabits a second, or 45,000 times faster than the typical modem. Ten sets of encyclopedias can move over fiber-optic lines supplied by Qwest in one second flat. The companies and universities are spending $500 million a year on the project.

The Abilene network, headquartered in Indianapolis, came into existence in 1999 and is growing rapidly. Now, university researchers again have an electronic playground where they can use supercomputers to manipulate vast amounts of information about, say, upper atmospheric disturbances. The hope is that as they develop visualization, holography, and other techniques, their knowledge will spread to the commercial Internet.

All these technologies are obviously too expensive for today's mass market. However, as costs fall and demand picks up, Internet2 executives figure the new technologies could trickle onto the commercial Net. It may never be realistic for individuals to own their own holographic caves, but they one day may be able to afford *telecubicles* that produce similar experiences. Internet2 execs even think that by 2004, they will be able to transmit holographic images via Abilene, à la *Star Trek*.

If just a few new technologies make it onto today's Net, the impact could be huge. At the University of North Carolina—Chapel Hill, for example, scientists are learning how to transmit the sense of touch. The implication is that someday, shoppers might be able to feel the fabric of an article of clothing they want to buy online. How does it work? Powerful 3D devices collect data about an object and shoot it across Abilene. At the receiving end, an Intel-made controller converts the data and applies electronic force to the human finger in a way that replicates the original object. Scientists have demonstrated the technique for only tiny objects (atoms) so far, but it could be just a matter of time before more computing power allows it to handle the fabric of a dress, for example.

Other technologies, such as full-screen video, are closer to fruition. IBM plans to open a facility in Illinois where it can help customers learn how to distribute full-screen video over the Net. Today's Internet will gradually morph into what the academic and research world has created for itself. If I'm right, you ain't seen nothing yet.

End Notes

1. Paul E. Keller and Arthur F. Gmitro (Pacific Northwest National Laboratory, K1-87, P.O. Box 999, Richland, WA 99352, USA), "Operational Parameters of an Opto-Electronic Neural Network Employing Fixed Planar Holographic Interconnects," Holographic Dimensions, Inc., 16115 Southwest 117th Avenue, #21-A, Miami, FL 33177, USA, 2001.
2. Ibid.
3. Ibid.
4. Ibid.
5. Myriam Meyer Matievic and M. B. Sayir (Institute of Mechanics, ETH Zentrum Swiss Federal Institute of Technology), "Holographic Measurement of Plastic Deformations," Holographic Dimensions, Inc., 16115 Southwest 117th Avenue, #21-A, Miami, FL 33177, USA, 2001.
6. Ibid.
7. Ibid.
8. "The Digital Holography System," Voxel, Inc., 5314 North 250 West, Suite 350, Provo, UT 84604, USA, 2001.
9. Ibid.
10. Ibid.
11. Ibid.
12. Ibid.
13. Joe Michael, "Fractal Shape Changing Robots," Robodyne Cybernetics, Ltd., 23 Portland Rise, London N4 2PT, United Kingdom, 2001.
14. Ibid.
15. Ibid.
16. Ibid.

PART VII

APPENDIXES

This part of the book contains the following 10 appendixes and a Glossary of holographic terms and acronyms:

A. International Hologram Manufacturers Association (IHMA) and List of Holographic-Related Sites
B. Holographic Research Condensed Report
C. Hologram Reference Wave Noise
D. Quantum Holography
E. Holographic Images
F. Frequently Asked Questions
G. A Generalized Geometry for ISAR Imaging
H. Hologramic Theory
I. Holographic Games
J. A Guide to the Companion CD-ROM

  Glossary

APPENDIX A

INTERNATIONAL HOLOGRAM MANUFACTURERS ASSOCIATION (IHMA) AND LIST OF HOLOGRAPHIC-RELATED SITES

The International Hologram Manufacturers Association (IHMA) was founded in 1993 to represent and promote the interests of hologram manufacturers and the hologram industry worldwide. It is a not-for-profit membership organization registered in the United Kingdom as a company limited by guarantee.

The International Hologram Manufacturers Association (IHMA) is made up of more than 60 of the world's leading hologram manufacturers (see Table A-1). It is dedicated to promoting the interests of these quality hologram manufacturers worldwide and to helping their customers to achieve their security, packaging, graphic, and other objectives through the efficient use of holography.

Membership is open to all producers of holograms, suppliers of equipment and materials for the manufacture of holograms, and hologram converters and finishers.

The IHMA requires its members to abide by a strict code of ethics, publishes technical guidelines for the use of holography, and operates the Hologram Image Register (a registry designed to prevent the copying of holograms used for security and other purposes). Contact: http://www.ihma.org.

The Code of Practice

Within its mission to protect and promote the common interests of the worldwide holography industry, the International Hologram Manufacturers Association has drawn up a Code of Practice for its members. All members agree to abide by this Code, which is intended to promote ethical business practices and the highest standards by its members:

1. Members of the IHMA undertake to operate their business in a manner which enhances the reputation of the holography industry and the IHMA. In dealings with each other, with customers and suppliers, members will observe the highest standards of business integrity and business ethics.

2. All IHMA members are bound by the IHMA's Articles of Association which oblige all members to follow this Code of Practice. A member in id reach of this obligation, or otherwise operating in such a way as to bring the IHMA and the holography industry into disrepute, should be brought to the attention of the Board for notification to a General Meeting which is empowered to suspend the membership or expel from membership such companies.

3. All members will issue standard conditions of contract which will be available to all customers or potential customers. In dealings with customers or potential customers, members will not knowingly misrepresent the characteristics and functioning of their products or their capabilities. Members will use their best endeavors to ensure that orders are delivered to customers as specified and on time.

4. Where by reason of factors outside their control, such as fire, flood, industrial unrest, bankruptcy or other factors, a member is unable to meet the contractual obligations for quantity or date delivery, then by agreement with the customer the member will endeavor to place that work with another IHMA member and will co-operate fully with that member. If required or necessary, and in strictest confidence, the IHMA will undertake the finding of a member to complete the order.

5. IHMA members will respect the intellectual property rights of their suppliers, sub-contractors and other members of the holography industry. Members undertake to operate within the Berne Convention and to make clear to customers and contractors who owns the designs, drawings, artwork, master and sub-master plates of any given hologram. The IHMA publishes guidelines to cover these matters.

6. Where appropriate, members undertake to pay a royalty or a royalty waiver fee and to honor all contractual obligations to an artist or hologram originator who has supplied holograms for reproduction.
7. IHMA members will operate within the environmental health and safety legislation that applies to them as a minimum standard of care and concern for the environment and the health and safety of their staff, customers and end users.
8. Every member of the IHMA accepts the responsibility of maintaining this Code of Practice and striving to enhance the reputation of the IHMA so that membership is seen to stand for quality and customer satisfaction.[1]

IHMA LIST OF MEMBERS

Table A-1 contains a partial membership list:[2]

TABLE A-1: List of IHMA members

Country	Members	E-Mail Address
England	API Universal Foils Limited	lholden@api-foils.co.uk
England	Applied Optical Technologies Plc	dtidmarsh@applied-holographics.co.uk
England	De La Rue Holographics	
England	Light Impressions International	
England	Spatial Imaging Limited	rob@holograms.co.uk
Bulgaria	Demax PLC	demax@aster.net
Czech Republic	LightGate AS	light@login.cz
Czech Republic	Optaglio sro	czholo@holo.cz
Finland	Starcke OY	ari-veli.starcke@starcke.fi
Hungary	Hologram Varga Miklós	mail@holovm.hu
Poland	Polskie Systemy Holograficzne SC	stigma@pol.pl
Russia	Concern Russian Security	krypto@orc.ru
Switzerland	Lange AG	info@lange-ag.ch

(*continues*)

TABLE A-1: (Continued)

Country	Members	E-Mail Address
Turkey	MTM Holografi & Guvenlik Sistemleri	mtmholography1@superonline.com
Ukraine	SPEKL	spekl@ukrnet.net
Yugoslavia	System Intelligence Products	office@sip.co.yu
France	Hologram Industries	Sales@Hologram-Industries.com
Germany	Hologram Company Rako GmbH	
Germany	Holographie Systems München	dausmann.hsm@t-online.de
Germany	Leonhard Kurz GmbH & Co	Christian.Hermann@kurz.de
Germany	Papierfabrik Louisenthal GmbH	Anette.Sieben@GDM.DE
Greece	Cavomit Foils	secure@taurus.com.gr
Italy	Diaures SpA	
Italy	Holo 3D SpA	olivier.pitavy@com.area.trieste.it
Italy	Istituto Poligrafico E Zecca Dello	a.ricci@ipzs.it
Portugal	Imprensa	Nacional-Casa Da Moeda
India	Holostik India Limited	holostik@del2.vsnl.net.in
Indonesia	P.T. Pura Barutama	purahg@kudus.wasantara.net.id
Turkey	Hologram Etiket Aksesuar San Ve Tic	
USA	AD 2000 Inc	peter@AD2000.com
USA	American BankNote Holographics	
USA	CFC International	
USA	Crown Roll Leaf	
USA	Dimensional Arts Inc	arts@holo.com
USA	DuPont Holographic Materials	
USA	FLEXcon	
USA	Holman Technology, Inc	holmtec@erols.com
USA	Holo-Source Corporation	Rob@Holo-source.com

TABLE A-1: *(Continued)*

Country	Members	E-Mail Address
USA	Holographic Label Converting Inc	
USA	HoloPak	www.holopak.com Halotek@holopak.com
USA	International Holographic Paper	hologramer@aol.com
USA	James River Products	jrp@richmond.infi.net
USA	Krystal Holographics Intl Inc	MBaker@khiinc.com
USA	NovaVision Inc	mmessmer@novavisioninc.com
USA	Reconnaissance Int'l	ltkontnik@reconnaissance-intl.com
USA	Pacific Holographics	pacholo@ix.netcom.com
USA	PROMA Technologies	
USA	Spectratek Inc	
Japan	Dai Nippon Printing Co	nushi-s@mail.dnp.co.jp
Japan	Toppan Printing Co	Yasumasa.Kamata@toppan.co.jp
Korea	HoloTopia	holotopia@holotopia.co.kr
South Africa	Guillemot Business Forms (PTY) Ltd	
Taiwan	AHEAD Optoelectronics Inc	julie@ahead.com.tw / sales@ahead.com.tw

Holographic-Related Sites

Optical Society of America   http://www.osa.org/

The International Society for Optical Engineering
 http://www.spie.org/meetings/calendar/index.cfm?fuseaction=home

LEOS: Lasers and Electro-Optics Society
 http://www.ieee.org/leos

MIT Spatial Imaging Lab
 http://spi.www.media.mit.edu/groups/spi/

Search The Light
 http://www.holo.com/peper/search.html

Voxel: Holographic medical imaging
http://www.voxel.com/

Lazing: A 3D Web site for deep thoughts
http://www.lazing.com/

Warning!! URLs and e-mail addresses can change without notice.

End Notes

1. Ian M. Lancaster, "The Source of Quality Holograms," IHMA, 7105 E. Powers Ave., Greenwood Village, CO 80111-1730, USA; and Runnymede Malthouse off Hummer Road, Egham, Surrey, TW20 9BD, England, 2001.
2. Ibid.

APPENDIX B

HOLOGRAPHIC RESEARCH CONDENSED REPORT

A holographic camera on which this project is based was designed and built by the Optical Engineering Laboratory (OEL) at the University of Warwick for Rolls Royce. The camera is used to make holograms of the transonic airflow as it passes through a rotating front fan on a testbed-based gas turbine engine. Rolls Royce conducted a major series of holographic tests on the Trent engine in 1998. As a result of the tests, they have recently returned the camera to Warwick for an update and re-fit. This has given the OEL the opportunity to explore several aspects of holographic technology.

Objectives

The following objectives have been specified by the Rolls Royce Electronics and Instrumentation group:

- The camera should be modified to accept two rolls of 127mm AGFA holographic film, such that it will be possible for the camera to be used at its full 250mm film width.
- An update of the current software should be provided. This will allow the Laser Firing System (LFS) to take control of both the holographic camera and the laser supplied by Lumonics.

- An update should be made of the hardware within the LFS to make it compatible with the latest LFS firing modes and functions.
- The LFS should be interfaced with a local computer (to be supplied with the system).
- The communications between the LFS and the computer will be performed to Rolls Royce's specification using an RS422 interface. The interface will be required to operate over a distance of 100m.
- The software should be modified so that a history of the firing conditions can be logged, including the information and descriptive codes printed on the holographic film by the camera's laser diode marking system.[1]

A secondary objective for the group is to undertake a marketing exercise to see how much industrial interest there is in OEL's work with holography. This seems appropriate, as over 1,000 inquiries were made to the OEL over the summer break in 1996 after the publication of two technical papers. A third objective is for the group to analyze holographic materials to be sourced from a Russian supplier and comment on their possible use.

ADDITIONAL OBJECTIVES

Holoviz Software specification additions of diagnostic capabilities to ensure that various functions of the system are fully operable, are as follows:

- Laser firing
- Shutter operations
- Plate movements
- Film winding
- Error messages where possible to identify malfunctions
- Laser feedback capabilities to ensure the laser fires within a certain time limit (including override capabilities)
- Laser feedback to read in and record incoming pulse signal from energy
- Monitor into D/a within LFS
- Access to feedback data (above)
- Remote testing routine (via RS422) connections over 100 meters

- Ensure film does not wind between pulses during double pulse
- Development of Help screens.[2]

Technical Research: Holography

Holography is a specialized form of photography that stores a three-dimensional image of an object on a photographic plate, as opposed to the two-dimensional image achieved with standard photography. The image is a store of the scattered light reflected from a test object, combined with a reference light beam. These two light sources interfere with one another and consequently contain enough information for a virtual recreation of the object.

The image created consists of the recording of an interference pattern from the light scattered by the object and the reference wave from the same source on a high-resolution, light-sensitive material. This recorded pattern is called a *hologram*. To visualize the image, the hologram is illuminated by the reference beam. The light beam incident to the hologram diffracts on the recorded pattern as with a diffraction grating. The observer can see through the hologram as a three-dimensional image of the object in the same place as during the recording.

To create a hologram, the light source must be coherent and of a single wavelength; therefore, the source used is usually a laser. This also provides a very intense beam with very little scattering. The beam from the laser is split into two separate beams, usually by an optical glass prism or similar device. One of these beams forms the reference beam, and the other is directed onto the test subject. This light is subsequently scattered from the subject's surface, and thus, a small fraction of it falls onto the photographic film together with the full reference beam. Lenses are used to spread the beams, so they cover the desired areas of photographic film and subject, respectively.

There are also considerable restrictions on the photographic materials that can be used in holography. The resolution must be very high, typically 1000–4000 lines per mm. The diffraction efficiency must also be high, and the light energy required to mark it must be low. This requirement is necessary, because even though the laser light intensity can be quite high, the duration of exposure is very small.

There are a number of typical types of hologram, each with a different method of creation. They all contain information about the object that is stored by the interference of the two aforementioned light beams on the photographic material. However, they all have unique setups, and consequently have separate advantages and applications. There are four basic types:

- **Fresnel holograms**: These are the most common variety, as described earlier.
- **Fourier transform holograms**: The object image is formed with the photographic plate in the back focal plane. This is usually achieved by the use of converging lenses. This allows the hologram to be stored on less photographic film than an equivalent Fresnel hologram. Thus, this type is particularly useful for creating high-resolution images.
- **Image holograms**: In this case, the object is placed very near to the photographic material when the hologram is taken. The advantage of the similar length paths for the two light beams is that the light need not be so coherent. It is even possible to use white light sources to produce and reproduce the hologram. This type is particularly useful for display/advertising type holograms.
- **Lensless Fourier transform hologram**: Here, the object is placed in the same plane as the reference light beam. Consequently, this method is only practical for two-dimensional holograms.

The Holoviz group is concerned with a laser and camera system that can create any of the first three types of hologram. The lensless Fourier transform hologram is limited to producing only two-dimensional images, and is therefore inadequate for Rolls Royce's requirements. It is unlikely that the camera will be required to create image holograms, as it is to be used to holograph air flows in a Trent engine (and perhaps the surrounding structure). Thus, the camera could not be placed too near the test region, as it would be at risk of being damaged. Therefore, it is believed that the difference in producing the remaining hologram types will not affect the group's challenges. The positioning of the test subject, camera, and the use of lenses is entirely dependent upon the Rolls Royce testing team; and it is appreciated that the group is in no position to recommend test procedures and is therefore unlikely to be required to do so.

In the context of obtaining vibration information on turbine blades (or in understanding the airflow as it passes through a rotating front fan on a testbed-based gas turbine engine), holographic techniques give a unique insight into this field. As an example, a wind tunnel is filled with small particles produced by a particle nebulizer spray, and a series of holograms are taken of the particles passing through the blades of a turbine engine. Existing technical applications for holography include:

- Vibration analysis
- Nondestructive testing
- Airflow visualization
- Data storage media[3]

AREAS OF CURRENT DEVELOPMENT/INTEREST

The use of holography covers many areas and, consequently, the paths of investigation being pursued are numerous, but the best documented at the moment include real-time holography, tomographic reconstruction, and research on different recording materials.

In real-time holography, the results can immediately be seen. It is seen as an area of a great deal of development. This has gone hand in hand with advances in available computing power. The most common arrangement is for the traditional camera and film to be replaced by a C.C.T.V. camera with high resolution, and directly linked to a computer and VDU.

Closely linked to real-time holography is another area of current research: that of tomographic reconstruction, which is the production of software and algorithms that are capable of analyzing interferometric holograms quickly, and with a minimum of specialist assistance. These are particularly commercially viable applications, as they will allow holography to be more approachable by smaller, less well-resourced companies as they wouldn't need a specialist staff.

A final field that is linked to the development of real-time holography is the research of different recording materials. A search for high-quality, economical, fully erasable film material that requires little or no development has now been underway for a long time.

POSSIBILITY OF ENCODING HOLOGRAPHIC IMAGES

Encoding holographic images allows the automatic analysis software already developed by the OEL to process the holographic information. The possibility of high-density encoding of holograms is explained here.

The selection of a particular hologram from an array of holograms is made by an X-Y beam deflector. The real image reconstructed from the selected hologram is projected onto a matrix of photo-transistors. The outputs from the photo-transistors are connected to the electronic circuits of the computer. A high-density recording is achieved through the use of Fourier transform hologram.

HOLOGRAPHIC CAMERA

The camera itself comprises:

- **Film wind motor**: RS 1.8ohm/phase, 1.8 degree step, 3V, 1.7A 150 × L toothed belt.
- **Shutter motor**: RS 1.8 degree step, 5V, 1.A.
- **Largest hologram size**: 256mm × 240mm.
- **Removable top cover**.
- **Types of objects**: Transparent, diffusely reflecting (flat and 3D).
- **Film type** : 8E–75, red sensitive, 5" or 10".

Proposed Modifications to the Camera

A subgroup of the main project-group studied the camera and its operation; then, with their observations in mind, undertook an exercise to produce a series of possible design solutions to modify the camera so that it could use both 5 and 250mm film. There were three conceptually different designs that were proposed and are outlined next.

Initial problems with the camera modifications largely revolved around a lack of knowledge of the exact requirements of the film to be used. Two spools on which the film would be contained were sent, but were not received until after the Christmas vacation. The information provided verbally was not sufficient. Constantly changing requirements had to be finalized, and confirmed in writing by Rolls Royce before any modifications could be started. The modifications were discussed with the technicians and the Project Director, and the possible solutions were drawn up and put in writing.

The reply from Rolls Royce enabled the modifications to be set in motion immediately. Suggestions for improving the current solution and removing the need for silicon sealant were made by the technicians. A larger portion of the spool was duly machined away than initially anticipated, and an insert was used. This enabled the split washer to be retained by springs enclosed in the insert that fit into a locator in the washer. This simplified fitting and removal of the split washer when film size changes were required.

SOFTWARE DEVELOPMENT

At the project outset, the existing software programmed in GWBasic was inadequate, and a primary objective of the group was to update it. Visual Basic was selected as the most appropriate language for the work required, due to its user-friendly format and high flexibility.

Development of the LFS software was outsourced to Lumonics. Outsourcing was necessary due to the lack of expertise in programming within the group and the fact that it would be an inappropriate use of project time for a group member to develop programming skills for such a management-based project.

WIRING REQUIREMENTS

The system requires positioning of the camera, laser, and Laser Firing System (LFS) close to the turbine engine. The environment is not suitable for the computer, which will be situated in another room. The harsh and noisy environment requires that all wiring is armored for electrical and mechanical protection. The wiring requirements specifications are as follows:

- 100m of RS422 twisted pair/shielded wire with RS232/422 converters
- RS232/422 converters
- Model 485Fi converter with 25-pin connector
- 230V UK Power Supply Unit (PSU)
- Model 222NF converter
- 25- to 9-pin adapter[4]

Communications

The original system that was implemented to control the Laser Firing System (LFS) used an RS-232 signaling protocol. It is for this reason that the transmission techniques studied will be of a near compatible type, so as to keep the system fundamentally the same. RS-232 is a long-established method of communication that is proven to work very well within its specified operating range. This system has been around for a long time and new signaling technologies have since emerged. These boast higher board rates and signaling distances. The following documentation examines the original signaling system, and looks at new methods, including RS-422 communications and current loop circuits. Rolls Royce has indicated strongly that they would like an RS-422 type of operation.

RS-422 SYSTEM ADVANTAGES

The RS-422 adapter plugs into the RS-232 communications ports and immediately converts the signaling technique to the RS-422. This has multitudinal benefits, including replacement upon failure, some confidence in operation, and somewhere to *dispel blame*. The physical appearance of the converters is small neat packages that would not only look professional in appearance, but be compact and uncluttered in operation. This is vital in a system such as this, with excessive wiring already. If the LFS system were to be upgraded in the future, allowing more speed on the communication or distance transmitted, the RS-232 would quickly become rather useless (little upgrade potential).

RS-422 will meet the 100m distance criteria easily and should offer good noise resistance. By using an opto-isolated converter with its own non-earthed power supply, ground loops will be avoided and noise should be absolutely minimal. If the system needs to cater for possible increases in specification at a later date (such as distance or data rate), the RS-422 would allow this to take place without further modifications.

RS-422 is the protocol Rolls Royce actually specified. In many situations such as this, if the solution is reasonably simple and practical, the customer's wishes are respected.

Using converters that plug into the existing communications ports means that the existing LFS will not need to be changed internally; thus, a well kept standard. Many systems could be modified as necessary, or one system could be used with any existing LFS. This also implies that any given LFS can revert to an old system extremely easily if required (see the sidebar, "Laser Firing System (LFS): Hardware, Software, and Limitations").

■ LASER FIRING SYSTEM (LFS): HARDWARE, SOFTWARE, AND LIMITATIONS

Laser Firing System

The LFS provides complete remote operation of a pulsed laser system.

Hardware:

- Model 222NF converter
- Converts RS232 to RS422
- Bidirectional
- Transmits and receives data signals
- No additional power required
- Surface mount technology for reliability
- DCE/DTE switch selectable
- 100m of twisted-pair shielded RS422 wire
- Model 485Fi converter
- Plugs directly into RS232 port on LFS
- Opto isolated allowing unit to operate with common mode peaks up to 75V without data corruption
- Bidirectional
- Externally powered with PSU supplied

Hardware Specification:

The following list provides information on the equipment used to allow the successful completion of holographic pictures. The details of the hardware are then described:

- Laser control unit
- Capacitor bank
- Power packs
- Holographic camera
- Laser
- Control computer (laptop)
- RS232/422 converters (one isolated, the other externally powered)
- Cooling facility
- Digital oscilloscope
- LFS
- Accessories (examples include armouring and electrical shielding of data transmission cables)

General Hardware Features:

- Minion microcomputer with 1 MHz clock speed
- EPROM chip to store several firing conditions

- Fires the laser amplifier and discharge tubes
- Controls the laser charging voltages
- Controls the internal and external pockel cells
- Sets up firing delay times
- External communications facility to control holographic/photographic camera, video or laptop computer
- Interface to laptop computer
- RS232 communications port
- RS232/422 conversion for transmission up to 100m
- Isolation of system provided by breaking ground loops; i.e., earth connection is insulated

Software Issues:

- EPROM 2764 contains instructions for controlling camera and laser simultaneously.
- Software written in Visual Basic 4.0 for control of camera and laser.
- Software user friendly.
- Provides records of previous holographic shoots.
- Facilitates different modes of use of laser; e.g., double pulse, delayed, etc.

Limitations of LFS:

- Minion slow (8 bit, 1 MHz)
- LFS remote to control computer[5]

LASER (Light Amplification by Stimulated Emission of Radiation)

The laser provides a surge of energy by a quantum mechanical effect that facilitates the phenomenon of taking holograms. The output energy of the laser varies according to the wavelength of light used. In turn, the *fuel* used will provide the laser system with electrons that upon returning to their stable level emit radiation. The *fuel* in the case of the Rolls Royce system is a powerful class 4 ruby laser (see the sidebar, "Laser Output Characteristics").

■ LASER OUTPUT CHARACTERISTICS

The laser is excited by a ruby rod, where:

1. $l = 694nm$
2. Maximum output = 3 J
3. Maximum power output = 10MW (for a 30ns pulse)

Examples of different modes of use: (as controlled by the software):
- Double pulse
- Delay pulse
- Single pulse
- Continuous pulse[6]

End Notes

1. "HOLOVIZ Condensed Report," School of Engineering, University of Warwick, Coventry CV4 7AL, UK, 2001.
2. Ibid.
3. Ibid.
4. Ibid.
5. Ibid.
6. Ibid.

APPENDIX C

HOLOGRAM REFERENCE WAVE NOISE

Deficiencies in the coding process of a hologram using reference waves are considered, and a correction for these deficiencies has been devised that will eliminate unwanted components. The physical processes of recording a hologram on film using a reference wave is reviewed and the standard mathematical interpretation is derived. A test image is used to calculate a Fourier transform and displayed as an image. A transform for each term in the equation is shown to contribute to unwanted parts in the total image; and it is shown that by subtracting the inverse transform of the unwanted components from the main image, a reference wave may reconstruct the original image without noise components.

Holograms Recorded on Film

To record a hologram, an object is placed near a photographic plate and illuminated by coherent light. A plane wave of coherent light, to be called a *reference wave*, is arranged to fall on a photographic plate at an angle called the *reference wave angle*. After exposing, the film is developed. This hologram is called an *amplitude hologram* because the darkened portions of the plate act to restrict transmittance at places where the phase does not constructively interfere to contribute to an image.

A phase object distorts a wavefront either because of varying refractive index within the material or because of varying thickness of the material.

This is demonstrated in the making of analogue phase holograms. After a holographic exposure is made, the photographic plate is developed and then immediately bathed in a bleach. The bleached plate is transparent under normal light, but reveals a good image when illuminated at the appropriate angle with coherent light. The film contains intensity information and the phase information about the original image. Both amplitude and phase are coded in the intensity variations. Bleached holograms would not work if only the transmittance of a film was altered. The refractive index of film is coincidentally also changed by the exposure process. Next is a representation of a typical black-and-white film composition:

- Supercoat of gelatin for resisting abrasion
- Emulsion of gelatin with a suspension of silver halide emulsion
- Substrate layer to promote adhesion of emulsion to film base
- Film base of cellulose triacetate or related polymer
- Backing layer to reduce curling during exposure

Holography may use film of this type or sometimes, instead of a flexible film with a cellulose backing, a glass plate is used. The intensity recording takes place in the silver halide emulsion where exposure to light converts parts of the grains in the emulsion to metallic silver. Developing the film spreads this conversion throughout each grain. The film is then fixed whereby the film is immersed in a bath and the remaining silver halide removed. To make a phase hologram, the plate is immersed in a solution of bleach. This is usually a brew of salts that will oxidize the metallic silver.

For a bleached photographic image, silver halide grains have a refractive index of 2.25 and are imbedded in a gelatin with an index of about 1.535. A bleached unexposed plate coated with gelatin and unexposed grains produces a refractive index somewhere in between. The refractive index of points on a photographic plate is considerably more complex than a function of these refractive indexes. The relief variations on the surface of an exposed emulsion are caused in part by changes in the volume of the silver salts, but variations of thickness are also controlled by stretching and compression which are apparently large and represent the most significant contribution to the relief.

These forces may be observed to cause the curling of developed film while it dries after being rinsed. Some of this thickness change takes place during the drying process and may be observed in making a hologram. After developing a photographic plate, it is removed from the bleach bath, washed in water containing a water-repelling emulsion, and superficially dried. The plate may be placed back on the holographic table and observed under coherent light. The holographic image on the bleached plate may be seen to improve over a period of about an hour as the surface dries.

The efficiency of a hologram is defined as the ratio of incident light on the phase object and the amount of light that is used to reconstruct an image. The advantage of the phase hologram is that because there is no attenuation of light as it passes through the phase object, nearly 100 percent of the light contributes to the formation of an image. Amplitude holograms may be only 2 or 3 percent.

The Role of the Reference Wave

Gábor relates that his first holograms were degraded by two images superimposed over one another (Gábor, 1972). In 1962, Leith and Upatnieks were able to use the laser as a superior coherent light source, and by using an angled reference wave the two images could be separated, one as a real image and the other a virtual image. Analogue holograms commonly utilize a reference wave oriented at 45 degrees to the plate, allowing an object to stand illuminated in front of a plate and a reference wave to also strike the plate from the side. Illumination of the plate at the same angle reproduces the object in its original position.

The reference wave should not be necessary for a computer-generated hologram, but there is the problem of removing the reflected image. The question remains as to whether it is possible to devise a hologram that forms a single image.

The Fourier Transform

The Fourier transform may describe the disturbance of a wavefront in the far field. Although three-dimensional information is carried by the fresnel transform, it is a computationally more expensive algorithm. The far field case is explored because of its comparative simplicity, the existence of a fast Fourier calculation, and the possibility of using the Fourier transform to obtain a wave description of an object and then modifying this solution to approximate the fresnel situation.

If an object is assumed to be at a large distance from the recording plate, then depth information is lost and the object is regarded as flat. Such an object is often portrayed as a slide with a picture on it. The slide may be divided into a rectangular array of pixels, each with some amplitude. Each pixel may also have a varying phase, induced by the thickness of the emulsion on the slide; but here, the image is assumed to have a fixed phase of zero across the object. The Fourier transform of this array has complex components and represents the wave disturbance on the holographic plate. This array is called the spectral image since it mirrors the spatial frequency distribution of the image on the slide.

To simulate an amplitude hologram on the computer, the Fourier transform of a real image is applied to a Fast Fourier transform, giving a two-dimensional array of complex values representing the wave disturbance in the holographic plane. A reference wave is added to this complex file and multiplied by its conjugate to obtain the intensity value that would be recorded on the film. These values are also referred to as the transmittance of the holographic plate. The reconstruction process is modeled by multiplying the transmittance by a term representing a plane wave illuminating the holographic plate striking the plate at some angle. An image is only formed when this wave strikes the plate at the same angle as the reference wave. At this angle, the reconstruction wave has the same spatial frequency, or phase function, as the reference wave. This reconstructed wave is then applied to an inverse Fourier transform to obtain an image.

The phase hologram is simulated by treating the values of the transmittance function as phase variation. An incident wave is retarded by a phase angle equal in magnitude of the transmittance. This is mathematically expressed by taking the cosine and sine of the intensities at each point on the plate. The resulting complex array is multiplied by the reconstruction wave and then input to a Fourier transform.

The Triangular Aperture and Its Fourier Transform

A method has been devised to obtain an equation representing the Fourier transform of a triangle of any shape and position. Summing individual wavelets over many points within the aperture is one way to calculate the wave disturbance at a point in front of the aperture. Instead of this point-by-point summation, the Fourier transform gives the disturbance at distant points for the aperture as a whole. The equation representing the triangular aperture is used in a computer program to output a complex array representing the Fraunhofer wave disturbance that is applied to a point-by-point fast Fourier transform to reconstruct the hypothesized image. The correct reconstruction of the original image verifies the correctness of the equations and the program. The Fourier shift theorem is used to position a shape in a window, and then the addition theorem is employed to merge several images in a single window and compose an image of multiple shapes. The C code of the program suite that performs these tasks is supplied in the appendix.

The triangular Fourier transform equation is used in the Fresnel-Kirkoff approximation integral to obtain an expression for the 3D version of the triangle. This is used to generate a complex array representing a hologram of a 3D triangle. This is applied to a fresnel transform to create a three-dimensional array shown to contain a triangle inclined within the array.

A method of theoretically calculating the holograms of nonexistent three-dimensional objects was first reported by Lohman in 1967. These considerations were initially concentrated on two-dimensional objects with the main goal to create spatial filters. Shortly thereafter, three-dimensional objects were calculated. These approaches to the problem were restricted to objects represented as a set of three-dimensional points. Every point is considered to spherically reflect incident radiation and its influence on a point in a holographic plane; it is summed according to the law of superposition of waves.

Such integration is computationally intensive, involving a calculation matching every point on the defined object to every pixel in the calculated hologram. This also involves the complexity of the Fresnel transform expression in three dimensions.

An integration process rather than a point-by-point method may be used to reduce this computation. Instead of defining an object as a set of points, it may instead be represented as a reflecting surface or aperture with a defined shape that diffracts an incident coherent wave. The Fraunhofer expression for the electromagnetic disturbance created by a square aperture may be shown to reduce to a sinc function. This expression is not suitable for 3D image synthesis, but there is a method by which the simplicity of the Fourier transform for surfaces making up an object may be retained, and the three-dimensional aspects of the surfaces later added to the expression. This implies that the two-dimensional Fourier transform for surfaces may be the basis for a three-dimensional representation.

Research into improving holographic images using the two-dimensional Fourier transform was employed by a number of honors students at Basser Department of Computer Science. Images produced by the students were reported as poor, but stimulated many good ideas to get better results including the suggestion to calculate the Fourier representation of a triangle. The development of a triangle primitive is advantageous, since any three-dimensional object may be perfectly tessellated with triangles; however, this has been avoided because of the apparent complexity of the Fourier transform. Most holography images have been based on the use of the rectangle that, when centered on the origin of the real plane, possesses the sinc^2 function as its Fourier transform.

Even with the N log N efficiency of this algorithm, the rendering of objects with sample points close together can be computationally expensive. The cost of a point-by-point Fourier transform may be reduced by considering the influence of many points at once. Expressions for the influence of points arranged in slits and squares and rectangles is well known, and now also the triangle. If an expression is derived, and is believed to represent the Fourier transform of a surface patch, it may be tested by applying the inverse fast Fourier transform to regain its original image.

APPENDIX D

QUANTUM HOLOGRAPHY

Quantum holography is a novel information-processing paradigm defining the science of consciousness and the morphology and dynamics of the brain/mind as testable hypotheses. The viewpoint is the physical, not the mathematical foundations of computation. Quantum holography constitutes a novel paradigm of nonclassical information-processing machines of which functional magnetic resonance tomograph imaging is already a most impressive realization. Such imaging is the hottest topic in cognitive neuroscience, for what experimentally is a revealing noninvasive look at the activity of the living brain. *Holography* is used as a generic term. Such information processing may be self-organized—concerning analogue computation performed quantum mechanically; and proceeds by selection and learning non algorithms, using phase conjugate adaptive resonance. Since quantum holography concerns nonlocal quantum interference, it employs phase and not amplitude or "bit" gates and geometric; not logic encoding and decoding, so that arbitrarily complex experiential knowledge arriving at such machines' sensory apparatus in the form of object image bearing "illumination" may be transduced, stored, and transmitted as signals without loss of information, which now concerns patterns of energy and not representations of bits. That is, a "movie" of all forms of sensory experience concerning a set of objects in motion in three dimensions may be transduced, stored, and processed without the need of human intervention to produce a mathematical specification or to turn such a specification into a program. In this paradigm, where the ontology, epistemology, syntax, and semantics of information all coincide, knowledge, perception, cognition, memory, learning, intelligence, language and creativity are no longer simply dictionary definitions, but well-defined physical processes applicable to particular quantum holographic information processing morphologies. These morphologies are in good accord with the empirically falsifiable hypothesis that the actual brain/mind is such a machine, as is the neuron, etc. Quantum mechanics is science's most accurate, tried and tested theory. It is therefore adequate to define the science of consciousness because the transactional interpretation is explicitly nonlocal and thereby consistent with recent tests of Bell's inequality. Yet, it is relativistically invariant, fully causal, and allows a long-sought-after visualization of macroscopic quantum phenomena denied not by the abstract quantum mechanical formalism, but by the Copenhagen interpretation. Such a visualization in the format of a three-dimensional movie elicited by phase conjugate adaptive resonance constitutes the stream of consciousness in the brain/mind defined through the emission/absorption model of quantum holography. This takes place against a background of unconscious activity. For example, there exists the taxonomization and storage of sensory experience and mental activity in the form of a fully distributed frequency organized paged holographic associative memory; the very considerable benefits of this form of memory; and the associated image and signal processing. The model confirms the nature of personal experience as having a conscious mental self or ghost in the machine in accordance with the traditional dualist metaphysics of a physical brain, and a separate mind that interact. It provides the mind with a mathematical specification, and a known quantum physical basis.

APPENDIX **E**

HOLOGRAPHIC IMAGES

HoloBank® is one of the world's largest selection of stock image embossed holograms. All HoloBank products are available plain or imprinted.

Holograms add a huge impact to any promotion. Stock image holograms make low runs and quick delivery both possible and cost effective. The imprinting of a stock image through matte etch or holographic diffraction add on makes it virtually custom.

In addition to stickers and magnets, HoloBank images can be used on the following products:[1]

- Bookmarks, business cards, buttons
  http://www.ad2000.com/bkmks.htm
- Calculators, coasters, keychains
  http://www.ad2000.com/calcs.htm
- Notebooks
  http://www.ad2000.com/notebk.htm
- Business cards
  http://www.ad2000.com/bus.htm
- Coasters
  http://www.ad2000.com/coast.htm

- Buttons
  http://www.ad2000.com/buttons.htm
- Keychains
  http://www.ad2000.com/keych.htm
- Diffraction labels
  http://www.ad2000.com/diff.htm
- HoloNotePads
  http://www.ad2000.com/notepad.htm
- Stickers and Magnets
  http://www.ad2000.com/stickmag.htm
- Tee shirts and garment transfers
  http://www.ad2000.com/tees.htm
- Wall plates
  http://www.ad2000.com/wall.htm

 Warning!! URLs are subject to change without notice.

End Notes

1. Peter Scheir, "HoloBank® Stock Hologram Collection," AD 2000, 780 State St., New Haven, CT 06511, USA, 2001.

APPENDIX F

FREQUENTLY ASKED QUESTIONS

What's a Hologram?

Holography is the process by which three-dimensional visual information is recorded, stored, and replayed. A hologram refers to the flat *picture* that displays a multidimensional image under proper illumination. Unlike any photograph, a holographic image has *parallax* (the ability to see a scene from many angles) and depth—just like the way you see things in the real world!

In other words, holography, in short, is a science of recording the light waves of a laser reflected off an object. When re-illuminated (placed under an ordinary light), the hologram will *replay* the image of the original object. The observer sees the recreated light waves, perceiving the object as if it were still there! Thus, a hologram is a photograph of light wave interference

How Do Holograms Work?

When a beam of light strikes a hologram at the proper angle, a multidimensional image will appear! Direct sunlight or a single overhead spotlight is the best way to illuminate holograms with deep imagery. Embossed holograms, with their shallower imagery, can be viewed when lighting is less than ideal (under florescent lights, for example).

How Are Holograms Made?

Holograms are made using precision optical instruments and special photosensitive materials, which are exposed with laser light. After the first hologram is made in the laboratory, the holographic image can be copied repeatedly on a variety of formats, depending on the intended application.

What's an Embossed Hologram?

For large runs of holograms to be produced in a cost-effective manner, a method was developed to emboss very complex microscopic patterns onto rolls of very thin plastic or foil materials. Light interacts with these patterns to create the holographic image.

What Are the Different Types of Embossed Holograms?

3D

These holograms display a three-dimensional image that looks identical to a *solid* object. The viewer can look around the top and sides of the image as if the actual object were there. Advantages: Under proper illumination, these holograms look spectacular! People are fascinated by images that project right out of the hologram or that appear to be behind it.

2D/3D

These holograms display a unique multilevel multicolor effect. These images have one or two levels of flat graphics *floating* above, or at the surface of, the hologram. Background artwork seems to be *under* or *behind* the hologram, giving the illusion of depth. Advantages: These intriguing images can be viewed under a wide variety of light sources. They are extremely eye-catching and are especially enjoyed by children.

STEREOGRAMS

This type of hologram originates with cinematic footage that is processed holographically to create a multidimensional effect. Advantages: Special effects, computer graphics, animated subjects, and even outdoor scenes can all be utilized as elements of a stereogram (also called a multiplex).

How Does a Hologram Differ from a Photograph?

The holographic film records an infinite number of views of the object, whereas a photograph records only one view. Thus, when viewing a hologram, your left eye sees a different apparent view of the object than your right eye, and the image appears three dimensional.

Can Moving Holograms Be Made?

Yes. Specifically filmed motion pictures from 3 to 20 seconds in length can be used to create animated holographic images that are astounding! This format is usually referred to as *integral* holography.

Can Holograms Be Reduced or Enlarged?

Most holographic techniques produce images that are identical in size to the original object. There are new developments that allow some enlargement and reduction if necessary.

Can Holograms Be Made from Flat Artwork or Photographs?

Yes, but the laser recording will recreate only what it sees. A hologram created from flat art looks quite different from the astounding, fully dimensional image created when a hologram is made from an object or sculpture.

Can You Control the Color of a Hologram?

Yes, to some extent you can. Certain types of holograms allow *individual color positions* for varied objects in your *scene*, while other types allow partial color *blends* at particular viewing angles.

Can Holograms Be Made from Living Subjects?

Yes. Using a special *pulse* laser, moving subjects can be *frozen* in time!

Is the Right Word Hologram or Holograph?

The preferred word is *hologram*. The dictionary defines a *holograph* as a hand-written document, as in a holographic will or deed. A *holographer* is someone who makes holograms. *Holography* is the word for the technology and art form. According to Isaac Asimov (science fiction author), a *holographist* is a person who collects or studies holography but does not make holograms. Things pertaining to holography are said to be *holographic.*

Are Holograms Projections?

No, holograms are not projected. Light fills up a hologram like plaster would fill up a cast. Technically, they are reconstructions of the light that reflected off the object.

If a Hologram Breaks, Is the Whole Image Visible in Each Piece?

No, each broken piece would let you see the image from its own unique perspective. Think of a hologram as a window. Anywhere you look through a window you see what's on the other side. If you were to paint the window black and scratch a hole in the paint on the left side of that window just big enough to look through, you would see everything on the other side of the window. It's like looking through a peephole. If you then scratch another viewing peephole somewhere on the right side of the window, you still can see through, but from a different perspective. This is the same effect that each broken piece of a hologram would display. Just remember that if you have two broken pieces taken from opposite sides of the hologram, and you are looking at an object that looks differently from each side, one piece may let you see just one of those sides, while the other piece will let you view the other side. So, you might say that each piece of a hologram stores information about the whole image, but from its own viewing angle. No two pieces will give you a view that is exactly the same.

How Many Lasers Do You Need to Make a Hologram?

One. However, you can shoot several different holograms on the same piece of film. Each holographic exposure can be shot with a different color laser if, for example, you are making a multicolor image of red, green, and blue. A color hologram can also be made with a single laser using tricks of the trade like emulsion swelling or multiple reference angles.

What Does the Word Laser Mean?

It is an acronym or abbreviation of the first letters of Light Amplification through Stimulated Emission of Radiation.

What Is Color?

Light is a wave. You see different sizes of light waves as different colors. It's something like the sizes of the strings of a harp making different musical notes. The largest strings of a harp make the lowest pitch notes, and the shortest strings make the highest pitched musical notes. A rainbow is like a harp with strings of light. The largest visible light waves are called red. Those a little smaller are called orange. A bit smaller and you get yellow. Smaller still is green. Smaller once again and you have blue. And the tiniest visible light waves are violet. Light waves smaller than violet are invisible and are called ultraviolet. Light waves larger than red are also invisible and are called infrared. The visible spectrum is from 400 nm (nanometers—1 billionth of a meter) for violet to 700 nm for red.

Is There a Word to Describe Where an Image Appears in a Hologram?

Yes, there are a few common ones that are quite helpful. If an image appears to be on the other side of the hologram, like looking through a window, it is called *virtual*. If an image jumps right out of the hologram and appears in front of the film, it is called *real*, since it has left the *virtual* world inside the film and entered the *real* world. When you flip a hologram over, the image is inside out and called *pseudoscopic*. Flip it back over and view it normally, right side out, and it is called *orthoscopic*. An image can be orthoscopic and real, or orthoscopic and virtual. Or an image can be pseudoscopic and real, or pseudoscopic and virtual. An image can be both real and virtual, as in the case of an image that starts behind the film and then protrudes right out of it. Holograms can be made (especially by artists) that have both orthoscopic and pseudoscopic images in them. Any combination of these terms is possible. So, to quickly rehash, Real = in front; Virtual = behind; Orthoscopic = right side out; Pseudoscopic = inside out.

How Are Images Made to Jump Out in Front of the Holographic Film?

As just explained in the previous response, images that protrude out in front of a piece of holographic film are called *real* images. Virtual image holograms are used as the masters for real image holograms. Most real image holograms are holograms of holograms. The basic concept is like the idea that a negative of a negative is a positive. In effect, when you typically make a hologram, it is orthoscopic (right side out) and virtual (the image appears behind the film). If you turn this orthoscopic and virtual image hologram over, the image you see is both pseudoscopic (inside out) and real (in front), since the spatial relationship of where the image is seen has flipped. If you use this image to record a second hologram, that image will be pseudoscopic (inside out), because you are recoding the pseudoscopic image of the first hologram and virtual. If you then turn it over, it is orthoscopic (right side out); because an inside out image of an inside out image is right side out and real; because, each time you flip a hologram over, you reverse from virtual image to real image. Voilà!

APPENDIX G

ISAR IMAGING: A GENERALIZED GEOMETRY

Traditional range/doppler ISAR techniques have inherent geometric limitations. By using concepts of microwave holography and tomography, a vector-based k space approach allows a more generalized geometry of the sampled Fourier space. By constructing a complete annulus in the polar sampling space, arbitrary apertures up to 360 degrees can be processed for *full-body* two-dimensional images. This processing also typically exhibits better resolution. The algorithm relies on linear interpolation for polar-Cartesian conversion. The general geometric formulation is also readily adaptable to arbitrary antenna configurations.

Conventional 2D ISAR processing uses the *keystone* method of resampling. The interpolation from polar to Cartesian is performed by two sequential one-dimensional interpolations. Typically, the resample space is a subset of the sample space resulting in a reduction of bandwidth and some loss of information. This method yields best results at relatively short azimuthal apertures.

By considering the whole sample space using a two-dimensional interpolation, the limitation on azimuth aperture is avoided, allowing arbitrary apertures up to 360 degrees. Further, the resample space retains the full bandwidth of the original sample space, resulting in improved resolution.

Finally, the two-dimensional formulation can be viewed as a special case of a general three-dimensional geometry in k space. The methods used in the two-dimensional case can also be generalized to process three-dimensional data.

Two-Dimensional ISAR

Figure G-1 typifies the standard layout of *keystone* processing.[1] Evenly spaced data points lie on radial lines aligned relative to azimuthal orientation to the imaged object (chirps). All data is taken at *waterline* elevation (orthogonal to the axis of object rotation). Since the two-dimensional Fourier transform requires data to be orthogonally aligned and evenly spaced, it must be resampled to a Cartesian grid. By overlaying a rectangular grid over our sample space, one can predict where the horizontal lines intersect each chirp. Each chirp is resampled so that all points lie on these intersections. This one-dimensional interpolation can be performed in place sequentially and independent of any other chirps. Utilizing the same methods, data can then be interpolated again along these resampled horizontal lines so that all data is now evenly spaced and orthogonal. A two-dimensional Fourier transform places the data in image space.

This method has the advantage of being efficient both in computation and memory. However, this method generally requires that the resample space be inside the sample space. This results in a loss of some of the original bandwidth, and hence a loss of resolution. This loss of information gets progressively worse as the azimuthal aperture increases. At some point, the algorithm fails to take into account new information as the aperture size reaches 180 degrees.

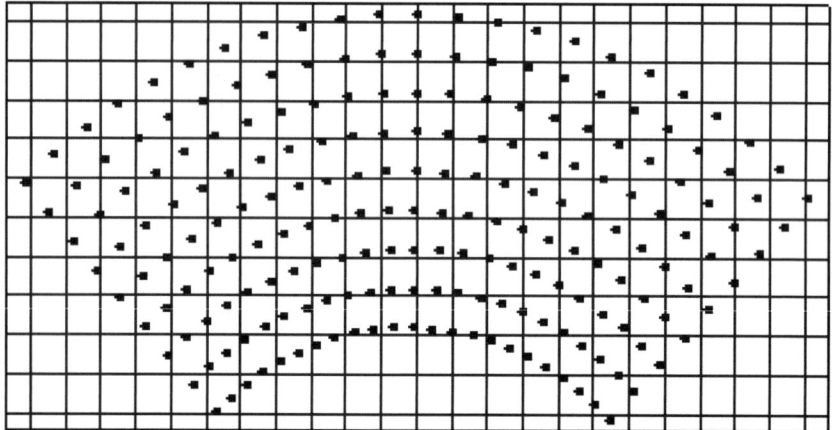

FIGURE G-1: Standard layout of keystone processing.

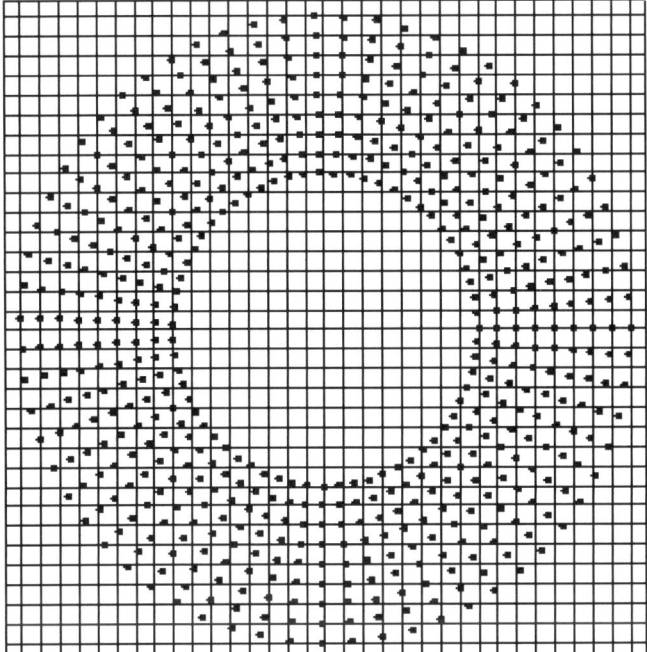

FIGURE G-2: Process any aperture up to a full 360 degrees.

These limitations can be avoided by considering the full data set and performing a two-dimensional interpolation. While one could perform this interpolation using the same inscribed resample grid, it is also possible to place the resample *box* so that it contains the whole sample space. Points not in the sample space are set to zero, effectively zero padding the resampled data. This results in some smoothing, but maximal resolution and information are retained. This method can be generalized to process any aperture up to a full 360 degrees (see Figure G-2).[2] Data is arranged as radial lines inside an annulus about the phase center of the imaged object. This method is similar to Fourier reconstructive tomography.

Several 2D interpolation schemes can be implemented. If the sample space is sufficiently dense, bilinear and nearest-neighbor techniques can be used. Typically, this involves locating the four nearest sample points that surround a grid point. Higher order methods such as bicubic splines or nonlinear time series methods may also be employed by using larger neighborhoods about resample points.

Three-Dimensional Geometry

The two-dimensional waterline case discussed in the preceding section can also be described as a special case of a generalized three-dimensional geometry utilizing holographic principles. If one measures data using elevation diversity as well as azimuthal diversity, one has data that occupies a three-dimensional sample k space.

Consider a standard mapping of spherical space as shown in Figure G-3.[3] The origin is the phase center of the illuminated object. The vectors Rt and Rr represent the range vectors of the transmit and receive antennae. Their vector sum determines the direction of the Fourier variable defined by k(Rt + Rr), where $k = 2pf/c$ is the freespace wavenumber of the sample. In the monostatic case, this reduces to $2k$, which is the distance variable r. However, r can simply be referenced as frequency, since k space maps linearly to frequency space. Therefore, one model's chirps in space as straight line vectors oriented toward the origin. At theta = 90 deg., this reduces to the

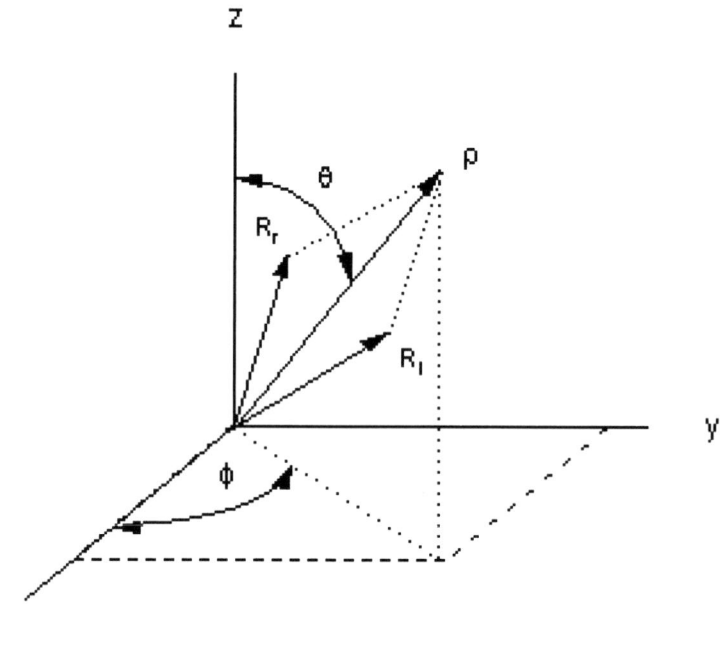

FIGURE G-3: Standard mapping of spherical space.

two-dimensional waterline case. For azimuth-over-elevation configurations, the sample space of constant elevation planes is described by Figure G-4.[4]

Displaying parametric surfaces of this model can be useful in gaining insight into the geometry of the sample/resample spaces. Since one is often physically constrained to short apertures, one generally processes only a small portion of the spherical sampling space. Figures G-5 and G-6 correspond to the sample space of the images presented at the end of this appendix (8-12 GHz, 67 deg-85 deg elevation, 167.5 deg-192.5 deg azimuth).[5] Data is typically collected as elevation planes, by holding elevation constant and varying azimuth.

Surfaces of constant elevation (see Figure G-5) result in sections of conical surfaces about the origin, gradually flattening into a plane at waterline. Because of this, one cannot use two-dimensional methods on individual elevation planes off of waterline, since the data is not planar. Distortion will increase due to curvature as one gets further away from waterline.

Surfaces of constant azimuth (see Figure G-6) result in flat planes about the origin. These surfaces are identical to waterline constant azimuth planes. This is not surprising, since this is the constant range spotlight SAR analogue of ISAR. As a result, these surfaces can be imaged to yield information in the orthogonal dimension to the waterline elevation plane (height).

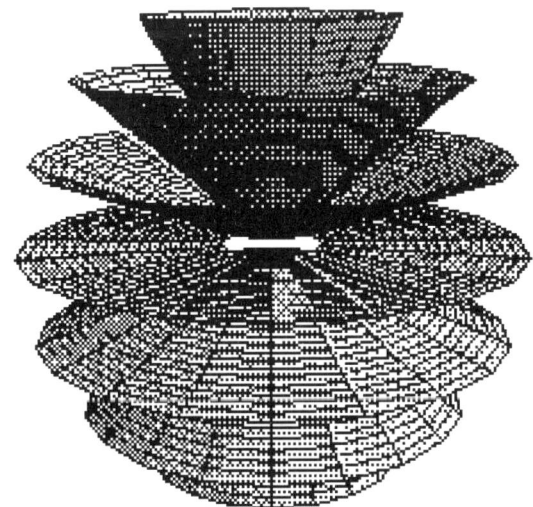

FIGURE G-4: Sample space of constant elevation planes.

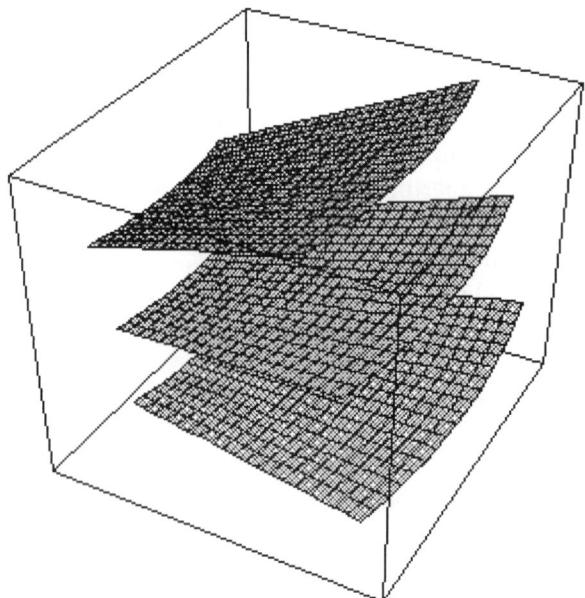

FIGURE G-5: Data is typically collected as elevation planes.

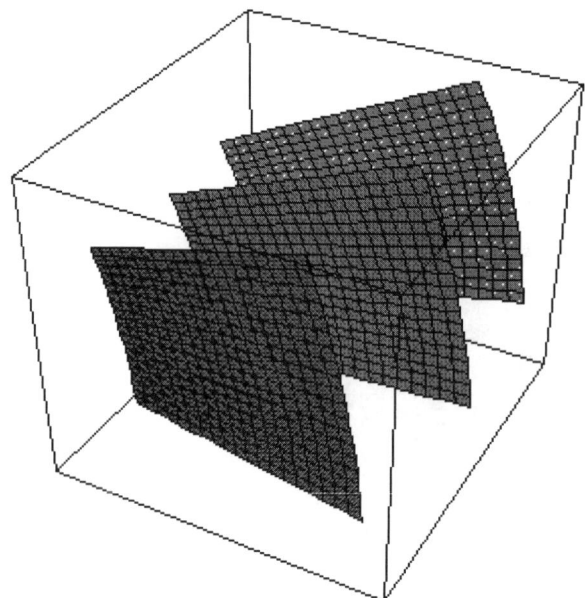

FIGURE G-6: Holding elevation constant and varying azimuth.

These visualizations of the sample space are useful in computing the limits of the resample volume by considering the sample volume extrema. For sample spaces that are approximately cubic, it is desirable to force the resample volume containing the sample space to be cubic (as in Figures G-5 and G-6). This bounding box is aligned relative to the axis of rotation and the waterline plane, then overlaid with a uniform lattice to simplify dimensioning and image processing.

While this example is a typical short angle case, the geometric model is general enough to accommodate any aperture. However, the sheer size of the data sets can make this impractical. As the apertures increase, the spherical null region in the middle can be extremely large relative to the sample space, forcing the resample grid to an impractical size.

Once the resample volume has been constructed, one can interpolate to a three-dimensional Cartesian grid using information from neighboring planes to plot points that will necessarily lie off the sampled surfaces. This is analogous to the two-dimensional case where points are resampled to locations that lie in between adjacent chirps.

2D Results

A test target of a 1/3 scale model C-29 (17.5' wingspan) was used to test the imaging algorithms based on the preceding discussion. The full body image was produced using 2-18 GHz with 10 MHz steps and 360 degrees aperture with 0.1 degree steps. Data was resampled to 2048 × 2048 square grid.

There is some loss of detail due to signal averaging. Scatterers that are specular for relatively short apertures will tend to be averaged out. This is due to shadowing from the target itself, or from the geometry of the individual scatterers.

3D Results

The 3D data set was 8-12 GHz with 10 MHz steps, 167.5 deg-192.5 deg azimuth with 0.1 deg. steps, 67 deg-85 deg elevation with 0.2 deg steps. Data was rcsamplcd to 256 × 256 × 256 cubic grid.

The result of the 3D processing is a cubic grid of the image space. For a two-dimensional display, one can take slices or tomograms of any of the three orthogonal planes, or consider any set of slices, projected onto an orthogonal plane. However, standard back-to-front spatial processing can result in weaker scatterers or sidelobes obscuring brighter ones. To avoid this problem, plots were generated taking the maximum value along a given line of sight. A vector-based, holographic/tomographic approach to ISAR processing offers a general model that can accommodate both two- and three-dimensional processing at arbitrary apertures and antenna/target configurations.

End Notes

1. Craig Malek (CompuQuest, Inc., 6564 Loisdale Court, Suite 425, Springfield, VA 22150), "A Generalized Geometry for ISAR Imaging," Holographic Dimensions, Inc., 16115 Southwest 117th Avenue, #21-A, Miami, FL 33177, USA, 2001.
2. Ibid.
3. Ibid.
4. Ibid.
5. Ibid.

APPENDIX H

HOLOGRAMIC THEORY

Theories are what make scientific revolutions happen, and we may just be in the early and uncertain rounds of one right now: the Hologramic theory. The abstract principles of the hologram form the basis of a theory to explain the brain's most elusive properties.

Hologramic theory was first proposed by physicists, not psychologists, physiologists, or anatomists. However, the theory has been used by a number of behavioral and biological scientists during the past few years to account for many diverse paradoxes about the brain. The theory has predicted incredible results in many laboratories, and has led to new and unsuspected possibilities of brain and mind. In addition, physical holograms may be used to model many features of mind that once seemed totally beyond the ken of science.

A generation ago, psychologist Karl Lashley found that he could dim the recollections of laboratory animals by destroying parts of their brains. However, it wasn't where he cut, it was how much. Lashley concluded that the engram, or memory trace, is distributed and repeated throughout the brain. He could never isolate the engram.

People's engrams have been equally elusive, although Lashley's doctrine may be too simplistic to cover the complex brains of humans, apes, and perhaps dolphins and whales. His basic idea works, however, even in us. Moreover, his beliefs have been sustained by many different kinds of experiments, some even conducted by his critics. A physiologist named E. Roy John and his colleagues have learned how to detect active memory signals in the living brains of rats and cats; the same signals recur in vastly different regions of the brain, just as Lashley's doctrine predicts. And, would you believe that shuffling a salamander's brain does not scramble the animal's mind? I might not.

The diffuse hologram makes no more common sense than the brain itself. Cut the hologram in half, and both halves still regenerate a whole scene. Try quarters, eighths—it's the same thing! The reason is that a whole code exists at every point in the medium. The reasons are mathematical, but basically, the hologramic code depends on ratios established within the medium, not absolute values. Like an angle, the code is relationships—abstract relationships, really.

At any rate, with very tiny pieces of a diffuse hologram, the regenerated image does blur, as did the memories of Lashley's animals. However, the information itself doesn't degenerate. It's in the communication of it; the loss of fidelity is an effect of *noise* on the very weak signal a tiny piece of hologram generates; it's like what static does to a weak radio signal or snow to a dim TV picture. Physiologist John believes that signal-noise differences account for specialized functions in particular parts of the brain.

A subtle dilemma pops up in work such as Lashley's. If engrams are distributed everywhere, how does the same brain house more than a single memory? Holographers have produced what are called multiple holograms—codes of many different scenes superimposed one on top of the other right in the very same medium! How can this be? In theory, the individual code can be made almost as small as a single geometric point. Thus, on any piece of medium, the number of individual codes may approach infinity, in theory.

In multiple holograms, the holographer keeps different scenes straight by manipulating such properties as color or the angle of the construction beam. Then, during decoding, by making comparable adjustments, the holographer can cause one whole scene to go off and another to come on—like an actor forgetting Othello so he can play Hamlet.

Consider, though, what happens if the holographer doesn't make adjustments during the coding process! During decoding, the holograms might act as though they had an imagination: scenes might merge that had never coexisted in physical reality at all. Or, depending on the scenes, the holograms might also act as though they were producing delirium tremens or were taking a trip on LSD.

Actually, some 400 reports exist alleging that memory may be transferred in chemical form from one rat to another—or even from a rat to a hamster! This is a very controversial subject, and it would take another book to do justice to both sides. But, one big imponderable used to be how a tiny molecule might store all the information required to make a memory. Hologramic theory doesn't settle the controversy. But holograms would have no trouble fitting on a molecule—in theory, at least.

There is evidence of memories on molecules. The most convincing case has come from studies of decision-making and memory in bacteria; believe it or not, organisms that don't even have brains. But, there's this major obstacle against reducing mind to molecules, and this extends to the chemical transfer of memory, too: the facts. Some physiologists have found no chemical changes associated with learning. Hologramic theory suggests reasons for such discrepancies; some mathematical, others that can be seen by way of physical simulation. Let's look at an example of the latter.

There are acoustical holograms—holograms made from sound instead of light. The information is gathered with a microphone and the code is displayed on a TV tube. It's possible to photograph the hologram for permanent storage, should someone pull the TV plug or change scenes.

Short- and long-term memory might work analogously. Or the transition from visual to some other mode might be modeled as follows. It is possible to shine a light through the photograph and visually regenerate a scene originally encoded by sound. But, consider this question: Where is the hologram? In the air near the microphone? In the vibrations of the microphone? In the electronics of the set? On the tube? In the photographic film? The answer is that it is in all these places. For the hologram is information. It is abstract relationships. The same code can exist in vastly different kinds of media and depend on different mechanisms—just as an angle of 30 degrees may be formed from ivory, mahogany, or extruded aluminum.

Finally, Hologramic theory does not free mind from media. It is no new variant of mind-body dualism. But, the theory's most far reaching implication is that many different mechanisms and media can store the same codes. Stored mind is not a thing. It is abstract relationships produced by things. In the sense of ratios, angles, and square roots, mind is a mathematic. No wonder it's hard to fathom.

APPENDIX I

HOLOGRAPHIC GAMES

3-D i*d™, a new patented holographic technology for CD developed by Nimbus and Applied Holographics, has been chosen to help promote a special edition of the PC Games disc at the Electronic Entertainment Expo (E3). Edge-to-Edge, or *full-face* application of the hologram technology, has been selected as an alternative to traditional, screen-printed label art on the magazine's monthly disc.

For the PC Games disc, Nimbus and Applied Holographics created a 3-D i*d hologram based on a *moonscape and planets* image. When viewed at various angles, the stunning three-dimensional scene *explodes* with amazing color and animation, a favorite theme among computer gamers. The whole purpose of the PC Games disc is to showcase the most exciting and cutting-edge computer games. With 3-D i*d, gamers can tell the disc by its cover. The 3D effect is amazing—there's nothing out there like it.

Nine thousand copies of the special edition disc are slated for distribution at E3 and will be packaged with PC Games. You can only get this disc at E3. It's a collector's item.

3-D i*d has been receiving a high level of attention from both the software and recorded music industries since its introduction. In addition to its use as an outstanding alternative to label art, 3-D i*d is one of the most effective deterrents against counterfeiting and piracy of CDs, a multibillion dollar problem in the software industry.

3-D i*d incorporates true 3D holograms onto CD without loss of disc capacity or playing time. In addition to graphic images, corporate logos, text, photographs, animated sequences, and unique covert images can be incorporated inside the 3-D i*d holograms. The covert images allow for authentication of the holographic images themselves, thus *raising the bar for even the most sophisticated counterfeiter.*

Nimbus is one of the world's leading, independent manufacturers of compact discs distributed throughout North America, the United Kingdom, and Continental Europe. Its operations service the software, entertainment, and recorded music industries. In 1984, Nimbus established one of the world's first CD pressing facilities in Europe and continues to be a front runner in offering state-of-the-art manufacturing innovations such as Enhanced CD (CD EXTRA), 3-D i*d holograms, and magazine-quality offset printing. In addition, the company was one of the first CD manufacturers to offer the new Digital Video Disc (DVD) format in the third quarter of 1996. Nimbus' corporate headquarters and East Coast manufacturing facilities are located in Charlottesville, Virginia, with additional facilities in Provo, Utah; Sunnyvale, California; and Cwmbran, South Wales. U.S. sales locations include Charlottesville, Virginia; Atlanta, Georgia; Gardena, California; Short Hills, New Jersey; and Sunnyvale, California.

APPENDIX J

GUIDE TO THE CD-ROM

The companion CD-ROM included with this book contains three folders, each containing unique images. The sets of images are found in the following folders:

- Book Figures
- Animation Files
- Web Sites

Note: This CD does not provide the necessary software for viewing these images, holograms and movies.

1) Book Figures

All of the figures included in the book to which this CD is a companion are included by chapter in this folder. These figures (.gif, .tif, .jpg, .vsd, .pcx, .esp, etc.) can be viewed with any graphics software, such as Paint Shop Pro, Visio or similar software products.

2) Animation Files

The animation illustrations (.qt) and movies (.mpeg, and .avi) were all composed using the Numerical Utility Displaying Ellipsoid Solids (NUDES) animation system, using figures composed by the author (Donald Herbison-Evans, Faculty of Applied Science, Central Queensland University,

Macleay Museum, University of Sydney, NSW 2006, Australia) and his students. They are appropriate for users with platforms with low horizontal resolution (PC, Mac, etc.), and for those with high resolution platforms (1200+ horizontal resolution (Silicon Graphics Indigo)). This is particularly critical for the images which are designed for parallel stereo viewing, as the dots at the top/bottom of these images should be about 3 to 5 centimetres apart for greatest comfort. The reader is urged to choose the images most suitable for the display being used. The animations are all 31 frame cyclic loops in the Macintosh quicktime format.

These files can be viewed with QuickTime, Windows Media Player, Real Player Plug-in, or similar viewing software. In QuickTime, movement of the figures can be achieved by manipulating the indicator underneath the figure back and forth. In video player software, the figures are moved by the software. See Central Queensland University, Macleay Museum web site at the following URL:

http://linus.socs.uts.edu.au/~don/fourd/fourb.html

to view more holographic images and videos; and, for more information about the preceding illustrations and movies just discussed.

3) Web Sites

This folder consists of the following:

- AD2000, Inc. Web Site (www.holobank.com)
- Reconnaissance International Web Site
- International Hologram Manufacturers Association Web Site
- Fractal Robot Web Site

VIEWING WEB SITES AND USE OF MATERIAL

The material in these web sites must be viewed with a Web browser. Open your Web browser first, then go to FILE/OPEN, and then find the appropriate folder on the CD-ROM. The Web browser, when opened, does not need to be connected to the Internet for viewing images. This material may be browsed on this CD and the Web Site owners must be contacted for permission to use any of the material found.

AD2000, Inc. Web Site

Contact: Peter Scheir

Company Name: AD2000, Inc.

Street Address: 780 State Street, New Haven, CT 06511

Phone: 203 624 6405

Web Site Address: www.holobank.com

E-mail Address: peter@ad2000.com

Description Of Web Site (This site best viewed on Microsoft Internet Explorer 4.5 or higher.)

Developed as the answer to short run or quick delivery projects, HoloBank is the world's largest selection of stock image holograms created specifically for advertising and promotional use. The HoloBank Web site is organized around three main sections: promotional products, stock images and security holograms. There are two simple approaches to finding what you need: identify a suitable product, then choose an image which fits the theme of your project from among those listed next to the product; or conversely, identify an image which fits your theme, and then choose from among the available products listed next to the image. If you are unable to find a product or stock hologram image which meets your requirements, please describe the project to your HoloBank sales representative, and we'll do our best to assemble an appropriate package. If you feel that fully-custom holography may be an option worth considering, refer to our section on custom holograms, then ask your HoloBank sales representative to provide you with further details.

Reconnaissance International Web Site

Contact: Lewis T. Kontnik

Company Name: Reconnaissance International

Street Address: 5650 Greenwood Plaza Blvd #225K, Greenwood Village, CO 80111

Phone: 303 779 1096

Web Site Address: www.Reconnaissance-Intl.com

E-mail Address: LTKontnik@Reconnaissance-Intl.com

UK Contact: Ian M Lancaster

Company Name: Reconnaissance International

Street Address: Runnymede Malthouse, Egham, Surrey, TW20 9BD, UK

Phone: (UK) +44 (0)1784 497008

E-mail Address: Ian.Lancaster@Reconnaissance-intl.com

Web site address: http://www.ihma.org

Description Of Web Site

Folder1: Hologram Manufacturers Association Web Site-The official web site of the International Hologram Manufacturers Association www.ihma.org

Folder 2: Reconnaissance Holography News Web Site-Web site for Holography News and Authentication News, published by Reconnaissance International.

Folder 3: Anti-counterfeiting News.pdf - Sample Issue Covering the Issues, Strategies and Technologies of Anti-counterfeiting, published by Reconnaissance International.

Folder 4: Holography News Dec 00.pdf-Sample Issue of the International Newsletter of the Holography Industry, published by Reconnaissance International.

International Hologram Manufacturers Association Web Site
(located in Reconnaissance International folder)

US Contact: Lew Kontnik

Comapany Name: International Hologram Manufacturers Association

Street Address: 5650 Greenwood Plaza Blvd #225K, Greenwood Village, CO 80111

Phone: 303 779 1096

E-mail Address: INFO@IHMA.org

Web site address: http://www.ihma.org

Description of Web Site

The IHMA is made up of more than 60 of the world's leading hologram manufacturers. We are dedicated to promoting the interests of these quality hologram manufacturers worldwide and to helping our customers to achieve their security, packaging, graphic and other objectives through the efficient use of holography.

Fractal Robot Web Site

Contact: Joseph Michael

Company Name: Robodyne Cybernetics Ltd.

Street Address: 23 Portland Rise, London N4 2PT, United Kingdom

Phone: (UK) (+44) 794 1100449

E-mail Address: JMichael@stellar.demon.co.uk

Web site address: http://www.stellar.demon.co.uk

Note: The data contained in the CD can be found at the following at web site URLs:

http://www.stellar.demon.co.uk/new/faqhtm

http://www.stellar.demon.co.uk/new/hmoviehtm

http://www.stellar.demon.co.uk/new/holodeck.htm

Description Of Web Site

The original information is taken from the Fractal Robot and Holodeck Information Web Site: http://www.stellar.demon.co.uk (this web site may be moved in the future to http://www.fractal-robots.com). The material has been developed at an amateur level (it is neither commercial nor intended to be academically rigorous material). It describes fractal robots and their futuristic applications such as the holodeck, Genesis Device, nanomachines, fractal tooling and other concepts. It is not in commercial production as yet (fractal robots have patent granted in many countries including USA and the author (Joseph Michael expects licensing and commercialization on a large scale soon). The web site is viewable with standard off the shelf web browsers.

CONTROL SOFTWARE

An original control software developed to move cubes around to make holodecks and other machines is included in the CD-ROM. The software has been tested on Windows platforms above Windows 3.11 (8Mb of RAM). The files are self contained (requiring no installation). This is beta code showing what can be done. No technical support is offered. Source code is available to download from Internet web site. Future demo versions of this program may become available for download from Internet web site.

VIDEO CLIPS

The video clips are in .AVI format (for Windows 3.11 and above) and show fractal robot 2nd generation prototypes. Despite appearances, the machines require considerable off-camera manual assistance to demonstrate, as there are no computers controlling the movement. Also included are computer animation video clips of theoretical possibilities, using fractal robots in 100% automated construction roles and other applications.

General System Requirements

If you wish to view all of the material on the CD-ROM, you'll need to have the following application software. It varies from folder to folder.

Graphics Viewing Software

Adobe Acrobat Reader

Web Browser

Media Viewer

QuickTime

Note: If you don't have viewing software, attempting to view the files will prompt you to download appropriate software.

DISCLAIMER!!

The information contained in this Appendix and CD-ROM is provided on the "as is" basis. Charles River Media, the author and third party contributors bear no responsibility whatsoever for any damage resulting from use of the software and system requirement recommendations contained in this Appendix or CD-ROM. Nothing in this Appendix or CD-ROM should be viewed as a commitment by Charles River Media and the author to release or maintain any product, version, feature, or performance level at any time. All brand names and product names mentioned in this book are trademarks or service marks of their respective companies. Any omission or misuse (of any kind) of service marks or trademarks should not be regarded as intent to infringe on the property of others. The publisher recognizes and respects all marks used by companies, manufacturers, and developers as a means to distinguish their products.

GLOSSARY OF TERMS AND ACRONYMS

3D viewing *Plain English*: Looking at a 3D picture and watching something pop out. *Technical English*: The act of viewing a 3D image with both eyes in order to experience stereoscopic vision and binocular depth perception.

absorption hologram A hologram that diffracts light by means of small patterns of exposed emulsion in the form of silver residue.

anaglyph A stereo image that requires glasses with red and green (or blue) lenses for 3D viewing. The two stereo images are printed on top of each other, but offset. To the naked eye, the image looks overlapping, doubled, and blurry. Traditionally, the image for the left eye is printed in red ink, and the right eye image is printed in green ink.

Angstrom Unit One 10-billionth of a meter; one-tenth of a nanometer, abbrev. Å.

AOM Acousto-optic modulator. A type of high-bandwidth SLM.

argon ion laser A continuous wave gas laser that is capable of emitting light in various wavelengths of both blue and green light, usually more powerful and more expensive than a helium-neon type laser.

autostereoscopic A 3D display type that presents left and right views of the imaged scene without special viewing aids. Examples include lenticular, parallax barrier, slice stacking, and holography. Some provide motion parallax by presenting more than two views.

basis fringe An elemental fringe precomputed to diffract light in a specific manner.

binocular Of or involving both eyes at once.

binocular depth perception *or* **stereoscopic depth perception** A visual skill that allows you to accurately judge relative distances between objects and perceive three-dimensional space. This is what you're playing with when you 3D view! The first computer-generated stereogram was invented by a vision scientist in 1960 specifically to test binocular depth perception. You don't have to have perfect BDP to see Magic Eye stereograms or other 3D illusions, but you have to have some.

binocular disparity The binocular depth cue affected by the slight differences between the two retinal images seen by each eye. The depth sensation caused by binocular disparity is called *stereopsis*.

binocular vision Vision as a result of both eyes working as a team; when both eyes work together smoothly, accurately, equally and simultaneously.

Cheops A digital image processing platform originally designed to explore scalable digital TV and real-time image encoding and decoding for the Television of Tomorrow (TVOT) consortium at the MIT Media Laboratory.

coherence volume That volume of space in which an object may be placed and be expected to make a successful hologram; is defined by the tolerable path difference of the reference and object beams; also referred to as *coherence length* or *depth*.

coherent Implies a definite phase relationship between wave of light or other radiation being emitted.

constructive interference The effects resulting from a superimposition of coherent wavelengths of light; for example, where a crest is superimposed on a crest.

Continuous Wave (CW) In referring to laser, this means that the energy emitted is continuous. A CW laser can be turned on and off like a normal light bulb, unlike a ruby laser, which emits its energy in pulses sometimes lasting less than a fraction of a billionth of a second.

convergence A binocular depth cue affected when the eyes rotate to align the retinal images seen by each eye.

CR Compression ratio.

cross-viewing *or* **cross-eyed viewing** A free viewing technique in which the lines of sight of the two eyes aim and meet at a point in front of the 3D image; the lines of sight of the eyes CROSS in front of the image.

Dennis Gábor A Hungarian-born physicist, the discoverer of the holographic technique and father of holography; proved that the phase information of light reflected or transmitted by an object could be captured by the interference of monochromatic, coherent light.

destructive interference The effects of the superimposition of a crest over a trough. The higher amplitude of the crest is cancelled by the lower amplitude of the trough. This occurs frequently when the waves of light are out of phase.

DMD Deformable micro-mirror device. A micromechanical SLM.

electromagnetic radiation EMR can be defined as waves of emitted energy characterized by the coexistence of electric and magnetic fields being propagated through air at approximately 186,000 miles per second. The visible light region is a miniscule portion of the entire EM spectrum.

electron A stable elementary particle having a negative charge. The electron orbits about the nucleus of the atom at a given distance from the nucleus. If an electron is raised to a higher energy level (i.e., more distant from the nucleus) by an energy input, then the electron will give off energy as it falls back to a lower energy level. This energy is sometimes visible light.

etalon An optical component made of fused quartz; sometimes used on a laser to filter out all other modes and ensure pure monochromaticity, thus greatly increasing the coherence volume.

excited state The condition of an electron when it has been raised to a higher energy level by some external force. In a laser, the force is electrical energy from the power supply.

free viewing Viewing 3D stereo images with the naked eye, without the aid of any special equipment such as stereoscopes, red/green glasses, polarized lenses, or VR equipment.

fringe The holographic pattern that is either recorded optically or generated computationally, and used to diffract light to form an image.

ground state The state of lowest energy of an atomic or molecular system.

helium-neon laser A continuous-wave gas laser emitting light in the visible red region at 6328 Å; also, the least expensive and most common laser.

Hertz A unit of frequency in the International System of Units. It is equal to one cycle per second—1 CPS.

hogel Holographic element. A small piece of hologram that has homogeneous diffraction properties.

hogel vector A sampled hogel spectrum, specifying the diffractive purpose of a hogel.

hologram A photograph that contains information about intensity and phase of light reflected by an object. When illuminated at the correct angle with a sufficiently coherent source, it will yield a diffracted wave that is identical in amplitude and phase distribution with the light reflected from the original object. The resultant three-dimensional image can be viewed or photographed.

holography The technique of capturing, on photosensitive material, the image of an object that contains the implitude, wavelength, and phase of the light reflected by that object. The result is a three-dimensional image of that object.

hololine A horizontal line of samples of a holographic fringe pattern.

holovideo An electronic interactive 3D holographic display system that uses holographic fringes to cause light to diffract and form a 3D image.

HPO Horizontal-parallax-only. A type of hologram that exhibits horizontal motion parallax horizontally but not vertically.

HVS Human visual system.

image hologram A type of hologram that uses a lens to focus the object information onto the film plane. A hologram is made of the object as imaged by the lens.

image resolution The number of resolvable image features in the lateral dimensions of an image.

image volume The space that may be occupied by a 3D image.

in-line hologram A type of hologram in which the reference beam is brought to the holographic film at the same angle or on axis with the object beam; a single-beam arrangement.

krypton-argon ion laser A continuous-wave gas laser that can laser in the blue, green and red region; is more powerful and more expensive that the helium-neon laser.

laser An acronym for light amplification by stimulated emission of radiation. It provides a source of light that can be phase coherent, and an intense beam can be attained by use of resonance techniques.

LCD Liquid crystal display. An electro-optic SLM.

lenticular Most people have seen this type of 3D image in a card shop (perhaps as a postcard of a religious figure). A lenticular is a thin, portable, full-color stereo picture. Thin plastic "lenses" placed over the photograph restrict the view of each eye to a particular part of the picture. Lenticulars can be composed on a computer, but (as far as this writer knows) cannot be viewed on a computer.

MB 1,048,576 bytes.

MAC Multiplication accumulation. A numerical calculation consisting of one multiplication and one addition.

mode One of several possible states of oscillation that may be sustained in a resonant system.

multiplex hologram A type of hologram that is formed by integrating a large number of photographs in a holographic manner; integral photography; usually only provides horizontal parallax.

nanometer One billionth of one meter, 1×10 meter.

noise Any undesirable disturbance or spurious signal.

occlusion, overlap Monocular depth cue affected when one part of an image is obstructed by another overlapping part.

ocular accommodation A monocular depth cue in which the eye senses depth by focusing at different distances.

off-axis reference beam A reference beam that travels a different path from the object or scene beam, and is brought to the holographic film from a different angle than the object or scene modulated beam; requires a twin-beam arrangement.

orthoscopic image The reconstructed image that maintains the same spatial relationships of the object(s) as they were when holographed; usually the virtual image.

oscillation A periodical change in the amount of energy contained in an electrical, atomic, or mechanical system.

parallax, motion parallax A (monocular) depth cue sensed from the apparent change in the lateral displacements among objects in a scene as the viewer moves. A display that provides parallax allows the viewer to move around the object scene.

parallax resolution The number of different perspective views available to the viewer.

parallel-viewing *or* **the parallel method** A free-viewing technique in which the lines of sight of the two eyes aim and meet at a point beyond and behind the 3D image; the eyes move outward (away from the nose) toward PARALLEL lines of sight. (NOTE: You view all Magic Eyes stereograms with the parallel method!)

phase That portion of a period or cycle through which a quantity (in this case, a wave of electromagnetic radiation) has proceeded from an arbitrary point; for example, the crest or highest point in amplitude of its wavelength, to the next crest after passing through zero.

phase hologram A hologram that refracts light by means of different thickness of a transparent substance; commonly, a bleached hologram.

photon A quantum or discrete package of electromagnetic radiation.

pictorial depth cues The monocular depth cues found in 2D images, including occlusion, linear perspective, texture gradient, aerial perspective, shading, and relative sizes.

pitch Sampling pitch. The number of samples per unit length in a discretized digital fringe pattern.

plane wave A wave of coherent laser light before it is changed or scattered by interfering with an object. A reference beam is essentially a plane wave.

population inversion The condition that occurs when the greater majority of electrons are in the higher energy level rater than the ground state.

pseudoscopic image That reconstructed image that reverses the spatial relationships of the objects from that which it was when holographed; usually the real image.

pulse ruby laser A solid-state laser the heart of which is a ruby rod comprised of aluminum oxide mixed with a small amount of chromium. When properly excited and amplified, the ruby emits light in the visible red region. When pulsed, it can emit powerful bursts of light energy commonly around 20 billionths of a second or 20 nanoseconds, thus virtually eliminating the concern with object movement in holography.

random dot stereogram The prototype for S.I.R.D.S. and Magic Eye stereograms was invented in1959 by Bela Julesz, vision scientist and recipient of the MacArthur Award (a.k.a., the "genius" award). With the random dot stereogram, Dr. Julesz created a computer-generated image that could be perceived only with binocular (two-eyed) depth perception. In one interview, Dr. Julesz was quoted as saying he dreams of inventing a tactile version of the random dot stereogram, so blind people could experience them. Is this person cool, or what? Of course, random dot stereograms are pretty cool, too.

real image The image that is projected out from the hologram toward the viewer.

reference beam That part of the laser beam that is not affected or changed by the object being holographed.

reflection hologram A type of hologram that is constructed by causing the object beam and reference beam to interfere from opposite sides of the holographic film or plate. In order to view the reconstructed image, incoherent light is reflected back toward the viewer from the hologram.

resonant cavity A chamber or enclosure whose design and physical characteristics permit only energy of a specific frequency to be propagated.

scene *or* **object beam** That part of the laser beam that is sent to the object being holographed, and is subsequently changed or modulated by the object before interfering with the reference beam on the photosensitive material.

Single Image Random Dot Stereogram (S.I.R.D.S.) A computer-generated stereogram in which the depth information is combined into a single image (a stereo pair is no longer visible to the naked eye). The first single image random dot stereogram was programmed on an Apple II computer in 1979 by Maureen Clarke and Christopher Tyler.

SLM Spatial light modulator. A device that modulates a beam of light.

spatial coherence That condition in which the light waves traveling through space are not only of the same frequency, but they are in phase in space.

stereo There's more to stereo than two speakers and a sound machine! According to *Webster's* dictionary, stereo comes from the Greek *stereos* for hard, firm, or solid, and it means combining form, solid, three-dimensional. With stereo sound or stereo vision, two inputs combine to create one unified perception of three-dimensional space.

stereogram Technically, any stereo picture, but in recent years this term has more commonly been used to refer to computer-generated Single Image Random Dot Stereograms. A long name, isn't it? That's why Single Image Random Dot

Stereograms are also called S.I.R.D.S. At this site, the term *stereogram* is used only to refer to S.I.R.D.S.

stereo pair The great-granddaddy of all 3D images! In 1838, Charles Wheatstone invented the first stereoscopic viewer for the 3D viewing of stereo pairs. Later in the nineteenth century, the viewing of stereo cards became a popular pastime. Perhaps your grandparents or great-grandparents had a hand-held stereoscope in their living room. To this day, all 3D pictures involve a pair of side-by-side stereo images. Even Magic Eye stereograms have stereo pairs hidden within them. To make things easier at this site, the term *stereo pair* will be used when an actual PAIR of images is visible to the naked eye.

stereo photographs Stereo photography was very popular in the nineteenth and early twentieth centuries (hello, Keystone). If you have never viewed a true stereo 3D photograph, you are missing a treat! The detail is amazing, and you really feel like you are immersed in the environment you are viewing. A good number of people still take stereo photographs.

stereoscopic A 3D display type that presents a left view of the imaged scene to the left eye, and a right view to the right eye. Examples include boom-mounted, head-mounted, and displays using polarizing glasses.

stereo vision *or* **stereoscopic vision** Two eye views combine in the brain to create the visual perception of one three-dimensional image.

TEM The lowest mode of oscillation in the laser; preferable because it gives the most uniform illumination and is the most stable mode of oscillation.

temporal coherence That condition in which the light waves are monochromatic (each cycle of the wave takes exactly the same time to pass a given point in space).

therapeutic 3D viewing Used in vision training, vision therapy, orthoptic therapy, behavioral optometry, or developmental optometry: 3D viewing for the sake of improving important visual skills such eye teaming, binocular coordination, and depth perception. Find out more about how stereo 3D images are used to improve vision.

transmission hologram A type of hologram that is constructed by causing the object beam and reference beam to interfere from the same side of the holographic film or plate. In order to view the reconstructed image, semi-coherent filtered light or very coherent laser light is transmitted to the viewer through the hologram.

transverse wave A wave motion in which the substance of the medium is displaced in a direction at right angles to the direction of propagation of the wave.

virtual image The image that appears in the space behind the hologram.

vision The act of perceiving visual information with the eyes, mind, and body.

volume hologram A type of hologram in which the angle difference between the object beam and reference beam is equal to or greater than 90 degrees. All reflection holograms are volume holograms.

wavelength The length of distance in the direction of travel of a wave motion between two arbitrary points in neighboring cycles having the same amplitude and phase.

INDEX

Abilene network, 595
Absorption holograms, 498–499
Acousto-optic addressing, 391–392
Acousto-optic modulators (AOMs), 127–128
    optical modulation and processing, 412–414
    SAW AOMs, 129, 413–414
Adhesives
    gluing Prismatic Illusions, 236
    self-adhesive labels, 231
Advertising applications, 5–6
Aerial perspective effects, 139–142
Aeronautical engineering
    combustion flow diagnostics, 180–182
    fracture mechanics, 569–672
    heads-up instrument panels, *xvi*
Affirm holograms, 227–228
Air copy counterfeiting, 539
Airy functions, 96
Alexander (English artist), 280–281
Amplifiers for Neodymium lasers, 94–95
Amplitude transmission holograms, 524, 617
Anaglyphs, 153
Angle multiplexing, 363
Animation
    aerial perspective effects, 139–142
    anaglyphs, 153
    artificial reality, 155
    autostereograms, 158–160
    background blurring techniques, 135–143
    binocular techniques, 150–163
    chiaroscuro techniques, 132–134
    cross polarized images, 155
    depth cueing, 143, 404
    double-beam display techniques, 162
    files on CD-ROM, 647
    hidden surface technique, 131–132
    lenticular film, 155–157
    monocular techniques, 131–150
    motion parallax and motion cues, 144–148
    multiplex (rainbow) holograms, 157–158
    perspective, 148
    polyocular techniques, 161
    projected holograms, 162–163
    Pulrich effect, 155–156
    shuttered glasses for stereoscopic viewing, 154
    texture gradients, 148–150
    vibrating mirror techniques, 162
    volume scanning, 161–163
    See also Motion picture holography
Anti-counterfeiting applications. *See* Counterfeiting
AOMs. *See* Acousto-optic modulators (AOMs)
Apertures, triangular, 620–622
Application methods for embossed holograms, 230–232
Applications
    flow visualization, 180–182
    interferometry, 59–81
    for micro-optics, 515
    museum exhibits, 100
    promotional, *xv*
    See also Art, holographic; Commercial / consumer applications; Security applications
Application to hologram substrates, 10, 230–232
Argon transfer holograms, 93
Arm's reach mode, 193, 414
Art, holographic
    aesthetic content and intent, 306–307
    Alexander (English sculptor and film maker), 280–281
    applications for holography, 104
    Berkhout, Rudi, 276–279
    business practices related to, 313–315
    commercial applications, 114
    commissioned works, 317
    computer holography and buoyant events, 302–303
    contemporary art scene as context for, 315
    critics, 317
    curators' concerns and expectations, 315–316
    defined and described, 305–306
    exhibitions of, 308–309, 311, 312–313
    galleries as outlets for, 312–315
    vs. giftware, 305–306
    Holographic Optical Elements (HOEs), 279
    HoloNet project, 354–355
    Holos Gallery of San Francisco, 286–288

664 INDEX

Art, holographic (*cont.*)
  illusionism, 295
  information storage used in, 275–276
  investment appreciation of, 307
  light in motion used in, 276–279
  market and sales, 285–288, 305–318, 306–307, 353
  museums, 314
  new technologies for holographic movies, 282–283
  price categories and sales, 310, 314
  promotion of, 316–318
  pseudoscopy phenomena, 297–299
  realism, 294, 296–297
  representation in holography, 292–297
  retinal rivalry effects, 299–300
  sales of, 305–318
  time asserted in, 273, 302–303
  university programs in, 309
Artifacts, off-axis electron holography
  Fresnel diffraction at biprism filament, 464–465
  geometric distortions, 466
  vignetting effects, 464–465
  windowing, 466
Artwork
  commercial hologram production, 111
  design specifications for, 199
Atomic theory, 487–488
Australian Holographics (AH), 92–93
Authentication. *See* Security applications
Author, e-mail address, *xxv*
Autonomous vehicle path following, 260–264, 266
Autostereograms, 158–160

Background, blurring of, 135–143
Backplanes, holographic optical backplanes, 219–220
Bandwidth compression, 165, 166–167
  diffraction-specific fringe computation, 439–440, 456–457
  fringlet bandwidth compression, 410
  spectral subsampling, 420
  See also Hogel-vector bandwidth compression
Basis fringes, 125, 420–421
  hogel-vector bandwidth compression, 167, 177
Beauty and the Beast (holographic movie), 280
Benton, Stephen A., xiv
Benton white-light transmission process, 38
Berkhout, Rudi (artist), 276–279
Binary optics, 514–515
Binocular techniques for animation
  crossview, 151–153
  freeview (parallel viewing), 150–151
Bipolar intensity summation, 446–447, 453–454
Bleaching, 499, 525–527
  bromine bleaching method, 525–527
  potassium dichromate bleaches, 93
  reflection holograms, 66–67
  silver halide emulsion, 499
Blurring
  background blurring techniques, 135–143
  hogel-vector bandwidth compression and, 175–176, 427–430
  image resolution and fringe manipulation, 167
  point spread and hogel-vector bandwidth compression, 427–430
Bohr, Niels, 487
Bragg's Law, 19–20
  Bragg-matching in dual-wavelength method, 372–375
Briefs, commercial hologram production, 111

Cameras, holographic, 610–611
CD-ROM contents, 657–659
Cheops image processing system, 173–175, 178–179, 403–404
  hogel-vector bandwidth compression implemented on, 434
  hogel-vector encoding implemented on, 425–426
Chiaroscuro, 132–134
Children's products, 247
Clear holographic film, 245
Code and code listings
  source code for hologram computation, 334
  translation and superposition of Fresnel zone plates, 328
Coherence curve, 509–510
Coherence length, 508–510
Coherence volume, 524
Collectibles, 247–249
Color
  advances in raman scattering, 98
  Bragg's Law and reflected light, 20
  color palettes for design specifications, 190
  controlling in holograms, 24, 629
  defined and described, 631
  disappearing in holograms, 32
  Hi-View 3D holograms, 182, 184, 186–187
  holovideo images, 129, 413
  holovideos + hogel vector arrays, 424
  papers and boards, colored, 244
  rainbow holograms, 15, 92–93
  saturated *vs.* pastel colors, 197

INDEX **665**

spectral harmonics and, 39
transmission holograms and color-
    recording, 52–53
Color transfer holography, 98–99
Combination holograms, 185
Comic books, 249, *xvii*
Commercial / consumer applications
    advertising, 5–6
    art, 114
    commercial hologram industry, 109–116
    design specifications for, 182, 186, 188–191
    design tips for hologram production,
        191–192
    display, 114
    entertainment, 114
    interactive graphics, 115
    labels, 114–115
    optimal holograms for, 182–186
    ornamental objects, 115
    packaging, 115
    products, 241–250
    pulsed laser holography, 10–18
    Ratcliffe's advances in, 92–104
    security and product authentication,
        115–116
    trade show exhibits, 11–14, 16
Compact phase conjugate holographic
    memory, 365–371
Comparison, visual
    design specifications, 190
    hologram mastering process, 202
Computational holography, 166
    basics, 416–417
Computational speed. *See* Speed,
    computational
Computation engines, 322, 330, 336
Computation techniques. *See* Diffraction-
    specific fringe computation; Hogel-vector
    bandwidth compression
Computer-generated holograms (CGH)
    binary optics, 514
    computation engines for, 322, 330, 336
    Fresnel zone plates, 322–328
    hologram programs *vs.* optical holography,
        337
    implementation of, 333–335
    look-up tables for interactive computation,
        442–444
    Nyquist theorem, 325
    rainbow holograms, 157–158
    translation and superposition of Fresnel zone
        plates, 327–328
    Web interface for, 329–333
    on World Wide Web, 321–337
Computer graphics, 120–130

electro-holography, 120–123
rendering with, 172
Computer holography, time and buoyant
    events, 302–303
Conoscopic holograms, 257–260
Consciousness
    engrams, 547
    hologramic theory, 641–643
    memory and, 641–642
Constructive interference, 30
Contact copy counterfeiting, 539, 540–541
Continuous Wave (CW) mastering, 97, 102
Converting process, 233–239
    commercial holograms, 112
    die cutting, 234–236
    equipment for, 237–239
    foil stamping, 236
    folding and gluing, 236
    laminating, 236
    printing phase, 239
    sheeting, 236–237, 239
Copyright issues, 552, 554
Costs
    application, 10
    custom hologram production, *xv–xvi*
    custom holograms, 534–535
    embossed holograms for security purposes,
        551
    helium-neon lasers, 489
    hologram security options, 533–534
    holovideos, 193
    for interferometry holography equipment,
        79–81
    1J system, 95
    metallized embossed holograms, 225
    origination, 9
    Polaroid Mirage holograms, 10
    production, 10
    3-D i*d holographic setup, 542–543
Counterfeiting
    air copy counterfeiting, 539
    anti-counterfeiting applications, 115–116,
        227–228
    complex imagery as countermeasure,
        547–548
    contact copy counterfeiting, 539, 540–541
    countermeasures, 540–541, 543–551
    Denisyuk holograms, 543–546, 550
    economics and counterfeiting cycles, 553
    embossed holograms, 543–544, 545, 550–551
    holograms, 537–541
    kinegrams, 546
    mechanical copying, 540, 547
    multiple connectivity as countermeasure,
        549–550

Counterfeiting (cont.)
  one-step copying, 547
  re-origination counterfeiting, 539, 546
  security benefits of holograms on products, 532–533
  special materials as countermeasure, 549
  techniques for, 538–540
Creating holograms. See Origination
The Creative Holography Index, 348–351
Credit cards, security and authentication holograms, 532
Cross-polarization of images, 155
Crossview binocular techniques, 151–153
CrystalEyes, 154
Custom holography
  described, 534–535
  security levels of, 534
Custom-made holograms
  production process for, 111–112
  types of, 113

Darkrooms
  construction of, 513
  darkroom techniques, 67–68
  troubleshooting the development process, 87–88
Data storage. See Memory, holographic
Deformable micro-mirror devices (DMDs), 127, 411
Deformation, measuring, 76
Denisyuk holograms, 98
  counterfeiting countermeasures, 543–546, 550
  manufacturing and use in security, 545–546
  variable information and processing parameters, 548–549, 558
Depth cueing, 143–147, 404
Depth of field, 524
Depth perception
  aerial perspective effects, 139–144
  background blurring techniques, 135–143
  chiaroscuro effects, 132–134
  depth cueing, 143–147, 404
  motion parallax and motion cues, 144–147
  perspective, 148
  texture gradients, 148–150
Design process, 182–191, also see design specifications
  inks for, 192
  tips for commercial holograms, 191–192
Design specifications, 182, 186
  artwork, 199
  color palettes, 190
  comparison, 190
  graphic effects, 189
  graphic software, 189
  registration, 190
  scanners, 188–189
  viewing software, 190
Destructive interference, 30
Developing, 203
  processing, 201–202
Die cutting, 234–236, 536
Diffraction gratings, 224
  in metallizing process, 226–227
Diffraction-specific fringe computation
  for electro-holography, 438–441
  equipment for, 125, 409–410
  hogel-vector bandwidth compression, 456–457
Diffraction tables, hogel-vector encoding, 421–423
Digital Holography System from Voxel, described, 572–573
Diode lasers, for playback, 52–56
Disparity, binocular depth cues, 145–146
Displays, holographic, 166, 417–418
  commercial applications, 114
  heads-up instrument panels in aircraft, *xvi*
  medical imaging, 575
  optical modulation and processing, 126–128
  point-of-purchase, 243, 249
  technologies for, 129–130, 455
Distance cues, 145–148
Division of amplitude transmission holograms, 525
Dot matrix holograms, 113
Dot matrix technology, master holograms and, 201
Dotz! holograms, 182–184
Double-exposure holography, problems encountered in, 78
Dual-wavelength storage method, 365, 372–378
Dust, mote diffraction, 65–66
DVDs, 360
Dynamic hologram refresher (DHR), 368
Dynamic refresh of memory, 368–371

E-commerce, 353
Electroforming
  defined, 212
  equipment for basic facility, 210–212
  jigs, 212
  in metallizing process, 226, 228–229
  shim production process, 205–210
  specifications for modules, 207–209
Electro-holography, 399–459
  applications for, 414–415
  described, 120
  diffraction-specific fringe computation for, 124–125, 438–441
  fringe computation for, 123–124, 405–410

optical modulation and processing, 405–407,
    411–414
  rendering, 436–437
  See also Holovideos
Electron holograms, off-axis, 462
  artifacts in, 464–466
  energy filtering, 474–475
  Fresnel diffraction at biprism filament,
    464–466
  interference experiments, 474–475
  neural nets, 467–469
  reconstruction scheme, 467–469
  vignetting effects, 464–465
Electron holography
  aberration correction, 463–464
  at atomic dimension, 462–464
  instrumentation, 462–464
  recording and reconstruction, 462–463
  Simplex Algorithmus for reconstruction,
    470–472
  transfer function, parameters, 472–474
  See also Electron holograms, off-axis
Electronics of lasers, 83–89
  reflection hologram setup, 85–86
  transmission hologram setup, 84
Embossed holograms, 7
  counterfeiting countermeasures, 543–544,
    550–551
  described, 628
  manufacturing and use in security, 545–546
  production of, 9–10
  types of, 628–629
  See also Embossing process
Embossed paper products, 231
Embossing process
  machines for, 218–223, 229
  manufacturing phase, 216
  metallized embossing, 225–232
  on oriented poly-propylene, 218
  origination phase, 216
  on paper, 217, 223
  on polyester, 217–218
  pre-origination phase, 216
  substrate transference, 216
Emulsions, 27–28
  See also Films
Encoding, optical rotation encoders, 220–222
Enlarging holograms, 24, 164–180
Entertainment applications, 114
Equipment
  computer generated holograms, 333–334
  for diffraction-specific fringe computation,
    125, 409–410
  for direct-encoding of hogel vectors,
    425–427
  electroforming process facility, 210–212

embossing machines, 218–223, 229
hardware to implement hogel-vector
    encoding, 173–175
for interferometry holography, 79–81
isolation tables, 510–513
light meters, 516
for onsite recording, 181–182
optical mounts, constructing, 513, 515–516
RS-232 Laser Firing System, 611–614
See also Lasers
Etalons, 96
Exposures
  lengths of, 33
  master holograms, 199, 203
  troubleshooting, 87, 88
Eyeball topographer, 257–260, 266

FAQ (Frequently Asked Questions), 627–632
Far-away mode, 193, 414
Film, holographic, lenticular film, 155
Film emulsion, 27–28
  as records of waveform patterns, 495
Films, holographic
  curling of, 618–619
  metallized films, 227–228
  removing development inhibiting coating,
    527
  resolution and selection of, 491
  typical composition of, 618
Finishing process, 112
Flow visualization techniques, 180–182
Foil stamping, in converting process, 236
Food products, 232, *xvi–xviii*
Fourier transform holograms, 608, 619–622
Fractal robots, 580, 582, 593
Freeview binocular techniques, 150–151
Frequently Asked Questions (FAQ), 627–632
Fresnel correlators, 253–257, 266
Fresnel diffraction at biprism filament, off-axis
    electron holography, 464–465
Fresnel zone plates, 322–328
  aligning images in UV space, 326
  calculation of, 324–325
  superposition and translation onto
    emulsion, 325–328
Fringe computation
  bipolar intensity summation, 446–447,
    453–454
  diffraction-specific approach, 409–410, 437,
    439–440
  electro-holography and, 405–410
  hardware for diffraction-specific
    computations, 125
  hogel-vector bandwidth compression,
    167–176, 414–415
  interference-based approach, 123–124, 408

Fringe computation (*cont.*)
  methods compared, 453–455
  point source summation, 444–446
Fringes
  basis fringes, 167, 177, 420–421
  diffraction-specific fringe computation, 409–410, 437, 439–440
  electro-holography and, 405
  fringe patterns, 120–121
  fringe printers, 194
  image resolution and manipulation of, 167
  look-up tables for precalculated, 447–453
  wave mechanics and, 484–486
Fringlets, 125–126, 194
  fringlet bandwidth compression, 410
  fringlet encoding, 437

Gábor, Dennis, 97–98, 389, 482, 501
Galleries
  for holographic art, 312–315
  Holos Gallery Web site, 345
  Light Fantastic Gallery, London, 313
Games, holographic, 647–648
  3-D i*d hologram technology, 645–646
  holodecks for, 588–592
General Optics Laboratory (GEOLA). *See* GEOLA (General Optics Laboratory)
GEOLA (General Optics Laboratory), 96–97
Geometric distortions, 466
Gerchberg-Saxton technique, 563–564
G5J system, 95
Glass plate holders, constructing, 32
Gluing, 236
Gold (24K) paper, 244
GRAM Archive project, 355
Graphic effects, 189
Graphics
  file formats for hologram output, 333
  output quality of hologram program, 335
Graphic software, design specifications for, 189
Graphics subsystems
  for encoding hogel-vectors, 175, 179, 427
  hogel-vector bandwidth compression implemented on, 434
  RE2 (SGI RealityEngine 2), 427, 434
Gratings, 495

Hardware. *See* Equipment
HEART (Holographic Electronic Authentication Recognition Tag) holograms, 228
Heerden, Pieter J. Van, 360
Helium-Neon lasers, 30
Helmholtz, Herman von, 297–298
Hidden surface algorithm, 132
High-density image encoding, 609–610
Hi-View 3D holograms, 182, 184, 186–187
Hiyama, Shigeo, 281–282
HOEs (Holographic Optical Elements). *See* Holographic Optical Elements (HOEs)
Hogels, 167
  encoded as fringlets, 125–126
  hogel-vectors and, 124
  See also Hogel-vector bandwidth compression; Hogel-vector decoding; Hogel-vector encoding
Hogel-vector bandwidth compression, 164–165, 167–176, 193–194, 415–436, 456–457
  basis fringes, 167, 170, 418
  color holograms, 173
  diffraction tables for direct encoding, 170–172
  encoding and decoding, 168
  imaging results, 176–177, 430–433
  point spread model for, 175–176, 427–430
  rapid linear superposition, 167, 418
  sampling and recovery, 168
  spatial coherence, 432–433
  spatial discretization, 167, 418
  spectral discretization, 167, 418
  spectral subsampling, 169–170
  speed, 177–180, 433–436
  3D computer graphics rendering for direct encoding, 172–173
  visual-bandwidth holography, 194
Hogel-vector decoding, 168, 410, 419–420, 421–425, 434–436
Hogel-vector encoding, 169–180
  equipment for, 425–427
  and holovideo, 401–402
  implementation, 173–176, 425–427
  sampling and recovery, 419–420
Hogel-vectors
  sampling and recovery, 419–420
  3D computer graphic rendering, 423–424
  See also Hogel-vector bandwidth compression; Hogel-vector decoding; Hogel-vector encoding
HoloBank
  products, 625–626
  security products, 536
  Web addresses, 625–626
Holocinema. *See* Motion picture holography
Holodecks, 579–594
  cinema systems, 586–587
  fractal robots, 580
  game applications of, 588–592
  holodeck movie production, 588–592
  holo-hotels, 593
  holo-objects, 593–594
  interaction, 583–584

INDEX   **669**

nano technology, 584
soundscaping in, 587–588
Star Trek series, 579–580
systems described, 580–594
telepresence and teleporting, 584–585
virtual design studios, 585–586
virtual reality helmets and suits, 581, 582–583
Holodisks, 185, 191
Holo-Gram, online publication, 348
Hologramic theory, 641–643
Holograms
  darkroom techniques, 67–68
  described and explained, 4, 27, 607, 627
  display options, 49–52
  enlarging or reducing, 24
  etymology of term, 288–289
  vs. holographic images, 291
  in-line, plane, transmission type, 500–501
  making, 628
  multiplex, 15, 157–158, 506–508
  vs. photographs, 629
  plane *vs.* volume, 502–503
  production process, 8–10
  redundancy, 4, 34–35, 630–631
  reflection holograms, 504–506
  resolution, 491
  theory and background, 17–20
  transmission holograms, 501
  types of described, 498–508
Holograph. *See* Holograms
Holographic Dimensions, Inc. (HDI), 273
Holographic Images, Inc. (HII), 273
Holographic industry, 109–116
Holographic optical backplanes, 219–220
Holographic Optical Elements (HOEs), 39–49
  artistic use of, 279
  copying large format holograms, 47
  dry photopolymer embossing (DPE), 48
  laser-Doppler velocimetry (LDV), 48
  ray tracing, 45–46
  types of, 45–48
Holographic process, test strips, 66–67
Holography
  conventional holography described, 573
  described, 3–4
  early experiments in, 37–39
  formats for, 6–8
  history of development, 37–39, 482, xiv
  multiple-exposure holography, 574
  photographic materials for, 607
  principles of, 23–35
  process basics described, 17–20
  public perceptions of, 101, 271–272
  theory, 641–643
Holography News, industry newsletter, 348, 349

Holo-hotels, 593
HoloNet, 354–355
Holo-objects, 593–594
Holoplex, 388
Holos Gallery of San Francisco, 286–288
Holovideos
  on CD-ROM, 654
  color holovideo images, 129
  costs of, 193
  displays, 120–123
  display systems, 401
  fringe computation, 123–124
  industry trends, 129–130
  resolution, 124
Horizontal-parallax-only (HPO) holograms, 457–458
  bandwidth savings and holovideos, 406
  computational holography and, 165–166
  imaging, 121–122
  look-up tables for interactive computation, 442–444
Hot-stamping, 10
  application method, 231
  integration/finishing process and, 112
How-to information. *See* Practical holography
Hybrid holograms, 228

Image holograms, 608
Images
  orthoscopic images, 632
  pseudoscopic (inside out) images, 34, 297–299, 506, 632
  real *vs.* virtual, 504–506
Industry trends, 109–116
Inks, design tips for commercial holograms, 192
Integral holography, 24, 269
  holocinema experiments, 280–281
Integral stereograms, 7–8
Interactive graphics, 115
  modes of, 193
Interactive holography, 399–401
  look-up tables, 440–455
  rendering images, 436–437
  3D holographic displays, 404–415
  3D imaging, information flow in, 407
  See also Holodecks; Holovideos
Interface design, parameters, and output, 329–333
Interference, wave, 28–30
  described and explained, 494–495
  diffraction-specific approach, 124–125
  fringe patterns, 120–121
  interference patterns, 32
  normal mode (nm) interference, 93
  standing waves, 29

INDEX

Interference holograms, 69–70
  defined and described, 75
  double-exposure method for, 75–76
  real-time method for, 75–76
Interference patterns, 32
  in interference holograms, 70
  master holograms and, 200
  See also Fringes
Interferograms. *See* Interferometry
Interferometry
  holders for holographic subjects, 70–73
  interference patterns, 70
  interferograms *vs.* phase extraction conoscopy, 261
  Michelson interferometry, 60–62
  optics of, 74–77
  photographic processes for, 77
  shear interferograms, 258–259
  spatial filtering and, 65–66
  used to test isolation tables, 512–513
International Hologram Manufacturers Association (IHMA), 532–533, 599–603
International Society for Optical Engineering (SPIE), 344
Internet, 339–356
  Abilene network, 595
  future trends in holographic Internet, 594–596
  interactivity and, 342–343
  on-line publications about holography, 348–351
  usage, 340–341
Internet holography
  early development of, 343–345
  World Wide Web and, 341–343
Internet Museum of Holography, 346
Iridescent paper, 245
ISAR imaging, 633–640
  keystone sampling method, 633–636
Ishikawa, Jun, 281–282
Isolation tables, constructing, 31

Jewelry, 115
Jigs, electroforming process, 209, 212
John, E. Roy, 641–642
Julesz, Bela, 662

Kinegrams, 546
Komar, Victor, 282

Labels
  commercial applications, 114–115
  as holographic products, 246–247
  pressure-sensitive, 10
  self-adhesive, 231
Laminating, 236

Lamination (application method), 231
Large-format holograms, 92–93
  laser-pulse techniques, 93–96
Laser-Doppler velocimetry (LDV), 48
Laser Firing Systems (LFS), 611–614
Laser-pulse holography
  commercial / consumer applications, 11–17
  vs. Continuous Wave (CW) mastering, 97
  described, 8, 11–17
  holographic cinematography and, 279
  large format and, 93–96
  material for large format holograms, 96
  requirements and specifications, 103
Lasers (Light Amplification by Stimulated Emission of Radiation)
  coherence curve, 509–510
  coherence length, 508–510
  described and explained, 489–491
  development and history, 486–487
  diode lasers for play back, 52–56
  electronics of, 83–89
  eye protection and, 62
  helium-neon lasers, 488–489
  laser light described, 482–483
  Neodymium *vs.* ruby, 94–95
  number required, 631
  ruby lasers, 94–95, 489, 611–615
  VCSEL arrays, 368–369
Lashley, Karl, 641
LCD stereograms, 113
LDV (laser-Doppler velocimetry), 48
Leith and Upatnieks, 336, 486
Lenses, 5, 495–496
Lensless Fourier transform holograms, 608
Lenticular film, 155
Light and lighting
  art, holographic, 276–279, 309–310
  chiaroscuro effects, 132–134
  direct light for replay, 112–113
  display options, 49–52
  laser light described and explained, 482–483
  of reflection holograms, 18
  sunlight and holograms, 34
  theories of, 484–486
  three-dimensional nature of, 25–26
  wave behavior described, 25–27
  white light illumination, 18
Light Fantastic Gallery, London, 313
Light meters, 516
Liquid-crystal displays (LCDs), 411
Liquid-crystal spatial light modulators, 126–127
Listserver for holographers, 344
Litho Canvas paper, 245
Litho Fuzz paper, 244
Litho Wood paper, 245

Living subjects, 24
Look-up tables, 458
  compared with other methods, 453–454
  data array reduction, 451–452
  for elemental fringes, 447–453
  interactive computation of holograms using, 440–455
  quantization, 450–451

Magnifying lenses in holograms, 5
Marketing, benefits of hologram security and authentication, 532–533
Marketing implications for holography, 92–104
Master holograms
  in commercial hologram production process, 111
  creation process, 197–203
  digital images, 198–199
  dot matrix technologies, 201
  exposures, 199, 203
  interference patterns, 200–201
  photoresists, 199
  processing of, 201–202
  3D imaging, 199–200
  visual comparison, 202
Mastering. See Master holograms
Mechanical counterfeiting, 540, 547
Medical applications
  advantages of holographic imaging, 577–578
  Digital Holography System from Voxel, 570–572, 575, 577–578
  eyeball topographer, 257–260, 266
  laser keratectomy, 260, 266
Mega-exhibitions, 308, 311
Memory, biological, 641–643
Memory, holographic
  advantages and disadvantages of, 360–361, 382–383
  art and nonlocal information storage, 275–276
  associative nature of, 389–390
  autonomous vehicle path following, 260–264
  compact phase conjugate holographic memory, 365–371
  crystal cube storage, 386–387
  data storage, 102, 104, 364–381, 386–387
  data transfer times, 393
  digital holographic storage systems, 388–389
  dual-wavelength storage method, 372–378
  holographic memory explained and described, 361–364
  lithium niobate crystals, 383–384
  master Fresnel zone plates and, 322–324
  memory crystals, 361–363
  multilaser holograms, 365, 378
  multiplexing developments, 386–389

  optical memories for data storage, 359–360
  vs. optical search, 391
  peristrophic memory systems, 260–264, 261–264
  photopolymer films, 383
  pixel-matching memory storage experiment, 365, 378–381
  playback accuracy, 363
  random rapid access to data, 382–383
  refreshing dynamic-read/write memory, 368–371
  signal strength developments, 381–382
  technological advances supporting, 385
Metallizing process
  described, 229–230
  electroforming phase, 226, 228–229
  origination phase, 226–228
Michelson interferometry, 60–62
Micro-optics, 514–515
Microscopes in holograms, 5, 102
Microwave holography and ISAR imaging, 633–640
Mirage holograms. See Polaroid Mirage Film
Mirror board, 243–244
MIT Spatial Imaging Lab, 399–459
  Web address, 603
Mixed media, 274–276
Moiré holograms, 569–571
Monocular techniques, hidden surfaces, 131–132
Mote diffraction, 65–66
Motion
  integral holography, 24
  pulsed laser holography techniques, 11
  See also Motion picture holography
Motion cues, 145–148
Motion parallax, 144–147
Motion picture holography, 271–283
  photonic cinema, 272–274
  technologies for production of, 273
  time and, 302
Multilaser holograms, 365, 378
Multiple images, 104
Multiplex holograms, 15, 157–158, 506–508
Multiplexing, holographic memory and, 260–264, 386–389, 392
Museums
  artifacts, archival recording of, *xvii*
  exhibits of holograms, 100
  Internet Museum of Holography, 346
  MIT Museum, 345, 347–348
  Primitive Lives exhibition, 100

Nano technologies, 584
National Geographic, holographic covers, 308, *xvii*

Negatives. *See* Shims
Neodymium lasers, 94–95
  YAG *vs.* YLF systems, 95–96
Neural nets, 467–469, 472–474, 561–562
  operational parameters, interconnectivity, 564–566
  opto-electronic neurons, 564, 565
  prototype-system results, 567–569
  systems architecture for, 562–564
  wave transfer functions, 472–473
New York Institute of Technology, Internet holography, 343
Nimbus CD International, Inc., 541–542
Normalization, 444
Normal mode (nm) interference, 93
Nyquist theorem, 325

Object beams, 32, 492–494
  multiple object beam transmission holograms, 524
Off-axis electron holography. *See* Electron holograms, off-axis
Onsite recording, 180–182
Optical Engineering Laboratory, research in holographic technology, 605–615
Optical holographic correlators, 251–252, 266
  target pattern recognition and tracking, 264–266
Optical modulation and processing, 126–128, 456
  acousto-optic modulators (AOMs), 412–414
  liquid-crystal displays (SLMs), 411
  micro-mirror devices (SLMs), 411
Optical mounts, constructing, 513, 515–516
Optical rotation encoders, 220–222
Optics
  binary optics, 514–515
  micro-optics, 514
  optical mount construction, 513, 515–516
Opto-Electronic Integrated Circuit (OEIC), memory refreshing, 368–369
Opto-electronic neurons, 564, 565
ORGEL (holographic movie), 281
Oriented poly propylene (OPP), 218
Origination of holograms, 8–9
Ornamental objects, 115
Orthoscopic images, 632

Packaging
  commercial applications, 115
  direct packaging of snack foods, 232
  Dotz! holograms, 182–184
Papers and boards
  colored, 244
  gold (24K) paper, 244
  iridescent paper, 245
  Litho Canvas paper, 245
  Litho Fuzz paper, 244
  Litho Wood paper, 245
  Prismatic Illusions, 242–243
  stock and specialty, 242–245
  as substrate, 217, 223
Parallax
  holovideos, 121
  motion parallax, 144–148
Parallel viewing, 150–153
Participate program of New York Institute of Technology, 343
Passivation booths, electroforming process, 206–209
Path following, 260–264, 266
Pattern recognition projects
  autonomous vehicle path following, 260–264, 266
  eyeball topographer, 257–260, 266
  Fresnel correlators, 253–257, 266
  optical holographic correlators, 251–252, 264–266
  target pattern recognition and tracking, 264–266
Patterns. *See* Fringes
Perception
  pseudoscopy, 34, 297–299, 506, 632
  psychological influences on, 295
  retinal rivalry, 299–300
  See also Depth perception
Peristrophic multiplexing, 260–264
Perspective, 148
  See also Depth perception
Phase holograms, 499, 617–618
  See also Bleaching
Photographic chemicals, safety precautions, 67–68, 86, 527
Photography
  vs. holography, 288–292, 491–497, 629
  stereoscopic photographs, 152–153
Photonic cinema, 272–274
Photopolymers, 383
Photo-processing for holograms, step-by-step directions, 77
Photoresists, 199
Piracy, anti-piracy technologies, 541–543
Pixel-matching memory storage experiment, 365, 378–381
Plane *vs.* volume holograms described, 502–503
Play back
  color-recording and transmission holograms, 52–53
  diode lasers for, 52–56
Pmp holograms, 227

Point-of-purchase displays, 114
Point source summation, 444–446
Point spread model for hogel-vector bandwidth
  compression, 175–176
Polarization of images, 155
Polaroid Mirage Film, 7
  production of holograms, 10
Polyester, as substrate, 217–218
Polymers, 383–384
Portraits
  holographic, 15–17, 269
  pulsed holograms, 8
Practical holography, 481–528
  division of amplitude transmission
    holograms, 524, 525
  experiments in, 524–525
  hologram basics, 516–517
  lasers for, 488–489
  object movement and object isolation,
    497–498
  reflection holograms, 522–523
  transmission holograms, 517–522
Precomputed elemental fringes, 447–453,
  447–454
Printers and printing
  fringe printers, 194
  on Prismatic Illusions, 239
Prismatic Illusions, 236
  stock holographic papers and boards, 242–243
Processing modules, shim production, 206
Production of holograms, 9–10
Production process, commercial holograms
  artwork, 111
  briefs, 111
  converting, 112
  integration/finishing, 112
  master holograms, 111
  production proofs, 111–112
Products
  custom manufacturing of, 247–249
  holographic papers and boards, 242–245
Promotional applications, *xv*
Proofs
  approval of, 185 n
  commercial hologram production, 111–112
Prototypes, embossed holograms, 219–222
Pseudoscopy, 34, 297–299, 506
  described, 632
Psychology
  hologramic theory, 641–642
  of perception, 295
Public perception of holography, 91, 101
  interest in holography, 345
  National Geographic cover's impact on, 308
  reactions to holograms, vx

Pulfrich effect, 155–157
Pulse holograms. *See* Laser-pulse holography
Puzzles, 247

Quantum Theory, 486, 487–488

Rainbow holograms, 15, 506–508
  computer-generated holograms (CGH),
    157–158
  large format, 92–93
Rainbow reflections, evidence of holographic
  process, 34
Raman scattering, 98
Random dot stereograms. *See* SIRDs
Rapid linear superposition, hogel-vector
  bandwidth compression, 167
Ratcliffe, David, 91–97
Real image projection transmission hologram,
  524
Real images
  described, 295–296, 632
  real image projection transmission
    hologram, 524
  vs. virtual images, 504–506
Real-time holography, 609
  tomographic reconstruction, 609
Reducing holograms, 24, 164–180
Redundancy of holograms, 4, 34–35
  explained and described, 496, 630–631
Reference angles, 31, 503
Reference beams, 31, 492–494
  angles for, 503
  collimated reference beams, 41–42
  in holographic memory systems, 362
  off-axis or twin beams, 501
Reference waves
  noise and, 617–618
  role of, 619
Reflection holograms, 7, 18–19
  described and explained, 504–506
  light and lighting of, 18
  making, 522–523
  reconstruction, 523
  scanning of, 73–74
  setup checklist, 522–523
  twin-beam reflection holograms, 63–64,
    508–510, 525
  viewing, 68
Registration, design specifications, 190
Re-origination counterfeiting, 539, 546
RE2 (SGI RealityEngine 2), 427
  hogel vector bandwidth compression
    implemented on, 434
Resolution
  bandwidth reduction and, 166–167

Resolution (*cont.*)
  hogel-vector bandwidth compression and, 175–176, 427–430
Resonance, 25
Retinal rivalry, 299–300
Retro-directive screens, 282
Rolls Royce Electronic and Instrumentation Group research, 605–616
Rotation encoder discs, 220–222
RS-232 Laser Firing System, 611–614

Safety considerations
  bromine bleaching, 527
  darkroom chemicals, 67–68, 86, 527
  lasers and eye protection, 62
Sales
  of holographic art, 305–318
  via e-commerce, 353
Scanned AOM systems, 127–128
Scanners, design specifications for, 188–189
Scanners, supermarket, *xvi*
Scene beams. *See* Object beams
Search The Light, 348–351, 352
  Web address, 603
Security applications
  anti-piracy technologies, 541–543
  benefits of security and authentication systems, 532–533
  combination holograms for, 182, 185
  commercial applications, 115–116
  counterfeiting countermeasures, 543–551
  data encryption signatures, 555
  3-D i*d, 542–543
  fundamentals, 531–532
  holographic counterfeiting, 537–540
  holographic locks, 389
  ID card systems, 556
  ideal security holograms, 551
  imprinting, custom, 535
  levels of security, 533–537
  machine-readable technologies, 540–541
  manufacturing and use of holograms in security, 545–546
  metallized films for, 227–228
  overt and covert devices in, 536
  product authentication systems, 557
  sequential numbering, 535
  special die cutting, 536
  stock holography options, customized, 535
  stock imagery, customized, 536
  tamper-evident adhesives, 535
  trends in anti-counterfeit security, 552–557
SGI RealityEngine 2, 427
Shadows, chiaroscuro effects, 132–134
Sheeting, 236–237

Shims
  defined and described, 205–210
  electroform modules, 206
  family tree of (mother, daughters, etc.), 208
  passivation booths / sink modules, 206–209
  processing module, 206
  production capacity of system, 210
  used in embossing, 228–229
Signal beams, in holographic memory systems, 362
Silver halide emulsion, 499
Simplex Algorithmus, 470–472
Single-beam reflection holograms
  in-line holograms, 500–501
  making, 30–34
Single-image random dot stereograms (SIRDs). *See* SIRDs (single-image random dot stereograms)
Sink modules, electroforming process, 206–209
SIRDS (single-image random dot stereograms), 146–147, 158–160, 662
Size
  art *vs.* giftware, 305–306, 311
  effected by color and wavelength of light source, 503
  enlarging or reducing holograms, 24
  large-format studio holograms, 92–98
  reconstruction and, 496
  reducing or enlarging holographic images, 164–180, 629
  and telepresence, 585
Slit holograms, 507–508
Soundscaping in holodecks, 587–588
Space-bandwidths, 165
  reducing, 166–167
Spatial discretization, 167
Spatial filtering, 76–77
  and interferometry, 65–66
Spatial light modulators (SLMs), 126–127
  holographic memory and, 366–368, 379–380
  liquid-crystal displays, 411
Spatial multiplexing, 386–387
Spectral discretization, 167
Spectral harmonics, 39
Spectral subsampling, 420
Speed, computational
  fringe computation times, methods compared, 453–454
  of fringe production, 440
  hogel-vector bandwidth compression, *vs.* interference-based methods, 177–180
  WWW interface performance, 335
SPIE (International Society for Optical Engineering), 344

INDEX **675**

Splotch engines, 403–404, 425
Standing waves, 29
Star Trek series, holodecks in, 579–580
Stereograms, 182, 185, 188–189
  autostereograms, 158–160
  computation, 408–409
  described, 628–629
  fringe computation, 452–453
  HPO stereograms, 124
  kinegrams, 546
  SIRDS (single-image random dot stereograms), 146–147, 158–160, 662
Stereoscopic images
  eyeware systems for, 154
  photographs, 152–153
Stereoscopic lustre, 298
Stickers, 246–247
Storage systems. *See* Memory, holographic
Storyboards for planning, 9
Subjects for holograms
  artwork for custom designs, 111, 116
  computer interface and selection of, 332–333
  holders for, 64–65, 70–73
  increasing reflectivity of, 65
  live subjects (digital imaging), 198
  selecting, 8, 32
Substrates
  oriented poly propylene (OPP), 218
  paper, 217
  polyester, 217–218
  quartz or glass, 221
  for security solutions, 227
Surface acoustic waves (SAWs), used in AOMs, 129
System requirements for CD-ROM, 654

Tables, isolation
  construction of, 510–513
  floating, 62–63
  interferometry to test, 512–513
  for large format holography, 102
  object movement and object isolation, 497–498
Tags, 246–247
Target pattern recognition and tracking, 264–266
TEM (Transverse Electromagnetic Mode), 30, 508
Texture gradients, 148–150
3D disks, 388, 392
360 degree transmission holograms, 525
3D holograms, 113
3-D i*d hologram technology, 542–543
  games development, 645–646

3D imaging, master holograms and, 199–200
Time (holographic temporal manifestations), 272–273, 280, 283
  buoyant events, 302–303
  discontinuity, 302
  fourth dimension as concealed truth, 293
  freezing, 301
  linearity, 301
  real-time, 302
  symultaneism, 301
  time-reversability, 302
  time suspension, 301
Tomographic reconstruction, 609
Tomography
  and ISAR imaging, 633–640
  tomographic reconstruction, 609
Toys, 247
Tracking tasks, optical holographic correlators and, 264–266
Trade shows
  display holography, 11–14, 114
  pulsed laser holograms in, 13–15
Trading cards, 245
Transmission holograms, 7
  color-recording and, 52–53
  commercial application examples, 14–15
  360 degree transmission holograms, 525
  described and explained, 501
  division of amplitude transmission holograms, 524, 525
  early experiments in, 38–39
  interferometry and, 73–74
  laser electronics, 84
  making, 517–522
  multiple object beam transmission holograms, 524
  real image projection transmission hologram, 524
  reconstruction, 521–522
  setup checklist, 518–521
  theory of, 17–18
Transverse Electromagnetic Mode (TEM), 30, 508
Troubleshooting
  anomalies caused by mote diffraction, 66–67
  developing process, 86–88
  double-exposure interferometry problems, 78
  hogel-vector bandwidth compression and spatial coherence, 177
  object movement during process, 497–498
  spatial filtering, 76–77
  viewing problems, 87, 88
  washout, 87

Twin-beam reflection holograms, 63–64, 508–510, 525
2D/3D holograms, 113, 187, 628

Underwater holography, 99
Upatnieks and Leith, 336, 486
User interface design, parameters, and output, 329–333

Vertical Cavity Surface Emitting Laser (VCSEL) arrays, 368–369
Vibrations
  holographic, 24–30
  resonance, 25
  See also Tables
Viewing software, design specifications for, 190
Vignetting effects, off-axis electron holography, 464–465
Virtual images
  described, 504–506, 631–632
  vs. real images, 504–506
Virtual reality. *See* Holodecks
Visual-bandwidth holography, 163, 193, 415
Visual phenomena
  pseudoscopy, 34, 297–299, 506, 632
  retinal rivalry effects, 299–300
  stereoscopic lustre, 298
Volume scanning, 161–163
Voxel Inc., 570–573
  voxgram production, 576–579
  Web address, 603

Washout, 87, 88
Wave mechanics
  light as transverse waveform, 484–486
  plane waves, 362
  wave behavior of light, 25–27
  wavelength described and explained, 483–484
  See also Interference, wave
Wave transfer functions, 472–474
Web addresses
  GEOLA Labs, 31
  HoloBank, 625–626
  holographic-related sites, 603–604
  Intergraph, 31
  MCM's Prehistoric Lives, 100
  Meredith Instruments, 31
  Search The Light, 603
White-light-transmission (WLT) holograms, 49
  origination phase of metallizing, 226
Windowing, 466
World Wide Web
  design considerations, 351–352
  e-commerce and hologram products, 353
  holography site development, 345–346
  interface design, parameters, and output, 329–333
  interface performance, 335

Young, Thomas, 484–486